PRINCIPLES OF HEATING, VENTILATION AND AIR CONDITIONING

with Worked Examples

PRINCIPLES OF HEATING, VENTILATION AND AIR CONDITIONING

with Worked Examples

Nihal E Wijeysundera

World Scientific

NEW JERSEY • LONDON • SINGAPORE • BEIJING • SHANGHAI • HONG KONG • TAIPEI • CHENNAI • TOKYO

Published by

World Scientific Publishing Co. Pte. Ltd.
5 Toh Tuck Link, Singapore 596224
USA office: 27 Warren Street, Suite 401-402, Hackensack, NJ 07601
UK office: 57 Shelton Street, Covent Garden, London WC2H 9HE

British Library Cataloguing-in-Publication Data
A catalogue record for this book is available from the British Library.

PRINCIPLES OF HEATING, VENTILATION AND AIR CONDITIONING WITH WORKED EXAMPLES

ISBN 978-981-4667-76-0

Printed in Singapore

To my grandchildren

Emiko Chrisanthi,
Sunil Hitoshi,
Isabella Anjali,
Amali Satomi, and
Helina Maya

Preface

Courses in Heating, Ventilation and Air Conditioning (HVAC) are usually offered in departments of mechanical engineering, civil engineering, architecture and building science. This book is written mainly with the interests of students and instructors in these departments in mind. However, a significant part of the contents may be used in courses such as, thermal systems and heat transfer, especially the worked examples. Practicing engineers could use this book to clarify the fundamental principles behind various design procedures recommended in professional handbooks.

A number of professional societies like the American Society of Heating, Refrigeration and Air Conditioning Engineers (*ASHRAE*) publish comprehensive handbooks and design guides for use by HVAC engineers. These handbooks are updated regularly to include the most recent design procedures, developed through sponsored research projects.

One of the main challenges for instructors in HVAC courses is to distill the materials available in professional handbooks, to a concise form to be included in regular undergraduate courses. This is often a time consuming task because the handbooks are intended for practicing engineers. This book tries to make the task easier for instructors by presenting the material in a directly useable format. For students the contents should appear as extensions and applications of the material covered in basic courses on thermodynamics, heat transfer and fluid mechanics.

Every effort is made to include simple derivations for most of the design parameters used in practice, without making the mathematical details unduly complicated. For instance, in chapter 9 a simple one-dimensional thermal network approach is used to derive the fenestration design parameter called the 'solar heat gain coefficient (SHGC)'. Likewise, in chapter 10 a lumped-capacity transient thermal model is used to clarify the physical meaning of 'the radiant time series (RTS)', and its application in cooling load estimation.

In chapter 9 a 'vector approach' is introduced to analyze complex three-dimensional geometrical design problems. These situations are encountered in computing incident angles of solar beams on inclined surfaces, and in determining the effectiveness of shading devices like overhangs.

Included in this book are the most up-to-date empirical models available in the *ASHRAE Handbook* - 2013 *Fundamentals*, that are relevant for design. In particular, in chapter 9, for computing the solar radiation absorption and transmission in building envelopes, the latest two-parameter model is used to estimate the 'clear-sky radiation' at different locations.

In design oriented courses such as HVAC, it is important for students to understand the fundamentals behind the recommended design procedures. Comprehensive worked examples provide an ideal means to present design concepts in a practically useful manner. With this objective in mind, about 15 worked examples are included in each chapter, carefully chosen to expose students to diverse design situations encountered in HVAC practice.

Computations required in worked examples illustrating basic principles are performed using a calculator. Worked examples involving more realistic design situations are done using MATLAB programs, included in the book.

At the end of each chapter there are additional problems for which numerical answers are provided. The format of the worked examples and problems is a novel feature of this book. For instructors, this should provide a useful source for problems to be included in courses, tutorials and examinations.

MATLAB programming is now taught routinely in most engineering and science courses. Therefore a number of MATLAB codes, for solving HVAC design problems requiring extensive computations, are also included. Computer codes are included for the following applications: (i) computation of psychrometric properties, (ii) design of cooling towers, (iii) design of wet-coil heat exchangers, (iv) computation of hourly diffuse and direct solar radiation intensities, (v) computation of sol-air temperature, (vi) estimation of hourly cooling load due to people, lights, roofs, and walls, (vii) design of overhangs, and (viii) design of duct and pipe systems.

I wish to thank ASHRAE for granting permission to extract representative design data from the *ASHRAE Handbook* - 2013 *Fundamentals*, for inclusion in this book.

I was fortunate to have had the opportunity to teach a number of courses in refrigeration, air conditioning, and thermal systems at the Department of Mechanical Engineering, National University of Singapore (NUS). The notes developed for these courses provided the framework and much of the material for this book. I am thankful to my colleagues in the energy and bio-thermal division at NUS, with whom I shared the teaching of these courses, for many valuable discussions on HVAC systems.

I am thankful to Dr. Raisul Islam for fruitful discussions on a number of practical design aspects of chiller systems and their energy efficiency. I wish to thank Dr. A. H. Jahangeer for providing valuable technical support on many occasions and Mr. Mahipala D. Fernando for fruitful discussions on heat pump systems.

Thanks are due to my sons Duminda and Harindra, and my daughters-in-law Sindhu and Sophia, for their constant encouragement.

Finally, my heartfelt thanks are given to my wife Kamani for her encouragement and generous support towards the completion of this project.

Nihal. E. Wijeysundera

Contents

Preface **vii**

Chapter 1
Introduction to Heating, Ventilation and Air Conditioning **1**
1.1 An Overview of HVAC Systems 1
1.2 Some Optional Designs of HVAC Systems 4
1.2.1 HVAC system using air as the energy transport medium 4
1.2.2 HVAC system using water as the energy transport medium 5
1.2.3 HVAC system using water and air as energy transport media 7
1.2.4 Packaged and unitary systems 9
1.2.5 Reversible heat pumps for heating and cooling 9
1.3 Overview of HVAC Design Procedure 11
1.4 Aims and Organization of the Book 13
References 15

Chapter 2
Heat Transfer Principles **17**
2.1 Introduction 17
2.2 Modes of Heat Transfer 17
2.3 One-dimensional Steady Heat Conduction 20
2.4 Thermal Network Analogy 21
2.4.1 Thermal resistances in series 22
2.4.2 Overall heat transfer coefficient 23
2.4.3 Thermal resistances in parallel 23
2.4.4 Boundary conditions 24
2.5 General Form of Fourier's Law 26
2.5.1 Cylindrical systems 26
2.6 Conduction with Internal Heat Generation 28
2.7 Convection Heat Transfer 30
2.7.1 Forced convection heat transfer 30
2.7.2 Correlations for the heat transfer coefficient 32
2.7.3 Natural convection heat transfer 33
2.8 Radiation Heat Transfer 34
2.8.1 Spectrum of electromagnetic radiation 35
2.8.2 Black surface 35
2.8.3 Emissive power of real surfaces 37
2.8.4 Emissivity of a gray surface 37

	2.8.5	Absorption, transmission and reflection	37
	2.8.6	Kirchhoff's law of radiation	38
	2.8.7	Radiation exchange between two black surfaces	38
	2.8.8	Radiation exchange between a black surface and a gray surface	39
	2.8.9	Radiation exchange between two gray surfaces	41
	2.8.10	Radiation exchange between a curved surface and a flat surface	42
2.9	Worked Examples		43
Problems			59
References			62

Chapter 3
Refrigeration Cycles for Air Conditioning Applications **65**

3.1	Introduction		65
3.2	Carnot Refrigeration Cycle Using a Vapor		66
3.3	Standard Vapor Compression Cycle		68
3.4	Analysis of the Standard Vapor Compression Cycle		71
3.5	Actual Vapor Compression Cycle		72
3.6	Modifications to the Standard Vapor Compression Cycle		74
	3.6.1	Two-stage compression with flash intercooling	74
	3.6.2	Two-stage compression with two evaporators	76
3.7	Refrigerants for Vapor Compression Systems		76
3.8	Vapor Compression Systems for Air Conditioning Applications		78
	3.8.1	Window-unit air conditioners	78
	3.8.2	Central air conditioning systems using chilled water	79
	3.8.3	Compressors of water chillers	80
	3.8.4	Reversible heat pump systems	82
3.9	Vapor Absorption Refrigeration Cycles		84
	3.9.1	Three-heat-reservoir model	85
3.10	Analysis of Actual Absorption Cycles		86
	3.10.1	Equilibrium of water–LiBr mixtures	87
3.11	Worked Examples		89
Problems			114
References			117

Chapter 4
Psychrometric Principles **119**

4.1	Introduction		119
4.2	Mixtures of Air and Water Vapor		119
4.3	Properties of Air–Water Mixtures		121
	4.3.1	Relative humidity, humidity ratio and degree of saturation	121
	4.3.2	Enthalpy of moist air	123
	4.3.3	Specific volume of moist air	126
	4.3.4	Adiabatic saturation and wet-bulb temperature	126
	4.3.5	Measurement of wet-bulb temperature	128
4.4	The Psychrometric Chart		129
	4.4.1	Constant dry-bulb temperature lines	130

4.4.2 Saturation curve and constant relative humidity lines 132
4.4.3 Constant wet-bulb temperature lines 133
4.4.4 Constant specific volume lines 133
4.4.5 Enthalpy–moisture protractor 134
4.4.6 Sensible heat ratio protractor 134
4.5 Worked Examples 136
Problems 153
References 156
Appendix A4.1 - MATLAB Code for Psychrometric Properties 157

Chapter 5
Psychrometric Processes for Heating and Air Conditioning **159**
5.1 Introduction 159
5.2 Basic Psychrometric Processes 159
5.2.1 Mixing of two moist air streams 160
5.2.2 Sensible heating or cooling 162
5.2.3 Dehumidification by cooling 163
5.2.4 Humidification of air 168
5.2.5 Evaporative cooling 169
5.2.6 Space condition line 170
5.3 Applications of Psychrometric Processes 173
5.4 Single-zone Air Conditioning Systems 174
5.4.1 Summer air conditioning systems 174
5.4.2 Summer air conditioning systems with reheat 176
5.4.3 Summer air conditioning systems with bypass paths 177
5.4.4 Winter air conditioning systems 178
5.4.5 Air conditioning systems using evaporative cooling 179
5.5 Multi-zone Air Conditioning Systems 180
5.5.1 Multi-zone systems with reheat 180
5.5.2 Dual-duct multi-zone air conditioning systems 181
5.5.3 Variable air volume (VAV) systems 183
5.6 Worked Examples 185
Problems 212
References 216

Chapter 6
Direct-Contact Transfer Processes and Equipment **217**
6.1 Introduction 217
6.2 Review of Mass Transfer Principles 218
6.2.1 Steady mass diffusion through a plane wall 218
6.2.2 Steady convection mass transfer 220
6.3 Simplified Model for Simultaneous Heat and Mass Transfer 221
6.4 Air Washers or Humidifiers 224
6.4.1 Analysis of air washers 225
6.4.2 Efficiency and number of transfer units (NTU) 228
6.5 Cooling Towers 229

6.5.1	Analysis of cooling towers	230
6.5.2	Enthalpy potential based model for cooling towers	232
6.5.3	Approach and range of cooling towers	233
6.6	Property Relations for Moist Air and Water	234
6.7	Worked Examples	235
Problems		258
References		260
Appendix A6.1 - MATLAB Code for Cooling Tower Design		261

Chapter 7
Heat Exchangers and Cooling Coils **265**

7.1	Introduction	265
7.2	Design–Analysis of Dry-Coil Heat Exchangers	266
7.2.1	Some common types of heat exchangers	267
7.2.2	Analysis of counter-flow heat exchangers	267
7.2.3	The LMTD method	270
7.2.4	The effectiveness–NTU method	270
7.2.5	Evaporators and condensers	272
7.2.6	Cross-flow heat exchangers	275
7.2.7	Efficiency of extended surfaces	278
7.2.8	Overall heat transfer coefficient for finned tubes	282
7.3	Wet-Coil Heat Exchangers or Cooling Coils	283
7.3.1	Physical processes in wet-coils	284
7.3.2	Analysis of wet-coil heat exchangers	285
7.3.3	Numerical model for wet-coils	288
7.4	Worked Examples	290
Problems		331
References		335
Appendix A7.1 - MATLAB Code for Design of Chilled Water Coils		335

Chapter 8
Steady Heat and Moisture Transfer Processes in Buildings **339**

8.1	Introduction	339
8.2	Steady Heat Transfer through Multi-Layered Structures	340
8.2.1	Parallel path method	341
8.2.2	Isothermal plane method	342
8.2.3	Zone method	343
8.2.4	Radiation heat transfer coefficient	344
8.2.5	Heat transfer in gas filled cavities	346
8.3	Steady Heat Transfer through Fenestrations	348
8.3.1	Windows and doors	348
8.3.2	Overall heat transfer coefficient	349
8.4	Below Grade Heat Transfer in Buildings	351
8.4.1	Heat transfer through basement walls	352
8.4.2	Heat transfer through basement floors	353
8.4.3	Heat transfer through surfaces at grade level	355

8.5 Infiltration in Buildings 355
8.5.1 Heating load due to infiltration 355
8.5.2 Infiltration air flow rates 356
8.5.3 Estimation of infiltration flow rates 360
8.6 Moisture Transport in Building Structures 361
8.6.1 Fick's law 362
8.7 Worked Examples 363
Problems 388
References 392

Chapter 9
Solar Radiation Transfer Through Building Envelopes 395
9.1 Introduction 395
9.2 Fundamentals of Solar Radiation 396
9.2.1 Beam and diffuse solar radiation 396
9.2.2 Direction of beam radiation 397
9.2.3 Angle of incidence of beam radiation on a surface 400
9.2.4 Total radiation incident on an inclined surface 402
9.2.5 Clear-sky model of direct and diffuse solar radiation 404
9.3 Absorption of Solar Radiation by an Opaque Surface 406
9.4 Transmission and Absorption of Solar Radiation 408
9.4.1 Effective properties of a single layer 408
9.4.2 Transmittance of a multi-layered fenestration 410
9.4.3 Radiation absorption in multi-layered fenestrations 411
9.5 Overall Energy Transfer through Fenestrations 412
9.6 Shading of Surfaces from Solar Radiation 417
9.7 Worked Examples 419
Problems 443
References 446

Chapter 10
Cooling and Heating Load Calculations 447
10.1 Introduction 447
10.2 Outdoor Design Conditions 450
10.3 Thermal Comfort and Indoor Design Conditions 452
10.3.1 Heat transfer from the human body 452
10.3.2 Indoor design conditions 455
10.3.3 Indoor air quality 457
10.4 Internal Heat Sources in Buildings 459
10.4.1 Heat gain from people 459
10.4.2 Heat gain from lighting 460
10.4.3 Heat gain from equipment 461
10.5 Transient Effects in Building Energy Transfer 462
10.5.1 Transient heat conduction through walls 463
10.5.2 Heat gain by a thin surface 465
10.6 Cooling Load Calculation Methods 468

10.6.1 Heat balance method (HBM) 468
10.6.2 Radiant time series (RTS) method 471
10.6.3 Application of the RTS method and the CTS method 475
10.7 Heating Load Calculation Methods 477
10.8 Worked Examples 479
Problems 507
References 513
Appendix A10.1 - MATLAB Code for Cooling Load due to People 514
Appendix A10.2 - MATLAB Code for Cooling Load due to Wall Conduction 515
Appendix A10.3 - MATLAB Code for Cooling Load due to Windows 519
Appendix A10.4 - MATLAB Code for Shading of Windows 523

Chapter 11
Air Distribution Systems **529**
11.1 Introduction 529
11.2 Total Pressure Distribution 530
11.3 Pressure Loss in Duct Networks 532
11.3.1 Pressure loss in straight ducts 532
11.3.2 Pressure loss in fittings 534
11.3.3 Total pressure loss in duct sections 539
11.4 Air Distribution Fans 539
11.4.1 Axial flow and centrifugal fans 540
11.4.2 Fan characteristics 541
11.4.3 Fan laws 542
11.5 Fan–Duct Network Interaction 543
11.6 Design Methods for Duct Systems 546
11.6.1 Equal friction method 547
11.6.2 Static regain method 549
11.7 Optimization of Duct Systems 551
11.8 Air Distribution in Zones 552
11.8.1 Air flow from diffusers 552
11.8.2 Air diffusion performance index 554
11.8.3 Design aspects of air distribution systems 555
11.9 Worked Examples 557
Problems 582
References 585
Appendix A11.1 - MATLAB Code for Pressure Loss in Circular Ducts 586
Appendix A11.2 - MATLAB Code for Equal Friction Design Method 587
Appendix A11.3 - MATLAB Code for Static Regain Design Method 588

Chapter 12
Water Distribution Systems **591**
12.1 Introduction 591
12.2 Energy Equation for Hydronic Systems 592
12.3 Head Losses in Hydronic Systems 593
12.3.1 Friction head loss in pipes 593

	12.3.2 Dynamic head loss in fittings	595
12.4	Pump Characteristics	596
12.5	System–Pump Interaction and Flow Control	599
12.6	Design of Water Distribution Systems	601
	12.6.1 Direct-return and reverse-return systems	602
	12.6.2 Design of pipe networks	603
12.7	Worked Examples	604
Problems		625
References		630
Appendix A12.1 - MATLAB Code for Head Loss in Pipe-Sections		630

Chapter 13
Building Energy Estimating and Modeling Methods **633**

13.1	Introduction	633
13.2	Degree–Day Method for Estimating Energy Use	634
13.3	Bin Method for Estimating Energy Use	638
	13.3.1 Generation of bin data	638
	13.3.2 Applications of the bin method	641
	13.3.3 Cycling of furnaces	642
	13.3.4 Air-source heat pumps	643
	13.3.5 Cooling towers	646
	13.3.6 Variable occupancy rates	647
13.4	Simulation Methods for Estimating Energy Use	649
	13.4.1 Central HVAC systems	649
	13.4.2 Simulation of multi-chiller systems	651
	13.4.3 Simulation of water-loop heat pump system (WLHPS)	654
13.5	Worked Examples	656
Problems		685
References		690
Appendix A13.1 - MATLAB Code for Bin Data and Degree–Days		691

Index **693**

Chapter 1

Introduction to Heating, Ventilation and Air Conditioning

1.1 An Overview of HVAC Systems

The commonly used acronym for heating, ventilation and air conditioning is HVAC. In its broadest sense, HVAC encompasses the means, the processes, and the technology used to maintain an indoor space at a desired set of physical conditions. Typically these conditions depend on the type of activity for which the space is used. For instance, if the space is the inside of a building occupied by people, the most pertinent conditions are the temperature, the humidity level, and the cleanliness of the air.

The indoor conditions for a health facility like a hospital or a clinic would be similar, but they would have to be controlled within more stringent limits compared to those of an ordinary home. In the case of a process plant, such as a food preparation facility or a paint-shop, the indoor conditions would have to satisfy those mandated by the industrial guidelines for these processes.

The conditions of all indoor environments are dynamic because they are subject to various time varying inputs, some of which are predictable while others are random or accidental. In the case of a residential building or a home, the indoor temperature, humidity, and cleanliness of the indoor air change due to a number of inputs which can be internally or externally generated.

The temperature difference between the indoor air and the ambient air causes heat to flow across the building envelope. During winter, when the temperature outside is lower than the indoor temperature, heat is lost to the outside. This makes the inside air colder, and therefore uncomfortable for occupants. The HVAC system needs to balance this heat loss by supplying the necessary heat input, using an external energy source. In addition, cold air leaking through any openings and cracks in the building envelope has to be heated to the indoor temperature by the HVAC system. If the indoor air is too dry and therefore uncomfortable for the occupants, moisture has to be introduced artificially. The total amount of energy supplied by the HVAC system per unit time, to maintain the space at the desired temperature and humidity, is referred to as the winter *heating load* of the building.

In the summer, the air outdoors is usually hotter and more humid than the typical indoor comfort conditions stipulated. In this case the heat flow across the building envelope occurs in the opposite direction. Moreover, the indoor air is heated indirectly by the solar radiation entering through the glass surfaces of the building envelope, such as, windows, glass doors, and skylights. The transmitted solar radiation is first absorbed by the interior surfaces of the building like the walls, the floor, and other items, such as furniture. This absorbed energy is later released to the indoor air when the latter surfaces get warmer.

People occupying the building, the indoor lights, and appliances, such as, computers and coffee makers, also release heat and moisture, which increases the temperature and humidity of the indoor air. If comfortable indoor conditions are to be maintained steadily, then all the aforementioned heat and moisture flows have to be balanced by the HVAC system. The amount of energy that needs to be removed by HVAC system per unit time is called the *cooling load* of the building. It is interesting to note that in the winter, the energy inputs, like absorbed solar radiation, energy from people, lights and equipment, tend to heat the indoor air, and thereby reduce the *heating load* of the building.

In section 1.2 below we shall present a few optional designs of HVAC systems including some of the equipment used. But first, it is instructive to highlight the interactions between the *conditioned space* and the HVAC system by referring to the conceptual diagram depicted in Fig. 1.1.

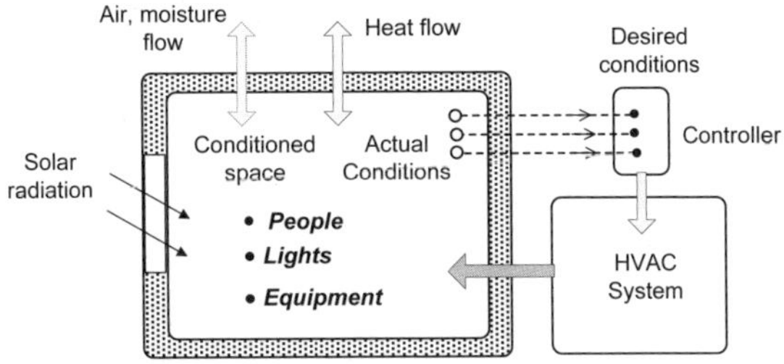

Fig. 1.1 Conceptual diagram of HVAC system

Within the indoor *conditioned space* shown in Fig. 1.1 energy is released by people depending on the type of activities they are engaged in. People also add moisture to the air and thereby increase the humidity of the air. The artificial lights inside the space and equipment, such as, computers, fax machines, coffee makers, and others produce heat that flows into the air. Solar radiation transmitted through the transparent sections of the enclosure, like the glass windows of a building, contributes indirectly to the heating of the air inside. Sensible heat is gained or lost through walls of the enclosure due to the inside-to-outside temperature difference. There may also be unwanted leakage of air and moisture through cracks and openings in the enclosure. As a consequence of these energy flows the conditions of the air inside the enclosure, such as, the temperature and humidity, change continuously.

The conditions of the space are monitored by sensors located inside, and the data is transmitted to a controller. This could be a simple thermostat in the case of a HVAC system serving a residence. In the case of a modern commercial building, the control system could be a computer based building management system (BMS).

The controller compares the actual conditions of the space with the desired conditions supplied to it by the designers of the HVAC system or its operators. Based on the discrepancy between the actual and desired conditions, the controller activates the necessary hardware items of the HVAC system to reestablish the desired conditions within the space. The HVAC system does this by either supplying or removing the appropriate amount of energy and moisture from the space. Careful control of the inside conditions, at the desired values, contributes both to the comfort of the occupants of the space and to the overall energy efficiency of the system.

1.2 Some Optional Designs of HVAC Systems

We shall now consider a few optional designs of HVAC systems that are being used for building related applications. However, these same designs could be modified for industrial and transportation applications where the requirements may be somewhat different.

1.2.1 *HVAC system using air as the energy transport medium*

Shown schematically in Fig. 1.2 is an HVAC system, commonly used for heating and cooling residences and small commercial buildings.

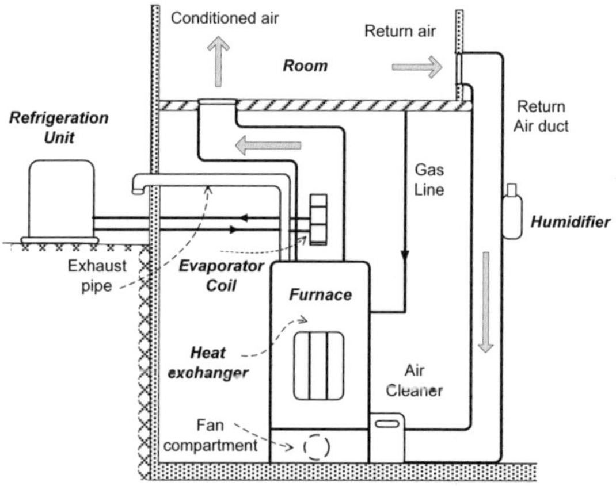

Fig. 1.2 Typical forced air heating and cooling system for homes

Air from the rooms or the conditioned spaces of the building is drawn by a fan, through a filter, to be processed by the HVAC system. In the heating mode of operation, the air passes over the tubes of a heat exchanger through which hot combustion gases flow in the opposite or ‘cross-flow’ direction. The combustion process occurs in a chamber that is usually supplied with a fuel, such as, natural gas. The heated air then flows through the supply duct network to the various spaces of the building. If necessary, the moisture content of the air can be increased by activating the humidifier located in the return air duct.

Typically, the temperature of the building is controlled by a thermostat located in one of the rooms. If the temperature of the room exceeds the preset temperature the thermostat switches off the combustion process, thus shutting down the furnace. The operation is reversed when the space temperature falls below the preset value.

In the cooling mode of operation, the furnace is shut down, and the compressor of the refrigerator is switched on. In most HVAC systems, the evaporator coil of the refrigerator is located in the supply air duct, as shown in Fig. 1.2. The condensing unit of the refrigerator, including the compressor, is usually placed outdoors. The air flowing over the finned tubes of the evaporator coil is cooled and dehumidified, and the condensate from the air is drained by gravity, to a sump in the plant room. When the temperature of the space falls below the preset temperature, the thermostat switches off the compressor. In this type of simple system the thermostat has to be manually set to either the heating or cooling mode of operation, depending on the outdoor conditions.

1.2.2 *HVAC system using water as the energy transport medium*

An HVAC system using water as the energy transfer medium is shown schematically in Fig. 1.3.

This system is equipped with a boiler, for producing hot water to be used in the heating mode of operation, and a water chiller, to be used in the cooling mode of operation. The water chiller is essentially a refrigerator, where the evaporator coil is used to produce chilled water, typically at a temperature of about 3 to 5°C. The system depicted in Fig. 1.3 is called a *two-pipe* arrangement [2]. As in the case of the HVAC system described in section 1.2.1, the *all-water, two-pipe* system either operates in the heating mode or the cooling mode at any one time.

In the heating mode of operation, the two header pipes that circulate water are connected to the boiler through valves A and B. The hot water from the supply header pipe flows through heat exchanger tubes in the fan coil units. Air from the room is circulated over the finned tubes by means of a fan located inside the fan coil unit. The flow rate of water through the fan coil is usually controlled by valves at the inlet, which could either be operated manually, or by means of a thermostat, to maintain the desired room temperature.

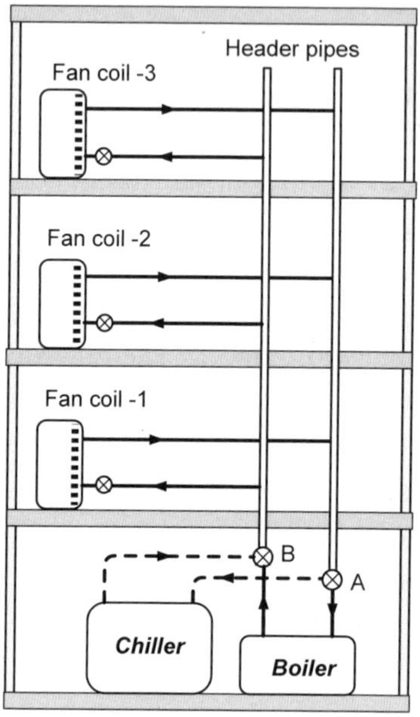

Fig. 1.3 All-water, two-pipe heating and cooling system

In the cooling mode of operation, the same header pipes are connected to the water chiller by repositioning the valves A and B. The chilled water flowing through the fan coils, from the supply header pipe, cools and dehumidifies the room air circulated through them by the fans. Any condensate produced within the fan coil units is piped out to a sink.

The two-pipe system shown in Fig. 1.3 cannot handle simultaneous heating and cooling needs of different rooms. If some rooms served by the system have high heat loads, and therefore require cooling, while others require heating, then the two-pipe system is not a suitable design option. However, two-pipe systems are less complicated and require fewer pipes, fan coils, valves, and controls.

A more versatile all-water system is the *four-pipe system* [2,4]. It has fan coil units which incorporate separate heating and cooling coils. These are connected to separate hot water and chilled water header pipes, similar to those in shown Fig. 1.3. The two sets of header pipes, in turn, are connected separately to the boiler and the chiller. When the boiler and the chiller are both operating, a fan coil unit can be rapidly switched between heating and cooling modes, by means of the valves at the inlets of the heating and cooling coils. Therefore these systems can provide heating to some rooms, while simultaneously providing cooling to other rooms of the building. The disadvantage, however, is that they requires more pipes, heat exchangers, and controls.

1.2.3 *HVAC system using water and air as energy transport media*

A central HVAC system using both air and water as energy transport media is shown schematically in Fig. 1.4. Return air from the conditioned space is drawn into the *air handling unit* (AHU) by the return air fan. As the air passes through the AHU a fraction of it is discharged to the outside ambient through the exhaust port EA, and replaced with an equal amount of fresh ambient air drawn through the inlet port OA, for hygienic reasons. Dampers are used to control this process. The mixture of return air and fresh air then passes through a filter before entering the cooling and dehumidifying coil.

In the cooling mode of operation, the air passing over the cooling coil is cooled and dehumidified, and the condensate produced is drained out from the AHU. The supply air fan then distributes the cold air through the supply duct network to the conditioned space.

Finally, the desired quantity of air is discharged to each conditioned space or room through flexible ducts connected to *ceiling diffusers* in the space.

In the heating mode of operation, the cooling coil is inactive, while the air is heated by water flowing through the hot water coil, and hot air is distributed to the different spaces as described above.

The air handling units (AHU), the air distribution system, the heating and cooling coils, and the liquid distribution network are commonly called *secondary components* of the HVAC system as indicated in Fig. 1.4.

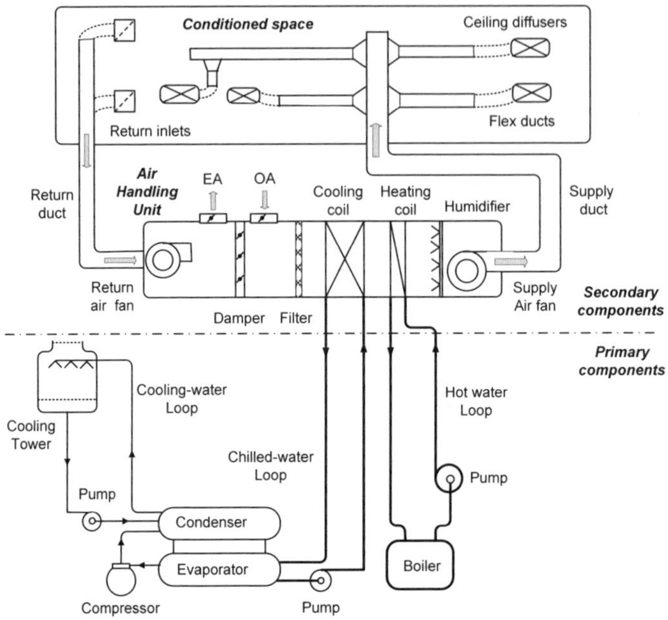

Fig. 1.4 Central heating and air conditioning system using air and water

The cooling coil receives chilled water pumped from the chiller, which is essentially a refrigerator, where the evaporator cools water to a temperature of about 3 to 6°C. The heat rejected by the condenser of the refrigerator is carried away by cooling water, pumped through the tubes of the condenser. This cooling water finally discharges heat to the atmosphere in a cooling tower, before being circulated back to the condenser, by the *cooling tower pump*.

The heating coil of the AHU receives hot water pumped from a fuel-fired boiler. The boiler and the chiller that convert fuel or electrical energy to heating and cooling effects respectively, are usually called the *primary components* of the system. Since the hot water and chilled water circuits are independent, they could serve a number of separate AHUs, similar to that depicted in Fig. 1.4, some supplying hot air, and others supplying cold air, at the same time.

1.2.4 *Packaged and unitary systems*

The HVAC systems described above can be adopted to serve different arrangements of conditioned spaces in a building by designing the duct and pipe networks accordingly. However, there are HVAC systems, commonly called *packaged systems* [4], that incorporate a vapor compression refrigeration unit, and a fuel-fired or electrical heating unit, in a single compact package. Such packaged systems are usually installed on flat roofs of commercial buildings, and connected through supply and return ducts to the conditioned space below.

Smaller air conditioning units, designed to serve a single space like an office room, are called *unitary systems* or *window units*. These are installed in a window or a wall opening, with the controls on the inside. Room air is cooled and dehumidified by circulating it across the finned tube coils of the evaporator using a fan. The condenser coil of the unit, facing the outside, is cooled by a fan blowing ambient air over it.

1.2.5 *Reversible heat pumps for heating and cooling*

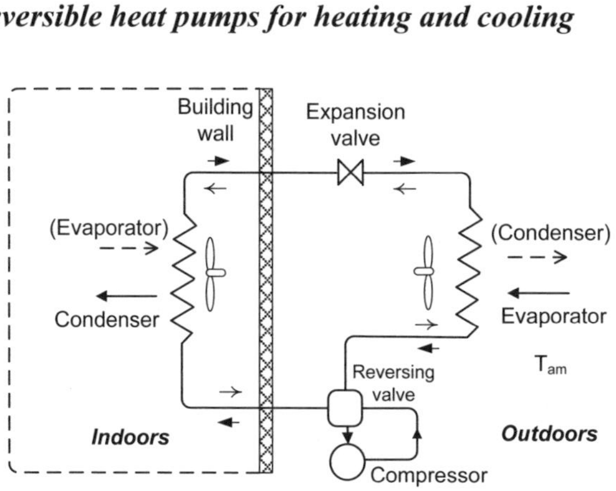

Fig. 1.5 Reversible heat pump for heating and cooling

A reversible heat pump is an HVAC system that can be used for cooling or heating an indoor space. A simplified schematic diagram illustrating its principle of operation is shown in Fig. 1.5. It is essentially a vapor compression refrigeration system, consisting of a compressor, a reversing valve, an expansion valve, and two coils, which are able to function both as the evaporator and the condenser of the system.

During the cooling mode of operation, the coil located inside the conditioned space is the evaporator. The evaporating refrigerant absorbs heat from indoor air, thus cooling the space. The position of the reversing valve allows the compressor to suck refrigerant from the evaporator and deliver the compressed vapor to the condenser, located outdoors. The refrigerant flowing through the condenser rejects heat to the ambient, and then passes through the expansion valve to enter the evaporator, thus completing the cycle.

In the heating mode of operation, the outdoor coil is the evaporator, and refrigerant passing through it absorbs heat from outdoor ambient air. The reversing valve is repositioned so that the compressor is now able to suck refrigerant from the outdoor evaporator, and deliver it to the condenser coil, located inside the space. The condensing refrigerant releases heat to the indoor air, thus heating it.

Reversible heat pump systems are being used to heat and cool homes and commercial buildings. Several variations of the basic system, described above, are now available commercially. In one of these, commonly called *ground-source heat pumps*, the refrigerant in the outdoor unit, shown in Fig.1.5, exchanges heat with a fluid circulating through a coil buried in the ground. Compared to the ambient air temperature, the fluctuation of the ground temperature over the seasons is much smaller, and therefore the variation of the performance of the heat pump is much less.

Another heat pump system, more suitable for large buildings like hotels, is called a *water-loop heat pump*. Here the outdoor coils of the individual heat pumps, located in different rooms, exchange heat with water passing through a common pipe-loop. For the rooms requiring heating, the heat pumps absorb heat from the common water loop and transfer it to the rooms. On the other hand, in the rooms being cooled, the heat pumps reject the heat absorbed from the room to the water-loop. The water loop temperature is typically maintained between 18°C and 32°C. A boiler is used to heat the water in the loop if the net heating demand becomes high, and a cooling tower is used to cool the water if the net cooling demand is high.

1.3 Overview of HVAC Design Procedure

The design, installation and commissioning of an HVAC system requires the inputs of specialists from several different areas.

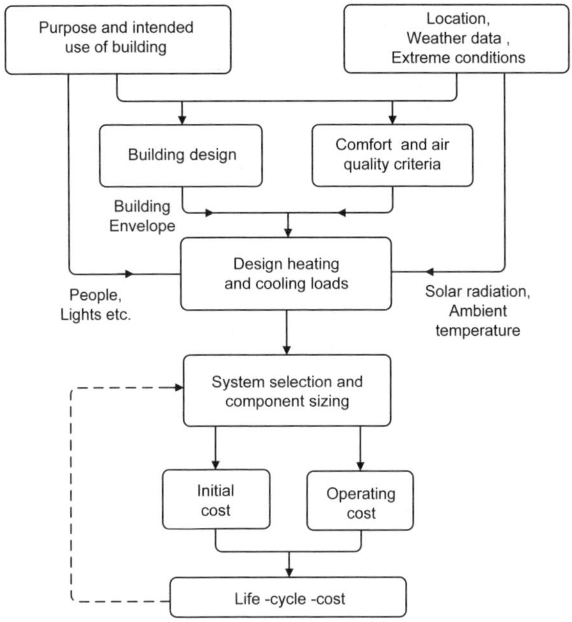

Fig. 1.6 Overview of HVAC design procedure

These include architects, civil engineers, HVAC-mechanical engineers, control engineers, equipment manufacturers and different contractors. The contribution of these specialists and the interactions between them can be best illustrated by referring to the block diagram in Fig. 1.6, which is an overview of the HVAC design process.

The overall design and construction of a building depends on the intended purpose and use of the building. Moreover, the location and the local weather conditions influence the design of the building. Hence the historical data on the variation of the solar radiation intensity, the ambient temperature, the humidity, and the wind speed are important inputs to the design. This data should also include the extreme weather conditions, such as, the highest and lowest ambient temperatures, and the highest wind speed at the location.

The orientation of the building, the types of construction material used, and the fraction of the building envelope area with transparent surfaces, are usually decided by architects and civil engineers. Although the aforementioned inputs are not within the purview of the HVAC engineer, they do affect the *heating and cooling loads* to a large extent. For example, the thermal properties and thicknesses of the construction materials used in the building envelope has a direct bearing on the *heating load* of the building. Moreover, the area of transparent surfaces in the envelope, and their thermal and optical properties, determine the contribution of the transmitted solar radiation to the *cooling load.* The materials used for items like doors and windows of the building, and the quality of their installation, influence the air *infiltration rate*, which affects the heating and cooling loads.

The cooling load also depends on the energy and moisture release rates by *people* occupying a building (see Fig. 1.1), which is a function of the type of activities they are engaged in, and their occupancy schedule in the building. These energy inputs to the cooling load would be very different for an office building, a sports hall, and a church.

Typically, the first step in the design of an HVAC system is to determine the design heating and cooling loads, based on the extreme weather conditions expected for the location. The heating and cooling loads are then used to select the appropriate system, and the sizes of the individual equipment. As we discussed in section 1.2, there are usually a number of design configurations to choose from. These include all-air systems (Fig. 1.2), all-water systems (Fig. 1.3), air–water systems (Fig. 1.4), packaged-systems, and others.

Once the type of HVAC system is selected, it is necessary to size the specific components that constitute the system. These sizes depend on the specific design parameters of the component which satisfy the intended performance requirements. For instance, for a chiller the appropriate performance parameters are the *cooling capacity* and the chilled water temperature range. For a water circulating pump or an air circulating fan (see Fig. 1.4), these are the pressure rise and the fluid flow rate.

Once the performance parameters have been determined, manufacturer's catalogues have to be consulted to select the appropriate unit that satisfies the performance requirements. The availability of units with similar capabilities from different equipment manufacturers makes this step of the design process somewhat complicated. Although most available equipment may satisfy the performance requirements, their indices of merit, like the efficiency, the coefficient of performance (COP), and others, could be quite different. More importantly, there could be considerable variation in the cost of the equipment offered by different manufacturers.

For instance, the first cost of a chiller with a high COP will usually be higher than that of a low COP chiller. However, the operating cost of the former chiller could be significantly lower than that of the latter. An appropriate life-cycle-cost (LCC) analysis may be needed to resolve the conflicting trade-offs with the two chillers. This situation could arise for most of the required equipment. Therefore the final decision on the selection of an equipment has to be made by an iterative process, as indicated in Fig. 1.6, where the LCC of owning and operating equipment from different manufacturers are compared.

1.4 Aims and Organization of the Book

In the preceding sections we presented a brief overview of the aims of HVAC, a few examples of specific HVAC systems, and a summary of the design and component selection procedure for HVAC systems. From this discussion it is clear that much of the fundamental knowledge required to understand the design and operation of HVAC systems is covered in basic courses in engineering. These courses include: thermodynamics, heat transfer, mass transfer, fluid mechanics, and control engineering.

A number of professional societies, like the American Society of Heating, Refrigeration and Air Conditioning Engineers (*ASHRAE*) publish comprehensive handbooks [1], and design guides, for use by HVAC engineers. These handbooks are updated regularly, to include the most recent design procedures, usually developed through sponsored research projects. One of the main challenges for students in HVAC courses is to distill the detailed content available in professional handbooks, to a concise form that could be easily understood and used.

The main aim of this book is to present the information available in professional handbooks as extensions and applications of the material covered in basic engineering courses mentioned above.

With the above goal in mind, we have tried to avoid presenting HVAC design procedures as simple recipes. Wherever possible we have included simple derivations of the various design parameters used in practice, without making the mathematical details unduly complicated. For instance, in chapter 9 we use a simple one-dimensional thermal network approach to derive the fenestration design parameter called the 'solar heat gain coefficient (SHGC)'.

Likewise, in chapter 10 we use a lumped capacity transient thermal model to clarify the physical meaning of 'the radiation time series (RTS)' and its application to cooling load estimation. We have introduced the 'vector approach' to analyze complex three-dimensional geometrical design problems in chapter 9. These situations are encountered in computing incident angles of solar beams on inclined surfaces, and in determining the effectiveness of shading devices like overhangs.

The book includes the most up-to-date empirical models available in the *ASHRAE Handbook* - 2013 *Fundamentals* [1], where these are relevant for design. In chapter 9, for computing the solar radiation absorption and transmission in building envelopes, the latest two-parameter model is used to estimate the 'clear-sky radiation' at different locations.

We have included in each chapter about 15 worked examples, carefully chosen to expose students to diverse design problems encountered in HVAC practice. The examples designed to illustrate the application of basic physical principles are solved using a calculator. The more comprehensive HVAC design examples are solved using the MATLAB codes included in Appendices in the different chapters. Comments are included in the computer codes to clarify the main steps of the computation procedure.

Using these codes judiciously, students could explore realistic design scenarios without having to perform tedious hand calculations. Computer codes are included for the following applications: (i) design of cooling towers, (ii) design of wet-coil heat exchangers using chilled water, (iii) computation of hourly diffuse and direct solar radiation intensities, (iv) computation of sol-air temperature, (v) estimation of hourly cooling load due to people, lights, roofs, walls, (vi) design of overhangs, (vii) design of air distribution systems, and (viii) design of water distribution systems.

References

1. *ASHRAE Handbook* - 2013 *Fundamentals*, American Society of Heating, Refrigeration and Air Conditioning Engineers, Atlanta, 2013.
2. Kuehn, Thomas H., Ramsey, James W. and Threlkeld, James L., *Thermal Environmental Engineering*, 3rd edition, Prentice-Hall, Inc., New Jersey, 1998.
3. Mitchell, John W. and Braun, James E., *Heating, Ventilation, and Air Conditioning in Buildings,* John Wiley and Sons, New York, 2013.
4. Stoecker, Wilbert F. and Jones, Jerold W., *Refrigeration and Air Conditioning*, International Edition, McGraw-Hill Book Company, London, 1982.

Chapter 2

Heat Transfer Principles

2.1 Introduction

The basic principles of heat transfer play an important role in the analysis and design of air conditioning systems. For example, the components of winter heating systems have to be sized to supply the heat needed to maintain the conditioned space at the desired temperature. Similarly, for summer air conditioning systems, the various subcomponents have to be designed to remove the heat inputs to the conditioned space from the surroundings. In order to estimate the above heat flow rates through the various structural components of a building envelope, like walls, roofs, windows, and doors we apply the basic principles of heat transfer.

Heat transfer is a well established engineering discipline on which the published literature is voluminous. This literature includes numerous textbooks, handbooks, and research journals. In most engineering courses heat transfer is covered as a separate subject. Therefore the main purpose of the present chapter is to briefly review the relevant principles of heat transfer to facilitate their application to air conditioning systems in later chapters. We have listed several text books [1-3] on heat transfer in the section on references for the benefit of readers who wish to pursue heat transfer in greater detail.

2.2 Modes of Heat Transfer

Heat is defined as the transfer of energy at the boundary between two systems due to a difference in temperature. From this point of view there are two modes of heat transfer: conduction and radiation. Heat transfer by *conduction* is usually associated with solids where heat is transferred from a region of higher temperature to a region of lower temperature by molecular vibrations. Often, the solid as a whole, remains stationary while heat conduction takes place. For example, the heat transfer through the wall of a metal pipe occurs by conduction.

All solids, liquids and gases have a tendency to emit energy in the form of electromagnetic waves. They also absorb such waves emanating from neighboring bodies. For example, the 'radiant energy' emitted by a room heater is absorbed by the walls, the ceiling, the floor, and any other items in the room. This type of energy transfer is known as *thermal radiation.*

Fluids like liquids and gases can transport momentum by virtue of their mass and velocity. Unlike the molecular vibrations in a solid, the flow of a fluid moves 'packets' of fluid physically from one place to another, thereby transporting energy with it. This type of energy transfer, due to fluid motion, is called *convection.* In engineering practice it is common to use the term convection broadly to describe heat transfer from a solid surface to a fluid moving over it.

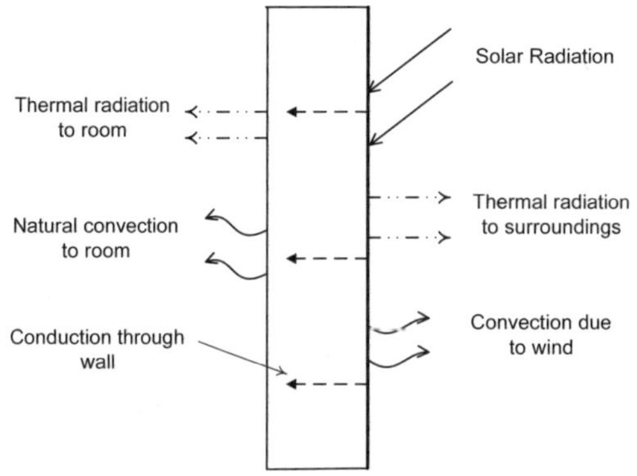

Fig. 2.1 Modes of energy transfer for wall

The distinction between the different modes of heat transfer is best illustrated by referring to a practical situation where several different modes of heat transfer are involved simultaneously.

Consider the external wall of a building, shown schematically in Fig. 2.1, where the various heat transfer processes are indicated. The outer exposed surface of the wall absorbs a portion of the solar radiation falling on it. As this surface gets warm heat flows through the wall by conduction. Heat is also transferred from the warm outer surface to ambient air due to *convection*, which is enhanced by wind impinging on the wall. Moreover, the warm surface exchanges thermal radiation with objects in its field of view, including part of the sky.

As the inner surface of the wall gets warm due to conduction, it transfers heat by *convection* to the air in the room. The heat flow into a thin air layer adjacent to the surface occurs by conduction. This causes the temperature of the air layer to increase, which in turn, decreases its density. The air circulation resulting from the density variation in the air layer is called *natural convection.* The inner surface of the wall also exchanges *thermal radiation* with surfaces inside the room, like the other walls, the ceiling, and the floor.

In general, a medium where heat transfer takes place has three dimensions. Depending on the shape of the medium, the temperature inside could vary along all three dimensions. Consider the wall, shown in Fig. 2.1, whose height and the breadth, are much larger than the thickness. Therefore the modes of heat transfer at different points on the surfaces are likely to be similar. Moreover, around the central area of the wall, it is reasonable to assume that the temperature varies only along the thickness. Therefore the heat flow through the central area of the wall is called *one-dimensional heat transfer*. However, at the edges and the corners of the wall the heat flow may depend on all three directions, namely, along the height, the breadth and the thickness. Hence, the heat flow at these locations is called *three-dimensional heat transfer.*

When the temperature distribution in a medium changes with time, the heat transfer process is called *transient heat transfer*. For example, as the solar radiation falling on the wall, shown in Fig. 2.1, becomes more intense, the temperature at different points inside the wall will increase. The temperature distribution is then a function of the three spatial dimensions as well as *time*.

When the temperature distribution in a medium remains constant with time, the heat flow in the medium is called *steady-state heat transfer*. For example, the heat flow to the surroundings from a well-insulated pipe carrying steam may be treated as a *steady-state* heat transfer process.

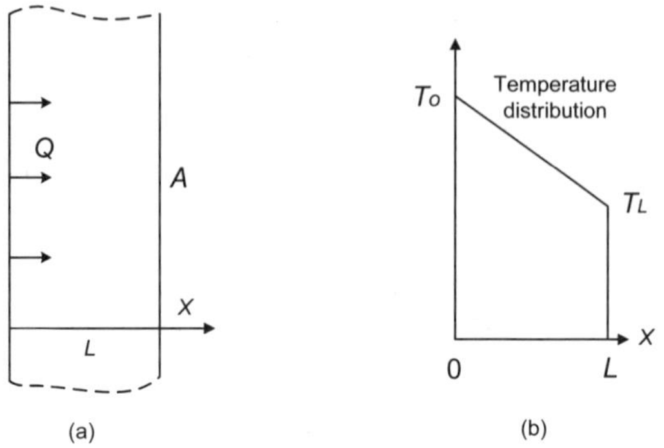

Fig. 2.2 Steady heat conduction in a slab: (a) the slab, (b) temperature distribution

2.3 One-dimensional Steady Heat Conduction

Consider the heat flow through the slab shown in Fig. 2.2, the length and breadth of which are much larger than the thickness. Therefore the heat flow in the slab may be treated as *one-dimensional*. Also assume that the heat transfer is *steady* and therefore the temperature distribution in the slab is independent of *time*. Under these conditions, the temperature distribution in the slab may be expressed in the form, $T = T(x)$ where x is the distance along the thickness.

The *steady, one-dimensional* heat transfer in the slab is governed by *Fourier's law* of heat conduction, which states that the rate of heat flow, Q is proportional to the area normal to the direction of heat flow, A and the temperature gradient across the thickness, L of the slab. We can express Fourier's law in the mathematical form

$$Q \infty \frac{A(T_O - T_L)}{L} \tag{2.1}$$

where T_o and T_L are temperatures at $x = 0$ and $x = L$ respectively, as indicated in Fig. 2.2(b).

We introduce the constant of proportionality k, called the *thermal conductivity*, to write the equation of heat conduction as

$$Q = \frac{kA(T_O - T_L)}{L} \tag{2.2}$$

The *thermal conductivity*, k $[Wm^{-1}K^{-1}]$, is a property of the material of the slab.

The thermal conductivity of a given substance depends on its microscopic structure and tends to vary somewhat with temperature. Metals, like copper and aluminum, have a high thermal conductivity, and are therefore *good conductors* of heat. On the other hand, materials, such as cork and fiberglass, have a very low thermal conductivity, and are therefore called *thermal insulators*. These insulating materials are applied on hot surfaces like the outside of steam and hot water pipes to reduce the heat loss to the surroundings.

2.4 Thermal Network Analogy

We encounter numerous situations involving one-dimensional, steady heat flow through multi-layered slabs when dealing with air conditioning design situations. In chapters 8 and 10 we shall consider the detailed analysis of heat flow through actual walls and roofs of buildings.

Here we develop a convenient method, based the *thermal network analogy*, to solve *steady*, *one-dimensional* heat transfer problems. Rearranging Eq. (2.2) we have

$$T_o - T_L = \left(\frac{L}{Ak}\right) Q \tag{2.3}$$

Compare Eq. (2.3) with Eq. (2.4), for the steady flow of electric current through a conductor subject to a voltage difference, given by Ohm's law as

$$V_1 - V_2 = RI \tag{2.4}$$

Since Eqs. (2.3) and (2.4) have the same mathematical form we could express Fourier's law in a form analogous to Ohm's law. The heat flow rate and the temperature difference are analogous to the electric current and the voltage difference respectively. The *equivalent thermal resistance* of the slab is then defined as

$$R_{th} = \frac{L}{Ak} \tag{2.5}$$

Hence Eq. (2.3) may be written in the form

$$T_o - T_L = \left(\frac{L}{Ak}\right) Q = R_{th} Q \tag{2.6}$$

Following the practice in electrical circuit analysis we adopt the equivalent thermal network element shown in Fig. 2.3(b), which is analogous to the electrical circuit element in Fig. 2.3(a), to solve *one-dimensional* heat transfer problems related to slabs.

Fig. 2.3 (a) Electrical circuit element, (b) Equivalent thermal network element

2.4.1 *Thermal resistances in series*

There are many air conditioning design applications where heat flow occurs through a slab consisting of parallel layers of different materials as shown schematically in Fig. 2.4. For instance, the walls of buildings usually have a layer of plaster or a siding on the outside. The wall itself could be made of brick or concrete blocks. The inside could have a layer of thermal insulation sandwiched between the wall and a finishing-layer of gypsum board.

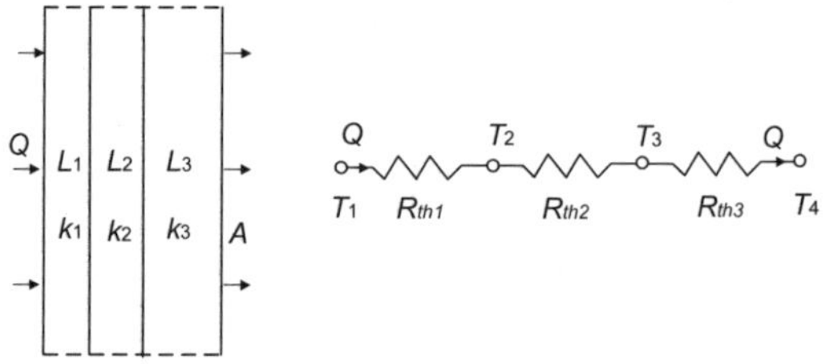

Fig. 2.4 Multi-layered slab and the corresponding thermal network

We now apply Eq. (2.6) to each element of the thermal network shown in Fig. 2.4 to obtain the following equations:

$$T_1 - T_2 = \left(\frac{L_1}{Ak_1}\right) Q = R_{th1} Q \tag{2.7}$$

$$T_2 - T_3 = \left(\frac{L_2}{Ak_2}\right) Q = R_{th2} Q \tag{2.8}$$

$$T_3 - T_4 = \left(\frac{L_3}{Ak_3}\right) Q = R_{th3} Q \tag{2.9}$$

Note that the *steady heat flow rate*, Q through the layers is the same due to energy conservation.

Adding Eqs. (2.7), (2.8) and (2.9) we have

$$T_1 - T_4 = (R_{th1} + R_{th2} + R_{th3}) Q = R_{tot} Q \tag{2.10}$$

We notice from Eq. (2.10) that the total thermal resistance of the composite slab is equal to the sum of the thermal resistances of the individual layers.

2.4.2 *Overall heat transfer coefficient*

The rate of heat flow through a series of parallel layers may also be expressed in terms of the *overall heat transfer coefficient, U*. For the slab shown in Fig. 2.4, this gives

$$Q = AU(T_1 - T_4) \tag{2.11}$$

Comparing Eqs. (2.10) and (2.11) we have

$$AU = \frac{1}{R_{tot}} \tag{2.12}$$

2.4.3 *Thermal resistances in parallel*

There many applications of air conditioning where conduction heat flow occurs through parallel paths. An example is the common wall section, shown in Fig. 2.5(a), where slabs of thermal insulation material like fiberglass are placed in cavities between vertical rectangular structural members, usually called studs. A portion of the heat entering through the surface of the wall section flows through the insulation while the rest flows through the studs. These heat conduction paths are parallel and therefore the thermal circuit representing overall heat flow includes two parallel thermal resistors as shown in Fig. 2.5(b).

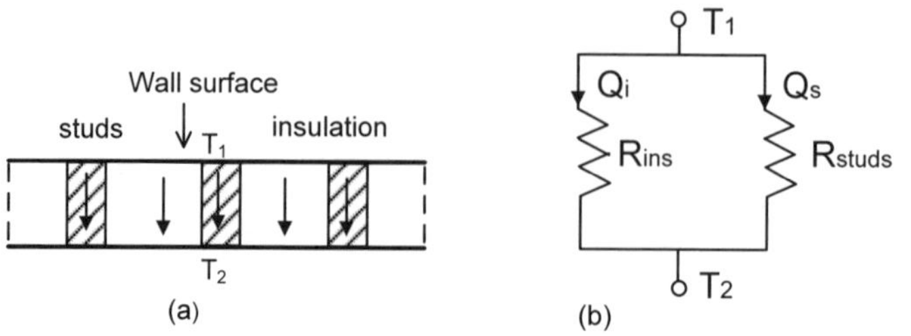

Fig. 2.5 (a) Heat flow through wall section, (b) Equivalent thermal network

The overall thermal resistance of two parallel resistors is given by

$$\frac{1}{R_{tot}} = \left(\frac{1}{R_{ins}} + \frac{1}{R_{stud}}\right) \tag{2.13}$$

The total heat flow, Q may be expressed as

$$Q = \frac{(T_1 - T_2)}{R_{tot}} \tag{2.14}$$

2.4.4 *Boundary conditions*

All heat conduction situations involve boundary conditions. These, in general, are the known or specified values of physical parameters at the boundary of the medium through which heat conduction occurs. We shall use the heat transfer process in a slab to illustrate some of the more common boundary conditions encountered in air conditioning applications.

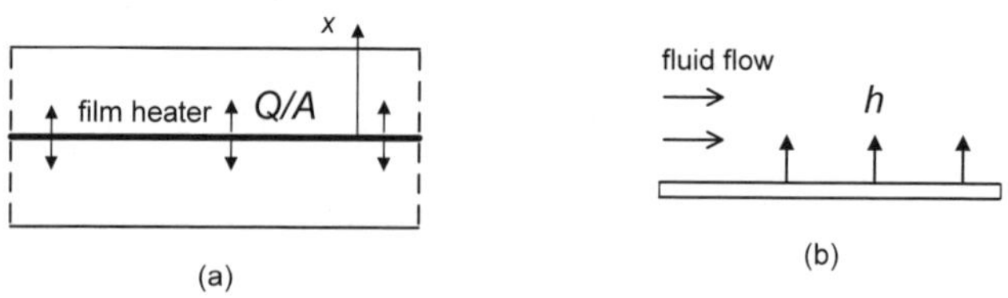

Fig. 2.6 Boundary conditions: (a) Specified heat flux, (b) Convection

(i) *Specified wall surface temperature*:

The simplest boundary condition is where the temperatures of the external surfaces of the slab are specified. For the multi-layered slab shown in Fig. 2.4 this would be the surface temperatures T_1 and T_4.

(ii) *Specified surface heat flux*:

For this boundary condition, the heat flux, defined as the rate of heat flow per unit area of the surface, (*Q*/*A*) is specified. A practical example is a flat air heater where a thin film electrical resistance element is sandwiched between two conducting slabs as shown in Fig. 2.6(a). The rate of heat flow into each slab at $x = 0$, is equal to half the electrical power input to the film heater.

(iii) *Convective heat flow at the surface*:

Heat is transferred to the fluid flowing over a warm surface of the plate, shown schematically in Fig. 2.6(b), by a process known as *forced convection.* A thin layer of fluid adjacent to the surface receives heat by conduction from the plate. The moving packets of fluid distribute this heat to the rest of the fluid. The convection process is governed by *Newton's law of cooling*, which states that the rate of heat transfer is proportional to: (a) the area of the surface, *A*, and (b) the difference between the surface temperature and the 'mean' temperature of the fluid. We can express Newton's law in the mathematical form

$$Q \infty A\left(T_{surface} - T_{fluid}\right) \tag{2.15}$$

Introducing a constant of proportionality, *h* we rewrite Eq. (2.15) as

$$Q = Ah\left(T_{surface} - T_{fluid}\right) \tag{2.16}$$

In Eq. (2.16), the constant *h* is called the *convective heat transfer coefficient.* Unlike the thermal conductivity, *h* is a complex function of a number of physical parameters, including the speed of the fluid and its physical properties like the density, the viscosity, the heat capacity and the thermal conductivity.

Here we have introduced convection mainly as a boundary condition. In section 2.6 of the present chapter we shall discuss convection heat transfer in greater detail.

We could develop a thermal network element to represent the convection heat flow process by rearranging Eq. (2.16) to the form

$$T_{surface} - T_{fluid} = (1/Ah)Q = R_{conv}Q \tag{2.17}$$

Therefore the thermal resistance for a *convective boundary* may be expressed as

$$R_{conv} = \frac{1}{Ah} \tag{2.18}$$

2.5 General Form of Fourier's Law

We considered Fourier's law for a slab of finite thickness in section 2.3. Although this form of the law is adequate to solve one-dimensional problems involving slabs, it is not directly applicable to other shapes like cylinders where the area changes in the direction of heat flow. For such situations we write Fourier's law in the following differential form:

$$Q = -kA\left(\frac{dT}{dx}\right) \tag{2.19}$$

We obtain Eq. (2.19) by applying Eq. (2.2) to a slab of infinitesimal thickness *dx* across which the temperature changes by *dT*. Note that the negative sign is introduced because heat flow occurs in the direction of decreasing temperature.

2.5.1 *Cylindrical systems*

Radial heat conduction through long cylindrical systems are encountered in many practical applications. The heat flow through insulations applied to hot water pipes and chilled water pipes are common examples from air conditioning systems. The analysis of condensers and evaporators also require knowledge of the heat flow through cylindrical tube walls.

Consider the radial heat flow through a long cylindrical annulus as shown in Fig. 2.7. When the section of the annulus of interest is far from the ends of the cylinder, the temperature distribution is a function *only of the radius*. The heat flow is therefore one-dimensional, for all practically purposes. The uniform temperatures of the inner and outer surfaces are T_i and T_o respectively. Let Q be the total heat flow rate across the cylindrical surface of length L and area, $A = 2\pi rL$, at a radius r.

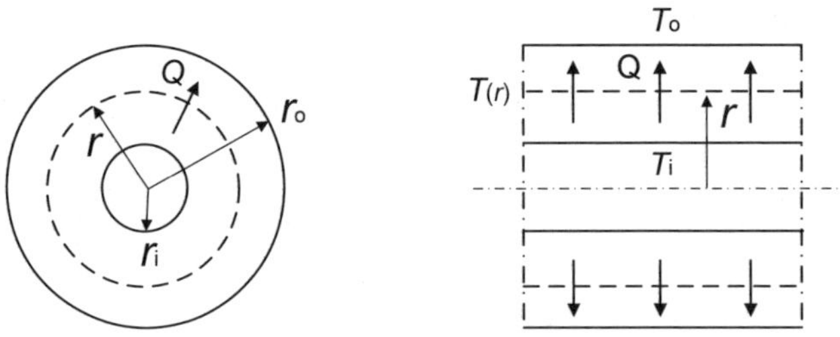

Fig. 2.7 Radial conduction in a cylindrical system

Since the heat conduction is steady, the total heat flow rate at any radius r is constant due to energy conservation. Therefore it follows from Fourier's law (Eq. (2.19)) that

$$Q = -(2\pi rL)k\left(\frac{dT}{dr}\right) = c_1 \tag{2.20}$$

where c_1 is a constant. Rearranging Eq. (2.20) we have

$$\frac{dT}{dr} = -\left(\frac{c_1}{2\pi krL}\right) \tag{2.21}$$

Integrating Eq. (2.21) we obtain

$$T(r) = -(c_1/2\pi kL)\, ln(r) + c_2 \tag{2.22}$$

where c_2 is a constant of integration. We now apply as boundary conditions the specified surface temperatures at the inner and outer radii to evaluate the constants of integration in Eq. (2.22).
Hence at $r = r_i$, $T = T_i$ and at $r = r_o$, $T = T_o$
Substituting the above boundary conditions in Eq. (2.22), we first obtain the constants c_1 and c_2 and then the following expression for the temperature difference:

$$T_i - T_o = \left[\frac{ln(r_o/r_i)}{2\pi kL}\right] Q \tag{2.23}$$

We observe that Eq. (2.23) has a form analogous to Ohm's law. Hence we obtain the thermal resistance of the cylindrical system as

$$R_{cyl} = \frac{ln(r_o/r_i)}{2\pi kL} \tag{2.24}$$

2.6 Conduction with Internal Heat Generation

There are important practical applications of heat transfer where energy entering the medium in a different form is converted to heat within a conducting medium. The most common example is an electrical heater in which electricity flowing through the conductor is converted to heat due to the resistance of the heater. Another example is the heating of glass windows, particularly the thick tinted type in large buildings, due to the absorption of solar radiation in the glass. We shall consider heat flow through such windows in chapter 9.

Consider the steady one-dimensional heat conduction in the slab of thickness L and area A, shown in Fig. 2.8, in which heat is generated internally at the rate of G per unit volume. Let the heat flow rate per unit area, that is the heat flux, be q_x at a section, distant x from one end.

Applying the energy conservation equation to an elemental layer of thickness dx at x we obtain

$$Aq_{(x+dx)} = Aq_x + GAdx \tag{2.25}$$

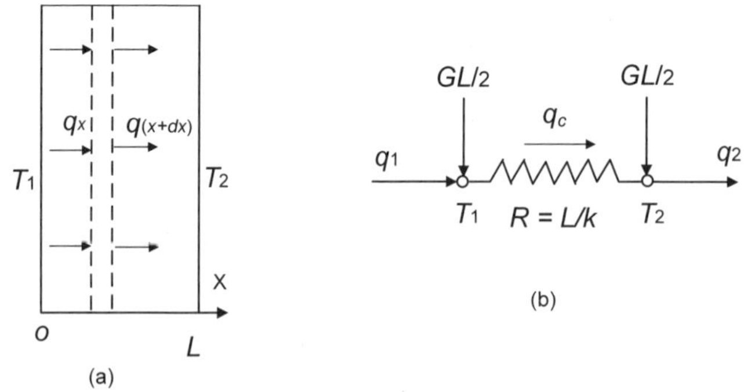

Fig. 2.8 (a) Slab with internal heat generation, (b) Equivalent thermal network

From Eq. (2.25) it follows that

$$\frac{q_{x+dx} - q_x}{dx} = \frac{dq}{dx} = G \tag{2.26}$$

Substituting for the heat flux, q from Eq. (2.19) in Eq. (2.26) we have

$$\frac{d}{dx}\left(-k\frac{dT}{dx}\right) = G \tag{2.27}$$

As an application of Eq. (2.27) we shall consider a simple case where heat is generated uniformly in the slab, and hence G is constant.

The solution of Eq. (2.27) for this condition is obtained by integrating it twice. This gives

$$-kT(x) = \frac{Gx^2}{2} + c_1 x + c_2 \tag{2.28}$$

To determine the constants of integration c_1 and c_2 we assume that the surface temperatures are specified. Hence the boundary conditions are: at $x = 0$, $T=T_1$ and at $x=L$, $T= T_2$. Substituting these conditions in Eq. (2.28) we have

$$c_1 = \frac{k(T_1-T_2)}{L} - \frac{GL}{2} \tag{2.29}$$

and

$$c_2 = -kT_1 \tag{2.30}$$

Substituting the above expressions for c_1 and c_2 in Eq. (2.28) we obtain the temperature distribution within the slab, which has the following parabolic form:

$$T(x) = T_1 - \frac{(T_1-T_2)x}{L} - \frac{G(x^2-Lx)}{2k} \tag{2.31}$$

We obtain the heat fluxes at the two outer surfaces of the slab by differentiating Eq. (2.31) and substituting in Fourier's law, given by Eq. (2.19). Hence we have

$$q_1 = \frac{k(T_1-T_2)}{L} - \frac{GL}{2} = q_c - \frac{GL}{2} \tag{2.32}$$

$$q_2 = \frac{k(T_1-T_2)}{L} + \frac{GL}{2} = q_c + \frac{GL}{2} \tag{2.33}$$

It is interesting to note that the form of Eqs. (2.32) and (2.33) allows us to represent the heat conduction process in the slab by the equivalent thermal network [5] shown in Fig. 2.8(b). The energy balance at the two nodes 1 and 2, representing the surfaces, gives the respective heat fluxes at the surfaces. We shall illustrate the application of the thermal networks developed thus far using the worked examples to follow in this chapter.

2.7 Convection Heat Transfer

2.7.1 *Forced convection heat transfer*

In the section 2.3 we introduced convection merely as a boundary condition for the analysis of conduction problems. The heat transfer coefficient, h was defined as the constant of proportionality in *Newton's law of cooling*. In this section we shall consider in greater detail the physical processes involved in convection.

Consider the ideal situation of a 'static' layer of warm fluid, of thickness y_f, resting on a large cold plate of uniform temperature T_w, as shown in Fig. 2.9(a). Since there is no fluid motion, in the steady state, the fluid would behave like a slab with a heat input at the top and a heat output to the plate at the bottom.

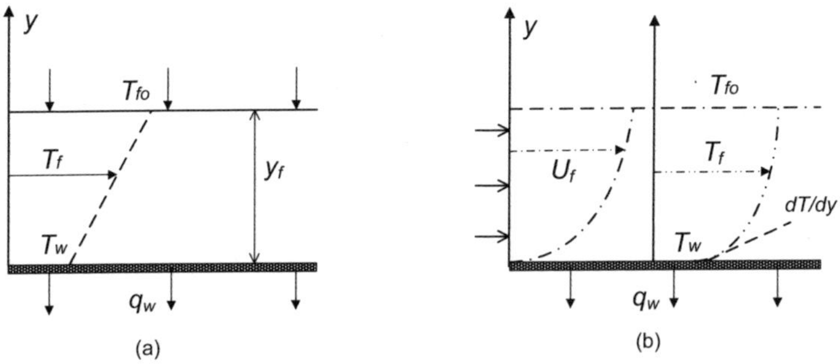

Fig. 2.9 (a) Stationary fluid layer on plate, (b) Fluid moving over a plate

The heat flow rate per unit area at the plate is given by Fourier's law as

$$q_w = \frac{k_f(T_{fo}-T_w)}{y_f} \tag{2.34}$$

The temperature distribution in the fluid is linear as indicated in Fig. 2.9(a).

Figure 2.9(b) shows a warm fluid layer, moving from left to right over a large cold plate. The fluid motion, usually brought about by applying a pressure gradient, is called *forced convection*. The temperature of the plate is maintained constant by removing heat from the bottom.

Due to the frictional force between the surface of the plate and the fluid, and the viscosity of the fluid, the different layers of fluid move at different speeds. This is seen from the graph of fluid speed, U_f versus distance y from the plate. However, a 'thin' layer of fluid adjacent to the plate surface will be at rest. The temperature distribution in the fluid is no longer linear, as in the case of the stationary fluid in Fig. 2.9(a), but will have a curved shape as indicated by the fluid temperature versus distance graph in Fig. 2.9(b).

In addition to heat conduction across thin layers of fluid, there is heat being transported horizontally by the moving layers of fluid. Therefore we cannot apply Fourier's law of heat conduction as we did for the stationary fluid layer in Fig. 2.9(a). However, for the infinitesimally thin fluid layer, at rest adjacent to the plate, we can apply the differential form of Fourier's law, given by Eq. (2.19). Hence we obtain the following expression for the rate of heat flow per unit area *into* the plate from the fluid

$$q_w = k_f \left(\frac{dT}{dy}\right)_{plate} \qquad (2.35)$$

where k_f is the thermal conductivity of the fluid.

Applying Newton's law of cooling we have

$$q_w = h_c\left(T_{fluid} - T_w\right) \qquad (2.36)$$

where T_{fluid} is the fluid temperature 'far away' from the plate, commonly called *the free stream temperature.* The *forced convective heat transfer coefficient*, denoted by h_c, can be expressed in the following form using Eqs. (2.35) and (2.36):

$$h_c = \frac{k_f (dT/dy)_{plate}}{(T_{fluid} - T_w)} \qquad (2.37)$$

It is clear from Eq. (2.37) that the temperature distribution is an important quantity that needs to be determined. Knowing the temperature distribution across the fluid we can compute the temperature gradient at the plate, and hence obtain h_c by substituting in Eq. (2.37).

Analytical expressions for the heat transfer coefficient have been derived for a number of forced convection flow situations, including flow over plates, flow in ducts and flow over concave surfaces. These solutions are available in Refs. [1] and [3].

2.7.2 *Correlations for the heat transfer coefficient*

The forced convection heat transfer coefficient for flow over a plate depends on the following quantities: (a) the length of the plate, L(b) the velocity of the fluid, V and (c) the properties of the fluid such as, the thermal conductivity (k), the specific heat capacity (c_p), the density (ρ), and the viscosity (μ). Therefore we could write the following functional relationship for the heat transfer coefficient, h_c:

$$h_c = F(L, V, k, c_p, \mu) \tag{2.38}$$

For many heating and cooling applications of engineering importance, the relationship given by Eq. (2.38) has been determined experimentally. The data have been correlated using dimensionless quantities leading to compact expressions that are convenient to use in design calculations.

For example, for flow over a flat plate, the local heat transfer coefficient, h_c at a distance x from the entrance is given by [3]

$$Nu_x = \frac{h_c x}{k} = 0.029\, {Re_x}^{0.8} Pr^{0.43} \tag{2.39}$$

where the heat transfer coefficient, h_c is included in the dimensionless group called the Nusselt number, Nu_x. The other two dimensionless groups are the Reynolds number, Re_x and the Prandtl number, Pr, defined by

$$Re_x = \frac{\rho V x}{\mu} \quad \text{and} \quad Pr = \frac{c_p \mu}{k} \tag{2.40}$$

where V is the free stream velocity. In Eqs. (2.39) and (2.40), k is the thermal conductivity, c_p is the specific heat capacity, ρ is the density, and μ is the viscosity.

A comprehensive list of heat transfer correlations for a variety of flow situations and flow geometries are available in Refs. [1–3]. The reader is referred to these sources for a complete treatment of forced convection heat transfer.

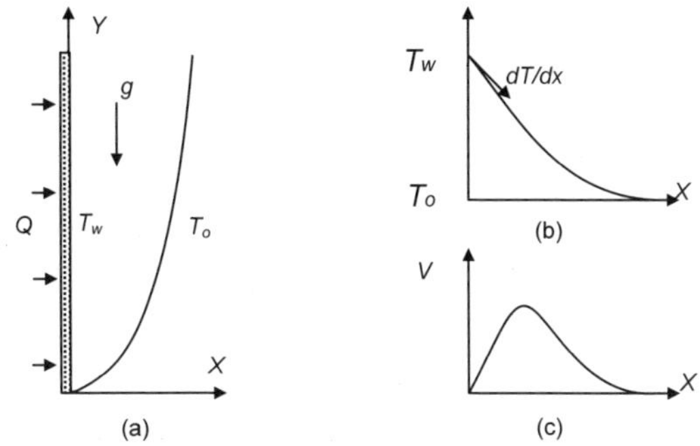

Fig. 2.10 (a) Natural convection from a heated wall, (b) Temperature profile, (c) Velocity profile

2.7.3 *Natural convection heat transfer*

For forced convection, considered in the preceding section, the fluid motion is caused by an external pressure gradient imposed on the fluid by a device like a fan or a pump. In *natural convection or free convection* the fluid flow is caused by the effect of *buoyancy*. This effect is distributed throughout the fluid, and it is the result of the general tendency of fluids to expand when heated at constant pressure.

Figure 2.10(a) shows a vertical wall, heated from the inside. The thin fluid layer adjacent to the hot surface at temperature, T_w receives heat from the wall by conduction and its temperature becomes higher than the rest of the fluid as indicated by the temperature profile in Fig. 2.10(b). The fluid expands with the increased temperature causing it to rise vertically due to buoyancy. The heat received from the wall is thereby distributed throughout the fluid.

However, a 'thin' fluid layer adjacent to the wall remains at rest due to friction. The fluid layers far away from the wall also remain stationary at a uniform temperature, T_o because this region is hardly influenced by the presence of the heated wall. These counteracting effects result in the velocity profile indicated in Fig. 2.10(c) where the fluid velocity reaches a maximum value at some distance from the wall. This heat transfer process is called *free convection*.

Equations (2.35), (2.36) and (2.37), derived earlier for forced convection, may be applied to free convection to obtain an expression for *free convection heat transfer coefficient.*

Numerous dimensionless heat transfer correlations for free convection are available in the published literature on heat transfer [1-3]. As a representative example we include the following correlation [1] for the average Nusselt number, which incorporates the *average heat transfer coefficient*, for flow adjacent to a vertical wall of height y (Fig. 2.10):

$$\overline{Nu}_y = \frac{\bar{h}_y y}{k} = 0.671 {Ra_y}^{1/4} \left(\frac{Pr}{Pr + 0.986 Pr^{1/2} + 0.492}\right)^{1/4} \tag{2.41}$$

The dimensionless groups are defined as follows:

$$\text{Rayleigh number, } Ra_y = \frac{g\beta(T_w - T_o)y^3}{\alpha\vartheta} \tag{2.42}$$

$$\text{Prandtl number, } Pr = \frac{c_p \mu}{k} \tag{2.43}$$

where g is the acceleration due to gravity, β is the coefficient of cubical expansion and α is the thermal diffusivity.

A detailed list of free convection heat transfer correlations for a variety of flow situations and geometries are available in Refs. [1-3].

2.8 Radiation Heat Transfer

In the previous section, we discussed two modes of heat transfer, namely conduction and convection. These types of heat transfer take place in a material medium. Conduction in a solid is caused by the vibration of the atoms constituting it. In convection, energy is transported by moving packets of fluid. In this section we shall consider a third mechanism of heat transfer called *thermal radiation transport* that does not require a material medium.

The phenomenon of radiation transfer occurs because of the propagation of *electromagnetic waves.* All bodies emit radiation at the expense of stored energy. When these waves of radiation fall on other bodies, a fraction of the radiation is absorbed and the rest is reflected. The absorbed energy is converted to stored energy in the receiving body. For instance, it is radiation falling on us that makes us feel warm when we stand some distance away from a hot object like a radiant space heater.

An important example of energy transport by radiation is the flow of energy from the sun to the earth. Radiation from the sun travels through empty space to be partially absorbed and partially reflected by the atmosphere and objects on the surface of the earth.

2.8.1 *Spectrum of electromagnetic radiation*

The electromagnetic waves emitted by a body have a characteristic distribution of wave lengths and frequencies. The portion of the electromagnetic spectrum between the wave lengths of 0.1 to 100 micrometers (μm) is broadly called the *thermal radiation* spectrum. The visible portion of the radiation spectrum, that can be sensed by the human eye, has wave lengths from 0.4 to 0.7 μm.

2.8.2 *Black surface*

A *black surface* (or a *black body*) is an ideal surface. It is defined as a surface that absorbs all radiation falling on it, irrespective of the wave length or the angle of incidence; no radiation is reflected. As will be shown later, no surface can emit more radiation than a black surface at the same temperature. Moreover, a black surface has no preferred direction for radiation emission. The quantity of radiant energy flowing into a solid angle at the point of emission, is independent of the direction of the solid angle. This type of radiation emission is called *diffuse emission.*

A black body is also called an ideal radiator because it emits the maximum possible radiation at a given temperature. The *spectral emissive power* of a blackbody at a given temperature is the total rate of energy emission per unit area per unit wavelength interval. The variation of the spectral emissive power with wavelength and temperature follows the Planck distribution shown in Fig. 2.11. The expression for the Planck distribution is

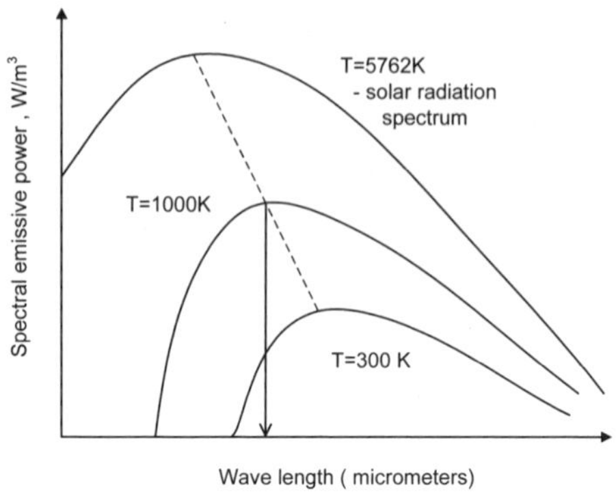

Fig. 2.11 Effect of wave length and temperature on the blackbody spectral emissive power

$$E_{\lambda b} = \frac{c_1 \lambda^{-5}}{exp(c_2/\lambda T) - 1} \tag{2.44}$$

where the constants are given by: $c_1 = 3.742 \times 10^8$ $W\mu m^4/m^2$ and $c_2 = 1.4389 \times 10^4$ μmK. The absolute temperature of the blackbody is T.

The wavelength, λ_{max} at which the spectral emissive power, $E_{\lambda b}$ of a blackbody at a given temperature, T is a maximum may be obtained by differentiating Eq. (2.44) with respect to λ. This gives the following relationship, called *Wien's displacement law.*

$$\lambda_{max} T = 2897.6\ \mu m\text{K} \tag{2.45}$$

The *total emissive power* of a surface is defined as the total rate of radiant energy emission in all directions over all the wavelengths per unit area of the surface. The emissive power of a blackbody is obtained by integrating the expression in Eq. (2.44) over all the wavelengths from zero to infinity. The resulting relationship, called *Stefan-Boltzman law,* may be expressed in the form

$$E_b = \sigma T^4 \tag{2.46}$$

where the constant, $\sigma = 5.67 \times 10^{-8}$ WK^{-4} m^{-2} is the Stefan-Boltzman constant.

2.8.3 *Emissive power of real surfaces*

For real surfaces the emissive power, E is less than the emissive power of a black surface, E_b at the same temperature. The ratio of E to E_b is called the *emissivity* of the surface. In general, the emissivity of a surface depends on the wavelength of the emitted radiation and the direction of emission.

2.8.4 *Emissivity of a gray surface*

A *gray surface* is an ideal surface that is less restrictive than an ideal black surface. It is a surface whose emissivity is independent of the wavelength of radiation emitted and the direction of emission. The emissivity of a *gray surface* is constant and it is defined by the expression,

$$\varepsilon = \frac{E_{gray}}{E_{black}} = \frac{E_{gray}}{\sigma T^4} \tag{2.47}$$

2.8.5 *Absorption, transmission and reflection*

Consider a beam of radiation of intensity G that is incident on a thin flat body as shown schematically in Fig. 2.12.

In general, a fraction, G_r of the incident radiation is reflected, a fraction G_a is absorbed in the material of the body, and a fraction G_t is transmitted through the body.

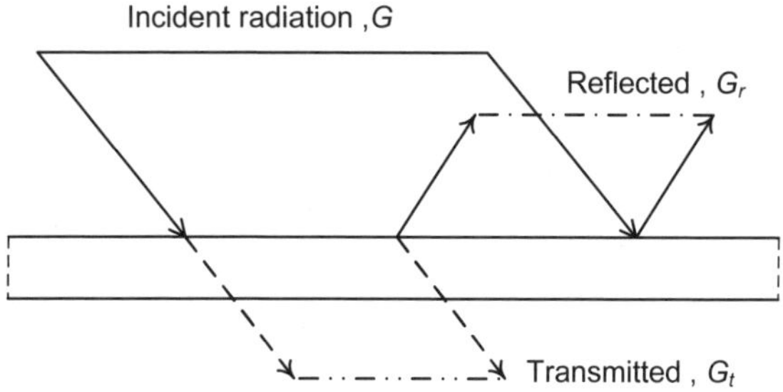

Fig. 2.12 Reflection, transmission and absorption of radiation

We define the *reflectivity* ρ, the *absorptivity*, α and the transmissivity, τ by the expressions,

$$\rho = G_r/G, \ \alpha = G_a/G \quad \text{and} \quad \tau = G_t/G \tag{2.48}$$

From energy conservation it follows that,

$$\rho + \alpha + \tau = 1 \tag{2.49}$$

We define an *opaque surface* as a surface whose transmittance is zero.

Therefore $$\rho + \alpha = 1 \tag{2.50}$$

It is clear that for a *black surface*:

$$\alpha = 1 \quad \text{and} \quad \rho = 0 \tag{2.51}$$

2.8.6 *Kirchhoff's law of radiation*

Kirchhoff's law of radiation states that the *emissivity* of a surface is equal to the *absorptivity*.

Therefore $$\varepsilon = \alpha \tag{2.52}$$

For a black surface:

$$\varepsilon = \alpha = 1 \tag{2.53}$$

2.8.7 *Radiation exchange between two black surfaces*

Two large parallel plates, shown in Fig. 2.13(a), are maintained at *constant* temperatures by external energy interactions. Assume that the space between the plates is evacuated so that conduction and convection heat transfer processes between the plates are absent. The inner surfaces of the two plates are *ideal black* surfaces.

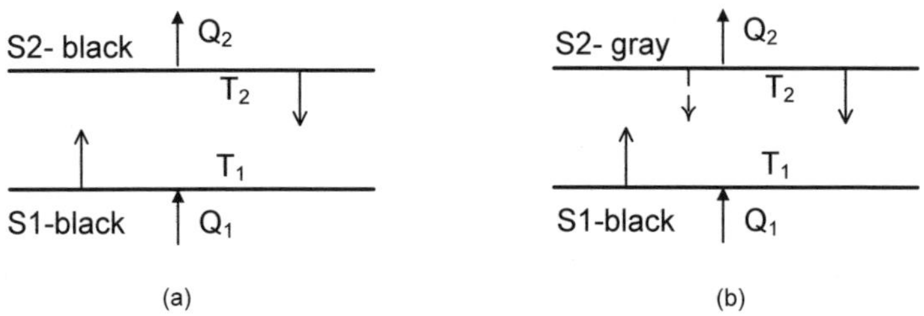

Fig. 2.13 Radiation exchange between surfaces: (a) two black surfaces, (b) a black surface and a gray surface

The steady rate of emission of radiant energy per unit area from surface S1 is

$$E_{b1} = \sigma T_1^{\ 4} \tag{2.54}$$

The steady rate of emission of radiant per unit area from surface S2 is

$$E_{b2} = \sigma T_2^{\ 4} \tag{2.55}$$

Since the areas of the plates are large compared to the distance separating them, all the energy emitted by S2 falls on S1 and vice versa. Moreover, being an ideal black surface, S1 absorbs all the radiant energy incident on it from S2. The net energy exchange between S1 and S2 is obtained by applying the overall energy balance equation to S1. Hence we have

$$Q_1 + E_{b2} - E_{b1} = 0 \tag{2.56}$$

Substituting from Eqs. (2.54) and (2.55) in Eq. (2.56) we obtain the following expression for the *net* rate of energy exchange between the surfaces, Q_1 which is also equal to external energy input:

$$Q_1 = \sigma\left(T_1^{\ 4} - T_2^{\ 4}\right) \tag{2.57}$$

Also, note that for overall energy balance of the system, $Q_2 = Q_1$.

2.8.8 *Radiation exchange between a black surface and a gray surface*

The two plates shown in Fig. 2.13(b) are similar in all respects to the system shown in Fig. 2.13(a), except that surface S2 is now a *gray surface with an emissivity* ε_2.

The steady rate of emission of radiant energy per unit area from the black surface S1 is given by

$$E_{b1} = \sigma {T_1}^4 \tag{2.58}$$

The steady rate of emission of radiant energy per unit area from the gray surface S2 is given by

$$E_{g2} = \varepsilon_2 \sigma {T_2}^4 \tag{2.59}$$

A fraction ρ_2 of the radiation falling on S2 from S1 is reflected. This fraction is given by

$$E_{ref,2} = \rho_2 \sigma {T_1}^4 \tag{2.60}$$

The total radiant energy emitted by S2, and the fraction reflected by S2, are completely absorbed by S1. Applying the *overall energy* balance equation to S1 we have

$$Q_1 + \varepsilon_2 \sigma {T_2}^4 + \rho_2 \sigma {T_1}^4 - \sigma {T_1}^4 = 0 \tag{2.61}$$

Now for the *opaque* surface S2, $\rho_2 = 1 - \alpha_2$. Also, from Kirchhoff's law, $\alpha_2 = \varepsilon_2$. These conditions are substituted in Eq. (2.61), to obtain the *net* rate of energy exchange between S1 and S2 as

$$Q_1 = \varepsilon_2 \sigma \left({T_1}^4 - {T_2}^4\right) \tag{2.62}$$

Also, note that for overall energy balance of the system, $Q_2 = Q_1$.

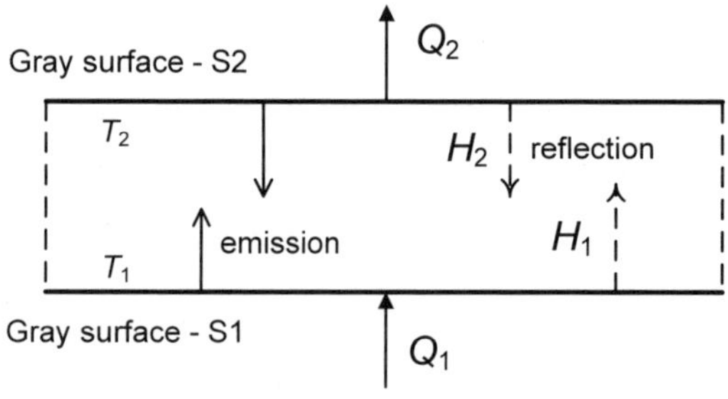

Fig. 2.14 Radiation exchange between two gray surfaces

2.8.9 *Radiation exchange between two gray surfaces*

Two large parallel gray surfaces S1 and S2, with emissivities of ε_1 and ε_2 respectively, are shown in Fig. 2.14. The rates of radiant energy emission by the two gray surfaces are given by

$$E_{g1} = \varepsilon_1 \sigma {T_1}^4 \quad \text{and} \quad E_{g2} = \varepsilon_2 \sigma {T_2}^4 \tag{2.63}$$

Since both surfaces have non-zero reflectivity, the radiation emitted by each surface will undergo multiple reflections at the two surfaces. At each of these reflections, a fraction of the incident radiation is absorbed by the surface. The net energy exchange between the surfaces may be found by tracing these multiple reflections and absorptions in detail.

Here we propose to the use an alternative approach called the *net radiation* method [4,6] to determine the rate of energy exchange. The rates of emission of radiation by the two surfaces due to their temperatures are given by the two expressions in Eq. (2.63).

Let the net radiation fluxes leaving the surfaces S1 and S2 be H_1 and H_2 respectively, as indicated in Fig. 2.14. These fluxes may be thought of as the readings of two radiation detectors facing the respective surfaces. These readings include two components: (i) the direct emission by the surface and (ii) the sum total of all the reflected radiation resulting from multiple reflections from the surfaces. Hence we can express the net radiation fluxes leaving the two surfaces S1 and S2 as

$$H_1 = E_{g1} + \rho_1 H_2 \tag{2.64}$$

$$H_2 = E_{g2} + \rho_2 H_1 \tag{2.65}$$

Note that in Eqs. (2.64) and (2.65), the LHS is the net radiation flux leaving the surface. The first term on the RHS is the direct emission rate by the gray surface and the second term is the reflected fraction from the net radiation flux incident on the surface.

Applying the overall energy balance equation to the *two plates* we have

$$Q_1 + \alpha_1 H_2 - E_{g1} = 0 \tag{2.66}$$

$$Q_2 - \alpha_2 H_1 + E_{g2} = 0 \tag{2.67}$$

where Q_1 and Q_2 are the external energy interactions as indicated in Fig. 2.14.

Applying Kirchhoff's law and the relationships between the various surface properties we obtain the following:

$$\alpha_1 = \varepsilon_1, \rho_1 = 1 - \alpha_1 = 1 - \varepsilon_1 \tag{2.68}$$

and

$$\alpha_2 = \varepsilon_2, \rho_2 = 1 - \alpha_2 = 1 - \varepsilon_2 \tag{2.69}$$

Eliminating H_1 between Eqs. (2.66) and (2.67) and substituting the relations in Eq. (2.63) we obtain

$$H_2 = \frac{\varepsilon_2 E_{b2} + \rho_2 \varepsilon_1 E_{b1}}{1 - \rho_1 \rho_2} \tag{2.70}$$

We substitute for H_2 from Eq. (2.70) in Eq. (2.66) and apply the conditions in Eqs. (2.68) and (2.69). Following some simple algebra, we obtain the energy exchange rate between the surfaces as

$$Q_1 = \varepsilon_{eff}\sigma\left(T_1{}^4 - T_2{}^4\right) \tag{2.71}$$

where the effective emissivity, ε_{eff} is defined by

$$\frac{1}{\varepsilon_{eff}} = \frac{1}{\varepsilon_1} + \frac{1}{\varepsilon_2} - 1 \tag{2.72}$$

Also, note that for overall energy balance of the system, $Q_2 = Q_1$.

2.8.10 *Radiation exchange between a curved surface and a flat surface*

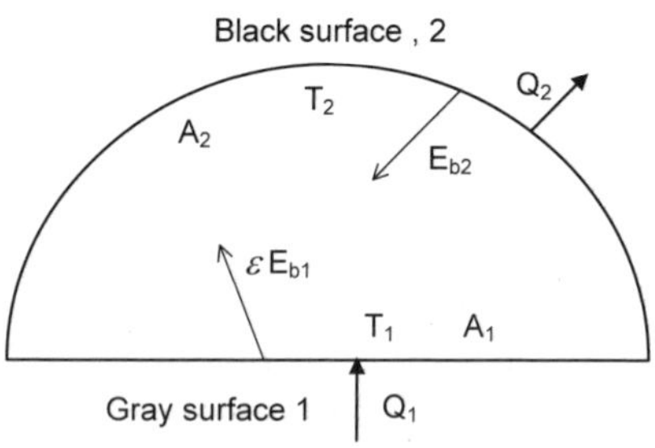

Fig. 2.15 Radiation exchange between a curved surface and a flat surface

The radiation exchange in an enclosure consisting of two surfaces is encountered in several air conditioning design applications. For instance, the ground and the sky may be idealized by considering an enclosure involving a flat surface and a curved surface, as depicted schematically in Fig. 2.15. The *flat gray* surface 1 has an area, A_1 and the *curved black* surface 2 is of area, A_2. The two surfaces are maintained at constant temperatures of T_1 and T_2 by external energy flows Q_1 and Q_2 respectively.

Applying the energy balance equation to surface 1 we obtain

$$Q_1 = A_1\varepsilon_1 E_{b1} - A_2 F_{21}\alpha_1 E_{b2} \quad (2.73)$$

where α_1 is the absorbtivity of surface 1.

Note that only a fraction F_{21} of the radiation emitted by the curved surface lands on the flat surface, while the rest lands on itself. However, all the radiation emitted by the flat surface lands on the curved surface. Substituting from Eqs. (2.54) and (2.55) in Eq. (2.73) we have

$$Q_1 = A_1\varepsilon_1\sigma{T_1}^4 - A_2 F_{21}\varepsilon_1\sigma{T_2}^4 \quad (2.74)$$

Note that in Eq. (2.73), $\alpha_1 = \varepsilon_1$, by Kirchhoff's law.

In order to determine the fraction, F_{21} we consider the situation as, $T_2 \rightarrow T_1$. The net energy exchange, Q_1 is then zero, and Eq. (2.74) takes the form

$$A_1\varepsilon_1\sigma{T_1}^4 - A_2 F_{21}\varepsilon_1\sigma{T_1}^4 = 0 \quad (2.75)$$

Hence we have, $A_1 = A_2 F_{21}$. Substituting in Eq. (2.74) we obtain the net energy exchange rate at any temperature as

$$Q_1 = A_1\varepsilon_1\sigma\left({T_1}^4 - {T_2}^4\right) \quad (2.76)$$

Also, note that for overall energy balance of the system, $Q_2 = Q_1$.

2.9 Worked Examples

Example 2.1 The exterior wall of a large industrial building has a layer of fiberglass insulation of thermal conductivity 0.035 $\mathrm{Wm^{-1}K^{-1}}$ and thickness 8 cm, sandwiched between two plywood sheets of thermal conductivity 0.11 $\mathrm{Wm^{-1}K^{-1}}$ and thickness 1 cm. The inner and outer surfaces of the wall are at 15°C and 32°C respectively. Calculate (i) the overall heat transfer coefficient and (ii) the steady heat flow rate through the wall.

Solution Consider unit area, $A = 1\text{m}^2$ of the wall (see Fig. 2.4). The thermal resistances of the plywood and fiberglass layers are

$$R_{pw} = \frac{L}{Ak} = \frac{0.01}{1\times0.11} = 0.0909 \text{ KW}^{-1}$$

$$R_{fg} = \frac{L}{Ak} = \frac{0.08}{1\times0.035} = 2.286 \text{ KW}^{-1}$$

The total thermal resistance is

$$R_{tot} = R_{pw} + R_{fg} + R_{pw} = 2.4678 \text{ KW}^{-1}$$

(i) The overall heat transfer coefficient, U is given by

$$U = \frac{1}{AR_{tot}} = \frac{1}{2.4678} = 0.405 \text{ WK}^{-1}\text{m}^{-2}$$

(ii) Applying the analogous Ohm's law we obtain the heat flow rate as

$$Q = \frac{(T_o - T_i)}{R_{tot}} = \frac{(32-15)}{2.4678} = 6.89 \text{ W}$$

Example 2.2 The inner surface of a wall of a furnace receives heat by convection from hot gases at 300°C. The convective heat transfer coefficient is 50 $\text{Wm}^{-2}\text{K}^{-1}$. The wall is made of firebrick of thermal conductivity 0.15 $\text{Wm}^{-1}\text{K}^{-1}$, and thickness 8 cm. The outer surface of the wall looses heat by free convection to ambient air at 35°C. The heat transfer coefficient is 20 $\text{Wm}^{-2}\text{K}^{-1}$. Calculate (i) the overall heat transfer coefficient, (ii) the rate of heat loss from the furnace to the ambient and (iii) the temperature at the outer surface of the wall.

Solution Consider unit area $A = 1\text{m}^2$ of the wall (see Fig. 2.4). The thermal resistances of the inner convection layer, the wall and the outer convection layer are respectively

$$R_{ci} = \frac{1}{Ah} = \frac{1}{1\times50} = 0.02 \text{ KW}^{-1}$$

$$R_w = \frac{L}{Ak} = \frac{0.08}{1\times0.15} = 0.533 \text{ KW}^{-1}$$

$$R_{co} = \frac{1}{Ah} = \frac{1}{1\times 20} = 0.05\ \text{KW}^{-1}$$

The total thermal resistance is

$$R_{tot} = R_{ci} + R_w + R_{co} = 0.603\ \text{KW}^{-1}$$

(i) The overall heat transfer coefficient is given by

$$U = \frac{1}{AR_{tot}} = \frac{1}{1\times 0.603} = 1.66\ \text{WK}^{-1}\text{m}^{-2}$$

(ii) Applying the analogous Ohm's law we obtain the heat flow rate as

$$Q = \frac{(T_i - T_o)}{R_{tot}} = \frac{(300-35)}{0.603} = 439.2\ \text{W}$$

(iii) Applying Ohm's law to the outer convection layer we have

$$Q = \frac{(T_{wo} - T_o)}{R_{co}} = \frac{(T_{wo}-35)}{0.05} = 439.2\ \text{W}$$

Hence we obtain the outer surface temperature as, $T_{wo} = 57°\text{C}$.

Example 2.3 A cold room maintained at −10°C has a wall made of two layers of different materials. The inner layer is 2 cm thick and has a thermal conductivity of 0.1 $\text{Wm}^{-1}\text{K}^{-1}$. The outer layer is 4 cm thick and has a thermal conductivity of 0.04 $\text{Wm}^{-1}\text{K}^{-1}$. The outside ambient air temperature is 30°C. The convective heat transfer coefficients on the outside and inside of the wall are 40 $\text{Wm}^{-2}\text{K}^{-1}$ and 20 $\text{Wm}^{-2}\text{K}^{-1}$ respectively. Calculate (i) the rate of heat flow through the wall and (ii) the temperature of the interface between the two layers making the wall.

Solution Consider unit area, $A = 1\text{m}^2$ of the wall (see Fig. 2.4). The thermal resistances of the inner convection layer, the two wall layers and the outer convection layer are respectively

$$R_{ci} = \frac{1}{Ah} = \frac{1}{1\times 20} = 0.05\ \text{KW}^{-1}$$

$$R_{wi} = \frac{L}{Ak} = \frac{0.02}{1\times 0.1} = 0.2\ \text{KW}^{-1}$$

$$R_{wo} = \frac{L}{Ak} = \frac{0.04}{1\times 0.04} = 1.0\ \text{KW}^{-1}$$

$$R_{co} = \frac{1}{Ah} = \frac{1}{1\times 40} = 0.025\ \text{KW}^{-1}$$

The total thermal resistance is

$$R_{tot} = R_{ci} + R_{wi} + R_{wo} + R_{co} = 1.275 \text{ WK}^{-1}$$

(i) Applying the analogous Ohm's law we obtain the heat flow rate as

$$Q = \frac{(T_o - T_i)}{R_{tot}} = \frac{(30-(-10))}{1.275} = 31.37 \text{ W}$$

(ii) Applying Ohm's law between the interface and the outside ambient we have

$$Q = \frac{(T_o - T_{wif})}{R_{wo} + R_{co}} = \frac{(30 - T_{wif})}{1.025} = 31.37 \text{ W}$$

Hence we obtain the interface temperature as, $T_{wif} = -2.15$°C.

Example 2.4 The inner section of a wall is made by placing slabs of fiberglass of thermal conductivity 0.038 $\text{Wm}^{-1}\text{K}^{-1}$ in the vertical spaces formed in a wooden frame of thickness 150 mm. The thermal conductivity of the framing material is 0.15 $\text{Wm}^{-1}\text{K}^{-1}$. The temperatures of the inner and outer surfaces of the wall section are 18°C and 6°C respectively. The area of the insulation is 75% of the total area of the wall. Calculate (i) the total heat flow rate through the wall per unit area and (ii) the heat flow rate through the insulation.

Solution Consider an area A m^2 of the wall. The insulation and the wooden framing occupy $0.75A$ and $0.25A$ respectively (see Fig. 2.5). The heat entering the wall takes parallel paths through the insulation and the frame. The thermal resistances of the insulation and the framing are respectively

$$R_{ins} = \frac{L_i}{A_i k_i} = \frac{150 \times 10^{-3}}{0.75A \times 0.038} = \frac{5.26}{A} \text{ KW}^{-1}$$

$$R_{fr} = \frac{L_f}{A_f k_f} = \frac{150 \times 10^{-3}}{0.25\text{A} \times 0.15} = \frac{4}{\text{A}} \text{ KW}^{-1}$$

The overall thermal resistance of the two parallel resistances is given by

$$\frac{1}{R_{tot}} = \left(\frac{1}{R_{ins}} + \frac{1}{R_{fr}}\right) = A\left(\frac{1}{5.26} + \frac{1}{4}\right) = 0.44A$$

(i) Therefore $R_{tot} = 2.27/A$. The heat flow rate per unit area is

$$Q = \frac{(T_i - T_o)}{AR_{tot}} = \frac{(18-6)}{2.27} = 5.29 \text{ Wm}^{-2}$$

(ii) Apply the analogous Ohm's law to the heat flow path through the insulation area. Hence we have

$$Q_i R_{in} = T_i - T_o$$

Substituting numerical values we obtain

$$Q_i = \frac{A(18-6)}{5.26} = 2.28A$$

Therefore the heat flow rate through the insulation per unit area of the wall is 2.28 Wm^{-2}.

Example 2.5 A steel pipe (k = 50 $\text{Wm}^{-1}\text{K}^{-1}$) of inner diameter 20 cm and thickness 5 mm carries chilled water at 5°C. The pipe is insulated on the outside with a layer of fiberglass (k = 0.035 $\text{Wm}^{-1}\text{K}^{-1}$) of thickness 6 cm. The convective heat transfer coefficient between the chilled water and the inner pipe surface is 100 $\text{Wm}^{-2}\text{K}^{-1}$. The heat transfer coefficient between the outer surface and the ambient air at 28°C is 20 $\text{Wm}^{-2}\text{K}^{-1}$. Calculate the rate of heat flow from the ambient to the chilled water over a length of 5 m of pipe.

Solution Consider a 5 m length of the chilled water pipe. The thermal resistances are as follows (see Fig. 2.7).
For convection from the inner pipe surface to chilled water

$$R_{cp} = \frac{1}{Ah_c} = \frac{1}{2\pi r_{pi} L h_c} = \frac{1}{2\pi \times 0.1 \times 5 \times 100} = 0.00318 \text{ WK}^{-1}$$

For conduction through the pipe wall

$$R_p = \frac{Ln(r_{po}/r_{pi})}{2\pi k_p L} = \frac{ln(10.5/10)}{2\pi \times 50 \times 5} = 3.1 \times 10^{-5} \text{ KW}^{-1}$$

For conduction through the insulation layer

$$R_i = \frac{Ln(r_{io}/r_{ii})}{2\pi k_i L} = \frac{ln(16.5/10.5)}{2\pi \times 0.035 \times 5} = 0.41 \text{ KW}^{-1}$$

For convection from outer insulation surface to ambient air

$$R_{ci} = \frac{1}{Ah_o} = \frac{1}{2\pi r_{ii} L h_o} = \frac{1}{2\pi \times 0.165 \times 5 \times 20} = 9.65 \times 10^{-3}\ \mathrm{WK^{-1}}$$

The total thermal resistance is

$$R_{tot} = R_{ci} + R_i + R_p + R_{cp} = 0.423\ \mathrm{WK^{-1}}$$

Applying the analogous Ohm's law we obtain the heat flow rate as

$$Q = \frac{(T_o - T_i)}{R_{tot}} = \frac{(28-5)}{0.423} = 54.37\ \mathrm{W}$$

Example 2.6 An electric cable has a 2 mm diameter copper wire encased in a 3 mm thick insulator of thermal conductivity 0.2 $\mathrm{Wm^{-1}K^{-1}}$. The heat transfer coefficient between the outer surface of the insulator and the ambient at 30°C is 40 $\mathrm{Wm^{-2}K^{-1}}$. If the maximum temperature limit for the insulator is 150°C, determine the maximum heat generation rate that can be allowed in the wire.

Solution The steady rate of heat generation in the copper wire due to Joule heating is given by, $Q_o = I^2R$. This heat will be conducted through the insulation layer and eventually transferred to the ambient by convection (see Fig. 2.7). Considering unit length of wire, the thermal resistances for these heat transfer processes are as follows.
For conduction through insulation layer:

$$R_i = \frac{Ln(r_{io}/r_{ii})}{2\pi k_i L} = \frac{ln(4/1)}{2\pi \times 0.2 \times 1} = 1.103\ \mathrm{KW^{-1}}$$

For convection from outer insulation surface to ambient air:

$$R_{ci} = \frac{1}{Ah_o} = \frac{1}{2\pi r_{io} L h_o} = \frac{1}{2\pi \times 4 \times 10^{-3} \times 1 \times 40} = 0.9947\ \mathrm{WK^{-1}}$$

The total thermal resistance is

$$R_{tot} = R_{ci} + R_i = 2.0977\ \mathrm{WK^{-1}}$$

In the steady state the rate of heat flow through the insulation is equal to the rate of heat generation in the metal wire. Applying the analogous Ohm's law we obtain the maximum heat generation rate as

$$Q_o = \frac{(T_i - T_o)}{R_{tot}} = \frac{(150-30)}{2.0977} = 57.2\ \mathrm{W}$$

Example 2.7 A concrete wall of thickness 8 cm and thermal conductivity 1.6 $Wm^{-1}K^{-1}$ absorbs solar radiation at a steady rate of 300 Wm^{-2}. The heat transfer coefficient between the outer surface of the wall and the ambient at 30°C is 25 $Wm^{-2}K^{-1}$. The heat transfer coefficient between the room air at 20°C and the inner surface of the wall is 10 $Wm^{-2}K^{-1}$. Calculate (i) the rate of heat flow into the room and (ii) the temperature of the outer surface of the wall.

Solution

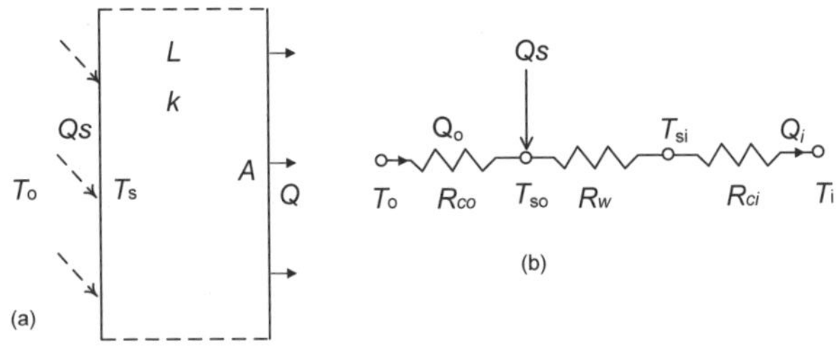

Fig. E2.7.1 (a) Heat flow through concrete wall, (b) Thermal network

Consider unit area of the wall shown in Fig. E2.7.1(a). The absorption of solar radiation at the outer surface produces a surface heat source. The corresponding thermal network is shown in Fig. 2.7.1(b). We introduce a heat source of 300 Wm^{-2} at the node representing the outer surface. The thermal resistances are as follows.

For convection at the outer surface

$$R_{co} = \frac{1}{Ah} = \frac{1}{1\times25} = 0.04 \text{ KW}^{-1}$$

For conduction through the wall

$$R_w = \frac{L}{Ak} = \frac{8\times10^{-2}}{1\times1.6} = 0.05 \text{ KW}^{-1}$$

For convection at the inner surface

$$R_{ci} = \frac{1}{Ah} = \frac{1}{1\times10} = 0.1 \text{ KW}^{-1}$$

Energy balance at the outer surface node gives

$$Q_i = Q_o + 300 \tag{E2.7.1}$$

Applying Ohm's law we have

$$Q_o = \frac{(30 - T_{so})}{R_{co}} = \frac{(30 - T_{so})}{0.04} \quad \text{(E2.7.2)}$$

$$Q_i = \frac{(T_{so} - 20)}{R_w + R_{ci}} = \frac{(T_{so} - 20)}{0.15} \quad \text{(E2.7.3)}$$

Substituting for Q_o and Q_i from Eqs. (E2.7.2) and (E2.7.3) in Eq. (E2.7.1) we obtain the outer surface temperature as, $T_{so} = 37.37°\text{C}$.

Substituting for T_{so} in Eq. (E2.7.3) the heat flow into the room is obtained as 115.8 Wm^{-2}. Note that some of the solar radiation absorbed is lost to the ambient air by convection because, $Q_o = -184.2\ \text{Wm}^{-2}$.

Example 2.8 A building has a large window made of clear plastic material of thickness 20 mm and thermal conductivity 0.173 $\text{Wm}^{-1}\text{K}^{-1}$. On a sunny day the window absorbs solar radiation at a uniform rate of 15 kWm^{-3} throughout the volume of material. The inner and outer surfaces of the window are at 20°C and 33°C respectively. Calculate the rate of heat flow at the inner surface and the outer surface per unit area.

Solution Consider unit area, $A = 1\ \text{m}^2$ of the window. We analyzed the steady heat conduction process with internal heat generation in section 2.6.

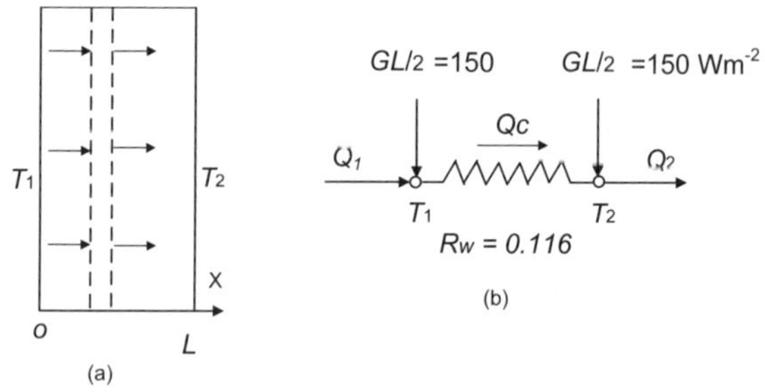

Fig. E2.8.1 Heat conduction in window with internal heat generation

The plastic window and the equivalent thermal network are shown in Figs. E2.8.1(a) and (b) above. The thermal resistance of the window is

$$R_W = \frac{L}{Ak} = \frac{20 \times 10^{-3}}{1 \times 0.173} = 0.116\ \text{KW}^{-1}$$

The conduction heat flux is

$$Q_c = \frac{(T_1 - T_2)}{R_w} = \frac{33-20}{0.116} = 112.1 \text{ Wm}^{-2}$$

Applying the energy balance equation at the node representing the inner wall surface we obtain the heat flux as (see section 2.6)

$$Q_2 = Q_c + \frac{GL}{2} = 112.1 + \frac{15\times10^3\times20\times10^{-3}}{2} = 262.1 \text{ Wm}^{-2}$$

Applying the energy balance equation at the node representing the outer wall surface we obtain the heat flux as (see section 2.6)

$$Q_1 = Q_c - \frac{GL}{2} = 112.1 - \frac{15\times10^3\times20\times10^{-3}}{2} = -37.9 \text{ Wm}^{-2}$$

For overall energy balance:

$$Q_2 - Q_1 = 300 \text{ Wm}^{-2}$$

Note that the thermal network analogy offers a convenient method for solving heat conduction problems with internal heat generation.

Example 2.9 Air at a temperature of 7°C flows over a heated flat plate maintained at a uniform temperature of 47°C. The temperature distribution of the air at a location 20 cm from the entrance section has been measured. This has the form:

$$T(y) = 7 + 40\left[1 - 1.5\left(\frac{y}{2.12}\right) + 0.5\left(\frac{y}{2.12}\right)^3\right]$$

where $T(y)$ is the temperature in °C at a location distant y mm from the surface of the plate. The thermal conductivity of air is 0.026 $\text{Wm}^{-1}\text{K}^{-1}$.

(i) Calculate fluid temperature at a distance of 1.5 mm from the surface.
(ii) Calculate the heat flux at the surface of the plate.
(iii) Calculate the convective heat transfer coefficient at the location.

Solution (i) The temperature at y = 1.5 mm is obtained by direct substitution in the measured temperature distribution given above. Hence we have

$$T(1.5) = 7 + 40\left[1 - 1.5\left(\frac{1.5}{2.12}\right) + 0.5\left(\frac{1.5}{2.12}\right)^3\right] = 11.63\text{°C}$$

Note that at a distance of 2.12 mm the temperature is 7°C. This distance over which the temperature changes from 47°C to 7°C is called the *thermal boundary layer.*

(ii) To obtain the heat flux at the wall surface we apply Fourier's law of conduction at the plate surface where, $y = 0$. Hence we have

$$-k\left(\frac{dT}{dy}\right) = -0.026 \times 10^3 \times 40\left[\frac{-1.5}{2.12} + 3 \times 0.5\frac{y^2}{2.12^3}\right] = 735.8$$

Therefore the heat flux at the plate is, $Q_o = 735.8$ Wm^{-2}.

(ii) We obtain the heat transfer coefficient, h_c by applying Newton's law of cooling. Hence we have

$$Q_o = h_c(T_{wall} - T_{air}) = h_c(47 - 7) = 735.8 \text{ Wm}^{-2}$$

Therefore the convective heat transfer coefficient, h_c is 18.4 Wm^{-2}K^{-1}.

Example 2.10 The external vertical wall if a room, made of thin metal, absorbs solar radiation at the rate of 480 Wm^{-2}. The wall looses heat to the air on the outside due to wind. On the inside the wall looses heat to room air by natural convection. The wind speed is 8 ms^{-1}. Assuming steady-state heat transfer, calculate the temperature of the wall.
The convective heat transfer coefficient (Wm^{-2}K^{-1}) at a wind speed V (ms^{-1}) is given by

$$h_w = 2.8 + 3V$$

The natural convection heat transfer coefficient, h_c (Wm^{-2}K^{-1}) for a vertical wall is given by

$$h_c = 4.2(T_s - T_a)^{1/4}$$

where T_s and T_a are the temperatures of the surface and the surrounding air respectively.

Solution Assume that (i) the metal wall is a good conductor of heat so that the temperature change across its thickness is negligible, (ii) radiation transfer is negligible, and (iii) the heat transfer processes are steady.

The energy balance equation for the wall may be expressed in the form

$$Q_{solar} = Q_{conv.} + Q_{wind}$$

Substituting the relevant expressions for the heat transfer rates in the above equation we have

$$480 = h_w(T_w - 31) + h_c(T_w - 29)$$

where T_w is the uniform wall temperature.
Substituting the given heat transfer correlations for h_w and h_c in the above equation we obtain

$$480 = (2.8 + 3V)(T_w - 31) + 4.2(T_w - 29)^{1/4}(T_w - 29)$$

For the purpose of solving the above equation for T_w we make the substitution, $\theta = T_w - 29$. Hence we have

$$480 = (2.8 + 3 \times 8)(\theta - 2) + 4.2\theta^{1.25}$$

A trial and error procedure gives the solution of the above equation as $\theta = 15.2$°C. Hence the temperature of the wall is

$$T_w = \theta + 29 = 44.2\text{°C}$$

Example 2.11 Air at 20°C flows with a speed of 2.2 ms^{-1} past a cylindrical heater generating 100 W per meter length. The diameter of the cylinder is 1.3 cm. The convective heat transfer coefficient ($Wm^{-2}K^{-1}$) for air flow past a cylinder is given by:

$$h_{air} = 2.5V^{0.46}D^{-0.54}$$

where V(ms^{-1}) is the air speed, and D(m) is the diameter of the cylinder. Calculate (i) the temperature of the heater and (ii) the air speed required to reduce the heater temperature to 75°C.

Solution (i) In the steady state all the heat generated within the cylinder flows out through its surface over which air flows. The convective heat transfer coefficient for this process is given by the correlation:

$$h_{air} = 2.5V^{0.46}D^{-0.54} = 2.5 \times 2.2^{0.46} \times 0.013^{-0.54} = 37.5\ \text{Wm}^{-2}\ \text{K}^{-1}$$

Apply the energy balance equation to the cylinder, neglecting radiation heat transfer from the surface. Hence we have

$$Q_{gen} = Ah_{air}\left(T_{surface} - T_{air}\right) \qquad \text{(E2.11.1)}$$

Substituting numerical values in Eq. (E2.11.1) we obtain

$$100 = \pi \times 0.013 \times 1.0 \times 37.5\left(T_{surface} - 20\right)$$

Hence the surface temperature of the cylinder is 85.3°C

(ii) When the surface temperature is 75°C, the heat transfer coefficient is obtained by substituting in Eq. (E2.11.1). This gives

$$100 = \pi \times 0.013 \times 1.0 \times h_{air}(75 - 20)$$

Hence the new heat transfer coefficient is 44.5 $Wm^{-2}K^{-1}$.

The air speed is obtained by substituting in the given heat transfer correlation. Hence we have

$$44.5 = 2.5 \times V^{0.46} \times 0.013^{-0.54}$$

Therefore the new air speed required is 3.2 ms^{-1}.

Example 2.12 An electric heater has a cylindrical shape with 2 cm diameter and 30 cm length. The outer surface of the heater is at 120°C and it can be treated as a black surface. Calculate (i) the wavelength of radiation for which the spectral emissive power is a maximum and (ii) the total energy emitted by the heater during 5 minutes.

Solution (i) We obtain the wavelength for the maximum spectral emissive power by applying Wien's displacement law (Eq. 2.45). Hence we have

$$\lambda_{max} T = 2897.6 \ \mu m\text{K}$$

Substituting numerical data in the above equation we obtain

$$\lambda_{max}(273 + 120) = 2897.6 \ \mu m\text{K}$$

Hence the wavelength is 7.37 μm.

(ii) We obtain the total energy emitted per unit area by the black surface by applying *Stefan-Boltzman law* (Eq. 2.46). Hence we have

$$E_b = \sigma T^4$$

where the constant, $\sigma = 5.67 \times 10^{-8}$ WK^{-4} m^{-2} is the Stefan-Boltzman constant. Hence we have

$$Q_b = \mathrm{A}\sigma T^4 = \pi \times 0.02 \times 0.3 \times 5.67 \times 10^{-8} \times 393^4 = 25.49 \text{ W}$$

The total energy emitted in 5 minutes is = $25.49 \times 300 = 7648$ J

Example 2.13 A large opaque gray surface maintained at 90°C, emits thermal radiation at the rate of 700 Wm^{-2}. (a) Calculate the emissivity, absorptivity and the reflectivity of the surface. (b) If a large black surface maintained at 30°C is placed parallel to the gray surface, calculate the net rate of exchange of thermal radiation between the two surfaces.

Solution (a) The emissive power of a gray surface is given by

$$E_g = \varepsilon\sigma T_2{}^4$$

Substituting numerical values in the above equation we have

$$\varepsilon = \frac{700}{5.67 \times 10^{-8} \times (273 + 90)^4} = 0.71$$

By Kirchhoff's law the absorptivity, α is equal to the emissivity, ε which is 0.71. The reflectivity is given by

$$\rho = 1 - \alpha = 0.29$$

(b) The net rate of energy exchange between the gray surface and the parallel black surface is given by Eq. (2.62) as

$$Q = \varepsilon_2\sigma\left(T_1{}^4 - T_2{}^4\right)$$

Substituting numerical values in the above equation we obtain the energy exchange rate as

$$Q = 0.71 \times 5.67 \times 10^{-8}(363^4 - 303^4) = 360 \text{ Wm}^{-2}$$

Example 2.14 Two large parallel black plates are maintained at 120°C and 40°C respectively. The space between the plates is evacuated. (a)

Calculate the net rate of radiation exchange between the two surfaces. (b) In order to decrease the radiation heat transfer between the two surfaces a radiation shield in the form of a thin perfectly conducting plate is placed between the two black surfaces. The emissivities of the two sides of the shield are 0.6 and 0.8 respectively. Calculate (i) the temperature of the radiation shield and (ii) the net rate of energy input to the black surfaces.

Solution (a) The net rate energy exchange between two black plates is given by Eq. (2.57) as

$$Q_b = \sigma(T_1{}^4 - T_2{}^4)$$

Substituting numerical values we have

$$Q_b = 5.67 \times 10^{-8}[(273 + 120)^4 - (273 + 40)^4] = 808.35\ \mathrm{Wm^{-2}}$$

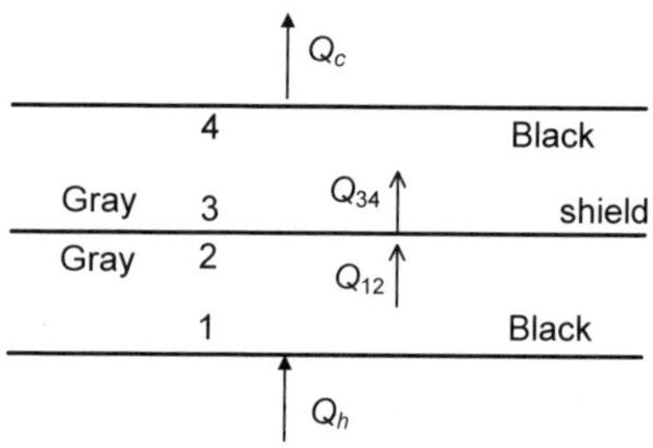

Fig. E2.14.1 Analysis of a radiation shield

(b) The two black plates with a gray radiation shield in between is shown in Fig. E2.14.1. The net rate of energy exchange between surfaces 1 and 2 is Q_{12}, and between surfaces 3 and 4 is Q_{34}. The system is in a steady state and there are no other modes of energy transfer between the surfaces. Therefore for overall energy balance

$$Q_{12} = Q_{34} = Q_h = Q_c \qquad \text{(E2.14.1)}$$

where Q_h and Q_c are the external energy interactions.

Now the net energy exchange rate between a black surface and a gray surface is given by Eq. (2.62). Apply this equation to the energy exchange between each gray surface of the shield and the black surface opposite to it and substitute the resulting expressions in Eq. (E2.14.1). Hence we have

$$Q_{12} = Q_{34} = \varepsilon_2\sigma\left(T_1{}^4 - T_2{}^4\right) = \varepsilon_3\sigma\left(T_3{}^4 - T_4{}^4\right) \quad \text{(E2.14.2)}$$

Let the uniform temperature of the shield be T_s. Therefore $T_2 = T_3 = T_s$. Hence Eq. (E2.14.2) may be written as

$$\varepsilon_2\sigma\left(T_1{}^4 - T_s{}^4\right) = \varepsilon_3\sigma\left(T_s{}^4 - T_4{}^4\right) \quad \text{(E2.14.3)}$$

Substituting numerical values in Eq. (E2.14.3) we have

$$0.6\sigma\left[(273 + 120)^4 - T_s{}^4\right] = 0.8\sigma\left[T_s{}^4 - (273 + 40)^4\right]$$

The solution of the above equation gives the temperature of the shield as $T_s = 354.0$ K. Substituting for T_s in Eq. (E2.14.2) we obtain the net heat exchange rate as, $Q_{12} = 277$ Wm^{-2}. Note that the radiation shield reduces the net energy exchange rate significantly.

Example 2.15 Two large vertical parallel gray surfaces A and B are maintained at 80°C and 20°C respectively. The emissivities of A and B are 0.9 and 0.7 respectively. There is convective heat transfer between the surfaces due to the movement of air in the space between the two surfaces. The convective heat transfer coefficient h_c (Wm^{-2}K^{-1}) is given by:

$$h_c = 0.95(T_A - T_B)^{1/3}$$

where T_A (K) and T_B (K) are the temperatures of the surfaces.

Calculate, (i) the convective heat transfer rate between the surfaces, (ii) the radiation heat transfer between the surfaces and (iii) the external energy input rate to plates A and B.

Solution (i) We assume that the thermal radiation transfer process and the convection process are independent (see Fig. E2.15.1). The rate of heat transfer per unit area by convection is given by

$$Q_{con} = h_c(T_A - T_B) \quad \text{(E2.15.1)}$$

Substituting the given correlation for the heat transfer coefficient, h_c in the Eq. (E2.15.1) we have

$$Q_{con} = 0.95(T_A - T_B)^{1/3}(T_A - T_B)$$

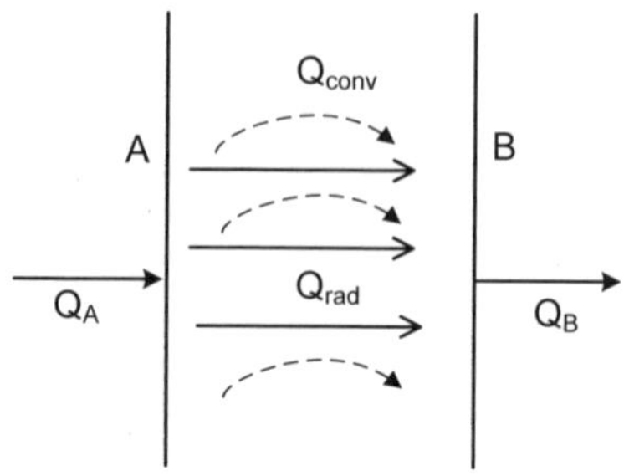

Fig. E2.15.1 Simultaneous radiation and convection heat transfer

Substituting numerical values in the above equation we obtain

$$Q_{con} = 0.95(T_A - T_B)^{4/3} = 0.95(80 - 20)^{4/3} = 223.2 \text{ Wm}^{-2}$$

(ii) The radiation heat transfer between the two gray surfaces per unit area is given by Eq. (2.71) as:

$$Q_{rad} = \varepsilon_{eff}\sigma\left(T_A{}^4 - T_B{}^4\right) \quad \text{(E2.15.2)}$$

where the effective emissivity, ε_{eff} is defined by

$$\frac{1}{\varepsilon_{eff}} = \frac{1}{\varepsilon_1} + \frac{1}{\varepsilon_2} - 1 \quad \text{(E2.15.3)}$$

Substitute numerical values in Eq. (E2.15.3) to obtain the effective emissivity. Hence we have

$$\frac{1}{\varepsilon_{eff}} = \frac{1}{0.9} + \frac{1}{0.7} - 1 = 1.54$$

Therefore the effective emissivity is 0.65. Substituting numerical values in Eq. (E2.15.2) we have

$$Q_{rad} = \varepsilon_{eff}\sigma\left(T_A{}^4 - T_B{}^4\right) = 0.65 \times 5.67 \times 10^{-8}(353^4 - 293^4)$$

This gives the radiation heat transfer rate as, $Q_{rad} = 300.6 \text{ Wm}^{-2}$.

(iii) Applying the overall energy balance equation to plates A and B we have

$$Q_A = Q_B = Q_{con} + Q_{rad} = 223.2 + 300.6 = 523.8 \text{ Wm}^{-2}$$

where Q_A is the rate of energy *input* to surface A and Q_B is the rate of energy *removal* from surface B as indicated in Fig. E2.15.1.

Problems

P2.1 The exterior wall of a building consists of 100 mm thick face brick (k = 0.9 $Wm^{-1}K^{-1}$), 40 mm thick polystyrene insulating board (k = 0.036 $Wm^{-1}K^{-1}$), 125 mm thick concrete block (k = 1.8 $Wm^{-1}K^{-1}$) and 15 mm thick interior gypsum board (k = 0.18 $Wm^{-1}K^{-1}$). The inside and outside convective heat transfer coefficients are 6.5 $Wm^{-2}K^{-1}$ and 22.5 $Wm^{-2}K^{-1}$ respectively. The outside air temperature is −5°C and the inside air temperature is 20°C. The wall is 3 m high and 15 m long. Calculate (i) the rate of heat loss through the wall, and (ii) the temperatures of the interior and exterior surfaces of the wall.
[*Answers*: (1) 715.9 W, (ii) 17.55°C, −4.3°C]

P2.2 Derive Eq. (2.13) for the overall thermal resistance of two parallel resistors, by applying Ohm's law to the equivalent thermal network.

P2.3 The section of a vertical wall is made up of fiberglass insulation slabs separated by wooden studs. The thermal conductivity of fiberglass and wood are 0.04 $Wm^{-1}K^{-1}$ and 0.18 $Wm^{-1}K^{-1}$ respectively. The thickness of the wall is 160 mm. The temperatures of the inner and outer surfaces of the wall section are 22°C and 4°C respectively. The ratio of the insulation area to the total wall area is 0.8. Calculate (i) the total heat flow rate through the wall per unit area and (ii) the rate of heat flow through unit area of studs.
[*Answers*: (i) 7.65 Wm^{-2}, (ii) 20.27 Wm^{-2}]

P2.4 A steel pipe (k = 48 $Wm^{-1}K^{-1}$) of a heating system carries wet steam at 120°C. The inner and outer diameters of the pipe are 15 cm and 16 cm respectively. The pipe is insulated on the outside with rockwool insulation (k = 0.05 $Wm^{-1}K^{-1}$) of thickness 8 cm. The ambient air temperature is 32°C. The outside heat transfer coefficient is 20 $Wm^{-2}K^{-1}$. The thermal resistance between the inner pipe surface and the steam is

negligible. Calculate (i) the rate of heat flow from the steam to the ambient over a 5 m length of pipe, (ii) the temperature of the insulation surface and (iii) the rate of condensation of steam over the 5 m length of pipe if the latent heat of steam is 2200 $kJkg^{-1}$.
[*Answers*: (i) 195 W, (ii) 33.9°C, (iii) 0.319 $kg.hour^{-1}$]

P2.5 The exterior wall of a building consists of concrete blocks (k = 1.6 $Wm^{-1}K^{-1}$) of thickness 80 mm, and an insulation layer (k = 0.12 $Wm^{-1}K^{-1}$) of thickness 15 mm. The exterior surface of the concrete absorbs solar radiation at the rate 400 Wm^{-2}. The outside and inside heat transfer coefficients are 20 $Wm^{-2}K^{-1}$ and 6 $Wm^{-2}K^{-1}$ respectively. The outside and inside air temperatures are 28°C and 18°C respectively. Calculate (i) the outside wall surface temperature and (ii) the rate of heat flow into the air inside.
[*Answers*: (i) 44.2°C, (ii) 76.5 Wm^{-2}]

P2.6 A clear plastic window of a building has a thickness of 22 mm and a thermal conductivity of 0.17 $Wm^{-1}K^{-1}$. It absorbs solar radiation at a uniform rate of 20 kWm^{-3} throughout the material. The outside and inside heat transfer coefficients are 25 $Wm^{-2}K^{-1}$ and 8 $Wm^{-2}K^{-1}$ respectively. The outside and inside ambient temperatures are 25°C and 15°C respectively. Calculate (i) the inside and outside wall surface temperatures, and (ii) the rate of heat flow into the room.
[*Answers*: (i) 38.8°C, 35°C, (ii) 190.4 Wm^{-2}]

P2.7 A heated flat plate is maintained at a uniform temperature of 50°C. The plate is cooled by air flowing over it with a free-stream temperature of 10°C. The measured temperature distribution of the air at a location 22 cm from the entrance section has the form:

$$T(y) = 10 + 40\left[1 - 1.4\left(\frac{y}{2.5}\right) + 0.4\left(\frac{y}{2.5}\right)^3\right]$$

where $T(y)$ is the temperature in °C at a location distant y mm from the surface of the plate. The thermal conductivity of air is 0.025 $Wm^{-1}K^{-1}$.

(i) Calculate air temperature at a distance of 2 mm from the surface.
(ii) Calculate the heat flux at the surface of the plate.
(iii) Calculate the convective heat transfer coefficient at the location.

[*Answers*: (i) 13.4°C, (ii) 560 Wm^{-2}, (iii) 14 $Wm^{-2}K^{-1}$]

P2.8 The temperatures of the outer and inner glasses of a triple-glazed (3 glass sheets) vertical window of a building are 8°C and 22°C respectively. The spaces between the glass sheets have air at atmospheric pressure. Assume that radiation transfer and the thermal resistances of the glass sheets are negligible. Calculate (i) the temperature of the glass sheet in the middle, and (ii) the rate of heat loss through the window. The convective heat transfer coefficient h_c ($Wm^{-2}K^{-1}$) between two vertical surfaces is given by:

$$h_c = 0.95(T_1 - T_2)^{1/3}$$

where T_1 (K) and T_2 (K) are the temperatures of the surfaces.
[*Answers*: (i) 15°C, (ii) 12.7 Wm^{-2}]

P2.9 The flat roof of a building has an area of 36 m^2. It may be treated as a gray surface with an emissivity of 0.6. The roof absorbs solar radiation at steady rate of 390 Wm^{-2}. The roof looses heat by convection due to wind. The convective heat transfer h_w ($Wm^{-2}K^{-1}$) at a wind speed V (ms^{-1}) is given by:

$$h_w = 2.8 + 3V$$

The roof exchanges thermal radiation with the sky which may be treated as a black hemispherical surface at the ambient temperature of 22°C. The wind speed is 5 ms^{-1}. The roof is well insulated on the inside so that the heat flow into the building is negligible. Calculate (i) the temperature of the roof, (ii) the rate of heat loss by convection and (iii) the rate of heat loss by radiation.
[*Answers*: (i) 40°C, (ii) 11.5 kW, (iii) 2.48 kW]

P2.10 Two large vertical parallel gray surfaces are maintained at 70°C and 10°C respectively. The emissivities of the surfaces are 0.9 and 0.6 respectively. The movement of the air in the space between the two surfaces causes convective heat transfer. The convective heat transfer coefficient h_c ($Wm^{-2}K^{-1}$) is given by:

$$h_c = 0.95(T_1 - T_2)^{1/3}$$

where T_1 (K) and T_2 (K) are the temperatures of the surfaces. Calculate, (i) the convective heat transfer rate between the surfaces, (ii) the radiation heat transfer rate between the surfaces and (iii) the external energy input rate to the two plates.
[*Answers*: (i) 223.15 Wm^{-2}, (ii) 236.9 Wm^{-2}, (iii) 460 Wm^{-2}]

P2.11 Two large parallel gray surfaces with emissivities of 0.9 and 0.8 are maintained at 90°C and 25°C respectively. The space between the plates is evacuated. (a) Calculate net rate of radiation heat transfer between the surfaces. (b) A thin meal radiation shield with gray surfaces of emissivity 0.3 is placed between the two plates. Calculate (i) the temperature of the shield and (ii) the external heat input to the two plates.
[*Answers*: (a) 395 Wm^{-2}, (b) (i) 63.3°C, (ii) 75.2 Wm^{-2}]

P2.12 The sky and the ground at a location may be idealized as a *black* hemispherical surface at temperature 18°C placed over a large horizontal *gray* surface of emissivity 0.75 and temperature 28°C. Calculate the steady rate of heat loss per unit area from the ground to the sky.
[*Answer*: 44.1 Wm^{-2}]

P2.13 For the two-surface enclosure shown in Fig. 2.15, obtain an expression for Q_2 by writing the radiation balance equation for surface 2.

References

1. Bejan, Adrian, *Heat Transfer*, John Wiley & Sons, Inc. New York, 1993.
2. Bejan, Adrian and Kraus, Allen D., *Heat Transfer Handbook*, John Wiley & Sons, Inc. New York, 2003.
3. Mills, Anthony F., *Heat Transfer*, Irwin, Richard D., Inc., Boston, MA, 1992.
4. Siegel, Robert and Howell, John R., *Thermal Radiation Heat Transfer*, Hemisphere Publishing Corporation, Washington, 1992.
5. Wijeysundera, N.E., 'Application of the network analogy to one-dimensional systems with internal heat generation', *Applied Energy*, 12, 1982, 229–236.

6. Wijeysundera, N.E., 'A net radiation method for the transmittance and absorptivity of a series of parallel regions'. *Solar Energy*, 17, 1975, 75–77.

Chapter 3

Refrigeration Cycles for Air Conditioning Applications

3.1 Introduction

The main function of refrigeration systems, or reversed heat engines, as these are sometimes called, is to transfer heat continuously from a low temperature region to a high temperature region using energy from an external source. In the more common applications of refrigeration, like food preservation and air conditioning, the aim of the refrigeration system is to maintain the temperature of the cold region below the local ambient temperature. The same system, however, may be used to transfer the heat extracted from a cold region to a high temperature region. Such systems are called *heat pumps*.

The functioning of conventional air conditioning systems rely critically on the refrigeration plant used to cool and dehumidify the air in the space. In the case of compact 'window-unit' air conditioners the air in the conditioned space like a room, is cooled directly by passing it over the tubes carrying cold refrigerant. A variation of this arrangement, called the 'split-unit', is used to cool the air in several separate spaces using the same refrigeration plant. For this purpose, the cold refrigerant is pumped to cooling coils located in the different spaces.

In the case of 'central air conditioning systems', used in large commercial buildings, the air in the different zones is cooled by chilled water produced by a refrigeration plant. The latter refrigeration plants are commonly known as 'chillers'.

The refrigeration plants of most air conditioning systems operate on the vapor compression refrigeration cycle using mechanical work or electricity as the energy input. In this chapter we shall analyze several refrigeration cycles used in practical air conditioning systems including vapor compression cycles, and vapor absorption cycles. Several applications of vapor compression cycles in HVAC systems will also be described.

3.2 Carnot Refrigeration Cycle Using a Vapor

In this section we shall review some fundamental aspects of refrigeration cycles which are usually covered in detail in most standard text books on engineering thermodynamics [2,7]. The ideal refrigeration cycle, with the highest coefficient of performance (COP) of any cycle operating between given heat absorption and rejection temperatures, is the *reversed-Carnot heat engine cycle* or *the Carnot refrigeration cycle.*

We shall now consider the operation of a Carnot refrigeration cycle where the working fluid is a vapor. A schematic diagram of the closed system that transfers heat from a low temperature region to a high temperature region is depicted in Fig. 3.1(a). The temperature-entropy (*T-s*) diagram of the Carnot refrigeration cycle in which all processes occur within the liquid-vapor region is shown in Fig. 3.1(b).

The working fluid (the refrigerant) enters the condenser as a saturated vapor at 1 and rejects heat isothermally to the high temperature region before exiting at 2 as a saturated liquid. The liquid undergoes an isentropic ($s_2 = s_3$) expansion 2-3 in an expander or turbine to produce a work output. During the evaporation process 3-4 the wet vapor absorbs heat isothermally from the low temperature region. Finally, the wet vapor is compressed from 4 to 1 in an isentropic ($s_1 = s_4$) process to complete the cycle. The work output of the expander (2-3) supplies a fraction of the work required by the compressor (4-1) while the rest is supplied from an external source.

We now apply the steady flow energy equation to each of the steady-flow processes of the cycle, neglecting the kinetic and potential energy of the fluid, to obtain the following expressions.
The heat rejection rate during the process 1-2 in the condenser is

$$\dot{Q}_{12} = \dot{m}(h_1 - h_2) \tag{3.1}$$

where $\dot{m}$ is the steady mass flow rate of the refrigerant. The fluid enthalpies at the entrance and exit are h_1 and h_2 respectively.

The heat absorption rate during the process 3-4 in the evaporator is

$$\dot{Q}_{34} = \dot{m}(h_4 - h_3) \tag{3.2}$$

The fluid enthalpies at the entrance and exit are h_3 and h_4 respectively.

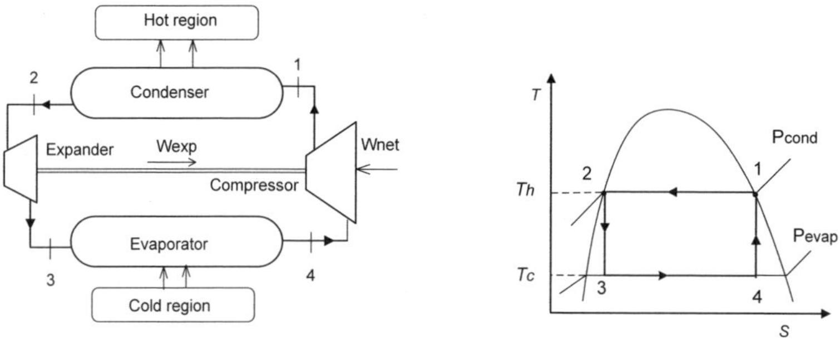

(a) (b)

Fig. 3.1 The Carnot refrigeration cycle

Applying the energy balance equation to the cycle, the net work input is

$$\dot{W}_{net} = \dot{Q}_{12} - \dot{Q}_{34} \tag{3.3}$$

The coefficient of performance of the *refrigeration* cycle 1-2-3-4, the purpose of which is to absorb heat from the *cold* region, is given by

$$COP_{ref} = \frac{\dot{Q}_{34}}{\dot{W}_{net}} \tag{3.4}$$

If the main purpose of the cycle is to supply heat to the hot region, then the cycle 1-2-3-4 is called a *Carnot heat pump* cycle and its coefficient of performance is given by

$$COP_{hp} = \frac{\dot{Q}_{12}}{\dot{W}_{net}} \tag{3.5}$$

Manipulating Eqs. (3.1) to (3.4) we have

$$COP_{ref} = \frac{h_4 - h_3}{(h_1 - h_2) - (h_4 - h_3)} \tag{3.6}$$

In order to express Eq. (3.6) in terms of temperatures, we invoke the well known thermodynamic relation [2]:

$$dh = Tds - vdP \tag{3.7}$$

The pressure and temperature during the condensation process 1-2 and the evaporation process 3-4 are constant. Therefore $dP = 0$ and, $T = T_{sat}$, the saturation temperature. Hence for these phase change processes, Eq. (3.7) takes the form

$$dh = T_{sat}ds \tag{3.8}$$

Integrating Eq. (3.8) and applying the resulting equation to processes 1-2 and 3-4 we obtain the following relations:

$$h_1 - h_2 = T_h(s_1 - s_2) \tag{3.9}$$

$$h_4 - h_3 = T_c(s_4 - s_3) \tag{3.10}$$

where T_c and T_h are the cold and hot fluid temperatures respectively.

For the *ideal reversible cycle* these fluid temperatures are the same as the respective cold and hot region (reservoir) temperatures. Due to the rectangular shape of the *T-s* diagram, $s_2 = s_3$ and $s_1 = s_4$.
Substituting from Eqs. (3.9) and (3.10) in Eq. (3.6) we have

$$COP_{ref} = \frac{T_c}{T_h - T_c} \tag{3.11}$$

Using Eq. (3.5) instead of Eq. (3.4), in the above analysis it is possible to show that

$$COP_{hp} = COP_{ref} + 1 \tag{3.12}$$

Therefore

$$COP_{hp} = \frac{T_h}{T_h - T_C} \tag{3.13}$$

We note that the COP of a Carnot refrigeration cycle depends *only* on the heat absorption temperature, T_c and the heat rejection temperature, T_h. It is independent of the properties of the refrigerant.

3.3 Standard Vapor Compression Cycle

The construction of a practical refrigerating plant operating on the Carnot refrigeration cycle with a vapor as the working fluid has been hampered by a number of practical difficulties. The isothermal processes 1-2 and 3-4 of the cycle, shown in Fig. 3.1(b), may be well approximated in practice. However, the development of a compressor to carry out the *wet compression* process 4-1, in a practical manner, poses several challenges.

In the case of a reciprocating compressor, liquid refrigerant may get trapped in the space between the head of the piston and the cylinder head, causing valve damage. Ideally, there should only be dry vapor at the end of the compression process. But this may not be the case in practice because droplet evaporation requires a finite time. Moreover, lubricant from the walls of the cylinder may be carried away by liquid refrigerant, accelerating wear. Due to these reasons, the compression process in actual refrigeration plants is carried out in the *dry* vapor region (4 -1), as shown in Fig. 3.2(a).

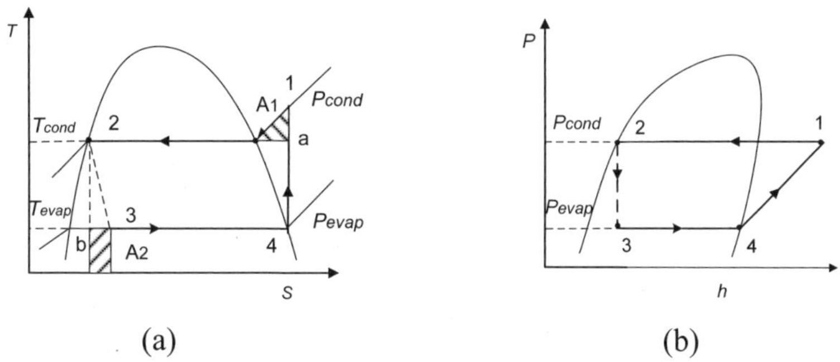

Fig. 3.2 Standard vapor compression cycle

In the Carnot refrigeration cycle, the expansion process 2-3 occurs isentropically in the vapor-liquid mixture region, as seen in Fig. 3.1(b). It is difficult to implement this process practically because of the effects of droplets, and the carryover of the lubricant. Also, the work output of the expander or turbine is only a small fraction of the work required by the compressor. For these reasons the expansion of the working fluid from the condenser pressure to the evaporator pressure is carried out in an *expansion valve*. Ideally, the enthalpy of the refrigerant is constant during this irreversible throttling process. We have indicated the expansion process by a broken line 2-3 in the *T-s* diagram. The increase in fluid entropy during 2-3 is due to the irreversible nature of the expansion process. Most practical refrigeration systems operate on the modified cycle 1-2-3-4, shown in Fig. 3.2(a), called the *standard vapor-compression* cycle. A schematic diagram of the corresponding vapor compression refrigeration system is shown in Fig. 3.3.

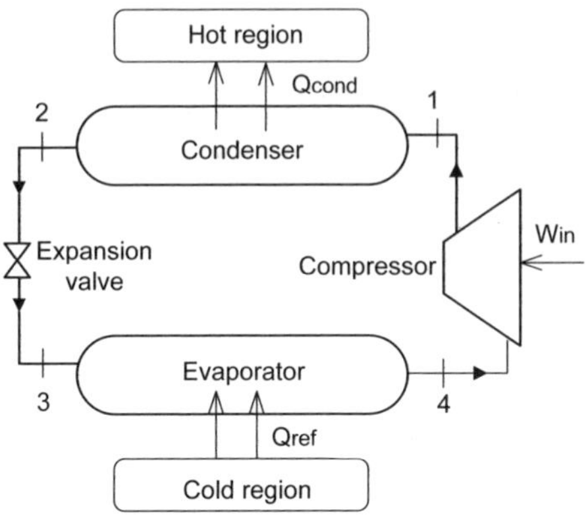

Fig. 3.3 Vapor compression refrigeration cycle

We shall now compare the performance of the *reversed Carnot cycle* 4-a-2-b, which has rectangular shape in the *T-s* diagram in Fig. 3.2(a), with the *standard vapor compression cycle* 4-1-2-3. As evident from Eq. (3.8), the area under a constant pressure line on the *T-s* diagram is equal to the difference in enthalpy of the states represented by the end points of the line. Applying this condition we obtain the following relations between the heat interactions of the two cycles, 4-a-2-b and 4-1-2-3,

$$q_{12} = h_1 - h_2 = q_{c,out} + A_1 \tag{3.14}$$

and

$$q_{34} = h_4 - h_3 = q_{c,in} - A_2 \tag{3.15}$$

where $q_{c,out}$ and $q_{c,in}$ are respectively the heat rejected and absorbed per unit mass in the reversed Carnot cycle 4-a-2-b. A_1 and A_2 are the shaded areas shown in Fig. 3.2(a). Applying the energy balance equation to the closed cycle 1-2-3-4 we have

$$w_{14} = q_{12} - q_{34} \tag{3.16}$$

Substituting from Eqs. (3.14) and (3.15) in Eq. (3.16) we obtain

$$w_{14} = w_c + A_1 + A_2 \tag{3.17}$$

where w_c is the work input in the reversed Carnot cycle 4-a-2-b.

The COP of the standard vapor compression cycle 1-2-3-4 is

$$COP_r = \frac{q_{34}}{w_{14}} = \frac{q_{c,in} - A_2}{w_c + A_1 + A_2} \tag{3.18}$$

The COP of the reversed Carnot cycle 4-a-2-b is

$$COP_c = \frac{q_{c,in}}{w_c} \tag{3.19}$$

From Eqs. (3.18) and (3.19) we have

$$\frac{COP_r}{COP_c} = \frac{1-A_2/q_{c,in}}{1+(A_1+A_2)/w_c} \tag{3.20}$$

The ratio of the COPs of the two cycles is a function of the areas A_1 and A_2. The area A_1, sometimes called the 'superheat horn', represents the additional work input required per unit mass due to superheating in the standard vapor compression cycle [3]. The area A_2 represents the loss in refrigerating effect due to the expansion valve. From Eq. (3.17), we could also interpret area A_2 as the work lost due to throttling. For actual refrigerants, the areas A_1 and A_2 depend on the shape of the saturation lines on the *T-s* diagram. Therefore we could use the magnitudes of these areas to compare graphically the impact of different refrigerants on the COP of the standard vapor compression cycle.

The pressure-enthalpy (*P-h*) diagram, shown Fig. 3.2(b), is a chart used commonly to represent refrigerant properties. It is used widely for the graphical analysis of vapor compression cycles. The heat rejection, process 1-2, the expansion process 2-3, and the heat absorption process 3-4, are represented by straight lines as seen in Fig. 3.2(b). Since the above heat interactions are proportional to the enthalpy differences, their magnitudes can be read off directly from the *P-h* chart of the refrigerant. The *P-h* charts used in practice also include constant temperature, constant entropy and constant specific volume lines, enabling these properties to be obtained directly from the chart. However, unlike the *P-v* and *T-s* diagrams, the area of the *P-h* diagram does not have a special physical significance.

3.4 Analysis of the Standard Vapor Compression Cycle

Consider the vapor compression cycle shown in Fig. 3.2. Apply the steady flow energy equation to each of the processes, neglecting the kinetic and potential energy of the fluid.

The *heat rejection rate* in the condenser during process, 1-2 is

$$\dot{Q}_{12} = \dot{m}(h_1 - h_2) \tag{3.21}$$

where $\dot{m}$ is the steady mass flow rate of the refrigerant.
The *heat absorption rate* in the evaporator during process, 3-4 is

$$\dot{Q}_{34} = \dot{m}(h_4 - h_3) \tag{3.22}$$

For the *adiabatic* throttling process 2-3 through the valve:

$$h_2 = h_3 \tag{3.23}$$

Applying the energy balance equation to the cycle, the net work input is

$$\dot{W}_{net} = \dot{Q}_{12} - \dot{Q}_{34} \tag{3.24}$$

The coefficient of performance of the *standard vapor compression refrigeration* cycle 1-2-3-4, which is used to absorb heat from the *cold* region, is given by

$$COP_r = \frac{\dot{Q}_{34}}{\dot{W}_{net}} \tag{3.25}$$

Substituting from Eqs. (3.21) to (3.24) in Eq. (3.25) we obtain

$$COP_r = \frac{h_1 - h_3}{(h_1 - h_2) - (h_4 - h_3)} = \frac{h_1 - h_3}{h_1 - h_4} \tag{3.26}$$

An additional design parameter of practical importance, called the *compressor displacement*, is the theoretical compressor swept volume per unit time. This is given by

$$\dot{V}_{com} = \dot{m}v_4 \tag{3.27}$$

where v_4 is the specific volume of the refrigerant at entry to the compressor.

3.5 Actual Vapor Compression Cycle

There are a number of differences between the *standard* vapor compression cycle, a-b-c-d, and the *actual* cycle 1-2-3-4, shown on the *T-s* diagram in Fig. 3.4. Some of these differences are due to practical reasons while others are intentional. The fluid pressure drops in the condenser and the evaporator are unavoidable. These pressure drops increase the overall pressure difference across the compressor, which in turn, increases the required work input to the compressor.

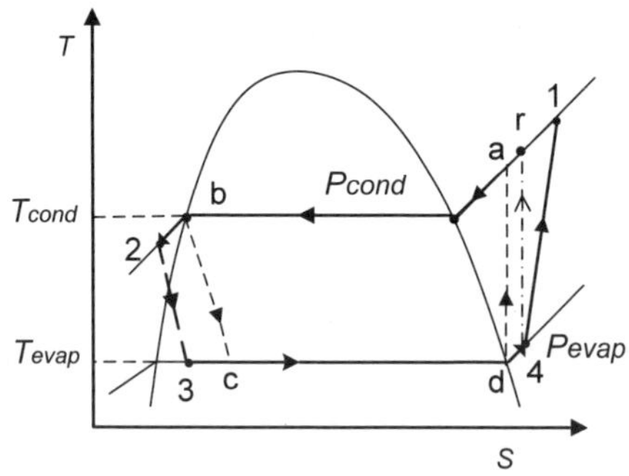

Fig. 3.4 Actual vapor compression cycle

In the standard cycle, the liquid entering the expansion valve at b is just saturated. In actual practice, however, subcooling the liquid slightly to 2 is found to be advantageous. In the case of the common form of expansion valve, known as the *capillary tube*, the presence of any vapor at the entrance to the tube could cause flow blockage. Subcooling ensures that only liquid enters the expansion valve. It is also desirable to slightly superheat the vapor to 4 to ensure that no liquid drops are present in the vapor entering the compressor.

In some practical refrigeration systems superheating of the vapor from d-4 is carried out in a counter-flow heat exchanger using the saturated liquid leaving the condenser at b. The liquid leaving the heat exchanger is subcooled to state 2.

The compression process in the actual cycle is not isentropic as assumed in the standard cycle. This difference could be readily accounted for by defining the *isentropic efficiency* of the compressor as the ratio of the isentropic work input to the actual work input. In Fig. 3.4 let r denote the state of the refrigerant at the end of the *isentropic* compression process, 4-r. Hence $s_r = s_4$. Assuming the actual compression process 4-1 to be *adiabatic*, we define the *isentropic efficiency* as

$$\eta_{is} = \frac{W_{ise}}{W_{act}} = \frac{h_r - h_4}{h_1 - h_4} \tag{3.28}$$

The differences between the standard cycle and the actual cycle mentioned above, relate mainly to the *internal* processes of the vapor compression cycle. There are also two important *external* factors that affect the performance of the actual refrigeration cycle. For the refrigerant to absorb heat from the cold space in the evaporator, and to reject heat to a heat sink in the condenser, there has to be finite temperature differences. These temperature differences constitute external irreversibilities, and they lower the COP of the actual cycle in comparison to the standard cycle.

Thus far we have used the COP as the main performance index of refrigeration cycles. However, for water chillers of HVAC systems, operating on the vapor compression cycle, it is more common to use a dimensionless efficiency index called the '*kW per Ton*'. This index is essentially the inverse of the COP. It is the amount of electrical energy (kW) consumed by the compressor of the cycle in producing one *Ton of refrigeration* (RTon), which is equal to 3.5168 kW. Hence we have

$$COP = \frac{3.5168}{kW\ per\ RTon}$$

3.6 Modifications to the Standard Vapor Compression Cycle

The standard vapor compression refrigeration cycle functions efficiently when the evaporating temperatures are relatively high. However, when the evaporator temperature is lowered, the required compressor power input increases significantly [3,6]. Consequently, the COP of the cycle decreases. Furthermore, as the evaporator temperature is lowered, the compressor displacement, which is proportional to the size of the compressor, and the discharge temperature, increase. Multi-stage compression with intercooling can mitigate some of these detrimental effects at low evaporator temperatures.

3.6.1 *Two-stage compression with flash intercooling*

A modified vapor compression cycle with two-stage compression and flash intercooling is shown schematically in Fig. 3.5(a). The *P-h* diagram of the cycle is depicted in Fig. 3.5(b).

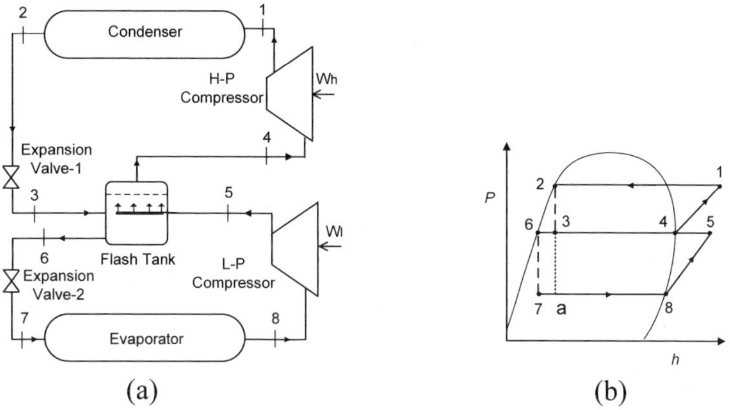

Fig. 3.5 Two-stage compression with flash intercooling

The saturated liquid leaving the condenser at 2 first expands through valve-1 to an intermediate pressure, before entering the flash intercooler at state 3. The liquid from the intercooler at state 6 then expands through valve 2 to the evaporator pressure. The low pressure vapor leaving the evaporator at state 8 is compressed by the low-pressure compressor to state 5, before entering the intercooler. The expansion valve 1 also functions as a float-valve to maintain a constant liquid level in the intercooler. Ideally, the vapor entering the high-pressure compressor at 4 is a saturated vapor at the intermediate pressure. This vapor is compressed to state 1 and enters the condenser, to complete the cycle.

If a single compressor was used with the same condenser and evaporator pressures the flow through the expansion valve would take place from 2 to a. We observe that during this expansion, the vapor fraction in the expanding mixture, usually called flash gas, increases progressively. Unfortunately, the flash gas makes no contribution to the heat absorption process in the evaporator, but the recompression of this gas back to the condenser pressure consumes additional work. Therefore, the compressor work input could be significantly reduced by continuously removing flash gas from the expanding mixture.

In the practical arrangement shown in Fig. 3.5(a), a single flash tank is used to effectively remove a fraction of the flash gas. Generally, the reduction in compressor work due to the flash intercooler depends on the type refrigerant used in the cycle. Furthermore, the higher liquid fraction at 7 increases the heat absorbed in the evaporator.

3.6.2 *Two-stage compression with two evaporators*

Refrigeration systems with two evaporators operating at different temperatures are common in industrial applications. The arrangement shown is Fig. 3.6(a) has two evaporators and a flash intercooler. The *P-h* diagram for the refrigeration cycle is shown in Fig. 3.6(b), and the analysis of the cycle is illustrated in worked example E3.12.

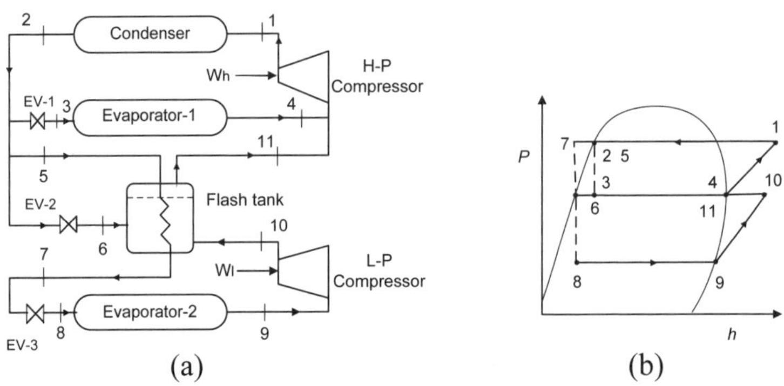

Fig. 3.6 Two-stage compression with two evaporators

3.7 Refrigerants for Vapor Compression Systems

Some of the desirable characteristics of a substance that could be used as a refrigerant in practical vapor compression systems are listed below:

(i) a positive gage pressure in the evaporator, to prevent leakage of ambient air into the system, (ii) a relatively high critical pressure, to enable operation in the liquid–vapor region of the phase diagram, (iii) a low freezing point, to operate at low evaporating temperatures, (iv) a high latent enthalpy of evaporation, to increase the heat absorbed per unit mass, (v) thermophysical properties that facilitate high heat transfer, (vi) inertness and stability, (vii) satisfactory oil solubility and low water solubility, (viii) non-toxicity and non-irritability (ix) non-flammability,

(x) low-cost, (xi) easy leakage detection, (xii) low ozone-depletion potential, and (xiii) low global-warming potential.

Substances considered as possible refrigerants for use in vapor compression systems include halocarbons, azeotropes, hydrocarbons and inorganic compounds.

(i) Halocarbons contain one or more halogens chlorine, fluorine and bromine. The names and chemical formulae of the more common halocarbon refrigerants are: Refrigerant 11 or R-11 ($CFCl_3$), Refrigerant 12 or R-12 (CF_2Cl), Refrigerant 22 or R-22 ($CHClF_2$), Refrigerant 134a or R134a (CH_2F-CF_3).

When first introduced, halocarbons were hailed as excellent refrigerants, because they had most of the desirable properties mentioned above, (i) to (xi). However, later studies found several halocarbon refrigerants to cause ozone layer depletion in the outer atmosphere, and also contribute to global warming. The use of these refrigerants has been discontinued by international agreement. The main environmentally safe halocarbon refrigerants are R134a and R22.

(ii) Mixtures of substances that evaporate and condense at a single temperature are called *azeotropes*. Refrigerants with desirable properties have been produced by mixing existing refrigerants. For example, refrigerant R-502, an azeotrpic mixture of R-22 and R-115, is used in industrial and commercial refrigeration systems.

(iii) Hydrocarbons such as ethane (R-50), methane (R-170), and propane (R-290) have been used as refrigerants in industrial applications where the safety procedures needed to deal with their flammability can be implemented satisfactorily.

(iv) Ammonia, an inorganic substance, has been widely used as a refrigerant in industrial applications. It has most of the desirable properties mentioned above. However, ammonia has not been used widely in domestic applications because it mixes easily with water, and it is a strong irritant.

3.8 Vapor Compression Systems for Air Conditioning Applications

In this section we briefly describe two practical air conditioning systems that use vapor compression refrigeration plants to cool and dehumidify air in a conditioned space. The first is a simple window air conditioner, and the second is a more complex central plant, commonly used in large commercial buildings.

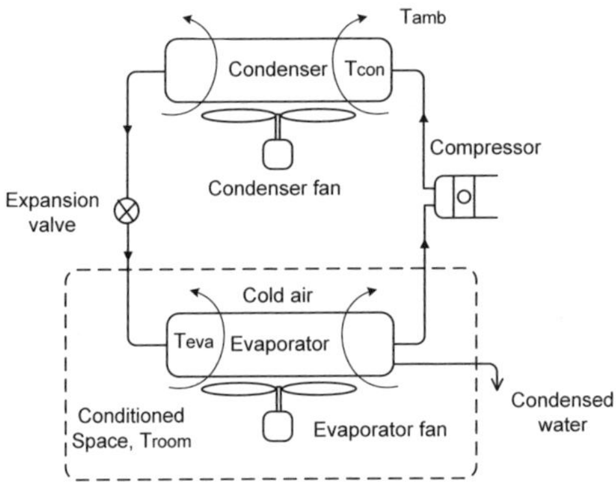

Fig. 3.7 Schematic of window-unit air conditioner

3.8.1 *Window-unit air conditioners*

A schematic diagram of a window-unit air conditioner is shown in Fig. 3.7. The refrigerant evaporator is located inside the conditioned space and the evaporator fan circulates indoor air across a bank of finned tubes carrying cold refrigerant. The air is cooled and dehumidified due to heat transfer to the cold refrigerant, and the moisture condensing from the air is discharged to the external ambient. The refrigerant is compressed by a reciprocating compressor, driven by an electric motor. The hot refrigerant flowing through a bank of finned tubes in the condenser is cooled by ambient air circulated by the condenser fan. The liquid refrigerant leaving the condenser expands through the expansion valve to complete the cycle.

For heat removal in the evaporator, the average refrigerant temperature T_{eva} in the evaporator has to be lower than T_{room}, the temperature of the conditioned space. Similarly, for heat rejection, the average refrigerant temperature, T_{cond} in the condenser has to be higher than the ambient temperature T_{amb}. The COP of the refrigeration system depends on the temperature differences, (T_{room} - T_{eva}) and (T_{cond} - T_{amb}). From the expression for the ideal COP, given by Eq. (3.11), we note that the COP can be increased by lowering the above temperature differences.

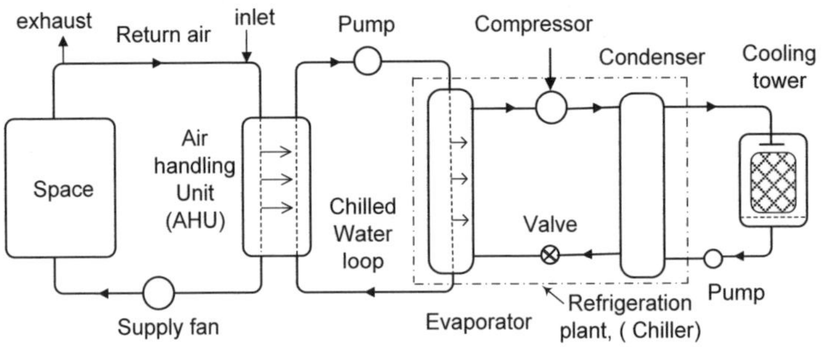

Fig. 3.8 Central air conditioning system

3.8.2 *Central air conditioning systems using chilled water*

The main subcomponents and fluid flow circuits of a typical central air conditioning system are depicted schematically in Fig. 3.8. The refrigeration plant, usually called a chiller, produces chilled water at a temperature of about 4°C to 10°C. The chilled water pump circulates the water through a heat exchanger, called the air handling unit (AHU), where the air from the conditioned space is cooled and dehumidified. The AHU is both a heat exchanger and a mass exchanger because, in addition to sensible cooling, the AHU also removes water vapor from the air by cooling it below the dew point temperature. The water condensing from the air is discharged to the ambient, and the conditioned air is delivered to the space by a fan. The detailed analysis of cooling and dehumidifying processes will be considered in chapter 7.

For health reasons, a fraction of the circulating air from the space is exhausted and replaced with an equal amount of fresh ambient air. These ventilation requirements are discussed in chapter 10.

On the condenser side of the chiller, cooling water is circulated between the condenser and a cooling tower, which eventually rejects heat to the atmosphere. The detailed analysis of cooling tower performance is given in chapter 6.

Figure 3.8 shows that there are several heat exchangers in the central air conditioning plant, each of which transfers heat across a finite temperature difference. These temperature differences constitute external irreversibilities of the subcomponents, and they lower the overall COP of the system. The design–analysis of dry and wet coil heat exchangers is presented in chapter 7.

3.8.3 *Compressors of water chillers*

The compressor is a key component of a water chiller on which its COP depends. The three main types of compressors used in commercial chillers are: (i) reciprocating compressors, (ii) centrifugal compressors, and (iii) rotary compressors. We shall now outline the salient features of these different compressors.

(i) *Reciprocating compressors*

Reciprocating compressors are positive-displacement devices, available as hermetically sealed units, and open units. In the more common hermitically sealed units, the motor and the compressor are directly coupled, and housed in a casing sealed from the atmosphere. In open units, the compressor and motor are housed separately, and their shafts are coupled through a drive.

For refrigeration capacities of 100 tons or less, reciprocating compressors are more cost effective than other types of compressors. Moreover, due to their higher condensing temperatures, they are better suited for applications where air-cooled condensers have to be used. The total cooling load to be met, and the total output of the chiller system, can be matched by installing a number of smaller reciprocating units that are switched on or off depending on the required cooling load.

A major disadvantage of reciprocating chillers is that they usually require a high level of maintenance, because they have more moving parts compared to other types of chillers. Reciprocating chillers are not suitable for cooling loads in excess of about 200 tons.

(ii) *Centrifugal compressors*

Centrifugal chillers are variable-volume units where the refrigerant is compressed by the centrifugal force created within the rotating impellor. The flow rate of the refrigerant through the compressor is controlled by adjusting inlet guide vanes on the impellor, which in turn, control the cooling capacity of the chiller. These chillers are also available as hermetically-sealed units and open units.

The capacities of commercial centrifugal chillers range from about 100 to 1000 tons. Their energy consumption per ton of cooling is, in general, less than that of other types of chillers.

A drawback of centrifugal chillers is that their COP decreases rapidly when the capacity is reduced by closing the inlet vanes. This method of control is effective until the load factor is about 20%. However, when the capacity is decreased below 25% of the rated output of the chiller, it is prone to a condition known as *surging* which can cause serious damage to the chiller.

The difficulties associated with part-load operation of centrifugal chillers have been overcome by the introduction of variable frequency drives (VFD). As the cooling load decreases the VFD alters the voltage and frequency input to the chiller, thus decreasing its speed. This method of capacity control is able to maintain the part-load COP of the chiller very close to its full-load value. Moreover, VFD units allow the chiller to operate at capacities as low as 10% without experiencing surging. Although there are some input energy losses in VFD units due to conversion inefficiencies, these are more than offset by the gains under part-load conditions.

(iii) *Rotary compressors*

Rotary compressors, or screw compressors, as these are sometimes called, are positive-displacement machines like reciprocating compressors. The refrigerant is compressed within the variable space between two meshing lobes, integral to two separate rotors. The rotors are driven by an electric motor. The contours of the lobes are such that the space between the lobes, occupied by the refrigerant, decreases continuously as the lobes move from the refrigerant-inlet to the refrigerant-outlet positions. The refrigerant is thus compressed.

The capacities of commercially available rotary chillers range from about 20 to 2000 tons. Their efficiencies are between about 0.7 and 0.8 kW per ton. For the same capacity, rotary chillers are more efficient than reciprocating chillers, but less efficient compared to centrifugal chillers.

The compact size and the light weight are two of the major advantages of rotary chillers. They also have fewer moving parts than reciprocating chillers and centrifugal chillers.

A complete description of the various types of chillers used in air conditioning systems, their analysis, and performance characteristics are available in Refs. [3] and [4].

3.8.4 *Reversible heat pump systems*

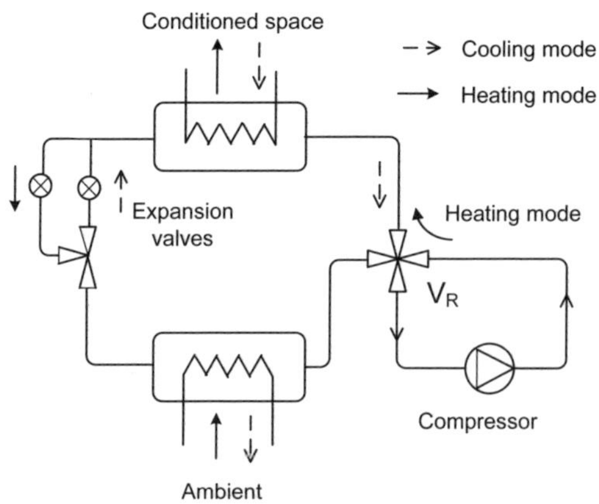

Fig. 3.9 Reversible heat pump

A reversible heat pump, used both for cooling and heating, is another common application of vapor compression refrigeration cycles in HVAC systems. A simplified schematic diagram illustrating its principle of operation is shown in Fig. 3.9. The cycle consists of a compressor, an expansion valve, and two coils which are able to function both as the evaporator and the condenser of the cycle.

During the cooling mode of operation, the coil located inside the conditioned space is the evaporator. The evaporating refrigerant absorbs heat from indoor air, thus cooling the space. The position of the reversing valve V_R allows the compressor to suck refrigerant from the evaporator, and deliver the compressed vapor to the condenser located outdoors. The refrigerant flowing through the condenser rejects heat to the ambient, and then passes through the expansion valve to enters the evaporator, thus completing the cycle.

During the heating mode of operation, the outdoor coil is the evaporator, and refrigerant passing through it absorbs heat from ambient air. The reversing valve is repositioned so that the compressor is now able to suck refrigerant from the outdoor evaporator and deliver it to the coil located inside the space, which is now the condenser. The condensing refrigerant releases heat to the indoor air, thus heating it.

Reversible heat pump systems are being used to heat and cool homes and commercial buildings. Several variations of the basic system described above are now available commercially. In one of these, commonly called *ground-source heat pumps* (GSHP) [3,4], the refrigerant in the outdoor unit exchanges heat with a fluid circulating through a coil buried in the ground. In the heating mode, the heat absorbed from the ground is transferred to the indoor air. The reverse occurs during the cooling mode of operation. Compared to the ambient air temperature, the fluctuation of the ground temperature over the seasons is much smaller, and therefore there is less variation in the performance of the heat pump.

Another heat pump system, more suitable for large buildings like hotels, is called a *water-loop heat pump* (WLHP) [4,6]. Here the outdoor units of individual heat pumps, located in different rooms, exchange heat with water circulating through a common pipe loop. In the rooms being heated, the heat pumps absorb heat from the common water loop and transfer it to the rooms. On the other hand, in the rooms being cooled, the heat pumps reject heat absorbed from the rooms to the water loop. The water loop temperature is typically maintained between 18°C and 32°C. A boiler is used to heat the water in the loop if the net heating demand becomes high, and a cooling tower is used to cool the water if the net cooling demand is high. Analysis of water-loop heat pumps is presented in section 13.4.3.

3.9 Vapor Absorption Refrigeration Cycles

Vapor absorption refrigeration systems use heat as the main energy input compared to the vapor compression systems that require work to drive the compressor. In Figs. 3.10(a) and (b) the vapor compression cycle and the vapor absorption cycle are shown side-by-side for easy comparison. The main difference is that the compressor in Fig. 3.10(a) is replaced by the unit within the dotted boundary in Fig. 3.10(b), consisting of the absorber, the liquid pump, the generator, and the pressure reducing valve (PRV). As in the vapor compression cycle, the heat absorption from the cold space occurs in the evaporator. The vapor leaving the evaporator then enters the absorber where it is absorbed in a liquid called the *absorbent*. The absorbent solution, rich in refrigerant, is pumped to the generator which is at the condenser pressure. In the generator the solution is heated using an external heat input to boil off the refrigerant. The vapor is condensed in the condenser and returned to the evaporator through the expansion valve, as in the vapor compression cycle. In the meanwhile, the solution weak in refrigerant, flows back from the generator to the absorber through a pressure reducing valve (PRV), to complete the cycle.

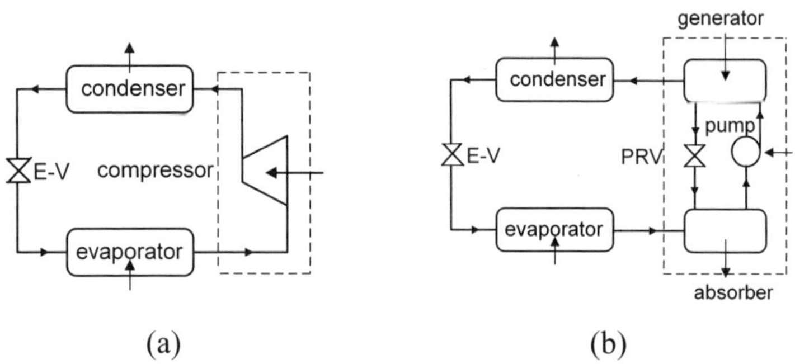

Fig. 3.10 Comparison of compression and absorption cycles

If the absorption of vapor occurs adiabatically in the absorber, then the heat of absorption would increase the solution temperature, impeding further vapor absorption. Therefore the absorber is cooled using an external source of cold fluid.

The main advantage of the absorption cycle is that the work input to the solution pump is only a small fraction of the work required by the compressor in the vapor compression cycle. Moreover, for the absorption system the heat input required in the generator may be provided with a gas burner, a waste-heat stream, or solar energy.

3.9.1 *Three-heat-reservoir model*

The absorption refrigeration system may be modeled as a cyclic device exchanging heat with three reservoirs, as shown schematically in Fig. 3.11. The refrigerated cold space is the low temperature reservoir at T_c, while the generator constitutes the high temperature reservoir at T_h. Since the condenser and the absorber are usually cooled by rejecting heat to the atmosphere, it is taken as the third reservoir at an intermediate temperature T_o.

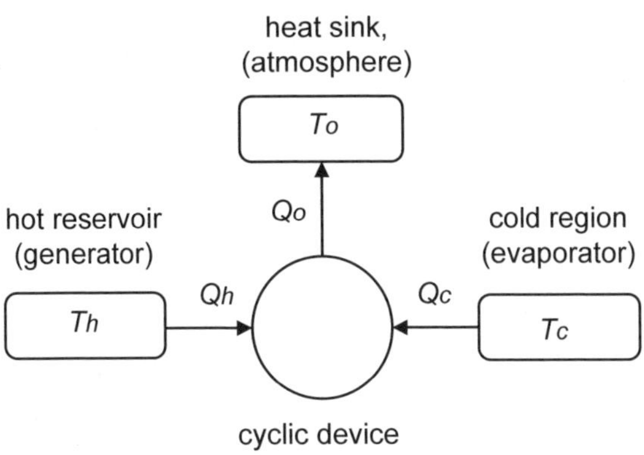

Fig. 3.11 Three-heat - reservoir model of the absorption refrigerator

We assume that the work input to the liquid pump is negligible. The *ideal absorption system* is a cyclic device where all the processes, including the heat interactions with the three reservoirs, are reversible. Applying the energy balance equation to the cyclic system operating in a steady manner we have

$$Q_h + Q_c = Q_o \tag{3.29}$$

Applying the second law of thermodynamics to the reversible cyclic system we obtain

$$\frac{Q_o}{T_o} = \frac{Q_c}{T_c} + \frac{Q_h}{T_h} \tag{3.30}$$

The COP of the absorption refrigeration system is defined as the ratio of heat absorbed from the cold reservoir Q_c, to the heat supplied in the hot reservoir Q_h. Hence we have

$$COP_{ab} = \frac{Q_c}{Q_h} \tag{3.31}$$

Manipulating Eqs. (3.29) to (3.31) the following expression is obtained for the COP:

$$COP_{ab} = \left(\frac{T_c}{T_o - T_c}\right)\left(\frac{T_h - T_o}{T_h}\right) \tag{3.32}$$

In Eq. (3.32), the first term within brackets is the COP of an ideal (reversed Carnot cycle) refrigeration cycle operating between the cold temperature T_c and the heat sink temperature T_o. The second term is the efficiency of a Carnot cycle heat engine operating between the high temperature T_h and the heat sink temperature T_o. Hence, in practical terms, we could consider the ideal absorption refrigerator, as a composite system where the work output of a Carnot heat engine is used to drive a Carnot refrigerator.

The above three-heat-reservoir model is a highly idealized model of the absorption refrigeration system, because all the processes involved are assumed to be reversible. However, in real absorption systems there are external irreversibilities due to heat transfer across finite temperature differences in the evaporator, the condenser, and the generator. Moreover, there are internal irreversibilities within the cyclic device due to property gradients.

A realistic three-heat-reservoir model that takes into the account the above irreversibilitis is described in Ref. [8]. This model could be used to simulate absorption cycles for design purposes.

3.10 Analysis of Actual Absorption Cycles

The most common, commercially available, absorption cycle air conditioning systems use water as the refrigerant and lithium bromide (LiBr) as the absorbent fluid. Water-LiBr systems are suitable for air conditioning applications because the required evaporator temperatures are above the freezing temperature of water. These systems are easier to analyze because LiBr remains in the liquid state during the entire operation and pure water vapor acts as the refrigerant. For low temperature refrigeration applications, ammonia has been used as the refrigerant and water as the absorbent. In the next section we shall review some basic thermodynamic conditions relating to the equilibrium of water-LiBr mixtures.

3.10.1 *Equilibrium of water–LiBr mixtures*

In the absorption refrigeration cycle shown in Fig. 3.10(b), the condenser and the generator are at different temperatures, but they are at the same pressure. Therefore the LiBr-water solution in the generator is in thermodynamic equilibrium with pure water vapor in the condenser. Similarly, the temperature of the solution in the absorber is different from that of the water vapor in the evaporator, but their pressures are equal.

In order to present the conditions of equilibrium for solutions and pure vapors at the same pressure, we consider the piston-cylinder apparatus shown in Figs. 3.12(a) and (b). In Fig. 3.12 (a) pure water and vapor are in equilibrium under the constant pressure applied by the load on the piston. For water vapor in equilibrium with liquid water, the temperature is a unique function of the applied pressure [2,7].

Figure 3.12(b) shows an identical arrangement containing a solution of water and the salt LiBr, in equilibrium with water vapor above it. The same pressure is applied on the system by the load on the piston. Unlike for pure water, the equilibrium of the salt solution and the vapor is dependent on an additional thermodynamic property called the *concentration*, X. It is defined as the mass of LiBr per unit mass of solution. For the same pressure P, the vapor and the salt solution can be in equilibrium for different combinations of the solution temperature T_s, and the concentration X. In Fig. 3.12(c), the solution temperature is plotted against the concentration, for three different values of the applied pressure. When the concentration is zero, the temperature is equal to the saturation temperature of pure water at the given pressure.

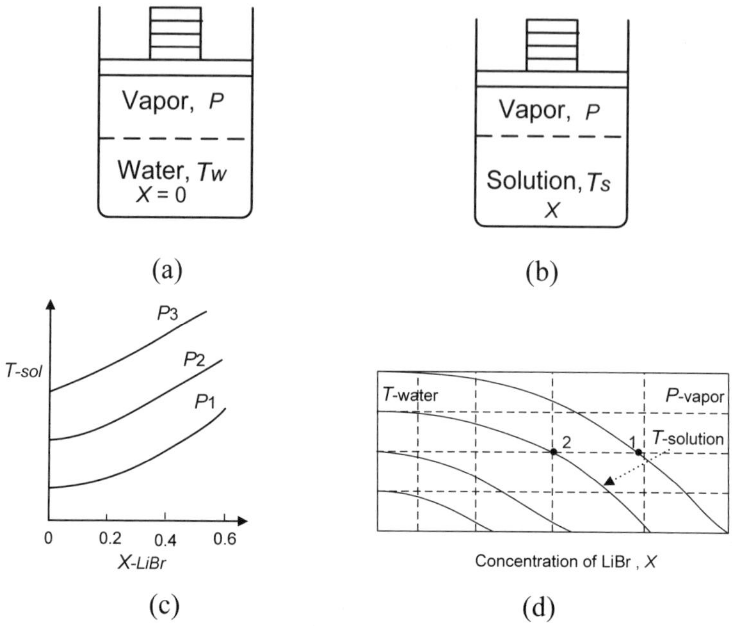

Fig 3.12 Equilibrium of water – LiBr mixtures

The equilibrium relation between the solution temperature, the solution concentration, and the water vapor temperature may be represented in the compact form shown graphically in Fig. 3.10(d). The vertical axis on the left is the saturation temperature of pure water and the corresponding saturation vapor pressure is on the right-side axis. This data may be directly obtained from the steam tables [5]. The horizontal axis gives the mass concentration, X of LiBr in the solution. The family of curves represent different solution temperatures.

Note that the states 1 and 2 of the solution, indicated in Fig. 3.12(d), are both in equilibrium with the same state of water vapor. The actual equilibrium diagram for LiBr-water is available in the *ASHRAE Handbook* - 2013 *Fundamentals* [1].

The specific enthalpy of the solution of LiBr-water is a function of the solution concentration and the temperature. Numerical values of solution enthalpy may be obtained from the *ASHRAE Handbook* - 2013 *Fundamentals* [1]. The absorption cycle can be analyzed in a straightforward manner by applying the mass balance equation and the steady flow energy equation to each component of the cycle. We shall illustrate the computational procedure in worked examples 3.14 and 3.15.

3.11 Worked Examples

Example 3.1 The heat absorption and rejection temperatures of a Carnot refrigeration cycle are −8°C and 36°C respectively. The rate of heat absorption is 28 kW. (a) Calculate (i) the COP of the refrigeration cycle, and (ii) the power input to the cycle. (b) If a Carnot heat pump operating between the same temperatures absorb 18 kW from the cold region, calculate (i) the COP of the heat pump, and (ii) the rate of heat supply to the hot region. Draw the *T-s* diagram of the cycle.

Solution A schematic diagram of a Carnot cycle refrigerator operating between heat reservoirs is shown in Fig. E3.1(a). The corresponding *T-s* diagram is given in Fig. E3.1(b). Since the cycle is reversible the reservoir temperatures are equal to the corresponding refrigerant temperatures.

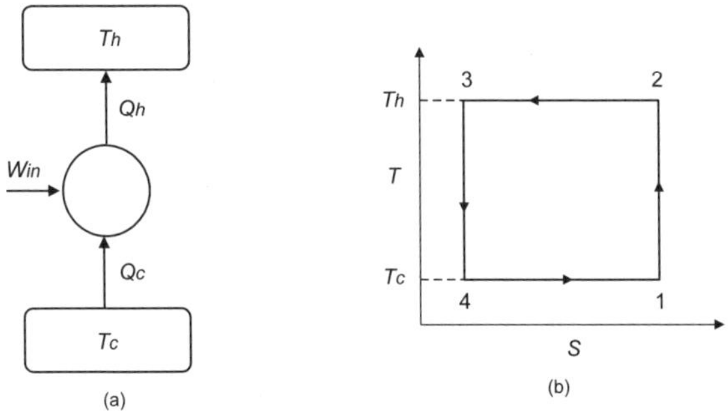

Fig. E3.1 (a) Carnot refrigeration cycle, (b) *T-s* diagram of the cycle

(a) From the given data we have

$$T_c = 273 - 8 = 265 \text{ K} \quad \text{and} \quad T_h = 273 + 36 = 309 \text{ K}$$

The COP of the Carnot refrigeration cycle is given by

$$COP_{ref} = \frac{T_c}{T_h - T_c} = \frac{265}{309 - 265} = 6.02$$

Also, $$COP_{ref} = \frac{Q_c}{W_{in}} = \frac{28}{W_{in}} = 6.02$$

Hence we obtain the work input to the refrigerator as 4.65 kW.

(b) Now the COP of the Carnot cycle depends only on the two reservoir temperatures which are the same as in (a). Therefore when the heat absorbed is 18 kW, the work input is 3kW.

The heat rejected to the high temperature reservoir is given by

$$Q_h = Q_c + W_{in} = 18 + 3 = 21 \text{ kW}$$

The COP of the *heat pump* is

$$COP_{hp} = \frac{Q_h}{W_{in}} = \frac{21}{3} = 7$$

Also, note that

$$COP_{hp} = COP_{ref} + 1 = 7$$

Example 3.2 The evaporating and condensing temperatures of a Carnot refrigeration cycle using R134a as the refrigerant are −5°C and 30°C respectively. The refrigerant flow rate is 0.2 kgs^{-1}. Calculate (i) the vapor quality at the beginning of the compression process and the end of the expansion process, (ii) the heat absorption rate, (iii) the net work input, and (iv) the COP of the cycle.

Solution The *T-s* diagram of a reversed Carnot cycle operating with a vapor is shown in Fig. 3.1(b). We obtain the following properties of R134a from the data tabulated in Ref. [5].

At 30°C, the entropies and enthalpies are:

$$s_g = 1.7142 \text{ kJK}^{-1}\text{kg}^{-1} \quad \text{and} \quad h_g = 414.74 \text{ kJkg}^{-1},$$

$$s_f = 1.1434 \text{ kJK}^{-1}\text{kg}^{-1} \quad \text{and} \quad h_f = 241.69 \text{ kJkg}^{-1}$$

At −5°C, the entropies and enthalpies are:

$$s_g = 1.7294 \text{ kJK}^{-1}\text{kg}^{-1} \quad \text{and} \quad h_g = 359.49 \text{ kJkg}^{-1},$$

$$s_f = 0.9754 \text{ kJK}^{-1}\text{kg}^{-1} \quad \text{and} \quad h_f = 193.32 \text{ kJkg}^{-1}$$

For the isentropic compression 4-1:

$$s_4 = s_1 = x_4 s_{g4} + (1 - x_4) s_{f4}$$

$$1.7142 = 1.7294 x_4 + 0.9754(1 - x_4)$$

Hence $$x_4 = 0.9798$$

$$h_4 = x_4 h_{g4} + (1 - x_4) h_{f4}$$

$$h_4 = 0.9798 \times 395.49 + (1 - 0.9798)193.32 = 391.4$$

For the isentropic expansion 3-2:

$$s_2 = s_3 = x_3 s_{g3} + (1 - x_3) s_{f3}$$

$$1.1434 = 1.7294 x_3 + 0.9754(1 - x_3)$$

Hence $$x_3 = 0.2228$$

$$h_3 = x_3 h_{g3} + (1 - x_3) h_{f3}$$

$$h_4 = 0.2228 \times 395.49 + (1 - 0.2228)193.32 = 238.36$$

Heat absorption rate is

$$Q_{34} = \dot{m}_r (h_4 - h_3) = 0.2(391.4 - 238.37) = 30.61 \text{ kW}$$

Heat rejection rate is

$$Q_{12} = \dot{m}_r (h_1 - h_2) = 0.2(414.74 - 241.69) = 34.61 \text{ kW}$$

The rate of work input is

$$W_{net} = Q_{12} - Q_{34} = 34.61 - 30.61 = 4 \text{ kW}$$

The COP of the refrigeration cycle is

$$COP_{ref} = \frac{Q_{34}}{W_{net}} = \frac{30.61}{4} = 7.65$$

The expression for the COP of a Carnot refrigeration cycle is

$$COP_{ref} = \frac{T_c}{T_h - T_c} = \frac{268}{303 - 268} = 7.66$$

As expected, the result from the detailed analysis agrees with the expression involving only the temperatures of the cold and hot reservoirs.

Example 3.3 A Carnot refrigeration system extracts heat from a cold reservoir at T_c and rejects heat to a heat sink reservoir at T_h. Determine which of the following changes to the reservoir temperatures will be more effective in increasing the COP of the refrigerator: (i) decrease T_h to $(T_h - \Delta T)$ keeping T_c constant, or (ii) increase T_c to $(T_c + \Delta T)$ keeping T_h constant.

Solution A Carnot refrigeration system operating between cold and hot reservoirs at absolute temperatures T_c and T_h are shown in Fig. 3.1(a). The COP of the refrigerator is given by

$$COP_o = \frac{T_c}{T_h - T_c} \tag{E3.3.1}$$

If T_h is decreased to $(T_h - \Delta T)$ keeping T_c constant, the COP becomes

$$COP_h = \frac{T_c}{(T_h - \Delta T) - T_c} \tag{E3.3.2}$$

If T_c is increased to $(T_c + \Delta T)$ keeping T_h constant, the COP becomes

$$COP_c = \frac{T_c + \Delta T}{T_h - (T_c + \Delta T)} \tag{E3.3.3}$$

From Eqs. (E3.3.2) and (E3.3.3) we have

$$\frac{COP_c}{COP_h} = \frac{T_c + \Delta T}{T_c} > 1$$

The above relation shows that increasing the cold reservoir temperature is more effective in increasing the COP than decreasing the hot reservoir temperature by the same amount.

Example 3.4 A standard vapor compression cycle using R134a as the working fluid has a condensing temperature of 40°C and an evaporating temperature of −5°C. The heat extraction rate is 25 kW. Calculate (i) the refrigerant flow rate, (ii) the compressor work input, (iii) volume flow rate of refrigerant at the compressor inlet, (vi) the COP, and (v) the COP of a Carnot refrigerator operating between the same temperatures.

Solution The *T-s* and *P-h* diagrams of a standard vapor compression cycle are shown in Fig. 3.2(a) and (b) respectively. We obtain the following data for R134a from Ref. [5].

At 40°C, the entropies and enthalpies are:

$$s_g = 1.7109 \text{ kJK}^{-1}\text{kg}^{-1} \quad \text{and} \quad h_g = 419.41 \text{ kJkg}^{-1},$$

$$s_f = 1.1903 \text{ kJK}^{-1}\text{kg}^{-1} \quad \text{and} \quad h_f = 256.38 \text{ kJkg}^{-1}$$

At −5°C, the entropies and enthalpies are:

$$s_g = 1.7294 \text{ kJK}^{-1}\text{kg}^{-1} \quad \text{and} \quad h_g = 359.49 \text{ kJkg}^{-1},$$

$$s_f = 0.9754 \text{ kJK}^{-1}\text{kg}^{-1} \quad \text{and} \quad h_f = 193.32 \text{ kJkg}^{-1}$$

For the isentropic compression 4-1:

$$s_1 = s_4 = 1.7294 \text{ kJK}^{-1}\text{kg}^{-1}$$

From the data in [5], it follows that the vapor is superheated at state 1. To calculate the enthalpy at state 1 we extract the following data from the tables in [5].

s	*h*
1.7109	419.41
1.7294	h_1 (?)
1.7460	430.55

By linear interpolation, $h_1 = 425.28$ kJkg^{-1}
For the throttling process 2-3:

$$h_3 = h_2 = 256.38 \text{ kJkg}^{-1}$$

The heat extraction rate is

$$Q_{34} = m_r(h_4 - h_3)$$

$$25 = m_r(395.49 - 256.38)$$

Hence the refrigerant flow rate, m_r is 0.1797 kgs^{-1}
The work input to the compressor is

$$W_{41} = m_r(h_1 - h_4)$$

$$\mathrm{W}_{41} = 0.1797(425.28 - 395.49) = 5.35 \text{ kW}$$

From the data in [5] the specific volume at 4 is, $v_4 = 0.08273$ m^{3}kg^{-1}
The volume flow rate of the refrigerant at 4 is

$$\dot{V}_4 = v_4 m_r = 0.08273 \times 0.1797 = 0.0149 \text{ m}^3\text{s}^{-1}$$

The COP of the cycle is

$$COP_{ref} = \frac{Q_{34}}{W_{41}} = \frac{25}{5.35} = 4.67$$

For comparison we calculate the COP of the Carnot refrigeration cycle operating between the same condensation and evaporation temperatures as

$$COP_c = \frac{T_c}{T_h - T_c} = \frac{268}{45} = 5.96$$

Example 3.5 The evaporator and condenser temperatures of a standard vapor compression cycle using R134a as the refrigerant are −5°C and 40°C respectively. The isentropic efficiency of the adiabatic compressor is 75% and the refrigeration capacity is 25 kW. Calculate (i) the refrigerant flow rate, (ii) the compressor work input, and (iii) the COP.

Solution The *T-s* diagram of the vapor compression cycle is shown in Fig. E3.5. For purposes of comparison we use the same condensing and evaporating temperatures as in worked example 3.4. For the irreversible adiabatic compression process 4-1 we define the isentropic efficiency as

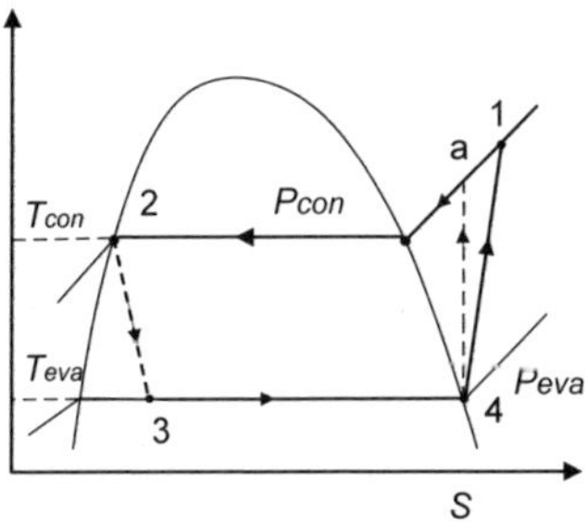

Fig. E3.5 *T-s* diagram

$$\eta_{iso} = \frac{h_a - h_4}{h_1 - h_4}$$

In worked example 3.4 we obtained the data for isentropic compression 4-a, which we shall use here. Substituting in the above equation we have

$$\eta_{iso} = \frac{425.28 - 395.49}{W_{41}} = 0.75$$

Hence the actual work input, W_{14} =39.72 kJkg^{-1}

Now the cooling capacity and the refrigerant flow rate are the same as in example 3.4 because the state points 2, 3 and 4 are unchanged. The work input to the compressor is

$$\dot{W}_{14} = \dot{m}_r W_{14} = 0.1797 \times 39.72 = 7.14 \text{ kW}$$

$$COP_{ref} = \frac{25}{7.14} = 3.5$$

As expected, the COP is decreased from 4.67 to 3.5 due to the irreversibility of the compressor, usually caused by frictional effects.

Example 3.6 A vapor compression cycle using R134a as the refrigerant operates with evaporating and condensing temperatures of −5°C and 40°C. Calculate (i) Carnot cycle compression work, (ii) excess work due to the 'superheat horn', (iii) the loss of work due to throttling, and (iv) the loss in refrigeration capacity due to throttling.

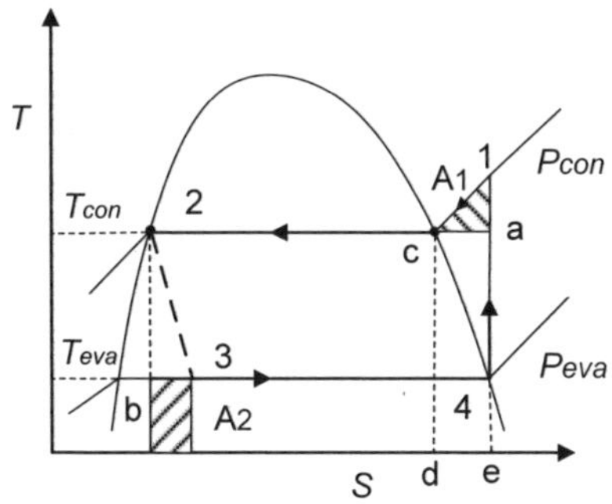

Fig. E3.6 *T-s* diagram

Solution The work loss due to the 'superheat horn' and the loss in refrigeration capacity due to throttling are indicated by the shaded areas of the *T-s* diagram shown in Fig. E3.6.

Now the area A_1 = area (1cde)-area (acde). From Eq. (3.8) it follows that the area under a constant pressure line in the *T-s* diagram is equal to the change in enthalpy.

Therefore area, 1cde = $(h_1 - h_c)$ and the rectangular area, acde = $T_c\,(s_a - s_c)$. Note that the evaporating and condensing temperatures of this example are the same as those in worked example 3.4. Hence we have

$$h_1 = 425.28 \text{ kJkg}^{-1}, \qquad h_c = 419.41 \text{ kJkg}^{-1},$$

$$s_a = s_4 = 1.7294 \text{ kJK}^{-1}\text{kg}^{-1}, \quad s_c = 1.7109 \text{ kJK}^{-1}\text{kg}^{-1}$$

$$T_c = 313 \text{ K}$$

Now the area, $A_1 = (h_1 - h_c) - T_c(s_a - s_c)$

$$A_1 = (425.28 - 419.41) - 313(1.7294 - 1.7109) = 0.0795 \text{ kJkg}^{-1},$$

which is the excess work due to superheating.
For the isentropic expansion 2-b, $s_2 = s_b$. Therefore

$$1.1903 = 0.9754(1 - x_b) + 1.7294x_b$$

Hence $x_b = 0.285$

$$h_b = 193.32(1 - x_b) + 395.49x_b = 250.9 \text{ kJkg}^{-1}$$

The loss of refrigeration capacity due to throttling is given by A_2 where

$$A_2 = h_3 - h_b = h_2 - h_b = 256.38 - 250.9 = 5.48 \text{ kJkg}^{-1},$$

which is also the loss of work due to throttling.

Example 3.7 The evaporator and condenser temperatures of a vapor compression cycle using R134a as the refrigerant are −5°C and 40°C. The refrigerant leaving the condenser is subcooled to 35°C while the vapor entering the compressor is superheated to 5°C. The heat absorption rate is 25 kW. (a) Calculate (i) the refrigerant flow rate, (ii) the compressor work input, and (iii) the COP (b) If the isentropic efficiency of the compressor is 75%, calculate the COP.

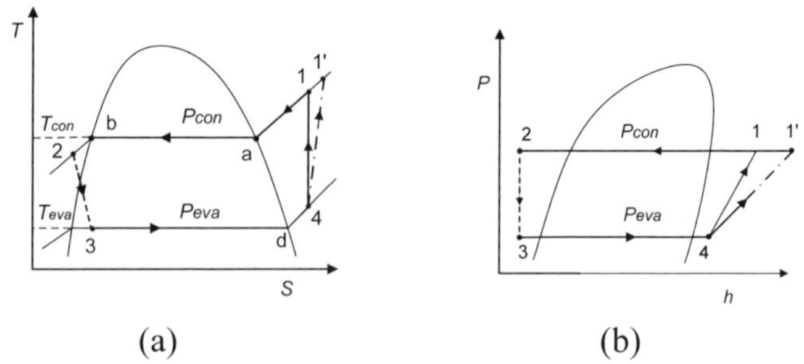

Fig. E3.7 (a) *T-s* diagram, (b) *P-h* diagram

Solution The *T-s* and *P-h* diagrams of the cycle are depicted in Fig. E3.7(a) and (b) respectively. We obtain the following properties of R134a from the tables in Ref. [5].

For the vapor at state 4 that is superheated by 10°C,

$$s_4 = 1.7614 \text{ kJK}^{-1}\text{kg}^{-1} \quad \text{and} \quad h_4 = 404.25 \text{ kJkg}^{-1}$$

For the sub-cooled liquid at 2, we ignore the effect of pressure and obtain the saturated liquid enthalpy at 35°C as, $h_2 = 248.98$ kJkg^{-1}.

Now for the isentropic compression 4-1, $s_1 = s_4 = 1.7614$

The vapor at 1 is superheated. To obtain the enthalpy at 1 we extract the following data from the table in [5]

s	*h*
1.746	430.55
1.7614	h_1 (?)
1.7788	441.32

By linear interpolation, $h_1 = 435.6$ kJkg^{-1}

For the throttling process 2-3:

$$h_3 = h_2 = 248.98 \text{ kJkg}^{-1}$$

The heat extraction rate is

$$Q_{34} = \dot{m}_r(h_4 - h_3)$$

$$25 = \dot{m}_r(404.25 - 248.98)$$

Hence the refrigerant flow rate is, 0.161 kgs^{-1}

The work input to the compressor is

$$W_{41} = \dot{m}_r(h_1 - h_4) = 0.161(435.6 - 404.25) = 5.05 \text{ kW}$$

The COP of the cycle is

$$COP_{ref} = \frac{Q_{34}}{W_{41}} = \frac{25}{5.05} = 4.95$$

(b) The non-isentropic compression is shown as 4-1′ in the Figs. E3.7(a) and (b). The work input during this process is given by

$$W_{act} = \frac{W_i}{\eta_{isen}} = \frac{5.05}{0.75} = 6.73$$

Therefore the COP of the cycle is

$$COP_{ref} = \frac{Q_{34}}{W_{act}} = \frac{25}{6.73} = 3.7$$

Example 3.8 A vapor compression system using R134a as the refrigerant includes a heat exchanger, as shown in Fig. E3.8(a), to superheat the vapor leaving the evaporator using the saturated liquid leaving the condenser. The condensing and evaporating temperatures are 40°C and −5°C respectively. The vapor leaving the evaporator is heated to 5°C. The compression process is isentropic. Calculate the COP of the cycle with and without the heat exchanger.

Solution The schematic diagram and the *P-h* diagram of the cycle are shown in Figs. E3.8(a) and (b) respectively. We obtain the following data for R134a from the tables in Ref. [5].
For the vapor at state 4, superheated to 5°C:

$$s_4 = 1.7614 \text{ kJK}^{-1}\text{kg}^{-1} \quad \text{and} \quad h_4 = 404.25 \text{ kJkg}^{-1}$$

For the saturated liquid state, a and the saturated vapor state, b:

$$h_a = 256.38 \text{ kJkg}^{-1} \quad \text{and} \quad h_b = 395.49 \text{ kJkg}^{-1}$$

Applying the steady flow energy equation to the heat exchanger, neglecting the kinetic and potential energy of the fluid, we have

$$\dot{m}_r(h_a - h_2) = \dot{m}_r(h_4 - h_b)$$

Substituting numerical values in the above equation we obtain

$$h_2 = 247.62 \text{ kJkg}^{-1}$$

For the throttling process 2-3

$$h_3 = h_2 = 247.62 \text{ kJkg}^{-1}$$

Now the heat extraction rate per unit mass is

$$Q_{ref} = h_b - h_3 = 395.49 - 247.62 = 147.87 \text{ kJkg}^{-1}$$

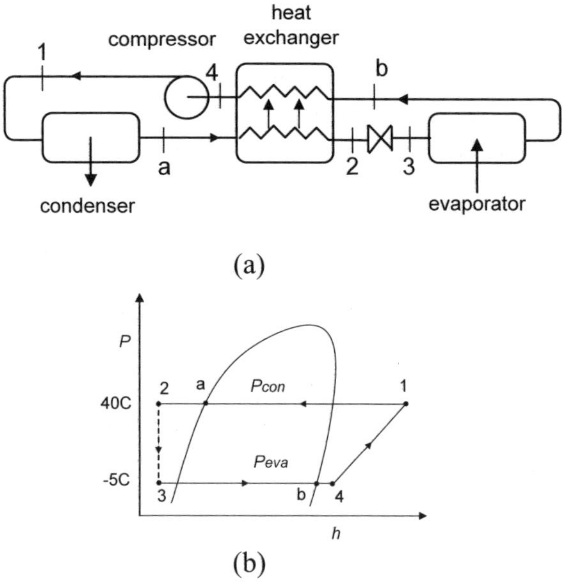

Fig. E3.8 (a) Refrigeration cycle with a heat exchange, (b) *P-h* diagram

Note that the energy needed for superheating the refrigerant from b to 4 is supplied internally and therefore does not contribute to the heat extraction rate in the evaporator as in example E3.7. From the data in example E3.7 we have, $h_1 = 435.6$ kJkg^{-1}.

The work input to the compressor per unit mass is

$$W_c = h_1 - h_4 = 435.6 - 404.25 = 31.35 \text{ kJkg}^{-1}$$

The COP is given by:

$$COP_{ref} = \frac{Q_{ref}}{W_c} = \frac{147.87}{31.35} = 4.72$$

In worked example 3.4, we obtained the COP of the standard vapor compression cycle with the same temperatures, but without the heat exchanger, as 4.67. The increase in the COP due to the inclusion of the heat exchanger is marginal. However, the practical advantages of the heat exchanger are: (a) only liquid enters the expansion valve, and (b) the vapor entering the compressor is superheated and therefore has no liquid.

Example 3.9 A standard vapor compression cycle using R134a as the working fluid is used to produce chilled water in an air conditioning plant. The condensing and evaporating temperatures of the cycle are 40°C and 5°C respectively. The chilled water enters the evaporator at 18°C and leaves at 8°C. The flow rate of chilled water is 0.22 kgs^{-1}. The condenser is cooled with water entering at 22°C and leaving at 31°C. The isentropic efficiency of the compressor is 75%. Calculate (i) the flow rate of refrigerant in the cycle, and (ii) the flow rate of condenser cooling water.

Solution Figure E3.9(a) shows a schematic diagram of the chilled water producing refrigeration system. The chilled water flowing counter to the refrigerant in the evaporator is cooled by the evaporating refrigerant. In the condenser the cooling water is heated due to heat absorption from the condensing refrigerant. Figure E3.9(b) shows the *T-s* diagram of the vapor compression cycle where the distribution of chilled water and condenser cooling water temperatures are also indicated.

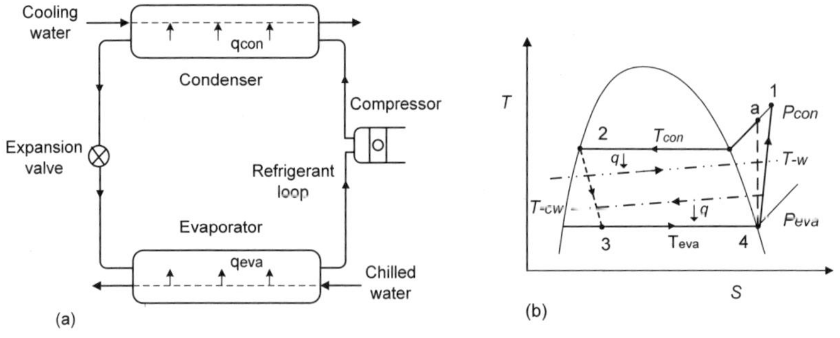

Fig. E3.9 (a) Chilled water producing refrigeration system, (b) *T-s* diagram

We obtain the following data for R134a from [5]. At 5°C:

$$h_4 = 401.33 \text{ kJkg}^{-1}, \qquad s_4 = 1.7238 \text{ kJK}^{-1}\text{kg}^{-1}$$

At 40°C: $$h_2 = 256.38 \text{ kJkg}^{-1}$$

For the expansion process, 2-3:

$$h_3 = h_2 = 256.38 \text{ kJkg}^{-1}$$

Applying the steady flow energy equation to the evaporator we have

$$\dot{m}_r(h_4 - h_3) = \dot{m}_{cw} c_w (T_{cwi} - T_{cwo})$$

The specific heat capacity of water is 4.2 kJK^{-1}kg^{-1}. Substituting numerical values in the above equation we obtain

$$\dot{m}_r(401.33 - 256.38) = 0.22 \times 4.2(18 - 8) = 9.24 = q_{abs}$$

Therefore refrigerant flow rate is, $\dot{m}_r = 0.0637$ kgs^{-1}.

Consider the isentropic compression, 4-a, for which

$$s_a = s_4 = 1.7238 \text{ kJK}^{-1}\text{kg}^{-1}$$

The vapor is superheated at state a. To determine the enthalpy at state a, we extract the following data from the tables in [5].

s	*h*
1.7109	419.19
1.7238	h_a (?)
1.7460	430.55

By linear interpolation: $h_a = 423.37$ kJkg^{-1}.

The isentropic efficiency of the compressor is 75%. Hence the actual work input is given by

$$W_{act} = \frac{\dot{m}_r(h_a - h_4)}{\eta_{iso}} = \frac{0.0637 \times (423.37 - 401.33)}{0.75} = 1.872 \text{ kW}$$

Applying the energy equation to the cycle, the heat rejection rate,

$$q_{rej} = q_{abs} + W_{act} = 9.24 + 1.872 = 11.1 \text{ kW}$$

Applying the energy equation to the condenser we have

$$q_{rej} = \dot{m}_{cl} c_w (T_{wo} - T_{wi})$$

Substituting numerical values we obtain

$$11.1 = \dot{m}_{cl} \times 4.2 \times (31 - 22)$$

Hence the mass flow rate of cooling water is 0.294 kgs^{-1}.

Example 3.10 A heat pump operates on the vapor compression cycle using R134a as the working fluid. The refrigerant flow rate is 0.18 kgs^{-1}. The evaporating and condensing temperatures are 10°C and 65°C respectively. The isentropic efficiency of the compressor is 75%.

The heat pump supplies heat at a constant rate to a tank of water of mass 1500 kg and specific heat capacity 4.2 kJK^{-1}kg^{-1}. (i) Calculate the time required to heat the water from 20°C to 45°C. (ii) If the heat pump is replaced with an electrical resistance heater using the same rate of energy input as the compressor of the heat pump, calculate the time taken to heat the water from 20°C to 45°C.

Solution The heat pump cycle is identical to the standard vapor compression cycle shown in Fig. 3.3. The *T-s* diagram of the cycle is shown in Fig. E3.10.

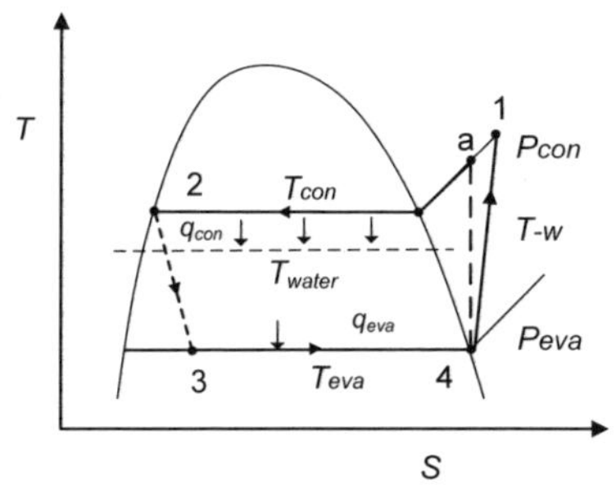

Fig. E3.10 *T-s* diagram of heat pump cycle

We obtain the following data for R134a from the tables in Ref. [5].

At 10°C: $s_4 = 1.7215$ kJK^{-1}kg^{-1}, $h_4 = 404.16$ kJkg^{-1}

At 65°C: $h_2 = 295.77$ kJkg^{-1}

Consider the isentropic compression 4-a, for which

$$s_a = s_4 = 1.7215 \text{ kJK}^{-1}\text{kg}^{-1}$$

The vapor is superheated at state a. To calculate the enthalpy at state *a* we extract the following data from the tables in [5]:

s	*h*
1.6995	427.89
1.7215	h_a (?)
1.7397	441.67

By linear interpolation, $h_a = 435.4$ kJkg^{-1}

For the expansion process 2-3:

$$h_3 = h_2 = 295.77 \text{ kJkg}^{-1}$$

Applying the steady flow energy equation to the evaporator we have

$$\dot{m}_r(h_4 - h_3) = q_{abs}$$

Substituting numerical values in the above equation we obtain

$$q_{abs} = 0.18(404.16 - 295.77) = 19.5 \text{ kW}$$

The isentropic efficiency of the compressor is 75%. Therefore the actual work input is given by

$$W_{act} = \frac{\dot{m}_r(h_a - h_4)}{\eta_{iso}} = \frac{0.18 \times (435.4 - 404.16)}{0.75} = 7.5 \text{ kW}$$

Applying the energy equation to the cycle, the heat rejection rate in the condenser is obtained as

$$q_{rej} = q_{abs} + W_{act} = 19.5 + 7.5 = 27 \text{ kW}$$

Apply the energy equation to the water tank, neglecting any heat losses, and assuming that the *rate of heat input is constant*. Hence we have

$$q_{rej}\tau = M_w c_w (T_{wf} - T_{wi})$$

where M_w is the mass of water, and c_w is the specific heat capacity of water. The time taken to heat the water is τ.
Substituting numerical values in the above equation we obtain

$$27\tau = 1500 \times 4.2 \times (45 - 20) = 157.5 \times 10^3 \text{ kJ}$$

Hence the time taken to heat the water is 1.62 hours.

The total energy consumed is, $7.5 \times 1.62 = 12.15$ kWh

When a resistance heater with the same electrical energy input is used, the time taken to heat the water is

$$\tau_{re} = 157.5 \times \frac{10^3}{7.5} = 5.84 \text{ hours}$$

The total energy consumed is

$$7.5 \times 5.84 = 43.8 \text{ kWh}$$

We note that the heat pump is a more energy efficient method of heating compared to the use of an electrical resistance heater with the same rate of electrical energy input.

Example 3.11 Shown schematically in Fig. E3.11(a) is a vapor compression refrigeration plant with a flash intercooler. It uses ammonia as the refrigerant. The saturation temperature of the vapor in the condenser is 30°C. The liquid in the flash tank is at −4°C. Vapor leaves the evaporator as a saturated vapor at −30°C. Liquid enters the expansion valve, EV-2 at −20°C. Calculate the COP of the refrigeration plant with and without the intercooler.

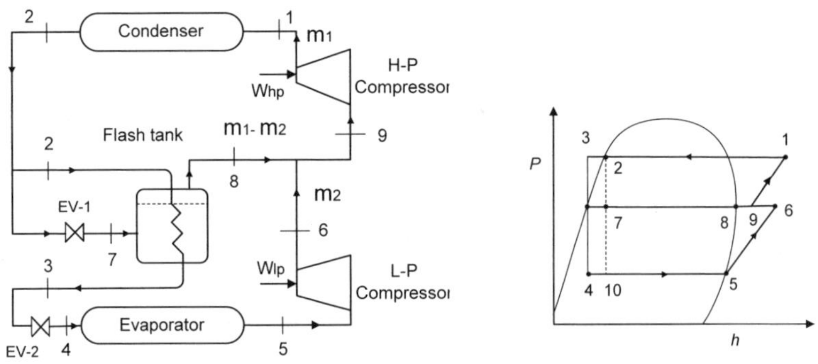

(a) Schematic diagram of refrigeration plant (b) *P-h* diagram

Fig. E3.11 Refrigeration system with flash intercooling

Solution Let the refrigerant flow rates through the high-pressure compressor and low-pressure compressor be $\dot{m}_1$ and $\dot{m}_2$ respectively. The following properties of ammonia are obtained from the data tables in Ref. [5].

At 30°C, $h_2 = 323.1$ kJkg^{-1}

Neglecting the effects of pressure, the subcooled liquid enthalpy at 3 is

$$h_3 = 89.8 \text{ kJkg}^{-1}.$$

At −4°C, $h_8 = 1439.9$ kJkg^{-1}

For the expansion process 2-7:

$$h_7 = h_2 = 323.1 \text{ kJkg}^{-1}$$

Applying the energy equation to the flash tank we have

$$\dot{m}_2(h_2 - h_3) = (\dot{m}_1 - \dot{m}_2)(h_8 - h_7) \qquad \text{(E3.11.1)}$$

$$\dot{m}_2(323.1 - 89.8) = (\dot{m}_1 - \dot{m}_2)(1439.9 - 323.1$$

Hence we have $$\dot{m}_1 = 1.209\dot{m}_2 \qquad \text{(E3.11.2)}$$

For the isentropic compression process 5-6:

$$s_6 = s_5 = 5.785 \text{ kJK}^{-1}\text{kg}^{-1}$$

The vapor at state 6 is superheated at a pressure of 3.691 bar. The saturation temperature at 3.691 bar is −4°C.
We obtain the superheated vapor enthalpy at state 6 by linear interpolation as

$$h_6 = 1554.68 \text{ kJg}^{-1}.$$

The work input in the low-pressure compressor is

$$W_{56} = \dot{m}_2(h_6 - h_5) = (1554.68 - 1405.6)\dot{m}_2 = 149.08\dot{m}_2$$

Applying the steady flow energy to the mixing process at the junction 6-8-9 we have

$$\dot{m}_2 h_6 + (\dot{m}_1 - \dot{m}_2)h_6 = \dot{m}_1 h_9$$

Substituting numerical values in the above equation we obtain

$$\dot{m}_2 \times 1554.68 + (\dot{m}_1 - \dot{m}_2) \times 1439.9 = \dot{m}_1 h_9$$

Substituting the condition in Eq. (E3.11.2) in the above equation we have

$$h_9 = 1534.84 \text{ kJkg}^{-1}$$

The pressure at state 9 is 3.691 bar. We obtain the superheated vapor entropy at state 9 by linear interpolation as

$$s_9 = 5.7169 \text{ kJK}^{-1}\text{kg}^{-1}$$

For the isentropic compression process 9-1,

$$s_1 = s_9 = 5.7169$$

We obtain the enthalpy at state 1 by linear interpolation as

$$h_1 = 1723.44 \text{ kJkg}^{-1}$$

The work input to the high-pressure compressor is

$$W_{91} = \dot{m}_1(h_1 - h_9) = (1723.44 - 1534.84)\dot{m}_1 = 188.6\dot{m}_1$$

For the expansion through the valve EV-2,

$$h_4 = h_3 = 89.8 \text{ kJkg}^{-1}$$

The heat extraction rate in the evaporator is

$$Q_{45} = \dot{m}_2(h_5 - h_4) = (1405.6 - 89.8)\dot{m}_2 = 1315.8\dot{m}_2$$

The COP of the refrigeration cycle is given by

$$COP_{ref} = \frac{Q_{45}}{W_{56}+W_{91}}$$

Substitute the relevant expressions for the heat extraction rate, Q_{45} and the work inputs, W_{56} and W_{91} in the above equation and use the condition in Eq. (E3.11.2) to obtain

$$COP_{ref} = \frac{1315.8\dot{m}_2}{149.08\dot{m}_2+1.209\times188.6\dot{m}_2} = 3.49$$

(ii) Consider the standard vapor compression cycle without the intercooler. The condensing and evaporating temperatures are 30°C and −30°C respectively. Then the expansion process in the valve is from 2 to 10, as indicated in Fig. E3.11(b).

Using the procedure in worked example 3.4 we obtain the following: The heat extraction rate in the evaporator,

$$q_{34} = 1082.5 \text{ kJkg}^{-1}$$

The work input to the compressor,

$$w_{41} = 343.48 \text{ kJkg}^{-1}$$

The COP of the standard cycle is given by

$$COP_{st} = \frac{q_{34}}{w_{41}} = \frac{1082.5}{343.48} = 3.15$$

Therefore the COP of the standard vapor compression cycle using a single compressor is 3.15. The COP is increased by about 11% due to the inclusion of a flash intercooler and two-stage compression.

Example 3.12 A vapor compression refrigeration system with two evaporators and an intercooler as shown in Fig. E3.12. The working refrigerant is ammonia. The heat extraction rate of the low-temperature evaporator at −40°C is 250 kW. The heat extraction rate of the high-temperature evaporator at 0°C is 150 kW. The vapor leaving the two evaporators is in a saturated state. The liquid in the flash tank is at 0°C. Dry saturated vapor enters the low-pressure compressor and the high-pressure compressor at −40°C and 0°C respectively. The condensing temperature is 35°C. Calculate (i) the work input to the two compressors and (ii) the COP of the cycle.

Solution A schematic diagram of the refrigeration plant and the *P-h* diagram of the cycle are shown in Fig. 3.12(a) and (b). Let the refrigerant flow rates through the low-temperature evaporator and the high-temperature evaporator be $\dot{m}_1$ and $\dot{m}_2$ respectively. The flow rate through the expansion valve, EV-2 is $\dot{m}_3$. The following data are obtained from the tables in Ref. [5].

At 35°C, $h_2 = 347.4$ kJkg^{-1}

At 0°C, $h_6 = 181.2$ kJkg^{-1}

$$h_4 = h_{10} = 1444.4 \text{ kJkg}^{-1}$$

At -40°C, $h_8 = 1390$kJkg^{-1}

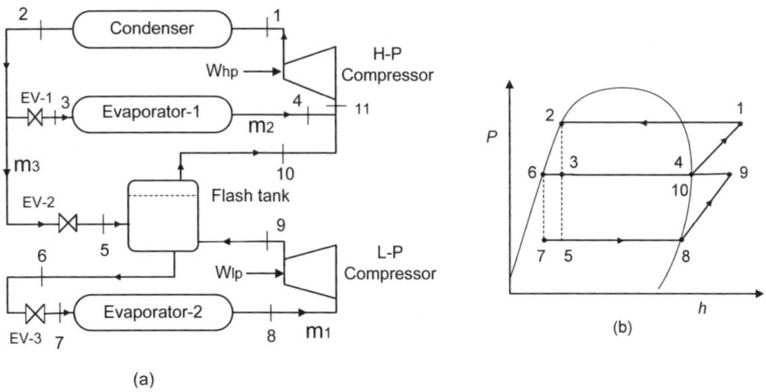

Fig. E3.12 Refrigeration system with two evaporators (a) Schematic diagram, (b) *P-h* diagram

For the expansion process 2-3, 2-5 and 6-7 we have

$$h_3 = h_2 = 347.4$$

$$h_5 = h_2 = 347.4$$

$$h_7 = h_6 = 181.2$$

The heat extraction rate in the low-temperature evaporator is

$$Q_{78} = \dot{m}_1(h_8 - h_7)$$

Substituting numerical values we have

$$250 = \dot{m}_1(1390 - 181.2)$$

Hence $$\dot{m}_1 = 0.2068 \text{ kgs}^{-1}$$

The heat extraction rate in the high-temperature evaporator is

$$Q_{34} = \dot{m}_2(h_4 - h_3)$$

Substituting numerical values we have

$$150 = \dot{m}_2(1444.4 - 347.4)$$

Hence $$\dot{m}_1 = 0.1367 \text{ kgs}^{-1}$$

For the isentropic compression 8-9 in low-pressure compressor we have

$$s_9 = s_8 = 5.962 \text{ kJK}^{-1}\text{kg}^{-1}$$

The vapor at 9 is superheated at a pressure of 4.295 bar. By linear interpolation using data from [5], the enthalpy at 9 is obtained as

$$h_9 = 1639.4 \text{ kJkg}^{-1}$$

Applying the steady flow energy equation to the flash-tank we have

$$\dot{m}_3 h_5 + \dot{m}_1 h_9 = \dot{m}_1 h_6 + \dot{m}_3 h_{10}$$

Substituting numerical values in the above equation we obtain

$$\dot{m}_1(1639.4 - 181.2) = \dot{m}_3(1444.4 - 347.4)$$

Hence $$\dot{m}_3 = 0.2748 \text{ kgs}^{-1}$$

For the isentropic compression process 4-1:

$$s_1 = s_4 = 5.34 \text{ kJK}^{-1}\text{kg}^{-1}$$

The vapor at 1 is superheated at a pressure of 13.3 bar. By linear interpolation using data from [5] we find the enthalpy at 1 as

$$h_1 = 1607.3 \text{ kJkg}^{-1}$$

The work input to the low-pressure compressor is

$$W_{89} = \dot{m}_1(h_9 - h_8)$$

$$W_{89} = 0.2068(1639.4 - 1390) = 51.58 \text{ kW}$$

The work input to the high-pressure compressor is

$$W_{41} = (\dot{m}_2 + \dot{m}_3)(h_1 - h_4)$$

$$W_{41} = (0.2748 + 0.1367)(1607.3 - 1444.4) = 67 \text{ kW}$$

The total work input to the cycle is

$$W_{tot} = W_{89} + W_{41} = 51.58 + 67 = 118.58 \text{ kW}$$

The COP of the cycle is

$$COP_{ref} = \frac{Q_{34}+Q_{78}}{W_{tot}} = \frac{150+250}{118.58} = 3.37$$

Example 3.13 An ideal absorption cycle has a cyclic device operating between three heat reservoirs. The heat extraction rate from the cold reservoir at 8°C is 15 kW. The hot reservoir is at 100°C. Heat is rejected to the ambient reservoir at 30°C. (i) Calculate the COP of the ideal cycle, and the rate of heat supply from the heat source. (ii) If there is temperature difference of 4°C between each reservoir and the working fluid of the internally reversible cyclic device, calculate the COP of the cycle.

Solution The three-heat-reservoir model of the ideal absorption cycle is shown in Fig. E3.13. The temperatures of the refrigerated space, the hot reservoir, and the heat sink are denoted by T_c, T_h and T_o respectively. The corresponding temperatures of the working fluid of the cycle are T_{c1}, T_{h1} and T_{o1} respectively.

(i) Consider the case when there is no temperature difference between the working fluid and the reservoirs. The COP is given by Eq. (3.32) as

$$COP_{abs} = \left(\frac{T_h - T_o}{T_h}\right)\left(\frac{T_c}{T_o - T_c}\right)$$

Substituting numerical values in the above equation we have

$$COP_{abs} = \left(\frac{373-303}{373}\right)\left(\frac{281}{303-281}\right) = 2.4$$

$$COP_{abs} = \frac{Q_C}{Q_h} = \frac{15}{Q_h} = 2.4$$

Therefore the rate of heat supply from the hot reservoir is, $Q_h = 6.25$ kW.

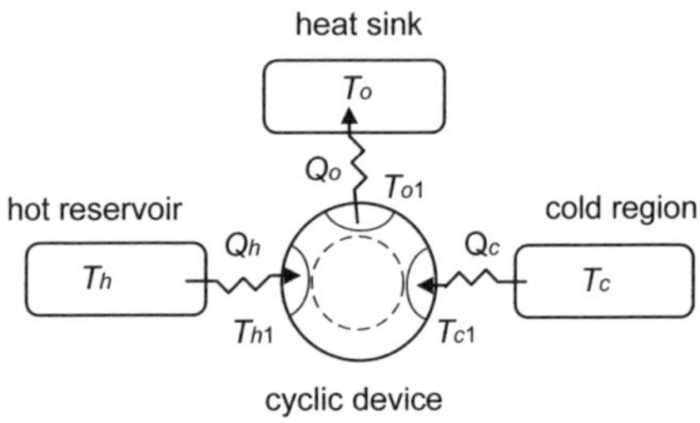

Fig. E3.13 Three-heat-reservoir model of the absorption refrigerator

(ii) Consider the situation when there is a temperature difference of 4°C between the three heat reservoirs and the working fluid of the cycle. This constitutes an external irreversibility. We assume that all internal processes of the cyclic device are reversible. The temperatures of the working fluid when it interacts with the different reservoirs are:

$$T_{c1} = 281 - 4 = 277 \text{ K}, \qquad T_{h1} = 373 - 4 = 369 \text{ K}$$

$$T_{o1} = 303 + 4 = 307 \text{ K}$$

The processes undergone by the working fluid of the cyclic device are *internally reversible*. Therefore the COP is given by

$$COP_{abs} = \left(\frac{T_{h1}-T_{o1}}{T_{h1}}\right)\left(\frac{T_{c1}}{T_{o1}-T_{c1}}\right)$$

Substituting numerical values in the above equation we have

$$COP_{abs} = \left(\frac{369-307}{369}\right)\left(\frac{277}{307-277}\right) = 1.55$$

$$COP_{abs} = \frac{Q_C}{Q_h} = \frac{15}{Q_h} = 1.55$$

The rate of heat input from the hot reservoir, Q_h is 9.68 kW.

Example 3.14 A schematic diagram of a vapor absorption refrigeration system is shown in Fig. E3.14. It uses water as the refrigerant and lithium-bromide as the absorbent. The temperatures of the evaporator, the absorber, the condenser, and the generator are 10°C, 25°C, 35°C and 90°C respectively. The solution flow rate through the liquid pump is 0.8 kgs^{-1}. Calculate (i) the refrigerant flow rate, (ii) the heat extraction rate in the evaporator, (iii) the heat rejection rate in the condenser, and (iii) the COP of the cycle.

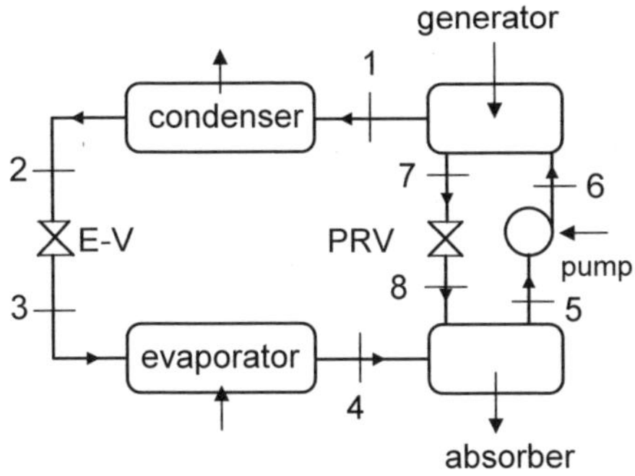

Fig. E3.14 Vapor absorption refrigeration cycle

Solution Let the refrigerant (water) flow rate through the condenser be $\dot{m}_1$ and the flow rate of weak solution leaving the generator be $\dot{m}_7$. Now the saturated solution at 25°C, leaving the absorber at 5, is in equilibrium with the pure water vapor in the evaporator at 10°C, as seen in Fig. E3.14.

From the equilibrium diagram for water-LiBr in [1] we obtain the concentration of the solution (kg of LiBr per kg of solution) at 5, as $x_5 = 0.451$.

The saturated solution at 90°C, leaving the generator at 7, is in equilibrium with pure water at 35°C in the condenser. From the equilibrium diagram we obtain the concentration at 7 as, $x_7 = 0.653$.

For mass balance of LiBr in the generator

$$\dot{m}_7 x_7 = \dot{m}_6 x_6$$

$$\dot{m}_7 = 0.8 \times \frac{0.451}{0.653} = 0.553 \text{ kgs}^{-1}$$

For total mass balance in the generator

$$\dot{m}_1 = \dot{m}_6 - \dot{m}_7 = 0.8 - 0.553 = 0.247 \text{ kgs}^{-1}$$

We obtain the following data for pure water from the steam tables in [5].

For the saturated liquid at 35°C at 2, $h_2 = 146.55$ kJkg^{-1}.

For the saturated vapor at 10°C at 4, $h_4 = 2519.2$ kJkg^{-1}

For saturated vapor at 90°C at 1, $h_1 = 2659.7$ kJkg^{-1}

For the expansion process, 2-3 we have

$$h_3 = h_2 = 146.55$$

The heat extraction rate in the evaporator is

$$Q_{34} = \dot{m}_1(h_4 - h_3) = 0.247(2519.2 - 146.55) = 586 \text{ kW}$$

The heat rejection rate in the condenser is

$$Q_{12} = \dot{m}_1(h_1 - h_2) = 0.247(2659.7 - 146.55) = 620.7 \text{ kW}$$

We obtain the enthalpy of the water-LiBr solution from the chart in [1]. For the solution leaving the absorber at 5,

$$T_5 = 25°\text{C}, \quad x_5 = 0.451, \quad h_5 = -162 \text{ kJkg}^{-1} \text{ (solution)}.$$

For the solution leaving the generator at 7,

$$T_7 = 90°\text{C}, \quad x_7 = 0.653, \quad h_7 = -69 \text{ kJkg}^{-1} \text{ (solution)}.$$

Neglecting the work input to the solution pump,

$$h_6 = h_5 = -162 \text{ kJkg}^{-1} \text{ (solution)}.$$

Applying the steady flow energy equation to the generator we obtain the heat input rate as

$$Q_{gen} = \dot{m}_1 h_1 + \dot{m}_7 h_7 - \dot{m}_6 h_6$$

$$Q_{gen} = 0.247 \times 2659.7 + 0.553 \times (-69) - 0.8 \times (-162) = 748 \text{ kW}$$

The COP of the absorption refrigeration cycle is

$$COP_{abs} = \frac{Q_{34}}{Q_{gen}} = \frac{586}{748} = 0.78$$

Example 3.15 The vapor absorption refrigeration system shown in Fig. E3.15 incorporates a heat exchanger in which the weak liquid flowing from the generator heats the strong liquid leaving the pump to 50°C. The temperatures of the evaporator, the absorber, the condenser, and the generator are 10°C, 25°C, 35°C and 90°C respectively. The solution flow rate through the liquid pump is 0.8 kgs^{-1}. Calculate (i) the heat input in the generator, (ii) the heat rejected in the absorber, and (iii) the COP of the cycle.

Solution The aim of the heat exchanger shown in Fig. E3.15 is to heat the solution entering the generator using the hot solution leaving it. This reduces the heat input required in the generator. The basic design parameters for this example are the same as those for worked example E3.14. Therefore we shall use the relevant fluid properties obtained in example E3.14.

The temperature and concentration at 9 are 50°C and 0.451 respectively. We obtain the enthalpy at 9 from the chart in Ref. [1] as $h_9 = -103$ kJkg^{-1} (solution).

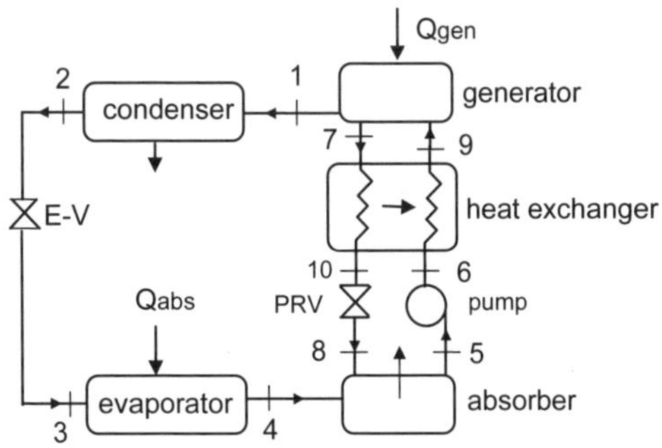

Fig. E3.15 Absorption cycle with heat exchanger

Apply the steady flow energy equation to the generator to obtain the heat input rate Q_{gen} as

$$Q_{gen} = \dot{m}_1 h_1 + \dot{m}_7 h_7 - \dot{m}_6 h_9$$

$Q_{gen} = 0.247 \times 2659.7 + 0.553 \times (-69) - 0.8 \times (-103) = 701.2$ kW

The COP of the absorption refrigeration cycle is

$$COP_{abs} = \frac{Q_{34}}{Q_{gen}} = \frac{586}{701.2} = 0.836$$

We note that the inclusion of the heat exchanger increases the COP of the refrigeration cycle by about 6.8 percent.
Neglecting the work input to the solution pump,

$$h_6 = h_5 = -162 \text{ kJkg}^{-1}$$

Applying the steady flow energy equation to the heat exchanger we have

$$\dot{m}_7(h_7 - h_{10}) = \dot{m}_5(h_9 - h_6)$$

Substituting numerical values in the above equation we obtain

$$0.553(-69 - h_{10}) = 0.8[-103 - (-162)] = 47.2$$

$$h_{10} = -154.5 \text{ kJkg}^{-1}$$

For the expansion process 10-8,

$$h_8 = h_{10} = -154.5 \text{ kJkg}^{-1}$$

Apply the steady flow energy equation to the absorber to obtain the heat rejection rate as

$$Q_{abs} = \dot{m}_7 h_8 + \dot{m}_1 h_4 - \dot{m}_5 h_5$$

$$Q_{abs} = 0.553 \times (-154.5) + 0.247 \times 2519.2 - 0.8 \times (-162)$$

$$Q_{abs} = 666.4 \text{ kW}$$

Problems

P3.1 A Carnot refrigeration cycle absorbs heat at −8°C and rejects heat at 55°C. (a) If the heat rejection rate is 30 kW, calculate the heat extraction rate, the work input, and the COP of the refrigeration cycle. (b) If the cycle operates as a heat pump calculate the COP of the cycle. [*Answers*: (a) 24.24 kW, 5.76 kW, 4.2, (b) 5.2]

P3.2 The heat absorption and rejection temperatures of a Carnot refrigeration cycle are −8°C and 55°C respectively. The heat rejection rate is 30 kW. There is a temperature difference of 6°C between the refrigerant and two heat reservoirs. The internal processes of the cycle are reversible. (a) Calculate the work input, and the COP of the

refrigeration cycle. (b) If the cycle is operated as a heat pump, calculate heat absorption rate and the COP.
[*Answers*: (a) 6.74 kW, 3.45, (b) 23.26 kW, 4.45]

P3.3 A Carnot cycle heat pump operates with R134a as the refrigerant. The condenser and evaporator temperatures are 50°C and 10°C respectively. The refrigerant flow rate is 0.3 kgs^{-1}. Calculate (i) the rate of heat input to the heat pump, (ii) the net work input, and (iii) the COP of the heat pump.
[*Answers*: (i) 40 kW, (ii) 5.64 kW, (iii) 8.075]

P3.4 The condensing and evaporating temperatures of a standard vapor compression cycle using R134a are 5°C and 35°C respectively. The heat absorption rate is 12 kW. Calculate (i) the refrigerant flow rate, (ii) the work input to the compressor, and (iii) the COP of the cycle.
[*Answers*: (i) 0.07875 kgs^{-1}, (ii) 1.52 kW, (iii) 7.89]

P3.5 A standard vapor compression cycle uses R134a as the refrigerant. The condensing and evaporating temperatures are 40°C and 0°C respectively. The isentropic efficiency of the compressor is 80%. The refrigerant flow rate is 0.35 kgs^{-1}. Calculate (i) the heat absorption rate, (ii) the work input to the compressor, and (iii) the COP of the cycle.
[*Answers*: (i) 49.7 kW, (ii) 11.33 kW, (iii) 4.39]

P3.6 A vapor compression cycle uses R134a as the refrigerant. The vapor condenses at 40°C and is subcooled to 35°C before entering the expansion valve. The evaporating temperature is −5°C. The vapor leaving the evaporator is superheated to 5°C at entry to the compressor. The heat absorption rate is 13 kW and the compression process is isentropic. Calculate (i) the refrigerant flow rate, (ii) the work input, and (iii) the COP.
[*Answers*: (i) 0.0837 kgs^{-1}, (ii) 2.63 kW, (iii) 4.95]

P3.7 The condensing and evaporating temperatures of a vapor compression cycle using R134a are 40°C and −5°C respectively. The vapor leaving the evaporator is heated to 5°C in a heat exchanger using

the saturated liquid leaving the condenser. The refrigerant flow rate is 0.3 kgs^{-1}. The compression process is isentropic. Calculate (i) the heat absorption rate in the evaporator, (ii) the work input to the compressor, and (iii) the COP.
[*Answers*: (i) 47 kW, (ii) 9.41 kW, (iii) 5]

P3.8 A vapor compression system using ammonia as the refrigerant has two evaporators and a flash intercooler as shown in Fig. 3.6(a). Refrigerant flows at the rate of 0.375 kgs^{-1} through the low-temperature evaporator at −30°C. The refrigerant flow rate through the high-temperature evaporator at −4°C is 0.225 kgs^{-1}. The liquid leaving the two evaporates is in a saturated state. The liquid in the flash tank is at −4°C. Dry saturated vapor enters the low-pressure compressor and the high-pressure compressor at −30°C and −4°C respectively. The liquid entering the expansion valve EV-2 is subcooled to 20°C. The condensing temperature is 40°C. Calculate (i) the heat absorption rates in the two evaporators, (ii) the work input to the two compressors, and (ii) the COP of the cycle.
[*Answers*: (i) 240 kW, 424 kW, (ii) 55.65 kW, 141.75 kW, (iii) 3.37]

P3.9 The temperatures of the three heat reservoirs representing the heat source, the refrigerated space, and the heat sink of an ideal three-heat-reservoir absorption cycle are 90°C, 6°C and 30°C respectively. The rate of heat absorption from the cold reservoir is 8 kW. (a) Calculate the COP of the cycle, and the heat rejection rate to the heat sink. (b) If the heat interaction between each reservoir and the working fluid of the internally reversible cycle requires a temperature difference of 5°C, calculate the COP of the cycle, and the rate of heat supply from the heat source.
[*Answers*: (a) 1.92, 12.1 kW, (b) 1.125, 7.1 kW]

P3.10 A water-LiBr vapor absorption system incorporates a heat exchanger as shown in Fig. E3.15. The temperatures of the evaporator, the absorber, the condenser, and the generator are 10°C, 25°C, 40°C and 100°C respectively. The strong liquid leaving the pump is heated to 50°C in the heat exchanger. The refrigerant flow rate through the condenser is

0.12 kgs^{-1}. Calculate (i) the heat input in the generator, (ii) the heat rejected in the absorber, and (iii) the COP of the cycle.
[*Answers*: (i) 346.7 kW, (ii) 328 kW, (iii) 0.81]

References

1. *ASHRAE Handbook* - 2013 *Fundamentals*, American Society of Heating, Refrigeration and Air Conditioning Engineers, Atlanta, 2013.
2. Jones, J. B. and Hawkins, G. A., *Engineering Thermodynamics,* John Wiley & Sons, Inc. New York, 1986.
3. Kuehn, Thomas H., Ramsey, James W. and Threlkeld, James L., *Thermal Environmental Engineering*, 3rd edition, Prentice-Hall, Inc., New Jersey, 1998.
4. Mitchell, John W. and Braun, James E., *Principles of Heating, Ventilation and Air Conditioning in Buildings*. John Wiley and Sons, Inc. New York, 2013.
5. Rogers G. F. C. and Mayhew Y. R., *Thermodynamic and Transport Properties of Fluids.* 5th ed. Blackwell, Oxford, U.K. 1998.
6. Stoecker, Wilbert F. and Jones, Jerold W., *Refrigeration and Air Conditioning*, International Edition, McGraw-Hill Book Company, London, 1982.
7. Van Wylen, Gordon J. and Sonntag, Richard E., *Fundamentals of Classical Thermodynamics,* 3rd Edition, John Wiley & Sons, Inc. New York, 1985.
8. Wijeysundera N. E., 'Performance of three-heat-reservoir absorption cycles with external and internal irreversibilities', *Applied Thermal Engineering*, 17(12) (1997), 1151–1161.

Chapter 4

Psychrometric Principles

4.1 Introduction

There are numerous processes of practical engineering importance, especially in HVAC systems, involving mixtures of gases and vapors. When subjected to certain processes, gas mixtures, and mixtures of gases and vapors, behave quite differently. In the case of a mixture of gas and vapor, the vapor could change phase during the process whereas a mixture of gases remains in a gaseous state during the process.

For instance, when ambient air passes through the cooling coil of an air conditioner, some of the water vapor in the air condenses to water on the cold surface of the coil. The air delivered to the conditioned space after passing through the cooling coil, is therefore much drier than the original ambient air supplied to the cooling coil. The reverse process occurs in a cooling tower where the air leaving the system becomes more moist due to the water vapor added to the air. In order to analyze these practical situations we need to develop a set of parameters to characterize a mixture of gases and vapors, and also obtain expressions for the relevant thermodynamic properties of the mixture.

4.2 Mixtures of Air and Water Vapor

The main focus of our discussion in this section is ambient air which is a mixture of water vapor in a superheated state and dry air. We shall refer to air that is free of any moisture as *dry air*. The pressure of water vapor in typical ambient air is relatively low. For example, at 30°C the pressure of water vapor in air that is fully saturated with water vapor is about 4.24 kPa (saturated pressure of water vapor at 30°C from the steam tables [3]), compared with a mixture pressure of about 100 kPa. Under these dilute conditions we shall assume that water vapor and dry air in the ambient behave as ideal gases. Furthermore, we shall treat them as independent pure substances where the properties of water vapor are not influenced by the presence of air. These assumptions are found to be reasonable for most practical calculations involving atmospheric air.

In order to illustrate the behavior of air–water vapor mixtures, we consider the piston cylinder arrangement shown in Fig. 4.1(a), where the cylinder contains *moist* ambient air initially. The pressure of the air in the cylinder is maintained constant by the fixed load on the piston. Subject of the above assumptions concerning ambient air, we may apply Dalton's rule [6] to express the pressure of the mixture in the form

$$P = P_a + P_v \tag{4.1}$$

where P_a and P_v are the partial pressures of the air and the water vapor respectively.

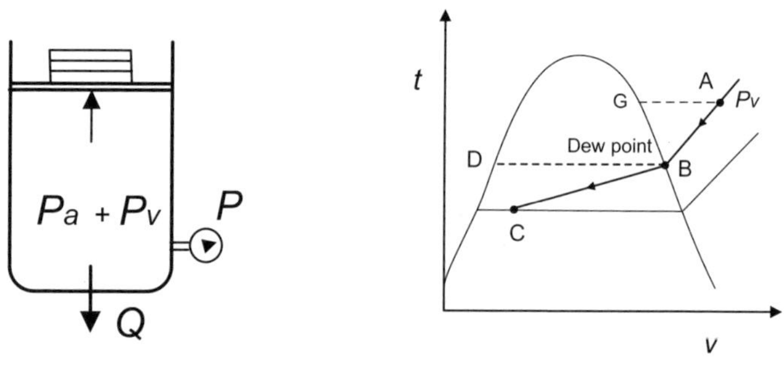

Fig. 4.1 (a) Piston-cylinder set-up with moist air, (b) *T-v* diagram for vapor

The initial state of the *water vapor* in the mixture is indicated by A in the temperature–volume (*t–v*) diagram for water depicted in Fig. 4.1(b). Imagine a quasi–static process in which the air in the cylinder is cooled at a *constant mixture pressure*. Since the mole fractions of air and water vapor are constant during the process, the partial pressure of the water vapor, also remains constant. Therefore the cooling process follows a constant pressure line on the *t–v* diagram for the vapor. However, when the state of the vapor reaches the point B on the saturated vapor curve in Fig. 4.1(b), the first liquid drops appear in the cylinder and condensation of water vapor just begins. The temperature at state B, where condensation just begins is called the *dew point* of the air, and the air is then referred to as *saturated air*.

As the cooling process continues more vapor would condense to water. Therefore, the mass of water vapor, and consequently the mole fraction of water vapor in the mixture, decrease. We note that the liquid water produced by condensation has a relatively small volume and therefore its effect on the air–vapor mixture in the cylinder can be neglected.

As condensation continues, the partial pressure of the vapor decreases due to the reduced mole fraction of the vapor. Since vapor is in a saturated state during condensation, its temperature, and consequently the temperature of the mixture, also decrease as shown by the line BC in Fig. 4.1(b). Because the mixture pressure is maintained constant by the fixed load on the piston, by Dalton's rule, the partial pressure of the air increases to compensate for the reduced vapor pressure.

It is noteworthy that in the absence of air in the cylinder the state of the pure vapor during condensation would follow the constant temperature line BD. We note that the presence of air alters the overall behavior of the vapor during condensation significantly.

4.3 Properties of Air-Water Mixtures

4.3.1 *Relative humidity, humidity ratio and degree of saturation*

We now introduce three parameters called the *relative humidity,* the *humidity ratio* and the *degree of saturation* that are commonly used to characterize the state of a mixture of dry air and water vapor. The humidity ratio is sometimes referred to as the *specific humidity.*

The following assumptions are made in obtaining analytical expressions for these parameters: (i) the air and vapor phases can be treated as a mixture of independent ideal gases, (ii) the conditions of equilibrium between the liquid and the vapor in the mixture are not influenced by the presence of air, (iii) there are no physical and chemical interactions between the liquid phase and the gaseous phase. When applied to air–water mixtures, these assumptions allow us to obtain the properties of the vapor from the steam table [3] at the mixture temperature and the partial pressure of the vapor in the mixture.

The *relative humidity*, ϕ is defined as the ratio of the partial pressure of the vapor, P_v in the mixture to the saturation pressure, $P_g(t)$ of the vapor at the mixture temperature, t. For air–water mixtures, the latter pressure is obtained directly from the steam table [3]. Therefore

$$\phi = \frac{P_v}{P_g} \tag{4.2}$$

The *humidity ratio*, ω is defined as the mass of water vapor in a given volume of mixture to the mass of *dry* air in the same volume. Therefore

$$\omega = \frac{\Delta m_v}{\Delta m_a} \tag{4.3}$$

where Δm_v and Δm_a are masses of vapor and dry air respectively in a volume ΔV. From Eq. (4.3) it follows that

$$\omega = \frac{\Delta m_v/\Delta V}{\Delta m_a/\Delta V} = \frac{\rho_v}{\rho_a} \tag{4.4}$$

In Eq. (4.4), ρ_v and ρ_a are the densities of vapor and dry air respectively.

Applying the density–form of the *ideal gas* equation to the vapor and air we have

$$P_v = \rho_v R_v T \tag{4.5}$$

$$P_a = \rho_a R_a T \tag{4.6}$$

where R_v and R_a are the respective gas constants of water vapor and air, and T is the absolute temperature. Substituting from Eqs. (4.5) and (4.6) in Eq. (4.4) we obtain

$$\omega = \frac{\rho_v}{\rho_a} = \frac{P_v R_a}{P_a R_v} \tag{4.7}$$

Applying Dalton's rule

$$P = P_v + P_a \tag{4.8}$$

From Eqs. (4.7) and (4.8) we have

$$\omega = \frac{P_v R_a}{(P - P_v) R_v} \tag{4.9}$$

From Eqs. (4.2) and (4.9) it follows that

$$\omega = \frac{R_a \phi P_g(t)}{R_v[(P - \phi P_g(t)]} \tag{4.10}$$

The gas constants for dry air and water vapor may be expressed in terms of the universal gas constant $\bar{R}$, and their molecular masses M_a and M_v as

$$R_a = \bar{R}/M_a = \bar{R}/28.96 \tag{4.11}$$

$$R_v = \bar{R}/M_v = \bar{R}/18 \tag{4.12}$$

Substituting in Eq. (4.10) from Eqs. (4.11) and (4.12) we have

$$\omega = \frac{0.622\phi P_g(t)}{P - \phi P_g(t)} \tag{4.13}$$

The *degree of saturation*, μ is defined as the ratio of the mass of water vapor in a mixture of air and water vapor to the mass of vapor that would be present if the air was in a saturated state at the same *temperature* and *mixture pressure*. The saturated vapor state is indicated as G in Fig. 4.1(b). From the above definition it follows that

$$\mu = \omega/\omega_s \tag{4.14}$$

where ω_s is the specific humidity of *saturate*d air at the same temperature and mixture pressure.
From Eqs. (4.13) and (4.14) we obtain the relation

$$\mu = \phi \left[\frac{0.622 + \omega}{0.622 + \omega_s}\right] \tag{4.15}$$

When the air is fully saturated with water vapor at a given temperature, $\mu = 1$ and $\phi = 100\%$.

4.3.2 *Enthalpy of moist air*

The enthalpy h, of a mixture of dry air and water vapor is equal to the sum of the enthalpy of dry air, h_a and the enthalpy of superheated water vapor, h_g. Therefore

$$h = h_a + \omega h_g(t) = c_{pa} t + \omega h_g(t) \tag{4.16}$$

where t is the dry-bulb temperature and ω is the humidity ratio. The units of h are kJkg^{-1} of *dry* air. In the above equation c_{pa} is the specific heat capacity at constant pressure of *dry* air. For the typical temperature range of about 0°C to 60°C, of interest in psychrometric calculations, c_{pa} varies from about 1.004 to 1.007 kJK^{-1}kg^{-1}. Therefore for practical calculations a constant value of c_{pa} = 1.00 kJK^{-1}kg^{-1} may be used in Eq. (4.16) without causing significant errors in the computed quantities [4].

Now the water vapor in the ambient is in a superheated state at a low pressure. We observe from the data in the steam tables [3] that under low pressures, the enthalpy of superheated steam is approximately equal to the enthalpy of saturated steam at the same *temperature*. Therefore in Eq. (4.16) we shall use the saturated vapor enthalpy $h_g(t)$ from the steam tables.

An alternative approximate relation for the enthalpy of moist air is obtained by assuming that low pressure water vapor behaves like an ideal gas with a constant specific heat capacity. At low pressures the variation of enthalpy of water vapor with temperature is well approximated by the linear relation given below [2]:

$$h_g = c_{pw}t + h_{go} \tag{4.17}$$

where h_{go} = 2501 kJkg^{-1}, is the enthalpy of saturated water vapor at the reference temperature of 0°C and c_{pw} = 1.86 kJkg^{-1}K^{-1}, is the specific heat capacity water vapor.

Substituting in Eq. (4.16) from Eq. (4.17) we have

$$h = c_{pa}t + \omega(c_{pw}t + h_{go}) \tag{4.18}$$

$$h = c_{pm}t + \omega h_{go} \tag{4.19}$$

where the specific heat of moist air is given by

$$c_{pm} = c_{pa} + \omega c_{pw} \tag{4.20}$$

A useful approximation for the change in enthalpy of moist air, often used in the solution of air conditioning design problems, is based on the concept of sensible enthalpy change, Δh_s and latent enthalpy change, Δh_l. Consider a process from state 1 to state 2 where the change of enthalpy is given by

$$\Delta h = h_2 - h_1$$

Expressing the enthalpy in the form given in Eq. (4.18) we obtain

$$\Delta h = c_{pa}(t_2 - t_1) + h_{go}(\omega_2 - \omega_1) + c_{pw}(t_2\omega_2 - t_1\omega_1) \tag{4.21}$$

Now the third term on the RHS of the above equation can be expressed in the form [2]:

$$c_{pw}(t_2\omega_2 - t_1\omega_1) = c_{pw}\left[\frac{(\omega_2-\omega_1)(t_1+t_2)}{2} + \frac{(t_2-t_1)(\omega_2+\omega_1)}{2}\right] \quad (4.22)$$

If we define the mean temperature and mean humidity ratio for the process as

$$\bar{t} = (t_1 + t_2)/2$$

$$\bar{\omega} = (\omega_1 + \omega_2)/2$$

then Eq. (4.22) may be expressed as

$$c_{pw}(t_2\omega_2 - t_1\omega_1) = \bar{\omega}c_{pw}(t_2 - t_1) + \bar{t}c_{pw}(\omega_2 - \omega_1) \quad (4.23)$$

Substituting for the last term in Eq. (4.21) from Eq. (4.23) we obtain

$$\Delta h = (c_{pa} + \bar{\omega}c_{pw})(t_2 - t_1) + (\bar{t}c_{pw} + h_{go})(\omega_2 - \omega_1) \quad (4.24)$$

Eq. (4.24) may be written in the form:

$$\Delta h = c_{am}(t_2 - t_1) + h_{gm}(\omega_2 - \omega_1) \quad (4.25)$$

$$\Delta h = c_{am}\Delta t + h_{gm}\Delta\omega \quad (4.26)$$

where the mean specific heat capacity of air and the mean enthalpy of water vapor are defined by:

$$c_{am} = c_{pa} + \bar{\omega}c_{pw} \quad (4.27)$$

$$h_{gm} = \bar{t}c_{pw} + h_{go} \quad (4.28)$$

For psychrometric calculations involving ambient air in the temperature range from 0°C to 60°C, the following mean values may be used [2]: $c_{am} = 1.02$ kJkg^{-1}K^{-1} and $h_{gm} = 2555$ kJkg^{-1}.
We now write Eq. (4.25) in the form

$$\Delta h = \Delta h_s + \Delta h_l \quad (4.29)$$

where the total change in enthalpy Δh of moist air is expressed as the sum of the sensible enthalpy change, Δh_s and the latent enthalpy change, Δh_l. These quantities are defined respectively by

$$\Delta h_s = c_{am}(t_2 - t_1) \quad (4.30)$$

$$\Delta h_l = h_{gm}(\omega_2 - \omega_1) \quad (4.31)$$

4.3.3 *Specific volume of moist air*

The specific volume of moist air, v in m^3 per kg of *dry air* may be obtained by using the ideal gas equation for *dry air* as

$$v = R_a T / P_a \tag{4.32}$$

We may rewrite Eq. (4.13) in the form:

$$\omega = \frac{0.622(P - P_a)}{P_a} \tag{4.33}$$

where P_a is the pressure of *dry* air and P is the total mixture pressure. From Eqs. (4.32) and (4.33) we have

$$v = (R_a T/P)(1 + \omega/0.622) \tag{4.34}$$

4.3.4 *Adiabatic saturation and wet-bulb temperature*

The schematic diagram in Fig. 4.2 depicts a device that brings a stream of air passing steadily through it to a saturated state by a process known as *adiabatic saturation*. The walls of the device are perfectly insulated and the air flows over the surface of a pool of water whose level is maintained constant by a steady supply of make–up water.

Let the air temperature and humidity ratio at the inlet section 1 and the exit section 2 be t_1, ω_1 and t_2, ω_2 respectively. The constant mass flow rate of *dry* air is $\dot{m}_a$, and the mass flow rate of make–up water at temperature t_3 is $\dot{m}_w$.

Applying the mass balance equation for water flowing through the control volume 1-2-3 we obtain

$$\dot{m}_w + \dot{m}_a \omega_1 = \dot{m}_a \omega_2 \tag{4.35}$$

Applying the steady–flow energy equation (SFEE), neglecting the changes in kinetic and potential energy of the air, we have

$$\dot{m}_w h_{f3} + \dot{m}_a h_{a1} = \dot{m}_a h_{a2} \tag{4.36}$$

where h_{f3} is the enthalpy of water at section 3.

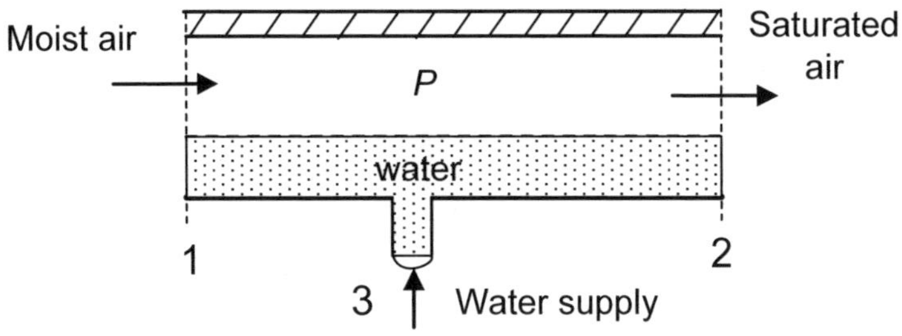

Fig. 4.2 Adiabatic saturator

From Eqs. (4.35) and (4.36) we obtain

$$h_{a1} + h_{f3}(\omega_2 - \omega_1) = h_{a2} \tag{4.37}$$

Substituting for the enthalpy of moist air from Eq. (4.16) in Eq. (4.37) we have

$$c_{pa}t_1 + \omega_1 h_{g1} + h_{f3}(\omega_2 - \omega_1) = c_{pa}t_2 + \omega_2 h_{g2} \tag{4.38}$$

Now h_{a2} and ω_2 are the enthalpy and humidity ratio respectively of saturated air at section 2. Since the air is *fully saturated* with vapor at section 2 we observe from Eqs. (4.10) and (4.16) that both ω_2 and h_{a2} are functions *only* of the temperature t_2.

We now introduce the important assumption that, under *ideal conditions*, the make–up water temperature t_3 is also equal to t_2, the exit air temperature of air. Substituting this condition in Eqs. (4.37) and (4.38) we have

$$h_{a1} = h_{a2} - (\omega_2 - \omega_1)h_{f2} \tag{4.39}$$

$$c_{pa}t_1 + \omega_1 h_{g1} + h_{f2}(\omega_2 - \omega_1) = c_{pa}t_2 + \omega_2 h_{g2}(t_2) \tag{4.40}$$

We note from Eq. (4.40) that, when *ideal conditions prevail*, air entering with different combinations, t_1, ω_1 will have the same exit saturation temperature t_2. The temperature, t_2 is therefore a property of the inlet air stream and is called the *thermodynamic wet-bulb temperature* (TWT) of the air.

The TWT provides an indirect method of determining the relative humidity of air. The instrument commonly used to measure the wet-bulb temperature is called the *wet-bulb thermometer*.

4.3.5 Measurement of wet-bulb temperature

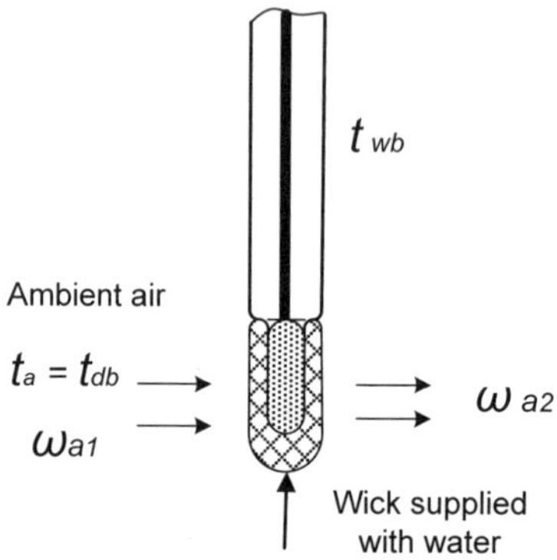

Fig. 4.3 Wet-bulb thermometer

A practical arrangement based on thermometry to measure the wet-bulb temperature is shown in Fig. 4.3. In its basic form the wet-bulb thermometer consists of an ordinary liquid–in–glass thermometer whose bulb is covered with a wet porous wick, continuously supplied with water. The air whose humidity is to be determined is blown over the wick usually with aid of a fan. The temperature indicated by the thermometer under steady conditions is called the *wet-bulb temperature,* t_{wb}.

The ideal conditions of adiabatic saturation, assumed in the development of the thermodynamic wet bulb temperature (TWT) as a property, may not be realized with the actual wet-bulb thermometer shown in Fig. 4.3. A complete analysis of the difference between TWT and t_{wb} is beyond the scope of this book. However, we shall briefly outline the main steps of a simplified analysis to make a comparison between the two types of wet-bulb temperatures.

Applying the SFEE to the control volume surrounding the wet wick in Fig. 4.3 we have

$$A_w h_c (t_a - t_{wb}) = A_w h_m h_{fg} (\omega_{a2} - \omega_{a1}) \tag{4.41}$$

The left hand side of Eq. (4.41) is the heat gained by the wick from the air due to forced convection and the right hand side is the energy absorbed by the water vapor evaporating from the wick. Note that we have neglected all other forms of energy interactions that usually affect the reading of a thermometer, such as radiation and conduction. The terms h_c and h_m are respectively the heat transfer coefficient and the mass transfer coefficient, and A_w is the area of the control surface just outside the wick. Rearranging Eq. (4.41) we obtain

$$\omega_{a1} = \omega_{a2} - (h_c/h_m h_{fg})(t_a - t_{wb}) \tag{4.42}$$

In Eq. (4.40) we may neglect the variation of, $(h_g - h_f) = h_{fg}$, over the sections 1 and 2 because of the narrow temperature range of practical interest. Hence we have

$$\omega_1 = \omega_2 - (c_{pa}/h_{fg})(t_1 - t_2) \tag{4.43}$$

From Eqs. (4.42) and (4.43) we observe that t_2 may be equal to t_{wb} if

$$(h_c/h_m c_{pa}) = Le = 1 \tag{4.44}$$

The non-dimensional quantity Le is called the *Lewis number.* Even though the Lewis number for ambient air in the temperature range of 10 to 60°C, is about 0.86, the TWT is found to be nearly equal to the temperature measured by the wet-bulb thermometer. In contrast, for most other mixtures of gas and vapor the equality of the two wet-bulb temperatures may not hold and therefore serious errors may result if they are assumed equal [2].

4.4 The Psychrometric Chart

In the preceding sections we developed several useful relations for the properties of moist air which could be used directly to analyze most air conditioning processes. However, their application often requires the use of data from property tables [3] and the iterative solution of nonlinear equations. The psychrometric chart is a nomograph where the various properties of moist air are represented in graphical form. The chart offers a convenient graphical means for the solution of most air conditioning design problems.

In constructing the psychrometric chart, two properties of moist air are taken as basic coordinates. The two common choices for this purpose are: humidity ratio and dry-bulb temperature (ω-t_{db}) or humidity ratio and enthalpy (ω-h). The latter choice has the advantage that a number of the more common air conditioning processes appear as straight lines on the chart. Moreover, on the (ω-h)-chart, constant dry-bulb temperature lines and the constant wet-bulb temperature lines are straight lines. The use of humidity ratio and enthalpy as the basic coordinates was pioneered by Richard Mollier in 1923. The widely used psychrometric chart, given in the *ASHRAE Handbook* - 2013 *Fundamentals* [1], is also based the same coordinates. We shall now outline the main steps in the development of the (ω-h) - psychrometric chart.

The vertical axis of the psychrometric chart, shown in Fig. 4.4, represents the humidity ratio (ω), and therefore the constant-ω lines are horizontal. The constant enthalpy (h) lines are parallel straight lines. These lines are inclined at an angle β to the horizontal, and the horizontal separation distance between them is L_h.

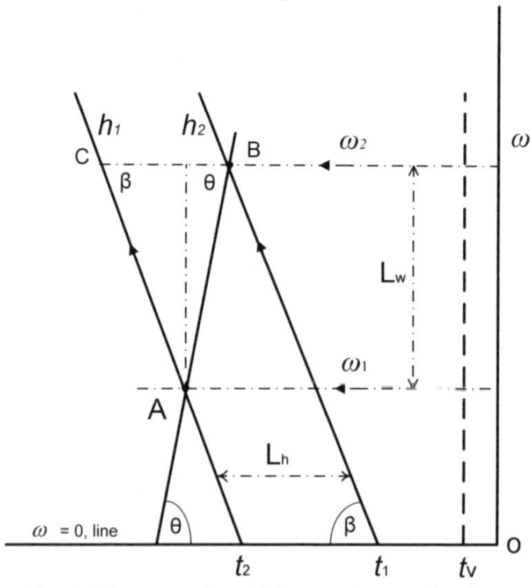

Fig. 4.4 Construction of the psychrometric chart

4.4.1 *Constant dry-bulb temperature lines*

We shall now illustrate the construction of the constant dry-bulb temperature lines on the psychrometric chart. Consider one such line (t) which intersects two constant enthalpy lines at A and B as shown in Fig.

4.4. The enthalpies at A and B may be obtained by applying Eq. (4.16) to the two points. Hence the change of enthalpy from A to B is

$$h_2 - h_1 = (\omega_2 - \omega_1)h_g(t) \tag{4.45}$$

From Eq. (4.45) we observe that on the (h-ω) chart a constant dry bulb temperature line is a straight line because $h_g(t)$ is constant on such a line. However, the lines at different dry-bulb temperatures are not parallel because the slopes of these lines depend on h_g, which is a function of the dry bulb temperature. The inclination, θ of a constant dry-bulb temperature line to the horizontal is obtained by considering the geometry of the triangle ABC. This gives

$$L_h = L_w cot\theta + L_w cot\beta \tag{4.46}$$

Now the scale factors for h and ω may be expressed as

$$s_h = \frac{h_2 - h_1}{L_h}$$

$$s_\omega = \frac{\omega_2 - \omega_1}{L_\omega}$$

From the two relations above it follows that

$$\frac{L_h}{L_\omega} = \frac{s_\omega(h_2 - h_1)}{s_h(\omega_2 - \omega_1)} = \frac{q}{S} \tag{4.47}$$

where the enthalpy-moisture ratio q, and the scale factor S are defined as:

$$q = \frac{h_2 - h_1}{\omega_2 - \omega_1} \tag{4.48}$$

$$S = \frac{s_h}{s_\omega} \tag{4.49}$$

From Eqs. (4.46) and (4.47) we have

$$cot\theta + cot\beta = q/S \tag{4.50}$$

From Eq. (4.45), for a constant dry-bulb temperature (t) line

$$\frac{h_2 - h_1}{\omega_2 - \omega_1} = q = h_g(t) \tag{4.51}$$

In order to fix the inclination of the constant enthalpy lines, β we *arbitrarily* select a temperature t_v for which the constant dry-bulb temperature line is drawn vertical ($\theta_t = 90^o$) as indicated in Fig. 4.4. Hence from Eqs. (4.50) and (4.51) it follows that

$$cot\beta = \frac{q}{S} = \frac{h_g(t_v)}{S} \tag{4.52}$$

The inclination of the constant temperature line, θ_t for temperature t is obtained by substituting in Eq. (4.50) for q from Eq. (4.51). This gives

$$cot\theta_t + cot\beta = h_g(t)/S \tag{4.53}$$

Substituting in Eq. (4.53) from Eq. (4.52) we have

$$cot\theta_t = \frac{[h_g(t) - h_g(t_v)]}{S} \tag{4.54}$$

From Eq. (4.16) we notice that at the points where the constant enthalpy lines intersect the horizontal axis ($\omega = 0$), the enthalpy is directly proportional to the dry-bulb temperature. Therefore this axis can be rescaled in terms of the dry-bulb temperature. Using Eq. (4.54) we then draw the constant dry-bulb temperature lines through the different temperatures marked on the horizontal axis. These are straight lines inclined at θ_t to the horizontal axis.

For many psychrometric processes it is useful to compare the inclination, α of the process line with the inclination, θ_t of the constant dry-bulb temperature lines. A general expression for α, applicable to a process with any q-value may be obtained by subtracting Eq. (4.53) from Eq. (4.50). Hence we have

$$cot\alpha - cot\theta_t = \frac{(q - h_g)}{S} \tag{4.55}$$

4.4.2 *Saturation curve and the constant relative humidity lines*

For any value of the dry-bulb temperature, the saturation vapor pressure P_g can be obtained directly from the steam table [3]. For a fixed mixture pressure P, the saturation humidity ratio ω_s is calculated using Eq. (4.13) with the relative humidity $\phi = 100\%$. The saturation curve is constructed by drawing a smooth curve through the points of intersection of the constant temperature line and the corresponding ω_s line. The constant relative humidity lines for other *fixed* values of ϕ are obtained using the same procedure.

4.4.3 *Constant wet-bulb temperature lines*

For the purpose of illustration, let us now assume the line AB in Fig. 4.4 to be a constant wet-bulb temperature (t_{wb}) line. Applying Eq. (4.39) to the two points of intersection A and B we obtain the following relations:

$$h_1 = h_{wb} - (\omega_{wb} - \omega_1)h_{fw}(t_{wb}) \tag{4.56}$$

$$h_2 = h_{wb} - (\omega_{wb} - \omega_2)h_{fw}(t_{wb}) \tag{4.57}$$

where h_{wb} and ω_{wb} are respectively the enthalpy and humidity ratio of saturated air at the wet-bulb temperature, and h_{fw} is the liquid enthalpy. From Eqs. (4.56) and (4.57) we obtain the change of enthalpy as

$$h_2 - h_1 = (\omega_2 - \omega_1)h_{fw}(t_{wb})$$

Hence

$$\frac{h_2 - h_1}{\omega_2 - \omega_1} = q = h_{fw}(t_{wb}) \tag{4.58}$$

From Eq. (4.58) we observe that on the (h-ω) chart a constant wet bulb temperature line is a straight line because h_{fw} is constant on such a line. However, the lines at different wet-bulb temperatures are not parallel because the slope of these lines depend on h_{fw}, which is a function of the wet bulb temperature. Note that Eq. (4.58) corresponds to Eq. (4.51) obtained earlier for a constant dry-bulb temperature line. Therefore by following the steps outlined above for the dry bulb temperature line we can show that the slope, θ_w of a constant wet bulb temperature line is given by

$$cot\theta_w = \frac{[h_{fw}(t_{wb}) - h_g(t_v)]}{S} \tag{4.59}$$

On the saturation curve the wet-bulb temperature t_{wb} is equal to the dry bulb temperature t. Therefore constant wet-bulb temperature lines can be constructed by drawing straight lines, inclined at θ_w to the horizontal axis, through a series of chosen points ($t = t_{wb}$) on the saturation curve.

4.4.4 *Constant specific volume lines*

To construct a constant specific volume line, a series of t–values are selected and the corresponding ω–values are calculated using Eq. (4.34), for a constant value of the specific volume, v. A smooth curve is drawn through the points on the chart whose coordinates are the chosen t–values and the calculated ω–values.

The procedure described above was used to construct the psychrometric chart shown in Fig. 4.5. The total pressure is assumed to be 101.325 kPa, the standard ambient pressure. The scale factor for the ω–scale is 2.0×10^{-3} kgkg^{-1}*per cm* and the scale factor for the h–scale is 1.8 kJkg^{-1} *per cm*. The constant dry-bulb temperature line for $t_{db} = 50°C$ is drawn vertically.

This chart could be used to solve most of the common air conditioning design problems where the pressures involved are close to that of the standard atmosphere. However, care should be exercised when the chart is used for other pressures as this may cause significant errors in the some of the computed quantities.

4.4.5 *Enthalpy–moisture protractor*

Many common psychrometric design calculations involve the enthalpy–moisture ratio, q. This ratio can be interpreted as the slope of a straight line on the psychrometric chart. The slopes θ for different values of q are obtained by substituting in Eq. (4.50) from Eq. (4.52). Hence we have

$$cot\theta = \frac{q}{S} - \frac{h_g(t_v)}{S} \tag{4.60}$$

The resulting values of q and θ are used to construct a protractor for easy use in graphical computations. This protractor is shown with the psychrometric chart in Fig. 4.5.

4.4.6 *Sensible heat ratio protractor*

The sensible heat ratio (SHR) is defined as the ratio of the change in sensible enthalpy Δh_s to the change in total enthalpy Δh. Therefore

$$SHR = \frac{\Delta h_s}{\Delta h} = 1 - \frac{\Delta h_l}{\Delta h} \tag{4.61}$$

where the change in latent enthalpy is given by

$$\Delta h_l = \Delta h - \Delta h_s$$

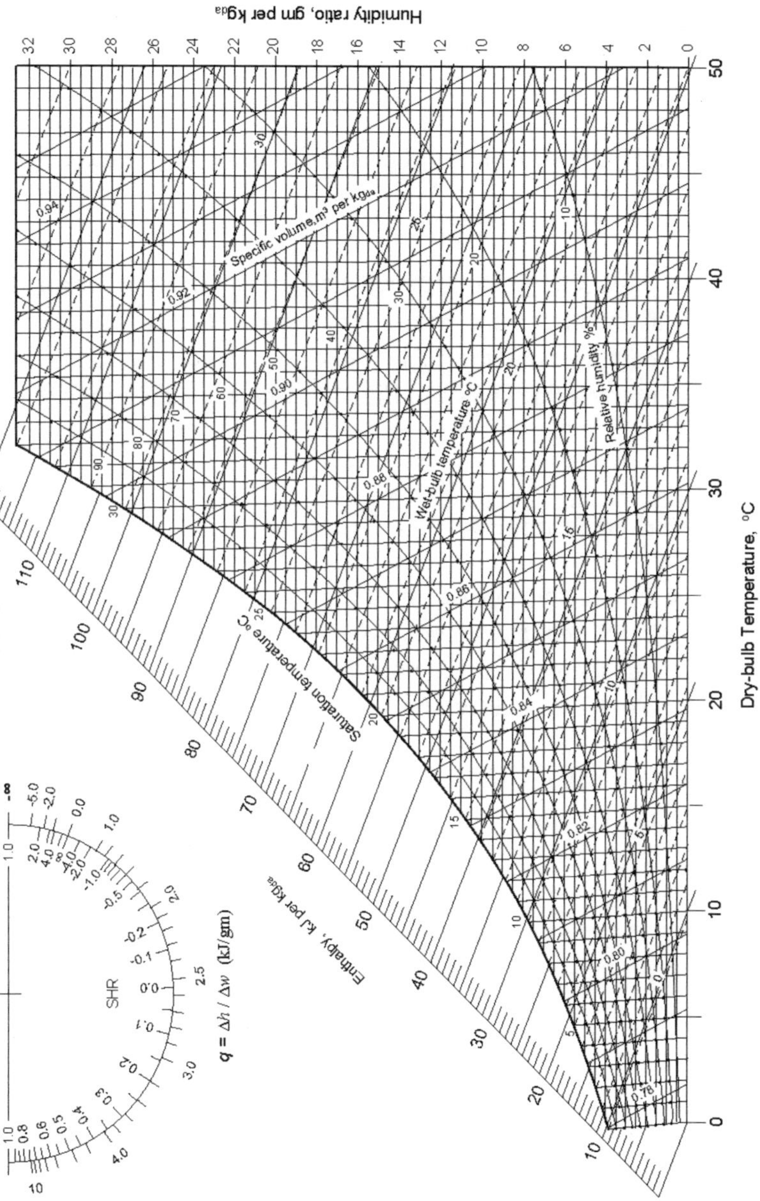

Fig. 4.5 **PSYCHROMETRIC CHART**
Pressure 101.325 kPa

The change in latent enthalpy is related to the change in humidity ratio by Eq. (4.31). Therefore

$$\Delta h_l = h_{gm}(\omega_2 - \omega_1) = h_{gm}\Delta\omega \tag{4.62}$$

Substituting in Eq. (4.61) from Eqs. (4.62) and (4.51) we have

$$SHR = \frac{\Delta h_s}{\Delta h} = 1 - \frac{h_{gm}\Delta\omega}{\Delta h} = 1 - \frac{h_{gm}}{q} \tag{4.63}$$

The SHR protractor is constructed using Eq. (4.63), where a constant value is assumed for the saturation vapor enthalpy, h_{gm}. For use in graphical computations the SHR–values are indicated on the enthalpy–moisture protractor as shown in Fig. 4.5, where a value of h_{gm} is taken as 2555 kJkg^{-1}.

4.5 Worked Examples

Example 4.1 Atmospheric air at a pressure of 95 kPa has a humidity ratio of 0.016 and a temperature of 27°C. Determine the following quantities: (i) the dew-point temperature, (ii) the relative humidity, (iii) the specific volume, and (iv) the enthalpy.

Solution We shall solve this problem by using the various equations derived above for the properties of moist air.

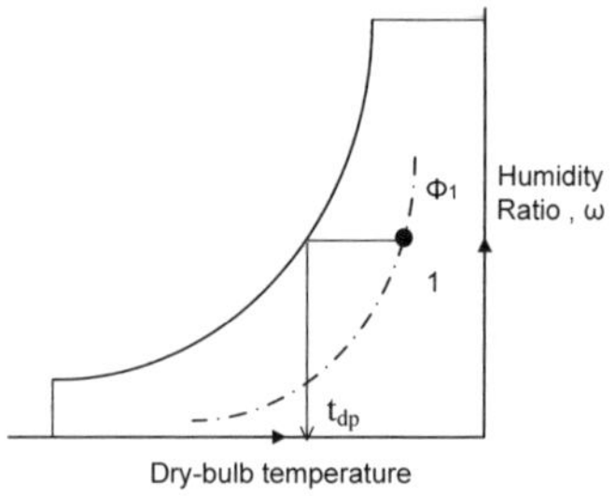

Fig. E4.1.1 Psychrometric chart

(i) At the dew-point temperature the air is in a saturated state with the same humidity ratio. Let the vapor pressure in this state be P_v. Since the relative humidity at the dew point is 100% we have from Eq. (4.13)

$$\omega = \frac{0.622P_v}{(P-P_v)} = 0.016$$

Substituting the total pressure, $P = 95$ kPa in the above equation

$$\frac{0.622P_v}{(95-P_v)} = 0.016$$

Hence $P_v = 2.382$ kPa. The saturation temperature at this pressure is obtained from the steam table [3] as 20.3°C, which is the dew-point temperature. This temperature is indicated in the psychrometric chart in Fig. E4.1.1.

(ii) The saturation vapor pressure P_s at the given temperature of 27°C is 3.564 kPa, from the steam table [3]. Therefore the relative humidity is

$$\phi = \frac{P_v}{P_s} = \frac{2.382}{3.564} = 66.8\%$$

(iii) The specific volume is given by Eq. (4.34) as

$$v = (R_a T/P)(1 + \omega/0.622)$$

Substituting numerical values in the above equation we have

$$v = [0.287 \times (27 + 273)/95](1 + 0.016/0.622) = 0.93$$

Note that the units of specific volume, v are [m^3kg^{-1} (*dry air*)]

(iv) The moist air enthalpy is given by Eq. (4.16) as

$$h = c_{pa}t + \omega h_g(t)$$

Substituting the relevant numerical values in the above equation we have

$$h = 1.0 \times 27 + 0.016 \times 2550.3 = 67.8 \text{ kJkg}^{-1}$$

Example 4.2 The dimensions of room are 10m by 6m by 3m high. The pressure, temperature and degree of saturation of the air in the room are 100 kPa, 25°C and 55 percent respectively. (i) Calculate the mass of air in the room. (ii) If the surface temperature of a window of the room is 10.5°C, will moisture condense out of the air?

Solution

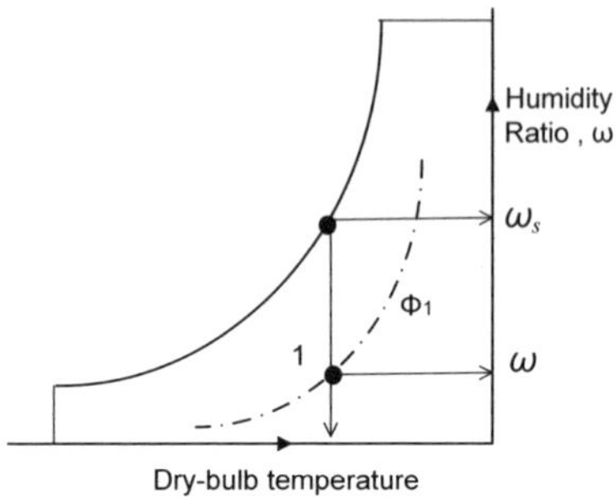

Fig. E4.2.1 Psychrometric chart

The saturation vapor pressure at 25°C is obtained from the steam table [3] as 3.166 kPa. The humidity ratio of saturated air at this temperature is obtained from Eq. (4.13) as

$$\omega_s = \frac{0.622P_v}{(P-P_v)} = \frac{0.622\times3.166}{(100-3.166)} = 0.02034$$

The degree of saturation is given by

$$\mu = \omega/\omega_s = 0.55 = \omega/0.02034$$

Therefore the humidity ratio, $\omega = 0.011187$

These quantities are indicated in the psychrometric chart in Fig. E4.2.1.

The specific volume (m^3kg^{-1}) of the air is given by Eq. (4.34) as

$$v = (R_aT/P)(1 + \omega/0.622)$$

Substituting numerical values we have

$$v = [0.287 \times (25 + 273)/100](1 + 0.011187/0.622) = 0.8706$$

(i) The volume of the room is

$$V = 10 \times 6 \times 3 = 180 \text{ m}^3$$

The mass of *dry air* in the room is obtained as

$$m_a = \frac{V}{v} = \frac{180}{0.8706} = 206.74 \text{ kg}$$

The mass of moist air is

$$m = m_a(1 + \omega) = 206.74 \times 1.011187 = 209.05 \text{ kg}$$

(ii) Now the saturation vapor pressure at the dew-point temperature of the air is given by Eq. (4.13) as

$$\omega = \frac{0.622P_v}{(100-P_v)} = 0.011187$$

Therefore $P_v = 1.7667$ kPa. The saturation temperature at this pressure is the dew-point, which is obtained from the steam table [3] as 15.6°C. The temperature of the window surface of 10.5°C, is below the dew-point of the air and therefore condensation of moisture will occur.

Example 4.3 The mass, pressure, dew-point temperature and relative humidity of a sample of moist air are 0.8 kg, 96 kPa, 20°C and 60 percent respectively. Determine (i) the dry-bulb temperature of the air, (ii) the humidity ratio of the air, and (iii) the volume of the sample.

Solution (i) Now at the dew-point temperature, the air is in a saturated state with the same humidity-ratio as that of the air sample (see Fig. E4.1.1). The saturation vapor pressure at 20°C from the steam table [3] is 2.337 kPa. The vapor in the air sample is at the same pressure. The saturation vapor pressure at the given temperature is

$$P_s = P_v/\phi = 2.337/0.6 = 3.895 \text{ kPa}$$

From the steam table [3] we obtain the saturation temperature at 3.895 kPa as 28.5°C, which is the dry-bulb temperature of the air sample.

(ii) The humidity ratio is given by

$$\omega = \frac{0.622 \times 2.337}{(96-2.337)} = 0.01552$$

(iii) The specific volume of the air is obtained from the equation

$$v = (R_a T/P)(1 + \omega/0.622)$$

Substituting numerical values in the above equation we have

$$v = [0.287 \times (273 + 28.5)/96](1 + 0.01552/0.622) = 0.924$$

Now the mass of dry air in the sample is

$$m_a = m/(1 + \omega) = 0.8/1.01552 = 0.7878 \text{ kg}$$

Hence the volume of the air sample is

$$V = m_a v = 0.7878 \times 0.924 = 0.728 \text{ m}^3$$

Example 4.4 The pressure and dry-bulb temperature of the air in a house are 100 kPa and 22°C respectively. The surface temperature of a window of the house is 6°C. What is the maximum relative humidity allowable in the house if no condensation is to occur on the surface of the window?

Solution Condensation of water will occur if the temperature of the window surface is below the dew-point temperature of the air in the room. Assume that the dew-point is 6°C. The saturation pressure at this temperature is obtained from the steam table [3] as 0.9346 kPa. This pressure is equal to the saturation vapor pressure of air with a dew-point of 6°C. The saturation vapor pressure at the air dry-bulb temperature of 22°C is 2.642 kPa [3]. The relative humidity of air with t_{db} = 22°C and t_{dp} = 6°C is given by

$$\phi = P_v/P_s = 0.9346/2.642 = 35.37\%$$

Therefore water vapor will condense on the surface of the window if the relative humidity of the air in the room exceeds 35.37%. The scenario is illustrated in the psychrometric chart in Fig. E4.4 1.

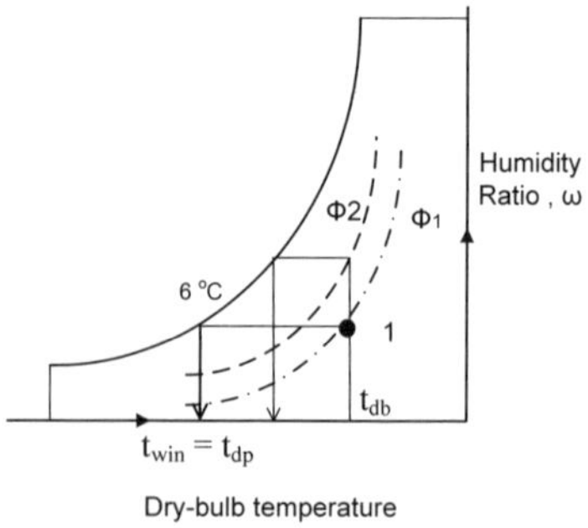

Fig. E4.4.1 Psychrometric chart

Example 4.5 The dry-bulb temperature and wet-bulb temperature of a sample of air are 30°C and 23°C respectively. The pressure of the air is 98 kW. Calculate (i) the humidity ratio if the air is adiabatically saturated, (ii) the humidity ratio of the air, (iii) the partial pressure of water vapor and dry air in the sample, and (iv) the relative humidity.

Solution Let the initial state of the air be 1 and final state after the adiabatic saturation process be 2. At state 2 the air is at its wet-bulb temperature. The following data are obtained from the steam table [3]:
At $t_1 = 30°\text{C}$, $h_{g1} = 2555.7\ \text{kJkg}^{-1}$.
At $t_2 = 23°\text{C}$, $P_{s2} = 2.808\ \text{kPa}$, $h_{g2} = 2543\ \text{kJkg}^{-1}$, $h_{f2} = 96.4\ \text{kJkg}^{-1}$

(i) The humidity ratio at the wet-bulb temperature is obtained from Eq. (4.13) as

$$\omega_2 = \frac{0.622 \times 2.808}{(98 - 2.808)} = 0.01834$$

The enthalpy of air at 23°C and 30°C are given by

$$h_2 = c_{pa}t_2 + \omega_2 h_{g2} = 23 + 2543 \times 0.01834 = 69.65\ \text{kJkg}^{-1}$$

$$h_1 = c_{pa}t_{21} + \omega_1 h_{g1} = 30 + 2555.7\omega_1$$

Application of the steady-flow energy equation to the adiabatic saturation process (see Fig. 4.2) we have

$$h_2 = h_1 + (\omega_2 - \omega_1)h_{f2}$$

Substituting numerical values in the above equation we obtain

$$69.65 = 30 + 2555.7\omega_1 + (0.01834 - \omega_1) \times 96.4$$

(ii) Solution of the above equation gives the humidity ratio, $\omega_1 =$ 0.0154

(iii) The partial pressure of water vapor, P_v in the air is obtained from Eq. (4.13) as

$$\omega_2 = \frac{0.622P_v}{(98 - P_v)} = 0.0154$$

Hence the partial pressure of water vapor is, $P_v = 2.368\ \text{kPa}$.
Applying Dalton's rule of partial pressures we have

$$P_a + 2.368 = 98$$

Therefore the partial pressure of dry air is, $P_a = 95.63\ \text{kPa}$.

(iv) The saturation vapor pressure at the dry-bulb temperature of 30°C is 4.242 kPa.
Hence the relative humidity is

$$\phi = P_v/P_s = 2.368/4.242 = 55.8\%$$

Example 4.6 A psychrometric chart is constructed with enthalpy and humidity ratio as the basic coordinates. The scale factor for the ω-scale is 1.5×10^{-3} kgkg^{-1}*per cm* and the scale factor for the h-scale is 1.8 kJkg^{-1} *per cm*. The constant dry-bulb temperature line for t_{db} = 60°C is drawn vertically. Calculate (i) the inclination of the constant enthalpy lines, and (ii) the inclination to the horizontal of the constant dry-bulb temperature lines at 30°C and 10°C.

Solution The overall scale factor, S for the psychrometric chart is given by

$$S = s_h/s_\omega = 1.8/0.0015 = 1200 \text{ kJkg}^{-1}$$

The geometrical details of the construction of the psychrometric chart are shown in Fig. 4.4. For the solution of the present problem we shall use the relations obtained earlier in section 4.4.1.
From Eqs. (4.46) and (4.47) we obtain the following relations:

$$\frac{L_h}{L_\omega} = \frac{s_\omega(h_2-h_1)}{s_h(\omega_2-\omega_1)} = \frac{q}{S} = \frac{q}{1200}$$

$$q = \frac{h_2-h_1}{\omega_2-\omega_1}$$

$$cot\theta + cot\beta = \frac{q}{S} = q/1200 \quad \text{(E4.6.1)}$$

The enthalpy of moist air is given by

$$h = c_{pa}t + \omega h_g(t) \quad \text{(E4.6.2)}$$

Hence for a constant dry-bulb temperature (t) line passing through the points 1 and 2 we have

$$\frac{h_2-h_1}{\omega_2-\omega_1} = h_g(t) = q \quad \text{(E4.6.3)}$$

From Eqs. (E4.6.1) and (E4.6.3) we obtain

$$cot\theta + cot\beta = h_g/1200 \qquad \text{(E4.6.4)}$$

Now the constant dry-bulb temperature line for 60°C is drawn vertically (i.e. θ = 0). The vapor enthalpy at this temperature is 2609 $kJkg^{-1}$[3]. Therefore from Eq. (E4.6.4) it follows that

$$cot\beta = \frac{q}{1200} = \frac{2609}{1200} = 2.174$$

Hence the inclination of the constant enthalpy lines, β = 24.7°.
The vapor enthalpy at 10°C is 2519.2 $kJkg^{-1}$[3]. Substituting in Eq. (E4.6.4) we have

$$cot\theta + 2.174 = 2519.2/1200 = 2.0993$$

$$cot\theta = -0.07466$$

The inclination of the constant dry-bulb temperature line (θ) at 10°C is 94.27°.
The vapor enthalpy at 30°C is 2555.7 $kJkg^{-1}$[3]. Substituting in Eq. (E4.6.4) we have

$$cot\theta + 2.174 = 2555.7/1200 = 2.1297$$

$$cot\theta = -0.04425$$

The inclination of the constant dry-bulb temperature line (θ) at 30°C is 92.54°. Note that the inclination, θ of the constant dry-bulb temperature lines decrease as the temperature increases. This is seen in the psychrometric chart in Fig. 4.5.

Now at the points where the horizontal axis (ω = 0) intersects the constant enthalpy lines, the enthalpy is directly proportional to the dry-bulb temperature according to Eq. (E4.6.2). Hence we have

$$h = c_{pa}t$$

We can now rescale the horizontal axis of the psychrometric chart in terms of dry-bulb temperature values as indicated in Fig. 4.5. The constant dry-bulb temperature lines are then drawn through these temperature points on the horizontal axis.

Example 4.7 A psychrometric chart is constructed with enthalpy and humidity ratio as the basic coordinates. The scale factor for the ω-scale is 10^{-3} $kgkg^{-1}$*per cm* and the scale factor for the h-scale is 2.4 $kJkg^{-1}$ *per* *cm*. The constant dry-bulb temperature line for t_{db} = 50°C is drawn vertically. Calculate (i) the inclination of the constant enthalpy lines and (ii) the inclination of the constant wet-bulb temperature lines at 40°C and 20°C to the horizontal.

Solution The overall scale factor, *S* for the psychrometric chart is given by

$$S = s_h/s_\omega = 2.4/0.001 = 2400 \ \text{kJkg}^{-1}$$

The geometrical details of the construction of the psychrometric chart are shown in Fig. 4.4.

We use the equations derived in Example 4.6 directly to obtain the inclination of the constant enthalpy line on the psychrometric chart. Thus we have

$$cot\theta + cot\beta = \frac{q}{S} = \frac{q}{2400} \tag{E4.7.1}$$

Now the constant dry-bulb temperature line for 50°C is drawn vertically (i.e. $\theta = 0$). The vapor enthalpy at this temperature is 2591.4 kJkg^{-1} [3]. Therefore from Eq. (E4.7.1) and Eq. (4.51) it follows that

$$cot\beta = \frac{q}{2400} = \frac{2591.4}{2400} = 1.07975$$

Hence the inclination of the constant enthalpy lines, $\beta = 42.8^\circ$.

The wet-bulb temperature is given by Eq. (4.39) as

$$h = h_{wb} - (\omega_{wb} - \omega)h_{fw}(t_{wb})$$

Hence for a *constant* wet-bulb temperature (t_{wb}) line passing through the points 1 and 2 we have

$$\frac{h_2 - h_1}{\omega_2 - \omega_1} = h_{fw}(t_{wb}) = q \tag{E4.7.2}$$

Substituting in Eq. (E4.7.1) from Eq. (E4.7.2) we obtain the equation for a constant wet-bulb temperature line as

$$cot\theta + cot\beta = h_{fw}/2400 \tag{E4.7.3}$$

The liquid water enthalpy at 40°C is 167.5 kJkg^{-1} [3]. Substituting in Eq. (E4.7.3) we obtain

$$cot\theta + 1.07975 = 167.5/2400 = 0.06979$$

$$cot\theta = -1.0099$$

The inclination of the constant wet-bulb temperature line (θ) at 40°C is 135.28°.
The liquid water enthalpy at 20°C is 83.9 $kJkg^{-1}$ [3]. Substituting in Eq. (E4.7.3)

$$cot\theta + 1.07975 = 83.9/2400 = 0.03496$$

$$cot\theta = -1.04479$$

The inclination of the constant wet-bulb temperature line (θ) at 20°C is 136.25°. Note that the inclination of the constant wet-bulb temperature lines decrease as the temperature increases.

Example 4.8 It is intended to construct (a) the saturation curve, (b) the constant relative humidity line for, $\phi = 60\%$ and (c) the constant specific volume line for, $v = 0.86\ m^3\ kg^{-1}$ in the psychrometric chart described in worked example 4.6. The total pressure is 101.3 kPa. Obtain (i) the (ω-t) coordinates for (a) and (b) at t = 10°C and 30°C and (ii) the (ω-t) coordinates for (c) at t = 23°C and 30°C

Solution (i) On the saturation curve ω and t are related by Eq. (4.13).

$$\omega_s = \frac{0.622 P_s(t)}{[P - P_s(t)]} \tag{E4.8.1}$$

The saturation pressure P_s, at t = 10°C, is 1.227 kPa [3]. The total pressure, P is 101.3 kPa. Substituting these values in Eq. (E4.8.1) we obtain $\omega = 0.007626$.
Similarly, the saturation pressure P_s, at t = 30°C, is 4.242 kPa [3]. The total pressure, P is 101.3 kPa. Substituting these values in Eq. (E4.8.1) we obtain $\omega = 0.02718$.
We have thus obtained two points on the saturation curve whose coordinates are (10°C, 0.00762) and (30°C, 0.02718). In a similar manner we could obtain more points on the saturation curve and then draw a smooth curve through the points.

(ii) On the constant relative humidity curve Eq. (4.13) applies:

$$\omega = \frac{0.622\phi P_s(t)}{P - \phi P_s(t)} \qquad \text{(E4.8.2)}$$

For, t = 10°C, substitute, P_s = 1.227 kPa, P = 101.3 kPa and ϕ = 60% in Eq. (E4.8.2) to obtain, ω = 0.004553.
For, t = 30°C, substitute, P_s = 4.242 kPa, P = 101.3 kPa and ϕ = 60% in Eq. (E4.8.2) to obtain ω = 0.01603.
We have thus obtained two points on the ϕ = 60% - constant relative humidity curve whose coordinates are (10°C, 0.004553) and (30°C, 0.01603). In a similar manner we could obtain more points on the ϕ = 60% - curve. We then draw a smooth curve through these points.

(iii) The specific volume is given by Eq. (4.34) as

$$v = (R_a T/P)\left(1 + \frac{\omega}{0.622}\right)$$

$$\omega = 0.622\left(\frac{Pv}{R_a T} - 1\right) \qquad \text{(E4.8.3)}$$

We shall now obtain the (ω-t) coordinates of the points on the constant specific volume curve for v = 0.86 m^3 kg^{-1}.
For t = 23°C we substitute the following values in Eq. (E4.8.3):
T = 296 K, R_a = 0.287 $kJkg^{-1}K^{-1}$, P = 101.3 kPa and v = 0.86 m^3 kg^{-1}.
This gives ω = 0.01586.
For t = 30°C we substitute the following values in Eq. (E4.8.3):
T = 303 K, R_a = 0.287 $kJkg^{-1}K^{-1}$, P = 101.3 kPa and v = 0.86 m^3 kg^{-1}.
This gives ω = 0.001123.
It is easy to verify that the various coordinates obtained above are in good agreement with the values in the psychrometric chart shown in Fig. 4.5.

Example 4.9 It is intended to construct the enthalpy/humidity ratio - protractor in the psychrometric chart described in worked example 4.7. Obtain the angles for $\Delta h/\Delta\omega$ ($kJkg^{-1}$) equal to 2000 and 5000.

Solution We shall use some of the relevant results obtained in worked example 4.7 to solve this problem. Now

$$cot\theta + cot\beta = q/2400$$

By using the fact that the constant dry-bulb temperature for 50°C is drawn vertically (i.e. $\theta = 0$), we obtained

$$cot\beta = 1.07975$$

From the two equations above it follows that

$$cot\theta = q/2400 - 1.07975$$

where $\Delta h/\Delta\omega = q$
Substituting in the above equation we obtain the following results:
(i) $q = 2000$, $\theta = -76.15°$ and (ii) $q = 5000$, $\theta = 44.9°$

Angles are measured positive from the horizontal as shown in Fig. 4.4.

Example 4.10 It is intended to construct the sensible heat ratio (SHR) - protractor in the psychrometric chart described in example 4.7. Obtain the angles for the SHR equal to 0.4 and −2.0.

Solution The sensible heat factor, SHR is defined by Eq. (4.63) as

$$SHR = \frac{\Delta h_s}{\Delta h} = 1 - \frac{h_{gm}\Delta\omega}{\Delta h} = 1 - \frac{h_{gm}}{q}$$

In example 4.9 we obtained

$$cot\theta = q/2400 - 1.07975$$

From the two equations above it follows that

$$cot\theta = \frac{h_{gm}}{2400(1-SHR)} - 1.07975$$

We assume a mean value of the vapor enthalpy, h_{gm} = 2555 kJkg^{-1} and substitute the given values of the SHR to obtain the following results:
(i) SHR = 0.4, $\theta = 55.2°$, and (ii) SHR = −2, $\theta = -54.06°$

Example 4.11 Moist air exists under conditions of 24°C db temperature and relative humidity 50 percent. The pressure is 101.3 kPa. Using the psychrometric chart, determine (i) the wet-bulb temperature, (ii) the dew-point, (iii) the humidity ratio, (iv) the enthalpy, and (v) the specific volume.

Solution The given state is located on the psychrometric chart by the point of intersection of the constant dry-bulb temperature line for 24°C and the constant relative humidity line for $\phi = 50\%$.
The required quantities are then read directly from the chart.
This gives the following results:
(i) wet-bulb temperature = 17°C, (ii) dew-point temperature = 13°C, (iii) humidity ratio = 9.45 gmkg^{-1}, (iv) enthalpy = 47.7 kJkg^{-1}, and (v) specific volume = 0.854 m^3kg^{-1}.

Example 4.12 A sample of moist air initially at 15°C dry-bulb temperature and 10°C wet-bulb temperature undergoes a process to a state with a dew-point of 18°C and a dry-bulb temperature of 35°C. Use the psychrometric chart to determine (i) the enthalpy-humidity ratio for the process, and (ii) the sensible heat ratio, SHR for the process.

Solution

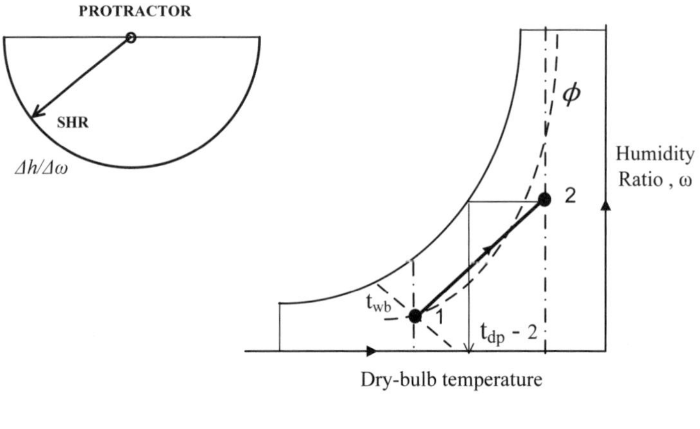

Fig. E4.12.1 Psychrometric chart

The process 1-2 undergone by the air is shown in the psychrometric chart in Fig. E4.12.1. The state 1 is located by the intersection of the 15°C dry-bulb temperature line and the 10°C wet-bulb temperature line. The state 2 is located by the intersection of the 35°C dry-bulb temperature line and the horizontal line drawn through the point on the saturation curve where the dry-bulb temperature is 18°C (which is also equal to the wet-bulb temperature of 18°C).

We then draw a straight line parallel to the line 1-2 through the center of the SHR-protractor. The following values are read directly from the scales of the protractor:

$$\Delta h/\Delta\omega\ (\text{kJgm}^{-1}) = 5.3 \quad \text{and} \quad \text{SHR} = 0.52$$

Example 4.13 Moist air undergoes a process from an initial state of 15°C dry-bulb temperature and 90 percent relative humidity to a final dry-bulb temperature of 40°C. The sensible heat ratio (SHR) of the process is 0.6. Use the psychrometric chart to determine the following quantities at the end of the process: (i) the relative humidity, (ii) the wet-bulb temperature, (iii) the dew-point temperature (iv) the specific volume, (v) the humidity ratio, and (vi) enthalpy.

Solution The initial state 1 is located in the psychrometric chart by the intersection of the 15°C dry-bulb temperature line and the 90 percent relative humidity line as indicated in Fig. E4.1.1 The line representing a SHR of 0.6 is drawn on the SHR-protractor. With the aid of two set squares a line is drawn through point 1, parallel to the above line on the protractor. The point of intersection of this line and the 40°C dry-bulb temperature line gives the final state 2 of the air.
The properties of air at 2 are read off directly from the chart. This gives the following results:
(i) relative humidity = 36%, (ii) wet-bulb temperature = 26.7°C, (iii) dew-point temperature = 21.8°C, (iv) specific volume = 0.912 m^3kg^{-1}, (v) humidity ratio = 0.0164, and (vi) enthalpy = 82.6 kJkg^{-1}.

Example 4.14 The dry-bulb temperature and humidity ratio of a sample of moist air are 30°C and 0.014 respectively. Calculate the following quantities for the two pressures, 80 kPa, and 101.3 kPa: (i) the dew-point temperature, (ii) the relative humidity, (iii) the enthalpy, and (iv) the specific volume. Compare the results obtained for the two pressures.

Solution Consider the pressure P = 80 kPa. The saturation pressure at 30°C is 4.242 kPa and $h_g = 2555.7\ \text{kJkg}^{-1}$. The humidity ratio is given by Eq. (4.13) as

$$\omega = \frac{0.622\phi P_s(t)}{P - \phi P_s(t)}$$

$$0.014 = \frac{0.622 \times 4.242\phi}{80 - 4.242\phi}$$

Hence the relative humidity, ϕ = 41.5%. Now at the dew-point temperature, the saturation pressure is equal to the prevailing vapor pressure, P_v which is given by

$$P_v = \phi P_s = 4.242 \times 0.415 = 1.76\ \text{kPa}$$

Therefore the dew-point temperature is 15.5°C. The enthalpy is given by

$$h = c_{pa}t + \omega h_g(t)$$

Substituting numerical values in the above equation we have

$$h = 1.0 \times 30 + 0.014 \times 2555.7 = 65.78\ \text{kJkg}^{-1}$$

The specific volume is given by Eq. (4.13) as

$$v = (R_a T/P)(1 + \omega/0.622)$$

Substituting numerical values we have

$$v = (0.287 \times 303/80)(1 + 0.014/0.622) = 1.11\ \text{m}^3\text{kg}^{-1}$$

We now repeat the above calculations for pressure, P = 101.3 kPa, which is the pressure for the psychrometric chart in Fig. 4.5. The relative humidity is obtained from Eq. (4.13).

$$0.014 = \frac{0.622 \times 4.242\phi}{101.3 - 4.242\phi}$$

Hence the relative humidity, ϕ = 52.57%. At the dew-point temperature, the saturation pressure is equal to the prevailing vapor pressure, P_v which is given by

$$P_v = \phi P_s = 4.242 \times 0.5257 = 2.23\ \text{kPa}$$

Therefore the dew-point temperature is 19.24°C. The enthalpy is given by

$$h = 1.0 \times 30 + 0.014 \times 2555.7 = 65.78\ \text{kJkg}^{-1}$$

The specific volume is given by Eq. (4.34) as

$$v = (0.287 \times 303/101.3)(1 + 0.014/0.622) = 0.878\ \text{m}^3\text{kg}^{-1}$$

We note that the relative humidity, the dew-point temperature and the specific volume are dependent on the total pressure while the enthalpy is independent of the pressure for a given dry-bulb temperature and humidity ratio. The psychrometric chart shown in Fig. 4.5 is constructed for a standard ambient pressure of 101.325 kPa. Therefore care should be exercised when this chart is used for other pressures.

Example 4.15 Moist air undergoes a process from an initial state 1 with t_{db1} = 10°C and ω_1=0.005 to a final state 2 with t_{db2} = 40°C and ω_2 = 0.021. Obtain the change in enthalpy, (h_2 - h_1) using (i) the basic expression for enthalpy of moist air, (ii) the expressions for the change in sensible enthalpy and latent enthalpy, and (iii) the psychrometric chart.

Solution (i) We use the following expression for enthalpy including the temperature dependence of the specific heat capacity. The saturation vapor enthalpy is obtained from the steam table [3].

$$h = c_{pa}t + \omega h_g(t)$$

For state 1

$$h_1 = 1.004 \times 10 + 0.005 \times 2519.2 = 22.64 \text{ kJkg}^{-1}$$

For state 2

$$h_2 = 1.0056 \times 40 + 0.021 \times 2573.7 = 94.27 \text{ kJkg}^{-1}$$

Therefore the change in enthalpy, (h_2 - h_1) is 71.63 kJkg^{-1}.

(ii) The change in enthalpy may be written in the form

$$\Delta h = \Delta h_s + \Delta h_l$$

where the total enthalpy change, Δh of moist air is expressed as the sum of the sensible enthalpy change, Δh_s and the latent enthalpy change, Δh_l. These quantities are defined respectively by Eqs. (4.30) and (4.31) as

$$\Delta h_s = c_{am}(t_2 - t_1)$$

$$\Delta h_l = h_{gm}(\omega_2 - \omega_1)$$

For the temperature range from 0°C to 60°C of a typical psychrometric chart the following mean values are used for the specific heat capacity and the enthalpy[2]: $c_{am} = 1.03\ \text{kJkg}^{-1}\text{K}^{-1}$ and $h_{gm} = 2555\ \text{kJkg}^{-1}$. Substituting the given numerical data in the above equations we have

$$\Delta h_s = 1.03(40 - 10) = 30.9$$

$$\Delta h_l = 2555(0.021 - 0.005) = 40.88$$

$$\Delta h = 30.9 + 40.88 = 71.78\ \text{kJkg}^{-1}$$

Therefore the change in enthalpy, $(h_2 - h_1)$ is 71.78 kJkg^{-1}.

(iii) We locate the states 1 and 2 of the air on the psychrometric chart by the intersection of the constant dry-bulb temperature lines and the constant humidity ratio lines for the two states. The enthalpy for the two states are read directly from the chart. This gives the change in enthalpy, $(h_2 - h_1)$ as 71.5 kJkg^{-1}. Note that the results obtained by the three methods agree closely.

Example 4.16 In many air conditioning design problems the state of air is defined by specifying either (a) the dry-bulb temperature and the relative humidity, or (b) the dry-bulb temperature and the wet-bulb temperature. Write a MATLAB program to determine the other important properties of the air for cases (a) and (b).

Solution The MATLAB program is developed by employing 'curve-fit' expressions for the properties of air and water involved in the various psychrometric equations derived in this chapter.

(i) The variation of the *saturation* vapor pressure of water, p_{sat} [kPa] with absolute temperature, T [K] is given by the expression in Ref. [5] as

$$p_{sat} = Exp\left[A + \frac{B}{T-C}\right]$$

where the constants are: $A = 16.577$, $B = -4023.05$ and $C = 37.2$

(ii) Listed below are analytical expressions for the enthalpy of water obtained by fitting polynomials to data tabulated in the steam tables [3] for the temperature range from 0°C to 60°C.
Enthalpy of *saturated* water vapor, h_g (kJkg^{-1}) is given by:

$$h_g = 2500.7 + 1.854t - 5.0 \times 10^{-4}t^2 - 6.0 \times 10^{-6}t^3$$

where t (°C) is the vapor temperature.
Enthalpy of *saturated liquid* water, h_w (kJkg^{-1}) is given by

$$h_w = 0.002 + 4.198t - 3.0 \times 10^{-4}t^3$$

(iii) The variation of enthalpy of *saturated* moist air, h_s (kJkg^{-1}) with temperature, at 101.3 kPa pressure is given by [4]

$$h_s = 9.3625 + 1.786t_s + 1.1135 \times 10^{-2}{t_s}^2 + 9.8855 \times 10^{-4}{t_s}^3$$

where t_s (°C) is the air temperature.

(iv) The variation of the specific heat capacity, c_p (kJK^{-1}kg^{-1}) of *dry air* with temperature (°C) is given by [3]

$$c_p = 1.0038 + 3 \times 10^{-5}t + 4.0 \times 10^{-7}t^2$$

(a) In the MATLAB computer program, given in Appendix A4.1, for case (a), the humidity ratio, the air enthalpy and the specific volume are computed directly using Eqs. (4.13), (4.16) and (4.34) respectively. However, the computation of the wet-bulb temperature, t_{wb} requires an iterative procedure using Eq. (4.40). We first assume an initial guessed value for t_{wb} and adjust the value iteratively until the LHS and RHS of Eq. (4.40) are equal.

(b) In case (b) where the db-temperature and wb-temperature are specified, Eq. (4.40) is first solved to determine the humidity ratio of the air. Then the relative humidity, the enthalpy and the specific volume are computed using Eqs. (4.13), (4.16) and (4.34) respectively. (See problem 4.13.)

Problems

P4.1 A sample of moist air with a volume of 3.5m^3 has a dry bulb temperature of 40°C and a relative humidity of 50%. The total pressure is 90 kPa. Calculate (i) the humidity ratio, (ii) the degree of saturation, (iii) the specific volume, (iv) the mass, and (v) the enthalpy. Show how the state of the air is located on a psychrometric chart.

[*Answers*: (i) 0.0266, (ii) 47.9%, (iii) 1.0407 $m^3\ kg^{-1}$, (iv) 3.45 kg, (v) 108.46 $kJkg^{-1}$]

P4.2 A quantity of moist air of mass 3 kg has a humidity ratio of 0.015 and a relative humidity of 60%. The pressure is 98 kPa. Calculate (i) the dry-bulb temperature, (ii) the volume, (iii) the dew-point temperature, and (iv) the enthalpy. Show how the state of the air is located on a psychrometric chart.
[*Answers*: (i) 28.38°C, (ii) 2.67 m^3, (iii) 19.8°C, (iv) 66.67 $kJkg^{-1}$]

P4.3 A sample of air has a dry-bulb temperature of 30°C and a wet-bulb temperature of 20°C. Calculate the relative humidity, the specific volume, and the enthalpy, if the pressure is: (i) 85 kPa and (ii) 101 kPa. Show how the state of the air is located on a psychrometric chart.
[*Answers*: (i) 42.2%, 1.045 $m^3\ kg^{-1}$, 64.25 $kJkg^{-1}$, (ii) 39.8%, 0.875 $m^3\ kg^{-1}$, 57.38 $kJkg^{-1}$]

P4.4 A fixed mass of air has a volume of 2.2 m^3 and a pressure of 98 kPa. The dry-bulb temperature and degree of saturation of the air are 28°C and 55% respectively. Calculate (i) the relative humidity, (ii) the partial pressure of dry air, (iii) the partial pressure of water vapor, (iv) the mass of dry air, and (v) the mass of water vapor. Show how the state of the air is located on a psychrometric chart.
[*Answers*: (i) 55.9%, (ii) 95.9 kPa, (iii) 2.11 kPa, (iv) 2.44 kg, (v) 0.033 kg]

P4.5 A sample of air has a dry-bulb temperature of 32°C and a relative humidity of 60%. The pressure is 101.3 kPa. Calculate (i) the wet-bulb temperature, and (ii) the degree of saturation. Compare the results with data obtained from the psychrometric chart.
[*Answers*: (i) 25.6°C, (ii) 58.9%]

P4.6 A psychrometic chart is constructed for a pressure of 95 kPa. The overall scale factor, S is 2100 $kJkg^{-1}$. The constant dry-bulb temperature line at 55°C is drawn vertically in the chart. Calculate (i) the inclination of the constant enthalpy lines, (ii) the inclination of the 25°C

constant dry-bulb temperature line, and (iii) the inclination of the 15°C constant wet-bulb temperature line.
[*Answers*: (i) 38.9°, (ii) 91.45°, (iii) 140.4°]

P4.7 A psychrometic chart, constructed for a pressure of 100 kPa, has an overall scale factor S of 1800 $kJkg^{-1}$. The constant dry-bulb temperature line at 60°C is drawn vertically in the chart. Calculate the angles of (i) the enthalpy-humidity ratio protractor for $\Delta h/\Delta\omega$ ($kJkg^{-1}$)-values of 4000 and 1400, and (ii) the SHR-protractor for the SHR-values of 0.4 and −1.2.
[*Answers*: (i) 52.3°, 123.9° (ii) 47.5°, 128.8°]

P4.8 Moist air undergoes a process from an initial state with a dry-bulb temperature of 32°C and a relative humidity of 60% to a final state with a wet-bulb temperature of 10°C. The sensible heat ratio, SHR for the process is 0.3. The pressure is 101.3 kPa. Use the psychrometric chart to obtain the following properties of air at the final state: (i) the dry-bulb temperature, (ii) the relative humidity, and (iii) the enthalpy.
[*Answers*: 17.5°C, 37%, 29.5 $kJkg^{-1}$]

P4.9 Air, initially with a dry-bulb temperature of 35°C and a wet-bulb temperature of 25°C, undergoes a process to a dry-bulb temperature of 20°C and a wet-bulb temperature of 10°C. The pressure is 101.3 kPa. Calculate the sensible heat ratio, SHR, and the enthalpy-humidity ratio for this process. Compare the results with those obtained from the psychrometric chart.
[*Answers*: 0.33, 3800$kJkg^{-1}$]

P4.10 A sample of moist air has a relative humidity of 50% and a dew-point temperature of 15°C. The pressure is 101.3 kPa. The air undergoes a process for which the sensible heat ratio is −2.0, and the air is saturated with water vapor at the end of the process. Calculate the following properties of air in the final state: (i) the dry-bulb temperature, (ii) the wet-bulb temperature, and (iii) the humidity ratio. What is the enthalpy-humidity ratio, ($\Delta h/\Delta\omega$) for the process?
[*Answers*: (i) 19.8°C, (ii)19.8°C, (iii) 14.6 $gmkg^{-1}$; 852 $kJkg^{-1}$]

P4.11 (a) Saturated air at 60°C and a pressure of 105 kPa undergoes a throttling process to a pressure of 80 kPa. Calculate the relative humidity at the final equilibrium state. (b) Saturated air at a pressure of 110 kPa undergoes a constant pressure process from 30°C to 50°C. Calculate the relative humidity in the final state. Assume that air and water vapor behave like ideal gases.
[*Answers*: (a) 76%, (b) 34.4%]

P4.12 Derive Eq. (4.15) from first principles.

P4.13 Amend the MATLAB code listed in Appendix A4.1 to compute the humidity ratio, the enthalpy, and the specific volume of air when the dry-bulb and wet-bulb temperatures are specified.

References

1. *ASHRAE Handbook* - 2013 *Fundamentals*, American Society of Heating, Refrigeration and Air Conditioning Engineeers, Atlanta, 2013.
2. Kuehn, Thomas H., Ramsey, James W. and Threlkeld, James L., *Thermal Environmental Engineering*, 3rd edition, Prentice-Hall, Inc., New Jersey, 1998.
3. Rogers G. F. C. and Mayhew Y. R., *Thermodynamic and Transport Properties of Fluids*. 5th ed., Blackwell, Oxford, U. K. 1998.
4. Stoecker, Wilbert F. and Jones, Jerold W., *Refrigeration and Air Conditioning*, International Edition, McGraw-Hill Book Company, London, 1982.
5. Stoecker, Wilbert F., *Design of Thermal Systems*, McGraw-Hill Book Company, New York, International Edition, 1989, page 328.
6. Van Wylen, Gordon J. and Sonntag, Richard E., *Fundamentals of Classical Thermodynamics*, 3rd Edition, John Wiley & Sons, Inc. New York, 1985.

Appendix A4.1 - MATLAB Code for Psychrometric Properties

```
% computation of  psychrometric properties
pamb=101.325       % chart pressure in kPa
tdb = 30           % dry-bulb temperature (C)
tabdb=273+tdb      % tdb in degrees K
rh= 0.4            % relative humidity, fraction
ca=1;              % specific heat capacity of air, kJ/kg/K
cw=4.19            % specific heat capacity of water, kJ/kg/K
% coefficients A,B,C for the vapor pressure versus temperature curve-fit
% see worked example 4.16
g1=16.577     % A
g2=-4023.05   % B
g3=37.20      % C
% coefficients for vapor enthalpy vs temperature, cubic expression
a0=2500.7;
a1=1.854;
a2=-0.0005;
a3=-6.0e-06;
A=[a3 a2 a1 a0];
% coefficients for saturated air enthalpy vs temperature, cubic expression
b0=9.3625;
b1=1.7861;
b2=0.01135;
b3=9.8855e-04;
B=[b3 b2 b1 b0];
% coefficients for liquid enthalpy vs temperature, quadratic expression
c0=0.002;
c1=4.198;
c2=-0.0003;
C=[c2 c1 c0];
% compute humidity ratio of air
exop1=g1+ g2/(tabdb-g3)
pvp=exp(exop1)                    % saturated vapor pressure
wa=0.622*rh*pvp/(pamb-rh*pvp)     % Eq. (4.13) in textbook
```

```
% compute enthalpy of moist air
   hg=polyval(A,tdb)                    % saturated vapor enthalpy
   ha=ca*tdb+wa*hg                      % Eq. (4.16) in textbook
   hama=ha
% compute specific volume of moist air
Ra=0.287                                % gas constant for air, kJ/kg/K
vspc=(Ra*tabdb/pamb)*(1+wa/0.622)        % Eq. (4.34) in textbook
% compute wet-bulb temperature by iteration using Eq. (4.40)
twb=20        % initial guess of wet-bulb temperature
 lhs=ha
for i=1:10                      % number of iterations
hf2=polyval(C,twb);
hg2=polyval(A,twb)
tabwb=273 + twb
exop2=g1 + g2/(tabwb-g3)
 pvpw=exp(exop2)
 wa2=0.622*pvpw/(pamb-pvpw)
 ha2=ca*twb+wa2*hg2
 rhs=ha2-(wa2-wa)*hf2
 err=abs((1-rhs/lhs)*100)
 if err<=0.5   wbtem=twb
 % at the wet-bulb temperature, lhs=rhs
 twb=twb-0.02
 end
  % compute saturated air enthalpy and humidity ratio at same
  % temperature
  hs=polyval(B,tdb)   % only valid for pressure of 101.3 kPa
 ws=(hs-ca*tdb)/hg
end
```

Chapter 5

Psychrometric Processes for Heating and Air Conditioning

5.1 Introduction

In chapter 4 we developed a number expressions to evaluate the properties of moist air. These expressions were used to construct the psychrometric chart which provides a convenient means to obtain the properties of moist air directly.

Processes involving moist air are usually called psychrometric processes. These processes play an important role in the design of heating, ventilation and air conditioning (HVAC) systems. In the present chapter we shall analyze several basic psychrometric processes and consider their application in the design of HVAC systems. Since the operating pressures of most practical HVAC systems are close to standard ambient conditions (101 kPa) we shall use the psychrometric chart to analyze most practical situations.

5.2 Basic Psychrometric Processes

The main purpose of heating and air conditioning systems is to maintain a space at a desired temperature and relative humidity. This is achieved by supplying air with the appropriate conditions to the space. The processing units of typical air conditioning systems utilize psychrometric processes to produce air with the required temperature and relative humidity, to be supplied to the space. In this section we shall analyze several basic psychrometric processes. These include: mixing of two moist air streams, sensible heating and cooling of air, dehumidification of air by cooling, and humidification of air by adding moisture. In subsequent sections we shall discuss how these basic processes are used to design practical heating and air conditioning systems.

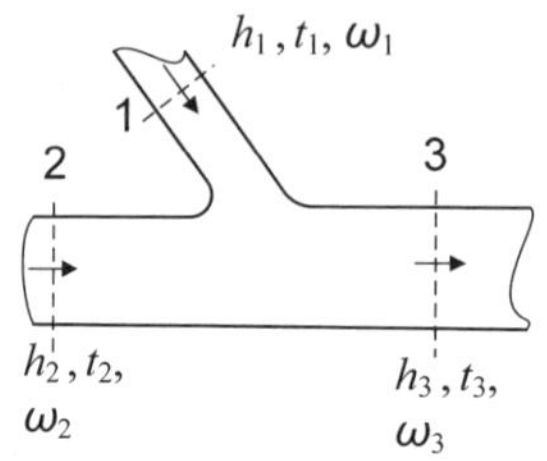

Fig. 5.1 (a) Mixing of two air streams

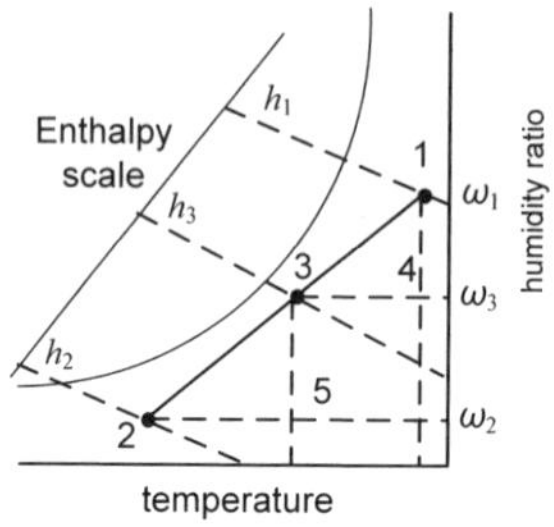

Fig.5.1. (b) Psychrometric chart

5.2.1 *Mixing of two moist air streams*

The mixing of two moist air streams, illustrated schematically in Fig. 5.1(a), is a common psychrometric process encountered in most heating and air conditioning systems. The inlet streams to the mixing section are indicated by 1 and 2 while 3 denotes the exit stream. For typical mixing processes, the changes in kinetic energy and potential energy of the air streams are negligible compared to the change in enthalpy. Moreover, any heat exchange between the air and the surroundings is negligible. For design purposes it is usually assumed that the mixing process occurs under steady-state conditions at constant pressure.

Consider a control volume with two inlet ports 1 and 2 and an exit port 3, surrounding the mixing section shown in Fig. 5.1(a). We now apply the conservation equations of mass and energy to the control volume, subject to the aforementioned assumptions.

For mass balance of air:

$$\dot{m}_{a1} + \dot{m}_{a2} = \dot{m}_{a3} \tag{5.1}$$

where $\dot{m}_{a1}$, $\dot{m}_{a2}$, and $\dot{m}_{a3}$ are the respective dry air mass flow rates of the three streams 1, 2 and 3.

For mass balance of water:

$$\dot{m}_{a1}\omega_1 + \dot{m}_{a2}\omega_2 = \dot{m}_{a3}\omega_3 \tag{5.2}$$

where ω_1, ω_2, and ω_3 are the respective humidity ratios of the three streams 1, 2 and 3.

Application of the steady-flow energy equation (SFEE) gives:

$$\dot{m}_{a1}h_1 + \dot{m}_{a2}h_2 = \dot{m}_{a3}h_3 \tag{5.3}$$

where h_1, h_2, and h_3 are the respective enthalpies of the three streams 1, 2 and 3.

Eliminating $\dot{m}_{a3}$ between Eqs. (5.1), (5.2) and (5.3) we have

$$\frac{\dot{m}_{a1}}{\dot{m}_{a2}} = \frac{h_3 - h_2}{h_1 - h_3} = \frac{\omega_3 - \omega_2}{\omega_1 - \omega_3} \tag{5.4}$$

Therefore

$$\frac{h_3 - h_2}{\omega_3 - \omega_2} = \frac{h_1 - h_3}{\omega_1 - \omega_3} \tag{5.5}$$

From Eq. (5.5) it follows that the slopes of the lines 2-3 and 1-2, which are also indicated on the $(\Delta h/\Delta\omega)$ - protractor of the psychrometric chart (see section 4.4.5), are equal. Since point 3 is common to both lines, it is clear that line 1-3-2 is a straight line. We recall from our construction of the psychrometric chart in chapter 4 that the constant humidity ratio, (ω) lines are horizontal straight lines. Notice that in Fig. 5.1(b), the right angle triangles 134 and 325 are similar. Therefore

$$\frac{\overline{23}}{\overline{31}} = \frac{\overline{53}}{\overline{41}} = \frac{\omega_3 - \omega_2}{\omega_1 - \omega_3} \tag{5.6}$$

From Eqs. (5.4) and (5.6) it follows that

$$\frac{\overline{23}}{\overline{31}} = \frac{\dot{m}_{a1}}{\dot{m}_{a2}} \tag{5.7}$$

Equation (5.7) suggests a convenient graphical method to locate the state 3 of the mixed air stream on the psychrometric chart, and thereby obtain the properties of the mixed stream. All we need to do is to draw the straight line connecting the state points 1 and 2, representing the two inlet streams, and then divide it in the inverse ratio of the *dry air* mass flow rates of the two streams.

It is important to note that the vertical lines 14 and 35, drawn through the points 1 and 3, in Fig. 5.1(b), are not constant temperature lines. However, we could perform an approximate analysis involving the dry-bulb temperatures and ideal gas relations to obtain the temperature of the mixed air stream. Using the expression for enthalpy given by Eq. (4.18), we rewrite the energy equation (5.3) in the form

$$\dot{m}_{a1}(c_{pa}+\omega_1 c_{pw})t_1+\dot{m}_{a1}\omega_1 h_{go}+\dot{m}_{a2}(c_{pa}+\omega_2 c_{pw})t_2 \\ +\dot{m}_{a2}\omega_2 h_{go}=\dot{m}_{a3}(c_{pa}+\omega_3 c_{pw})t_3+\dot{m}_{a3}\omega_3 h_{go}$$

Invoking Eq. (5.2), the above equation can be reduced to

$$\dot{m}_{a1}(c_{pa}+\omega_1 c_{pw})t_1+\dot{m}_{a2}(c_{pa}+\omega_2 c_{pw})t_2=\dot{m}_{a3}(c_{pa}+\omega_3 c_{pw})t_3$$

If in the above equation we assume that the specific heat capacities of the moist air streams 1, 2 and 3 are equal, then we obtain the following approximate relation for the temperature of the mixed air stream:

$$t_3=(\dot{m}_{a1}/\dot{m}_{a3})t_1+(\dot{m}_{a2}/\dot{m}_{a3})t_2 \tag{5.8}$$

For most practical mixing processes, Eq. (5.8) gives results that are in close agreement with those obtained by referring to the psychrometric chart.

5.2.2 *Sensible heating or cooling*

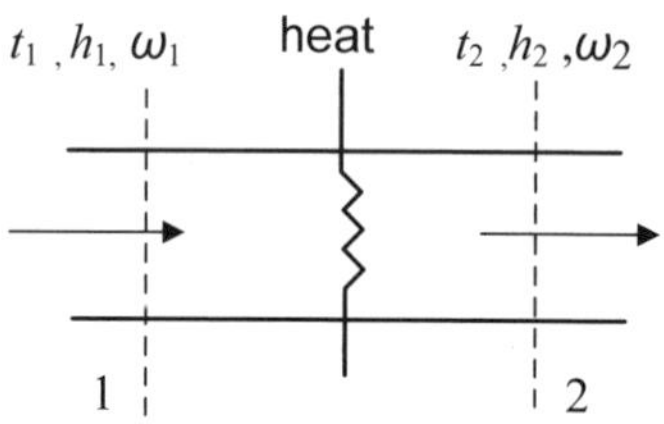

Fig. 5.2 (a) Sensible heating or cooling

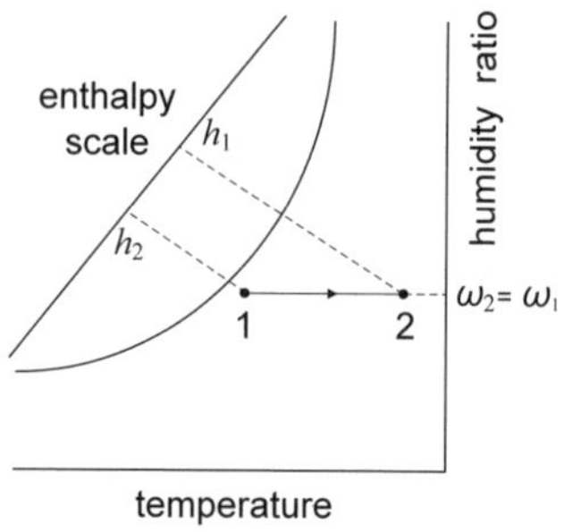

Fig. 5.2 (b) Psychrometric chart

An arrangement for the sensible heating or cooling of moist air is shown schematically in Fig. 5.2(a). Moist air enters the heating/cooling section at 1 and leaves at the exit section 2. An electrical resistance heater or a tubular coil carrying a hot fluid may be used to heat the air. The air could be cooled by passing a cold fluid like chilled water or a refrigerant through a similar coil.

During the *sensible* heating and cooling of air, the moisture content of the air remain constant. Therefore to prevent condensation of moisture during sensible cooling, we need to ensure that the surface temperature of the cooling coil is above the dew-point temperature of the entering air.

Consider the control volume surrounding the *heating* section shown in Fig. 5.2(a), with an inlet port 1 and an exit port 2. We now apply the conservation equations of mass and energy to the control volume.

For mass balance of dry air:

$$\dot{m}_{a1} = \dot{m}_{a2} = \dot{m}_a \tag{5.9}$$

For mass balance of water:

$$\dot{m}_{a1}\omega_1 = \dot{m}_{a2}\omega_2 \tag{5.10}$$

where ω_1 and ω_2 are the humidity ratios at 1 and 2 respectively.

Applying the steady-flow energy equation (SFEE), neglecting changes in kinetic and potential energy we have

$$\dot{m}_{a1}h_1 + \dot{Q}_{in} = \dot{m}_{a2}h_2 \tag{5.11}$$

where h_1 and h_2 are the air enthalpies at 1 and 2 respectively. The rate of heat input to the air is $\dot{Q}_{in}$.

From Eqs. (5.9) and (5.11) the rate of heat input is obtained as

$$\dot{Q}_{in} = \dot{m}_a(h_2 - h_1) \tag{5.12}$$

Assuming that the specific heat capacity of moist air is constant, Eq. (5.12) may be written as

$$\dot{Q}_{in} = \dot{m}_a c_{pm}(t_2 - t_1) \tag{5.13}$$

where t_1 and t_2 are the temperatures at 1 and 2 respectively.

For a *sensible* cooling process, Eqs. (5.9) to (5.13) may be used to determine the cooling rate.

5.2.3 *Dehumidification by cooling*

Dehumidification of moist air by cooling is one of most widely used processes in summer air conditioning systems. Three practical arrangements to carry out such a dehumidification process are shown schematically in Fig. 3(a)-(c). The corresponding process lines are depicted in the psychrometric chart in Fig. 5.4.

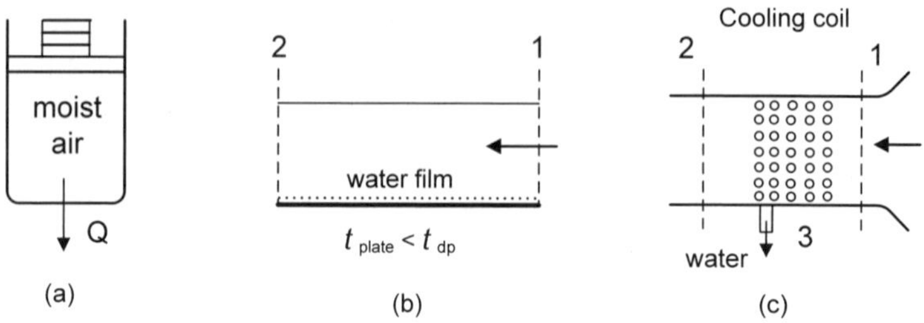

Fig. 5.3 (a) Ideal cooling process, (b) straight-line process, (c) actual cooling coil

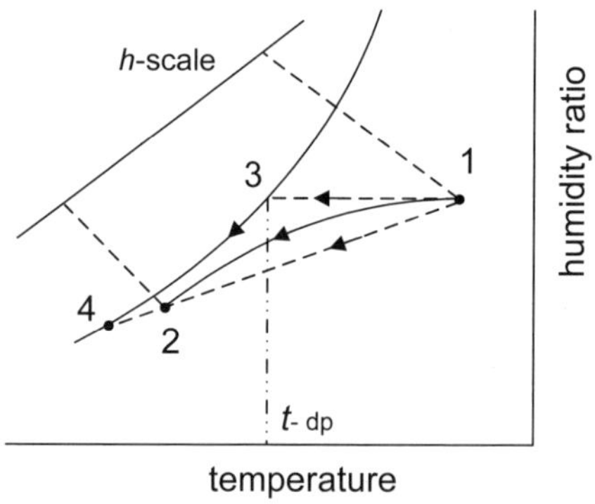

Fig 5.4 Psychrometric chart

The piston-cylinder set-up, shown in Fig. 5.3(a) is used to illustrate an *ideal, quasi-static* cooling and dehumidification process. The pressure of the moist air in the cylinder is held constant by the fixed load on the piston. Heat is removed in small steps allowing sufficient time at the end of each step for the air to attain an equilibrium state with spatially uniform properties. The path 1-3-4 on the psychrometric chart in Fig. 5.4 shows the state of the air during this ideal process. In the first phase from 1 to 3 the air undergoes sensible cooling with the humidity ratio remaining constant. At state 3 the air attains its *dew-point* temperature and the air is then fully saturated with water vapor. Further cooling results in vapor condensation and a reduction of the temperature. The state of the air then follows the saturation curve from 3 to 4 in a continuous manner.

The set-up shown in Fig. 5.3(b) is a more realistic arrangement to carry out the cooling and dehumidification process. Moist air flows over a cold plate, maintained at a *uniform temperature* below the dew-point temperature of the air at the entry section 1. The line 1-4 on the psychrometric chart in Fig. 5.4 shows the variation of the *mean* temperature and humidity ratio of the air at different sections of the flow stream. Since the plate temperature is below the *dew-point*, water vapor will condense on the plate from the air layer just adjacent to it. However, the *mean* temperature of the air at a section would be above the dew-point temperature for some distance along the plate.

A detailed analysis of the heat and moisture transfer between the air and the plate for this dehumidification process will be presented later in chapter 6. The main result from that analysis, pertinent to the present discussion is Eq. (6.16), which may be written in the form

$$\frac{dt}{d\omega} = Le\frac{(t_p - t)}{(\omega_p - \omega)} \tag{5.14}$$

where t_p and ω_p are respectively the *constant values* of the plate temperature and the *saturation* humidity ratio of the air adjacent to the water film. The mean temperature and humidity ratio of the air at a section are t and ω respectively.

An important conclusion from Eq. (5.14) is that if the dimensionless quantity *Le*, called the *Lewis number*, is *equal to one*, then the line 1-4, in Fig. 5.4, representing the state of the air is *approximately linear*. This is sometimes called the *straight line law* [5]. Moreover, at point 4 the air temperature approaches the plate temperature.

Figure 5.3(c) shows schematically some details of an actual cooling coil used to dehumidify moist air by cooling. A cold fluid, which could be either a refrigerant or chilled water, flows through the rows of tubes located in a duct. Moist air flows over the outside of the tubes. In typical applications the air is cooled sensibly as it passes over the first few tube rows and condensation of water vapor occurs over the rest of the tubes. Ideally, vapor condensation begins at the tube row where the *tube surface* temperature is just below the *dew-point* temperature of the air at the entry section 1. The condensed water drains to the bottom of the flow duct by gravity, and is discharged to the ambient through a tube at 3 as shown in Fig. 5.3(c).

It should be noted that unlike the ideal set-up in Fig 5.3(b), the tube surface temperature of the actual cooling coil is not uniform over the different tube rows. This results in the *non-linear* variation of the *mean* air temperature shown by the curve 1-2 in Fig. 5.4.

A simplified control volume analysis of cooling coils that enables us to perform psychrometric calculations is given below. Consider a control volume surrounding the cooling coil section shown in Fig. 5.3(c). Air flows in at 1 and exits at 2 and condensate (water) leaves the control volume at port 3. We now apply the conservation equations of mass and energy to the control volume.

For mass balance of dry air:

$$\dot{m}_{a1} = \dot{m}_{a2} \tag{5.15}$$

For mass balance of water:

$$\dot{m}_{a1}\omega_1 - \dot{m}_{a2}\omega_2 = \dot{m}_{w3} \tag{5.16}$$

where ω_1 and ω_2 are the humidity ratios at 1 and 2 respectively. The condensate flow rate at 3 is $\dot{m}_{w3}$.

Applying the steady-flow energy equation (SFEE) neglecting changes in kinetic and potential energy we obtain

$$\dot{m}_{a1}h_1 = \dot{m}_{a2}h_2 + \dot{m}_{w3}h_{w3} + \dot{Q}_{out} \tag{5.17}$$

where h_1 and h_2 are the air enthalpies at 1 and 2 respectively. The enthalpy of water leaving at 3 is h_{w3}. $\dot{Q}_{out}$ is the total rate of heat transfer from the air to the cooling fluid flowing through the tubes, commonly called the *refrigeration load* or *refrigeration capacity*.

A complete design model of a cooling coil including all the basic parameters is developed in chapter 7. However, such a model requires considerable computational effort. Therefore, we shall first present a semi-empirical model of a cooling coil, based on the *bypass factor*, that is much easier to use.

In Fig. 5.4, point 2 represents the state of the air after it passes through the actual coil shown in Fig. 5.3(c). The point 4 gives the state of the air after it passes over a cold plate maintained at *constant temperature*. We could visualize the real cooling process as a composite process, where a portion of the air bypasses the cooling coil, with its state unchanged, and the rest of the air undergoes cooling to the plate surface temperature in the set-up shown in Fig. 5.3(b). The two portions of air are then mixed *adiabatically* to obtain the final state 2 of the actual coil.

To develop a model based on the aforementioned processes we introduce two design parameters called the *bypass factor*, b and the cooling coil temperature or *apparatus dew-point*, t_d. Applying the equations of mass and energy conservation we obtain the following relations.

For mass balance of dry air:

$$\dot{m}_a = \dot{m}_{ab} + \dot{m}_{ac} \tag{5.18}$$

where $\dot{m}_{ab}$ and $\dot{m}_{ac}$ are the dry air flow rates of the bypass stream and the cooled stream.

The bypass factor is defined as

$$b = \frac{\dot{m}_{ab}}{\dot{m}_a} \tag{5.19}$$

Applying Eq. (5.8) to the adiabatic mixing process we obtain

$$t_2 = (\dot{m}_{ab}/\dot{m}_a)t_1 + (\dot{m}_{ac}/\dot{m}_a)t_d \tag{5.20}$$

where t_d (equal to t_4 in Fig. 5.4) is the saturation temperature of the air which is ideally equal to the cooling coil temperature or the *apparatus dew-point*. Manipulating Eqs. (5.18), (5.19) and (5.20) we obtain

$$b = \frac{t_2 - t_d}{t_1 - t_d} \tag{5.21}$$

The sensible cooling rate is

$$\dot{Q}_s = \dot{m}_a c_{am}(t_1 - t_2) = \dot{m}_a c_{am}(1 - b)(t_1 - t_d) \tag{5.22}$$

The total cooling rate is given by

$$\dot{Q}_t = \frac{\dot{Q}_s}{SHR} = \frac{\dot{m}_a c_{am}(1-b)(t_1 - t_d)}{SHR} \tag{5.23}$$

where *SHR* is the sensible heat ratio, which may be obtained directly from the protractor in the psychrometric chart for the process 1-4 shown in Fig. 5.4.

5.2.4 *Humidification of air*

Humidification of air by adding moisture to an air stream flowing steadily is a common psychrometric process, used in typical winter air conditioning systems. Shown schematically in Fig. 5.5 is a humidifier where a series of nozzles spray moisture directly to the air stream. Under ideal conditions, we assume that all the moisture sprayed is retained in the air stream.

Consider a control volume surrounding the humidifier, shown in Fig. 5.5(a), where the mass flow rate, the temperature, the humidity ratio and the enthalpy of the three streams are indicated.
Mass balance of dry air and water gives the following equations:

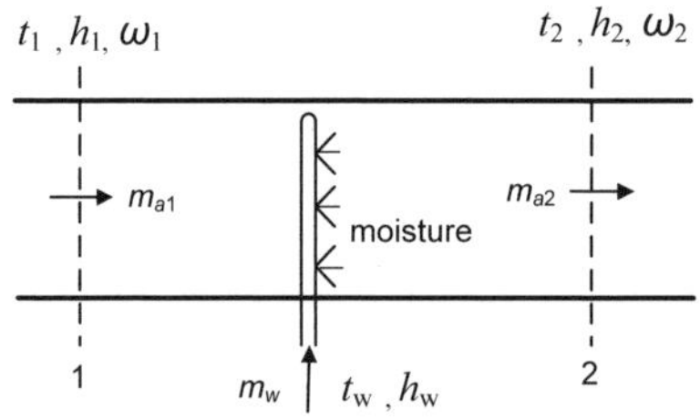

Fig. 5.5 (a) Humidification of air

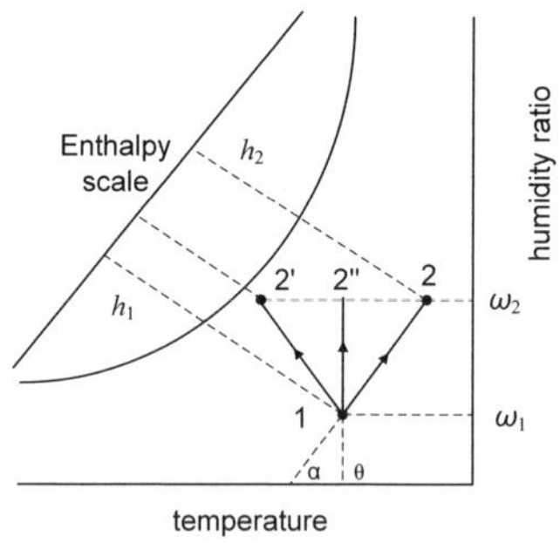

Fig. 5.5 (b) Psychrometric chart

$$\dot{m}_{a1} = \dot{m}_{a2} \tag{5.24}$$

$$\dot{m}_{a2}\omega_2 = \dot{m}_{a1}\omega_1 + \dot{m}_w \tag{5.25}$$

Applying the steady-flow energy equation (SFEE), neglecting changes in kinetic and potential energy of the air and any heat exchange with the surroundings, we obtain

$$\dot{m}_{a2}h_2 = \dot{m}_{a1}h_1 + \dot{m}_w h_w \tag{5.26}$$

The enthalpy-humidity ratio, q is obtained by manipulating Eqs. (5.24), (5.25) and (5.26) as

$$q = \frac{h_2 - h_1}{\omega_2 - \omega_1} = h_w \tag{5.27}$$

where h_w is the enthalpy of the sprayed moisture.

Now if h_w is equal to h_g, the enthalpy of water vapor at state 1, then by comparing Eq. (5.27) with Eq. (4.51) we conclude that the line 1-2$''$ is the constant dry-bulb temperature line passing through 1. For other values of the moisture enthalpy h_w, we apply Eq. (4.55) to obtain the inclination α of the line 1-2. Hence we have

$$cot\alpha - \ cot\theta = \frac{h_w - h_g}{S} \tag{5.28}$$

where S is the scale factor of the psychrometric chart and θ is the inclination of the constant dry-bulb temperature line as shown in Fig. 5.5(b).

From Eq. (5.28) we see that if $h_w > h_g$, then $\alpha < \theta$, and the air will be sensibly heated during humidification as indicated by line 1-2 in Fig. 5.5(b). However, if $h_w < h_g$, then $\alpha > \theta$, and the air will be sensibly cooled during humidification as indicated by line 1-2$'$.

5.2.5 *Evaporative cooling*

Evaporative cooling offers an effective and convenient method for cooling air in hot and humid climates. In practical evaporative coolers or air washers ambient air is passed through a porous structure, supplied with water from an external source, as depicted schematically is Fig. 5.6.

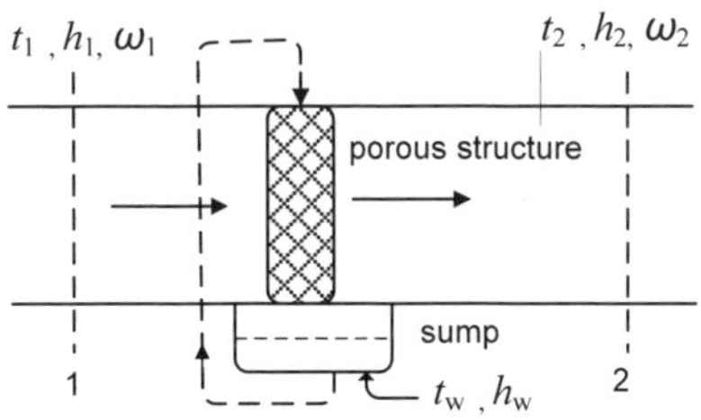

Fig. 5.6 (a) Evaporative cooler

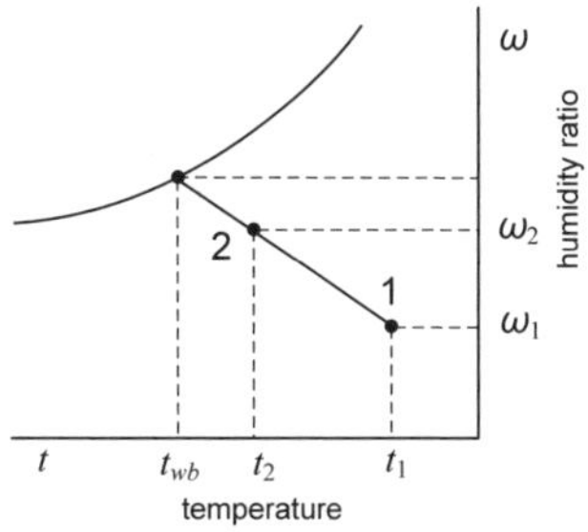

Fig. 5.6 (b) Psychrometric chart

The water that does not enter the air stream drips down to the sump from which it is recirculated to the top of the porous structure. A constant water level is maintained in the sump by supplying make-up water from an external source. The porous structure helps to distribute the water in the form of a thin film and thereby increase the contact area between air and water to facilitate evaporation. Since the incoming air is relatively dry the water evaporates readily, absorbing the latent heat of evaporation from the air. This process cools the air.

A control-volume analysis of the evaporative cooler could be carried out using the mass balance equations, (5.24) and (5.25) and the SFEE given by Eq. (5.26) in the foregoing section. A detailed design-analysis of the evaporative cooler is presented in chapter 6. It is found that under *ideal* conditions, the state of the air during *adiabatic evaporative cooling* follows a constant wet-bulb temperature line in the psychrometric chart as indicated in Fig. 5.6(b). However, in the practical system shown in Fig. 5.6(a) the moist air leaving the cooler at 2 is not cooled to the wet-bulb temperature, the lowest possible temperature to which the air could be cooled.

We define the ratio of the actual dry-bulb temperature drop of the air to the maximum possible temperature drop as the *saturation effectiveness* of the cooler, e_c. Hence we have

$$e_c = \frac{t_1 - t_2}{t_1 - t_{wb}} \tag{5.29}$$

5.2.6 *Space condition line*

The main purpose of air conditioning is to maintain the temperature and relative humidity of a space at desired values. There may also be a need to introduce a specified quantity of fresh ambient air to the space for ventilation purposes. These requirements and the stringency of their control, however, depend on the type of activities for which the space is used. We shall discuss these practical aspects of air conditioning applications in chapter 10.

The temperature and humidity of the air in a space change due to energy and moisture flows into and out of the space. For purposes of illustration, consider a typical summer air conditioning system used to remove heat and moisture from a space. It is customary to list all energy and moisture flows into the space which the air conditioning system has to remove, under two categories called the *sensible cooling load* and the *latent cooling load.*

The *sensible* cooling load is due to: (i) heat flow through walls, roofs, and windows, (ii) appliances generating heat, lighting, and occupants within the space, and (iii) unintended air infiltration. The *latent* cooling load is due to: (i) the moisture released by appliances and occupants within the space, and (ii) moist air infiltration. We shall discuss the various methods to estimate these cooling loads in chapter 10.

Under typical winter weather conditions, the air conditioning system (usually called a *heating system*) has to supply hot air to balance the heat loss from the space due to: (i) heat flow through the building envelope, and (ii) unintended cold air infiltration into the space. These energy flows constitute the *heating load* on the space. In chapter 8 we shall discuss the various methods to estimate these heating loads.

For psychrometric analysis of air conditioning systems, to be presented in the next few sections of the present chapter, we will assume that the cooling loads and the heating loads on the space have been estimated.

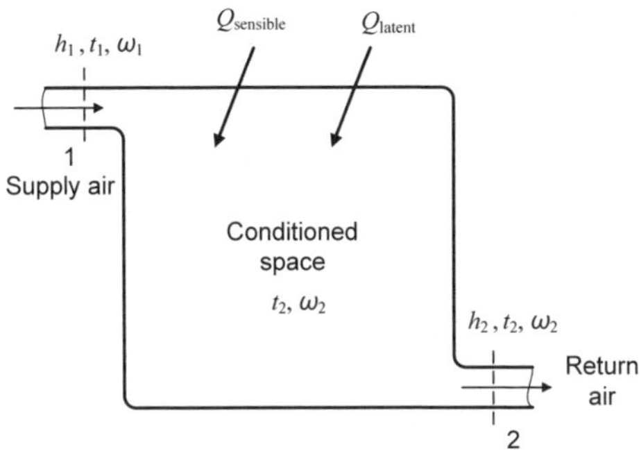

Fig 5.7 (a) Air conditioned space

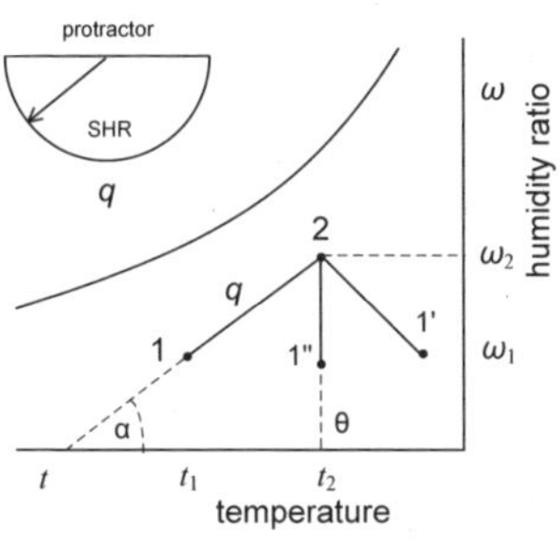

Fig. 5.7 (b) Psychrometric chart

The dry-bulb temperature t_2 and the humidity ratio ω_2 of the space, shown schematically in Fig. 5.7(a), are maintained at specified values by supplying conditioned air through the inlet at 1. We assume that the conditions of the space are steady and the air in the space is well mixed. Therefore the conditions of the air withdrawn from the space at the outlet

2, and returned to the air conditioning plant for processing, are the same as those of the air in the space.

Let the total rate of sensible heat gain (sensible cooling load) by the air in the space be $\dot{Q}_{st}$, the total rate of latent heat gain (latent cooling load) be $\dot{Q}_{lt}$, and the total rate of moisture gain by the air be $\dot{m}_{wt}$. Consider a control volume with a single inlet port 1 and a single outlet port 2, surrounding the space. Applying the SFEE to this control volume, neglecting changes in kinetic and potential energy of air, we have

$$\dot{Q}_{st} + \dot{Q}_{lt} = \dot{m}_a(h_2 - h_1) \quad (5.30)$$

Applying the dry air and water mass balance equations we obtain

$$\dot{m}_{a1} = \dot{m}_{a2} = \dot{m}_a \quad (5.31)$$

$$\dot{m}_{wt} = \dot{m}_{a2}\omega_2 - \dot{m}_{a1}\omega_1 \quad (5.32)$$

Now the enthalpy-moisture ratio, q for the inlet and outlet conditions is obtained by manipulating Eqs. (5.30), (5.31) and (5.32). Hence we have

$$q = \frac{h_2 - h_1}{\omega_2 - \omega_1} = \frac{\dot{Q}_{st} + \dot{Q}_{lt}}{\dot{m}_{wt}} \quad (5.33)$$

From Eq. (5.33) we seen that for fixed values of the sensible and latent cooling loads and the rate of moisture gain, the enthalpy-moisture ratio, q is constant. Moreover, the state point of the supply air at 1 on the psychrometric chart must lie on a straight line drawn through 2, parallel to the direction of q in the $(\Delta h/\Delta\omega)$ - protractor in Fig. 4.5. This straight line is called the *space condition line.*

Now if q is equal to h_{g2}, the enthalpy of vapor in the space, then by comparing Eq. (5.33) with Eq. (4.51) we observe that the direction of q coincides with the constant dry-bulb temperature line passing through 2, as indicated by the line 2-1$''$ in Fig. 5.7(b). For other values of q, we apply Eq. (4.55) to obtain the inclination α of the line 1-2. Hence we have

$$\cot\alpha - \cot\theta = \frac{(q - h_{g2})}{S} \quad (5.34)$$

where S is the scale factor of the psychrometric chart and θ is the inclination of the constant dry-bulb temperature line through 2.

From Eq. (5.34) we see that if, $q > h_{g2}$, then $\alpha < \theta$, and the supply air at 1 is cooler and less humid than the air in the space as indicated by line 1-2 in Fig. 5.7(b). This is the condition line for a typical summer air conditioning situation.

However, if $q < h_{g2}$, then $\alpha > \theta$ and the supply air is warmer and less humid than the air in the space as indicated by line 1'-2. This could be a typical winter air conditioning situation. The foregoing analysis is compatible with the (h-ω) - psychrometric chart (Fig. 4.5) and it could therefore be used to solve most air conditioning design problems graphically.

We now outline an approximate procedure based on the analytical expressions obtained earlier in chapter 4, for the changes in sensible enthalpy [Eq. (4.30)] and latent enthalpy [Eq. (4.31)]. Thus the sensible cooling load, $\dot{Q}_{st}$, the latent cooling load, $\dot{Q}_{lt}$ and the total cooling load $\dot{Q}_t$ may be expressed as

$$\dot{Q}_{st} = \dot{m}_a c_{am}(t_2 - t_1) \tag{5.35}$$

$$\dot{Q}_{lt} = \dot{m}_a h_{gm}(\omega_2 - \omega_1) \tag{5.36}$$

$$\dot{Q}_t = \dot{Q}_{st} + \dot{Q}_{lt} \tag{5.37}$$

For the temperature range from 0°C to 60°C of a typical psychrometric chart the following mean values may be used [3]: $c_{am} = 1.02$ kJkg^{-1}K^{-1} and $h_{gm} = 2555$ kJkg^{-1}.

The sensible heat ratio (SHR) is given by

$$SHR = \frac{\dot{Q}_{st}}{\dot{Q}_t} \tag{5.38}$$

When using the approximate Eqs. (5.35) to (5.38) in graphical procedures it is more convenient to use the SHR-protractor of the psychrometric chart. The space condition line is obtained by drawing a straight line through point 2 (Fig. 5.7b) in the direction of the SHR for the space, as indicated in the SHR-protractor.

5.3 Applications of Psychrometric Processes

Air conditioning systems are designed to maintain the temperature and humidity ratio of a space at desired values when the space is subjected heating loads or cooling loads. This is usually achieved by supplying conditioned air with the appropriate temperature and humidity ratio to the space. The desired conditions of the space, its heating/cooling loads, and the required supply air conditions are related by the *condition line* of the space.

In the preceding section we considered a number of basic psychrometric processes that could be used to change the temperature and humidity ratio of air. In real air conditioning systems these basic processes are combined appropriately in the air processing units to achieve the required conditions in the air being supplied to a space. The design of air conditioning systems is complicated by the dynamic nature of the space heating or cooling loads which vary during the day. Moreover, there may be large seasonal variations of the loads, especially in temperate climates.

In residential and commercial buildings, there would be the additional demand on the air conditioning system to maintain different sections or rooms of the building at different temperatures and humidity ratios. A space, such as a room, with a specified temperature, humidity ratio, and heating/cooling load is called a *zone*. Usually the conditions of the air in a *single-zone* will be controlled by its own thermostat. A large building will have a number of zones with different heating/cooling loads and different temperature and humidity ratio requirements. These are called *multi-zone* systems.

The next section presents several air conditioning systems where the basic psychrometric processes introduced in section 5.1 are configured to produce the desired supply air conditions for single-zone applications. Both winter heating systems and summer cooling systems are considered. Several multi-zone systems of practical interest are presented in subsequent sections.

5.4 Single-zone Air Conditioning Systems

5.4.1 *Summer air conditioning systems*

A schematic diagram of a basic summer air conditioning system is depicted in Fig. 5.8(a). The essential air processing components consisting of a filter, a cooling and dehumidifying coil, and a fan are arranged in series. These are connected using metal ducts through which air flows. For ventilation purposes, a portion of the air withdrawn from the conditioned space at 2 is discharged to the ambient at 3, and replaced with an equal mass of *fresh* air at 4.

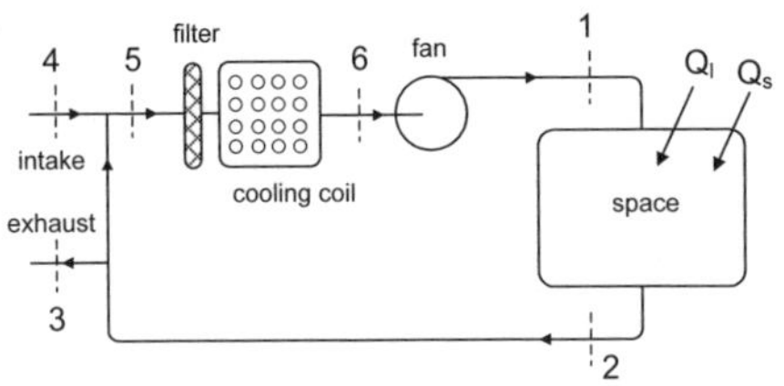

Fig. 5.8(a) Summer air conditioning system

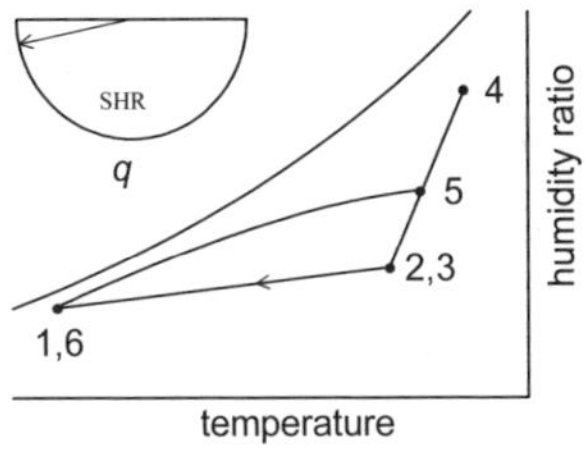

Fig. 5.8(b) Psychrometric chart

The state points of the air, and the processes undergone by the air as it passes through the different components of the system, are indicated on the psychrometric chart in Fig. 5.8(b). For psychrometric design-analysis, it is customary to assume that the system operates under steady conditions. By knowing the temperature and humidity ratio, or two other specified properties of the space, we locate point 2 on the psychrometric chart. If we assume that the sensible and latent cooling loads of the space have been estimated then we can calculate the SHR for the space. As was described in the foregoing section, the condition line of the space is obtained by drawing a straight line though state 2 in the direction of the line on the protractor pointing towards the SHR-value for the space as shown in Fig. 5.8(b).

For ventilation purposes, fresh ambient air is admitted to the system at 4. The state of fresh air is easily located on the psychrometric chart by knowing two of its properties like the temperature and the relative humidity. The dry air mass fraction of fresh air to be admitted to compensate for the return air discharged at 3 is usually specified.

The air entering the cooling coil through the filter at 5 is a mixture of the *fresh* air admitted at state 4 and the *return* air at state 2. Following the discussion in section 5.2.1, we obtain the state of air at 5, by dividing the straight line 2-4 in the inverse ratio of the dry air mass flow rates of the above two air streams.

The curved line 5-1, commonly called the *coil condition line*, gives the state of the air as it flows through the cooling and dehumidifying coil. There are several ways to obtain the coil condition line. For commercially available cooling coils, the manufacturer usually provides tabulated performance data from which the coil condition line may be constructed. Alternatively, we could use the semi-empirical model based on the bypass factor, outlined in section 5.2.3, to obtain the state of the air exiting the cooling coil. A computer-based heat and mass transfer model of the cooling coil, to be described in chapter 7, is also a possible option to determine the coil condition line. The point intersection of the space condition line and the coil condition line gives the state 1 of the supply air.

The air undergoes a slight increase in temperature as it flows through the fan which, for all practical purposes, could be neglected. However, we could add the energy input to the fan as a sensible cooling load to the space. In the basic air conditioning system, the cooling and dehumidifying coil is the sole air processing unit, and therefore only one property of the air could be controlled. In most systems this property would be the dry-bulb temperature.

5.4.2 *Summer air conditioning systems with reheat*

If the condition line of a space is very steep, then the coil condition line and the space condition line drawn on the psychrometirc chart may not intersect. In physical terms, this implies that the cooling coil is unable to produce the required state of the supply air to meet the cooling load of the space. The inclusion of a reheat coil after the cooling coil, as shown in Fig. 5.9(a), may help to remedy this situation.

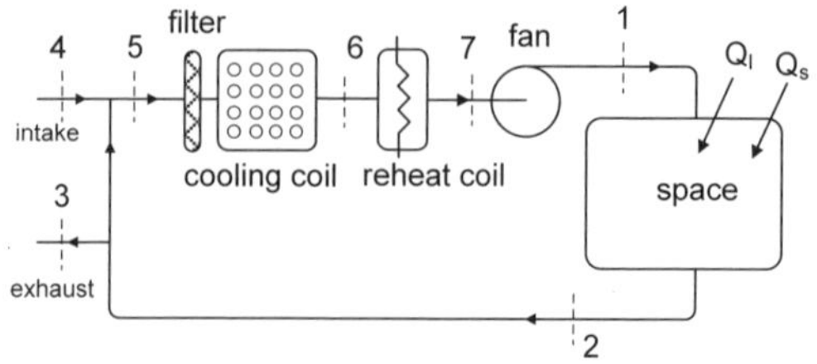

Fig.5.9 (a) Air conditioning system with reheat

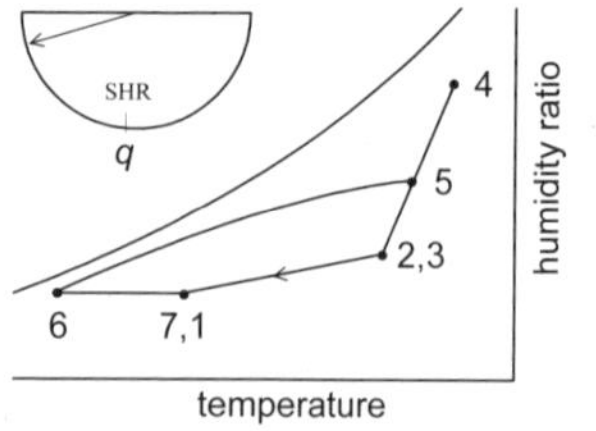

Fig. 5.9 (b) Psychrometric chart

The psychrometric chart for the system including the reheat coil is shown in Fig. 5.9(b). In the reheat coil, the air leaving the cooling coil at 6 undergoes sensible heating with the humidity ratio remaining constant. Therefore the intersection of the space condition line and the horizontal line through 6 gives the supply air state 1. The rest of the psychrometric chart is similar to that for the basic system shown in Fig. 5.8(b).

Since energy has been expended to cool the air to state 6, the energy input to the reheat coil is an additional energy input that lowers the overall energy efficiency of the system. This demonstrates that there is a trade-off between the desired comfort conditions in a space and the operating energy cost of the system (see worked example 5.4).

5.4.3 *Summer air conditioning systems with bypass paths*

When a basic summer air conditioning system operates under *part-load* conditions, a portion of the air returning from the space could be made to bypass the cooling coil and thereby maintain the desired conditions in the space. Such a system is shown schematically in Fig. 5.10(a) and the corresponding psychrometric chart is depicted in Fig. 5.10(b).

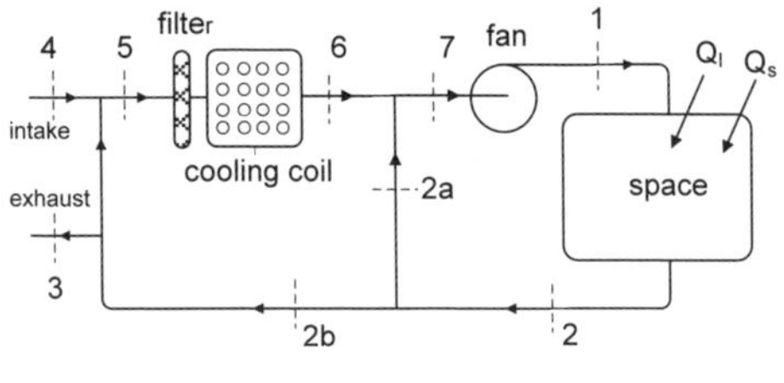

Fig. 5.10 (a) Air conditioning system with bypass

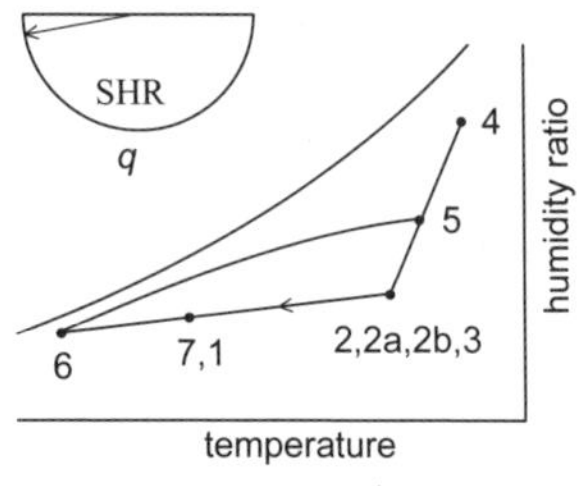

Fig. 5.10 (b) Psychrometric chart

In this system as the cooling load on the space decreases a larger proportion of the return air is sent through the bypass path without being cooled. Since the mass flow rate through the cooling and dehumidifying coil is now less, the air passing through it would be dehumidified to a greater degree.

The state 6 on the psychrometric chart is the point of intersection of the coil condition line, which depends on the air flow rate through the cooling coil, and the space condition line. The cooled air at state 6 is mixed with return air flowing through the bypass path to produce supply air at state 1. This mixed state is obtained by dividing the line 2-6 in the inverse proportion of the dry air mass flow rates of the two streams. When using the data from the psychrometric chart, we need to be aware that the air flow rates may be different along the different process paths as, for instance, on 5-6 and 1-2.

5.4.4 *Winter air conditioning systems*

Most winter air conditioning systems involve processing equipment that humidify and heat the air supplied to the space. The basic form of such a system consisting of a preheat coil, a humidifier, and a reheat coil is shown schematically in Fig. 5.11(a). The preheat coil ensures that the air entering the humidifier is above 0°C and therefore there is no danger of water freezing in the humidifier. Moreover, by controlling the temperature of the air entering the humidifier it is possible to control the rate of evaporation of water. The reheat coil is used to control the dry-bulb temperature of the air entering the space.

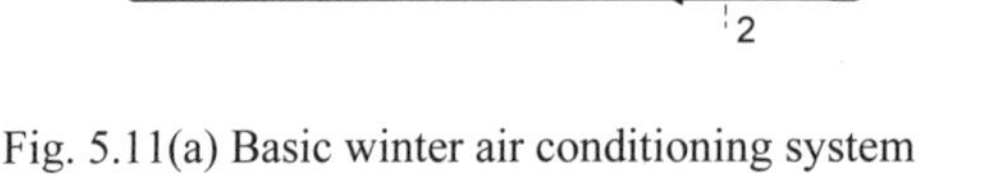

Fig. 5.11(a) Basic winter air conditioning system

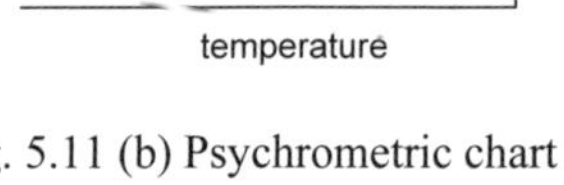

Fig. 5.11 (b) Psychrometric chart

The various processes are depicted on the psychrometric chart in Fig. 5.11(b). Fresh air at state 4 is mixed with a fraction of the return air to produce air at state 5. In the preheat coil the air is heated sensibly at constant humidity ratio to state 6. Ideally, the process line 6-7 in the humidifier follows a constant wet-bulb temperature line. However, as discussed in section 5.2.6, the state 7 of the air leaving the humidifier depends on the *saturation effectiveness* of the humidifier. In the reheat coil, the air is sensibly heated at constant humidity ratio and therefore the process is represented by a horizontal line through 7. The intersection of this line with the space condition line drawn through 2 locates the required supply air state 1 on the psychrometric chart.

5.4.5 *Air conditioning systems using evaporative cooling*

A relatively inexpensive summer air conditioning system that may be used in hot and dry climates is shown schematically in Fig. 5.12(a). The system uses a humidifier to cool the air. Because the air is dry the water evaporates readily in the humidifier extracting the latent heat of evaporation from the supply air, thereby lowering its dry-bulb temperature.

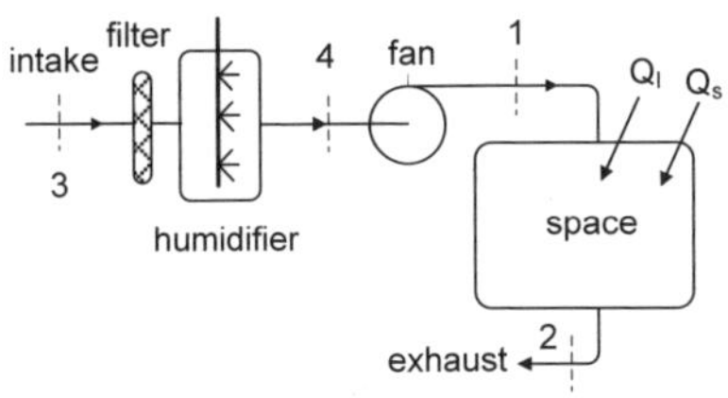

Fig. 5.12(a) Air conditioning system

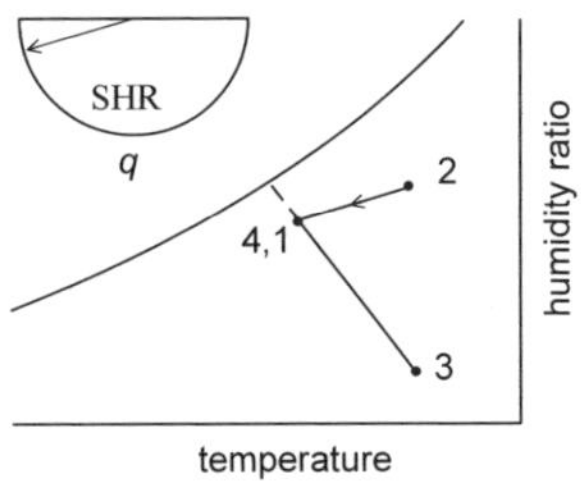

Fig. 5.12(b) Psychrometric chart

The psychrometric chart for the system is shown in Fig. 5.12(b). The air entering the system at 3 absorbs water in the humidifier with the state of the air following a wet-bulb temperature line, ideally. However, the exit state 4 of the air depends on the *saturation effectiveness* of the humidifier. The intersection of the space condition line and the wet-bulb temperature line through 3 locates the state of the supply air.

In the preceding sections we have presented several examples of single zone air conditioning systems. The overall energy efficiency of some of these systems, especially those requiring heat inputs, could be improved by utilizing any waste heat that is generated within the system. For example, in the system shown in Fig. 5.9(a), the reheat coil could make use of the heat rejected by the condenser of the refrigeration unit that provides the cooling fluid for the cooling coil.

In the winter heating system shown in Fig. 5.11(a), some of the energy in the warm air leaving at 3 could be transferred to the fresh air entering at 4 with the aid a simple *heat recovery* system. In the worked examples to follow in this chapter we shall illustrate the analysis of such systems using numerical design values.

5.5 Multi-zone Air Conditioning Systems

In most large buildings, different sections of the building usually have different heating/cooling loads and also different temperature and humidity ratio requirements. Such systems are commonly called *multi-zone* systems. In the next few sections we shall present examples of air conditioning systems, especially suitable for multi-zone applications.

5.5.1 *Multi-zone systems with reheat*

The multi-zone reheat system is a modification of the single-zone reheat system discussed earlier in section 5.4.2. Although an actual building could have a number of zones, for purposes of illustration of the principle of operation we shall use a system with two zones as depicted in Fig. 5.13(a).

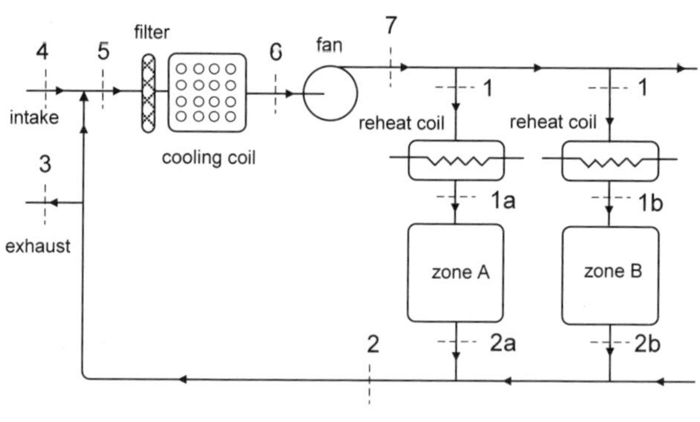

Fig. 5.13(a) Multi-zone system with reheat

The states of the return air from the two zones are represented by points 2a and 2b on the psychrometric chart shown in Fig. 5.13(b). The two streams of return air mix adiabatically to produce air with the state 2.

Fresh air at state 4, admitted for ventilation purposes, is mixed with the return air at 2 to produce the state 5 of the air entering the cooling coil.

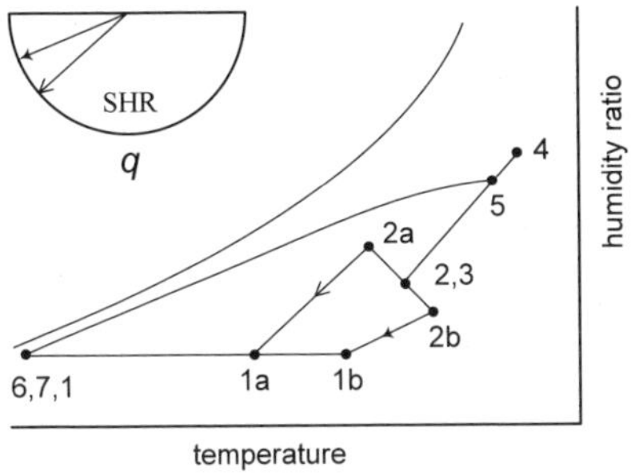

Fig. 5.13(b) Psychrometric chart

The air is cooled to state 6 before being distributed to the two zones A and B through the respective reheat coils. In the reheat coils the air streams are sensibly heated at constant humidity ratio. The points of intersection of the space condition lines of the two zones A and B and the horizontal line drawn through 6 locates the supply air states 1a and 1b for the two zones, respectively.

In practice, the dry-bulb temperatures of the zones are maintained by thermostats that control the heat inputs of the two reheat coils. Reheating tends to lower the overall energy efficiency of the system because the air is first cooled in the cooling coil by expending energy and later heated in the reheat coil with an additional energy input. However, the reheating process helps to control the required temperature and humidity of the zone more precisely, which is desirable for comfort air conditioning.

5.5.2 *Dual-duct multi-zone air conditioning systems*

In the dual-duct multi-zone air conditioning system shown schematically in Fig. 5.14(a), the supply air from the main fan is divided into two streams. One stream passes through a cooling and dehumidifying coil, and the other through a heating coil and a humidifier. The temperature of the air supplied to each zone is controlled by mixing air from the hot and cold streams in correct proportion. This is done in a mixing box which, in practice, responds to a signal from the thermostat in the zone.

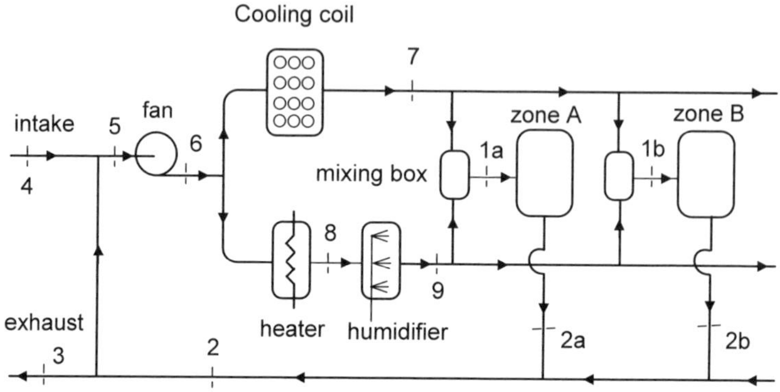

Fig. 5.14(a) A dual-duct system for multi-zone applications

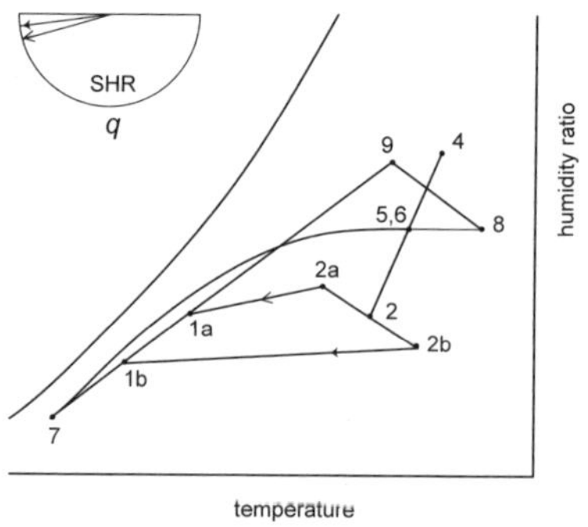

Fig. 5.14(b) Psychrometric chart

The dual-duct system is capable of dealing with zones with widely different temperature requirements. For instance, there could be zones that need only cooling and others requiring only heating. A disadvantage of the system is its high initial cost due to the need for two supply air ducts which should be capable of handling the entire air flow. Moreover, there could be situations where energy is expended both for heating and cooling air as in a reheat system, which lowers the overall energy efficiency of the system.

The psychrometric chart in Fig. 5.14(b) depicts the various processes of the dual-duct system. The return air from the two zones A and B, at states 2a and 2b respectively, mix adiabatically to produce air at state 2. The outdoor ventilation air at state 4 mixes with return air at 2 to produce the state 5 of the air entering the fan. The state of air passing through the cooling and dehumidifying coil changes from 5 to 7 along the coil condition line. The air passing through the heating coil is heated sensibly at constant humidity ratio from 5 to 8. Ideally, in the humidifier, the state of the air follows a wet-bulb temperature line. However, the final temperature at 9 of the air leaving the humidifier depends on the saturation *effectiveness* of the humidifier. The conditions 1a and 1b of the air supplied to the two zones A and B, through the mixing boxes, are located by the points of intersection of the line 7-9 and the respective space condition lines of the two zones.

5.5.3 *Variable air volume (VAV) systems*

A variable-air-volume (VAV) system used for cooling applications is shown schematically in Fig. 5.15(a). It supplies conditioned air to two zones A and B. In actual practice the system could be serving many more zones.

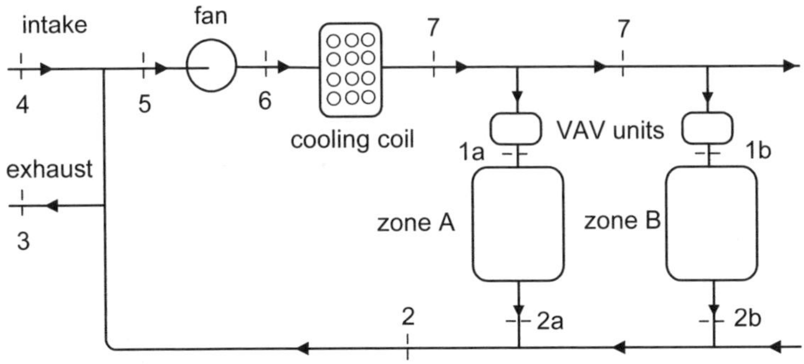

Fig 5.15(a) Variable air volume (VAV) system for cooling

In the VAV system the temperature of a zone is controlled by varying the supply air flow rate to the zone. This is achieved by changing the settings of the dampers in the VAV-unit mounted in the supply air duct to the zone.

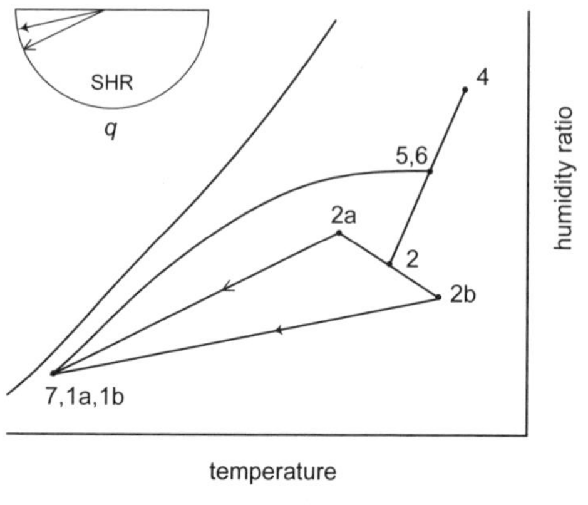

Fig. 5.15(b) Psychrometric chart

The main fan of the system is designed to supply the air flow rate required to meet the maximum total cooling load of the system. Usually the cooling loads on the different zones attain their maximum values at different times. Therefore it is possible to balance the air flow to the different zones to meet the required maximum flow rates of each zone. Typical VAV systems are used for summer cooling applications where only a cooling and dehumidifying coil is required. Therefore any reheat required in a zone has to be supplied locally by using separate heaters in the zone.

However, there are variations of VAV systems that incorporate reheat and dual-duct arrangements which were discussed earlier in sections 5.5.1 and 5.5.2 respectively. The inclusion of variable air flow in the latter systems greatly enhances their ability to control the temperature and humidity of the zone, and also improve the overall energy efficiency of the system.

The psychrometric chart depicted in Fig. 5.15(b) is for the two-zone, VAV cooling system shown in Fig. 5.15(a). The return air from the two zones A and B, at states 2a and 2b respectively mix adiabatically to produce air at state 2. The mixing of the outdoor ventilation air at state 4 and the return air at state 2 results in the state 5 of the air entering the fan. The state of air passing through the cooling and dehumidifying coil changes from 5 to 7 along the coil condition line. The condition lines for the two zones A and B are along the lines 7-2a and 7-2b respectively.

The cooling load of each zone is met by varying the mass flow rate of air to the zone through the VAV unit.

In the preceding sections we have discussed the principle of operation of a series of air conditioning systems for both heating and cooling applications. However, there are many details of these systems, of practical importance, that have not been included here. For a comprehensive presentation of the design and operation of air conditioning systems, the reader is referred to the *ASHRAE Handbook* - 2012 *HVAC Systems and Equipment* [2].

5.6 Worked Examples

Example 5.1 In an air conditioning system return air at 26°C dry-bulb temperature and 50% relative humidity is mixed with outdoor ambient air at 34°C dry-bulb temperature and 60% relative humidity. The dry air mass flow rate of outdoor air is 30% of the supply air mass flow rate to the space. The pressure is constant at 101.3 kPa. Calculate (i) the enthalpy, (ii) the humidity ratio, and (iii) the dry-bulb temperature of the supply air. Compare the results obtained using the ideal gas expressions with those obtained using the psychrometric chart.

Solution Let 1 and 2 denote properties of the two air streams and 3 the properties of the mixed stream. In order to calculate the required quantities using the expressions obtained in chapter 4 we extract the following data from the steam table [4].

For t_{db1} = 26°C, P_{g1} = 3.36 kPa, h_{g1} = 2548.4 kJkg^{-1}. For t_{db2} = 34°C, P_{g2} = 5.318 kPa, h_{g2} = 2562.9 kJkg^{-1}. The total pressure, P = 101.3 kPa. The humidity ratio is given by Eq. (4.13) as

$$\omega = \frac{0.622\phi P_g(t)}{P-\phi P_g(t)}$$

Substituting numerical values in the above equation we obtain

$$\omega_1 = \frac{0.622\times 0.5\times 3.36}{101.3-0.5\times 3.36} = 0.010489$$

$$\omega_2 = \frac{0.622\times 0.6\times 5.318}{101.3-0.6\times 5.318} = 0.02023$$

The air enthalpy is given by Eq. (4.16) as

$$h = c_{pa}t + \omega h_g(t)$$

Substituting numerical values in the above equation we have

$$h_1 = 1.0 \times 26 + 0.010489 \times 2548.4 = 52.73 \text{ kJkg}^{-1}$$

$$h_2 = 1.0 \times 34 + 0.02023 \times 2562.9 = 85.85 \text{ kJkg}^{-1}$$

The enthalpy of the air after the mixing process is given by Eq. (5.3) as

$$\dot{m}_{a1}h_1 + \dot{m}_{a2}h_2 = \dot{m}_{a3}h_3$$

Substituting numerical values we obtain the enthalpy as

$$h_3 = 0.7 \times 52.73 + 0.3 \times 85.85 = 62.67 \text{ kJkg}^{-1}$$

Applying the water mass balance equation we have

$$\dot{m}_{a1}\omega_1 + \dot{m}_{a2}\omega_2 = \dot{m}_{a3}\omega_3$$

Substituting numerical values we obtain the humidity ratio as

$$\omega_3 = 0.7 \times 0.010489 + 0.3 \times 0.02023 = 0.01341$$

Now the enthalpy at the exit 3 may be expressed as

$$h_3 = 1.0t_3 + 0.01341h_{g3}(t_3) = 62.67$$

To find the temperature we need to solve the above non-linear equation by trial-and-error. We make an initial guess of the temperature and then obtain the saturation vapor enthalpy from the steam table [4]. The guessed value is adjusted in an iterative manner until the above equation is satisfied. This gives the dry bulb temperature as 28.4°C.
The dry-bulb temperature is also given by Eq. (5.8) as

$$t_3 = (\dot{m}_{a1}/\dot{m}_{a3})t_1 + (\dot{m}_{a2}/\dot{m}_{a3})t_2$$

Substituting numerical values in the above equation we have

$$t_3 = 0.7 \times 26 + 0.3 \times 34 = 28.4°\text{C}$$

To obtain the results from the psychrometric chart (Fig. 4.5), we first locate on the chart the states 1 and 2 of the return air and the fresh air respectively (see Fig. E5.1.1 below). The state 1 of the return air is located by the intersection of the 26°C constant db-temperature line and the 50% - constant relative humidity (RH) curve. Similarly, the state 2 of the outdoor air is located by the intersection of the 34°C constant db-temperature line and the 60% constant RH curve.

Draw the line 1-2 connecting the two states on the chart. To locate the state 3 of the mixed air stream, divide the line 1-2 such that, length (1-3): length (2-3) is 3:7. The properties of air at 3 are read off directly from the chart. Thus we obtain the enthalpy, dry-bulb temperature and humidity ratio as 62.9 $kJkg^{-1}$, 28.4°C and 0.0134 respectively.

The results obtained by the two methods agree very closely. Computations using analytical expressions are somewhat tedious, but at the same time reading values accurately from the psychrometirc chart can be challenging. However, it is clear that for design calculations involving a number of psychrometric processes, the psychrometric chart undoubtedly is a more convenient tool to use.

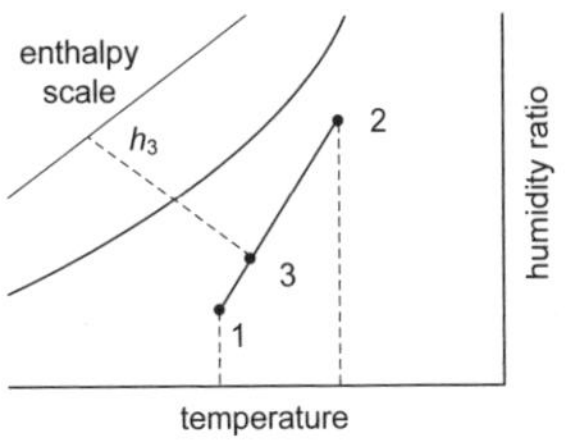

Fig. E5.1.1 Psychrometric chart

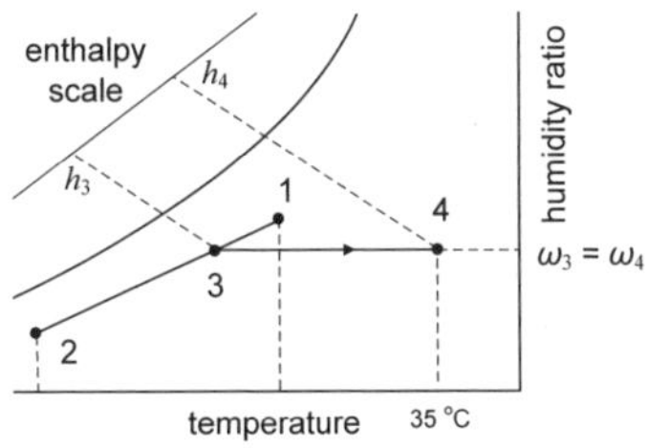

Fig. E5.2.1 Psychrometric chart

Example 5.2 In a winter air conditioning system, outdoor air at 2°C and 20% relative humidity is mixed with return air at 23°C and 40% relative humidity. The ratio of the dry air mass flow rates of outdoor air to supply air is 1:4. The supply air is heated sensibly to a dry-bulb temperature of 35°C before being supplied to the space at the rate of 30 kg of dry air per minute. The pressure is constant at 101.3 kPa. Calculate (i) the wet-bulb temperature of the air after the mixing process, (ii) the relative humidity of the air supplied to the space, and (iii) the rate of heat input to the heater.

Solution The processes are shown on the psychrometric chart Fig. E5.2.1. The state 1 of the return air is located by the intersection of the 23°C constant db-temperature line and the 40% constant relative humidity (RH) curve. Similarly, the state 2 of the outdoor air is located by the intersection of the 2°C constant db-temperature line and the 20% constant RH curve.

Draw the line 1-2 connecting the two states 1 and 2 on the chart. To locate the state 3 of the mixed air stream, divide the line 1-2 such that, length (1-3): length (2-3) is 1:3. The properties of the air at 3 are read off directly from the chart. (i) Thus the wet-bulb temperature at 3 is 11°C, and the db-temperature at 3 is 17.8°C.

The sensible heating of the air from 3 to 4 occurs with a constant humidity ratio. Therefore state 4 of the air at the end of the heating process is located by the intersection of the horizontal line through 3 and the 35°C constant db-temperature line. We read off the relative humidity at 4 as 16.5%. The sensible heat supplied to the air is given by

$$\dot{Q}_{in} = \dot{m}_a(h_4 - h_3) = \dot{m}_a c_{am}(t_4 - t_3)$$

Substituting numerical values in the above equation we obtain the heat input rate as

$$\dot{Q}_{in} = 0.5 \times 1.02 \times (35 - 17.8) = 8.77 \text{ kW}$$

Example 5.3 In a summer air conditioning system, 25% of the return air from a space at 30°C dry-bulb (db) temperature, and 22°C wet-bulb (wb) temperature is exhausted and an equal quantity of fresh air at 34°C db-temperature, and 28°C wb-temperature is mixed with the remaining return air. The mixture passes over a cooling coil whose coil-surface temperature (apparatus dew-point) is 8°C and the bypass factor is 0.25. The mass flow rate of dry air to the space is 0.8 kgs^{-1}. The pressure is constant at 101.3 kPa. Calculate (i) db-temperature and the relative humidity of the air leaving the cooling coil, (ii) enthalpy-humidity ratio of the cooling process, and (iii) the refrigeration capacity of the cooling coil.

Solution The processes are indicated on the psychrometric chart shown in Fig. E5.3.1. The state 1 of the return air is located by the intersection of the 30°C constant db-temperature line and the 22°C constant wb-temperature line. Similarly, the state 2 of the fresh air is located by the intersection of the 34°C constant db-temperature line and the 28°C constant wb-temperature line. To locate the state 3 of the mixed air we divide the line 1-2 such that length (1-3) / length (2-3) = 1/3. Hence we have $t_{db3} = 31°C$, $t_{wb3} = 23.5°C$.

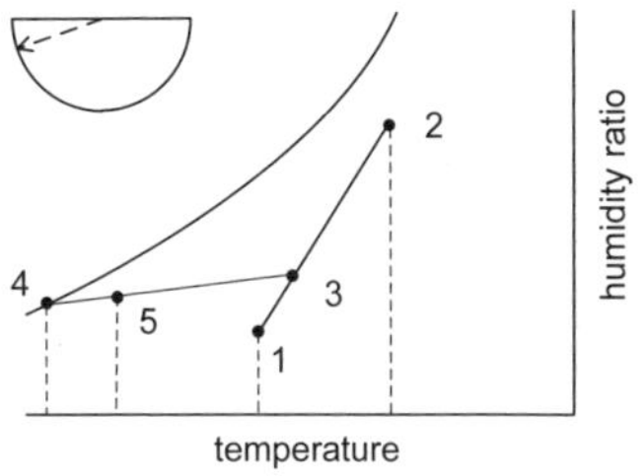

Fig. E5.3.1 Psychrometric chart

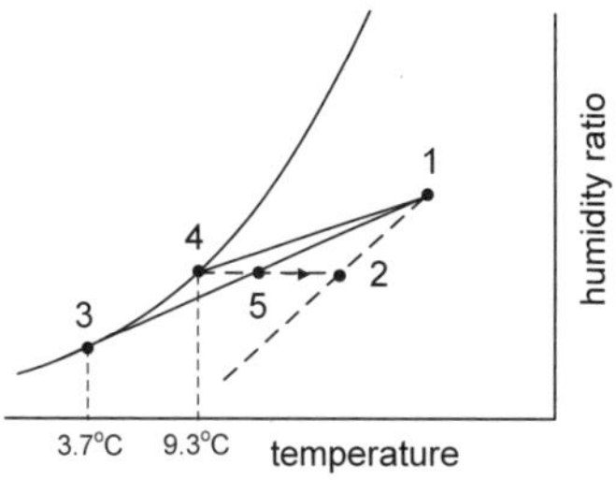

Fig. E5.4.1 Psychrometric chart

The mixed air at 3 now passes through the cooling coil to which we shall apply the bypass factor model developed in section 5.2.3. The bypass factor of the coil is 0.25, which means that only 75% of the air passes though the ideal straight line cooling process 3-4 in Fig. 5.4, to finally attain the coil surface temperature (apparatus dew-point) of 8°C. The cooled air at state 4 is mixed with the air that bypassed the coil (state 3) to produce the supply air at state 5. To locate 5 we divide the straight line 3-4 such that length (4-5) / length (3-5) = 1/3.

(i) We read off the db-temperature and relative humidity at state 5 as 13.8°C and 89%.

(ii) To obtain the enthalpy-humidity ratio of the cooling process, we draw a line parallel to line 3-4 through the centre of protractor as shown in Fig. E5.3.1. This gives the $(\Delta h/\Delta\omega)$-ratio as 5300 kJkg^{-1}.

(iii) The refrigeration capacity of the cooling coil is given by

$$\dot{Q}_{ref} = \dot{m}_a(h_3 - h_5)$$

$$\dot{Q}_{ref} = 0.8(71 - 36.5) = 27.6 \text{ kW}$$

Example 5.4 Return air from an air conditioned space is at 32°C db-temperature and 50% relative humidity. To satisfy the design requirements of the space, air has to be supplied at 16°C db-temperature and 65% relative humidity. The pressure is constant at 101.3 kPa. (i) Can a cooling and dehumidifying coil be used as the only processing unit of the system? (ii) What other processing units could be included to satisfy the required conditions of the supply air.

Solution The processes involved are depicted on the psychrometric chart in Fig. E5.4.1. The state 1 of the return air is located by the intersection of the 32°C constant db-temperature line and the 50% constant relative humidity (RH) curve. Similarly, the state 2 of the supply air is located by the intersection of the 16°C constant db-temperature line and the 65% constant RH curve.

We shall use the bypass factor model of the cooling coil, developed in section 5.2.3, to study the feasibility of the cooling process. We observe that the condition line 1-2 of the required *ideal* cooling coil does not intersect the saturation line on the psychrometric chart (Fig. 4.5). Therefore it is not possible to achieve the required state 2 of the supply air by a cooling coil alone. Now if we draw the line 1-3 that is tangential to the saturation curve, then 3 gives the lowest temperature to which air could be cooled using an *ideal* cooling coil.

We obtain this temperature from the psychrometric chart (Fig. 4.5) as 3.7°C. Air at state 5, lying on the horizontal line through 2, can be produced by cooling 75% of the dry air mass flow to state 3 and later mixing it with the 25% of the mass flow that bypassed the cooling coil. We obtain the temperature at 5 after mixing as 10.8°C. The air is now heated sensibly from 5 to 2 using a reheat coil to produce the required supply air state.

Alternatively, we could cool the entire air flow to the saturated state 4 with a temperature of 9.3°C. The air could then be heated sensibly from 4 to 2 using a reheat coil. By varying the mass fraction of air that bypasses the coil, we could produce any temperature between 3 and 4 in the cooled air stream. Mixing will then produce air with a corresponding state that lies between 4 and 5. This air can be reheated with the appropriate energy input to produce supply air at state 2.

Example 5.5 Ambient air at 10°C db-temperature and 20% relative humidity enters a steam humidifier with a dry air mass flow rate of 0.85

kgs^{-1}. Superheated steam at 110°C is sprayed into the air stream. The air leaves with a relative humidity of 70%. The pressure is constant at 101.3 kPa. Calculate (i) the db-temperature of the air leaving, and (ii) the rate of flow of steam.

Solution

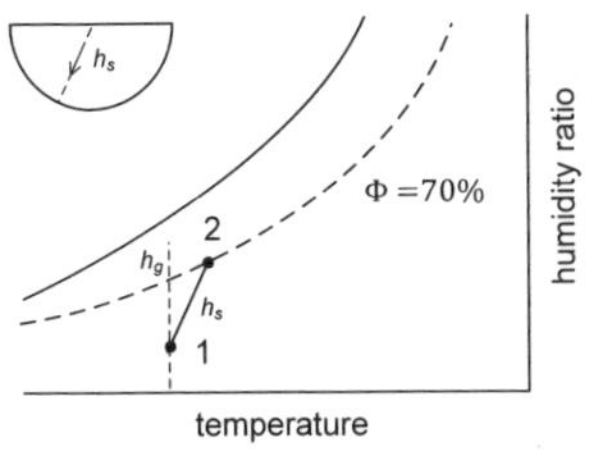

Fig. E5.5.1 Psychrometric chart

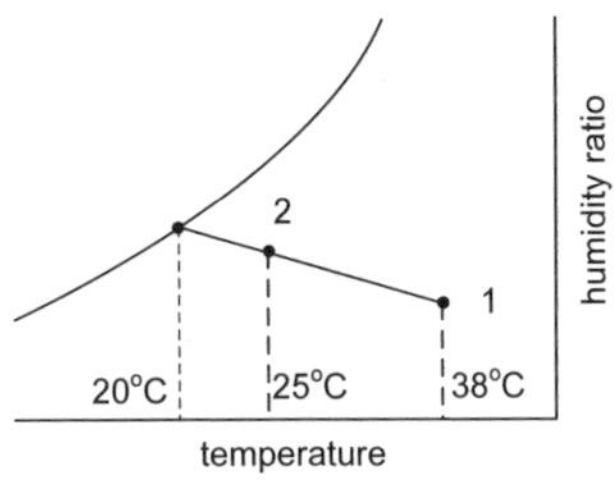

Fig. E5.6.1 Psychrometric chart

The state 1 of the ambient air entering the steam humidifier is located by the intersection of the 10°C dry-bulb temperature line and the 20% relative humidity curve. Applying the water mass balance equation and the SFEE to the humidifier we obtain the following equations:

$$\dot{m}_a\omega_2 = \dot{m}_a\omega_1 + \dot{m}_s \tag{E5.5.1}$$

$$\dot{m}_a h_2 = \dot{m}_a h_1 + \dot{m}_s h_s \tag{E5.5.2}$$

The enthalpy-humidity ratio, q is obtained from Eqs. (E5.5.1) and (E5.5.2) as

$$q = \frac{h_2 - h_1}{\omega_2 - \omega_1} = h_s \tag{E5.5.3}$$

The enthalpy of superheated steam at 110°C is obtained from the steam table [4] as h_s = 2696 kJkg^{-1}. We first draw the radial line on the enthalpy-humidity ratio protractor pointing to, q = 2696 kJkg^{-1}. To locate the state 2 of the air at the exit of the humidifier, we draw a line through 1, parallel to the line on the protractor to intersect the 70% constant relative humidity curve as indicated in Fig. E5.5.1.

The properties at the exit 2 are read directly from the chart as t_2 = 10.8°C and ω_2 = 0.0057. Now at the entrance, ω_1 = 0.0015. The mass flow rate of steam is given by Eq. (E5.5.1) as

$$\dot{m}_s = 0.85(0.0057 - 0.0015) = 0.00357 \text{ kgs}^{-1}$$

Note that the constant db-temperature line through 1 points in the direction of h_g at 10°C, which from the steam table [4] is 2519 kJkg^{-1}. Since $h_s > h_g$ the air is heated and humidified (see section 5.2.4).

Example 5.6 Ambient air at 38°C db-temperature and 20°C wb-temperature enters an evaporative cooler with a dry air mass flow rate of 0.75 kgs^{-1}. The pressure is constant at 95 kPa. The air leaves at a db-temperature of 25°C. Calculate (i) the relative humidity of the air at inlet, (ii) the relative humidity of the air at exit, (iii) the rate of flow of water to the cooler, and (iv) the saturation effectiveness of the cooler.

Solution A schematic diagram of the evaporative cooler is shown in Fig. 5.6(a). Since the pressure is 95 kPa we should not use the standard psychrometric chart, (Fig. 4.5) which is constructed for a pressure of 101.3 kPa, to solve this problem. Therefore we shall use the various relations derived in chapter 4 to compute the required properties of moist air. However, the cooling process is indicated on the psychrometric chart in Fig. E 5.6.1 above.

The following data are obtained from the steam table [4]:

At t_1 =38°C, P_{g1} = 6.624 kPa, h_{g1} = 2570 kJkg^{-1};

At t_{w1} = 20°C, P_{gw1}=2.33kPa, h_{gw1} = 2537.6 kJkg^{-1}, h_{fw1} = 83.9 kJkg^{-1};

At t_2 =25°C, P_{g2} = 3.166 kPa, h_{g2} = 2546.6 kJkg^{-1}

At the wb-temperature the relative humidity, ϕ_2 = 100%. The humidity ratio is given by

$$\omega_{w1} = \frac{0.622\phi P_{gw1}(t)}{P-\phi P_{gw1}(t)} = \frac{0.622\times 2.337}{95-2.337} = 0.01569$$

The enthalpy is

$$h_{w1} = 1.0 \times 20 + 0.01569 \times 2537.6 = 59.81 \text{ kJkg}^{-1}$$

The properties at the wb-temperature and the db-temperature are related by Eq. (4.39). Therefore

$$q = \frac{h_{w1}-h_1}{\omega_{w1}-\omega_1} = h_{fw1}$$

Substituting numerical values in the above equation we have

$$59.81 - (38 + 2570\omega_1) = 83.9(0.01569 - \omega_1)$$

Hence $\omega_1 = 0.00824$. Now the humidity ratio is given by

$$\omega_1 = \frac{0.622\phi P_{g1}(t)}{P - \phi P_{g1}(t)} = \frac{0.622 \times 6.624\phi_1}{95 - 6.624\phi_1} = 0.00824$$

(i) Therefore the relative humidity at the inlet is, $\phi_1 = 18.75\%$. Assume that the water sprayed in the humidifier is at the wb-temperature of the incoming air, that is at 20°C. The governing equation of the humidifier is given by Eq. (5.27) as

$$q = \frac{h_2 - h_1}{\omega_2 - \omega_1} = h_w$$

Substituting numerical values in the above equation we have

$$59.81 - (25 + 2546.6\omega_2) = 83.9(\omega_2 - 0.00824)$$

Hence $\omega_2 = 0.0135$. Now the humidity ratio is given by

$$\omega_2 = \frac{0.622\phi P_{g2}(t)}{P - \phi P_{g2}(t)} = \frac{0.622 \times 3.166\phi_2}{95 - 3.166\phi_2} = 0.0135$$

(ii) Therefore the relative humidity at the outlet is, $\phi_2 = 63.7\%$

(iii) The mass flow rate of water is given by

$$\dot{m}_w = \dot{m}_a(\omega_2 - \omega_1)$$

$$\dot{m}_w = 0.75(0.0135 - 0.00824) = 3.95 \times 10^{-3} \text{ kgs}^{-1}$$

(iv) The saturation effectiveness of the humidifier is given by Eq. (5.29) as

$$e_c = \frac{t_1 - t_2}{t_1 - t_{w1}} = \frac{38 - 25}{38 - 20} = 72\%$$

Example 5.7 The rate of sensible heat gain and the rate of moisture gain by a space are 23 kW and 0.0024 kgs^{-1} respectively. The space is maintained at 24°C db-temperature and 50% relative humidity. The air supplied to the space is at a db-temperature of 15°C. Assume that the moisture entering the space has an enthalpy of 2555 kJkg^{-1}. The pressure is constant at 101.3 kPa. Calculate (i) the relative humidity, the wb-temperature, and the dry air mass flow rate of air supplied, (ii) the refrigeration capacity of the cooling coil, and (iii) the bypass factor and the apparatus dew-point of the cooling coil.

Solution

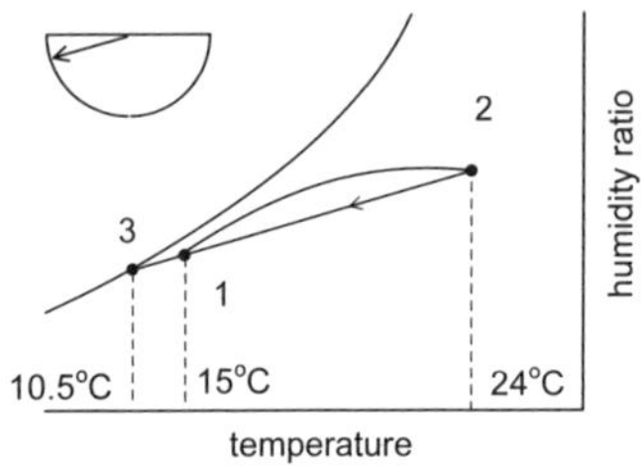

Fig. E5.7.1 Psychrometric chart

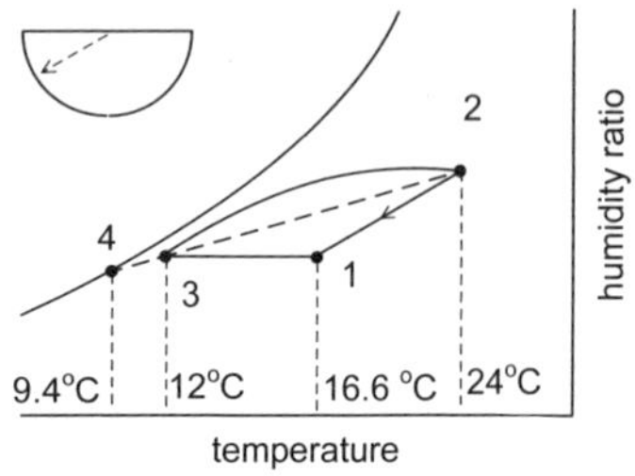

Fig. E5.8.1 Psychrometric chart

A schematic diagram of the system is shown in Fig. 5.7(a). The states of the air are indicted on the psychrometric chart in Fig. E5.7.1. The state 2 of the air in the space is located by the intersection of the 24°C constant db-temperature line and the 50% constant relative humidity (RH) curve. The sensible and latent heat loads are given as

$$\dot{Q}_s = 23 \text{ kW} \quad \text{and} \quad \dot{Q}_l = 0.0024 \times 2555 = 6.13 \text{ kW}$$

The sensible heat ratio is given by

$$SHR = \frac{\dot{Q}_s}{\dot{Q}_s + \dot{Q}_l} = 0.7895$$

We first draw a line on the SHR-protractor pointing towards the value of 0.7895. The space condition line is obtained by drawing a straight line through 2, parallel to the line drawn on the protractor, as indicated in Fig. E5.7.1. The state 1 of the supply air is located by the intersection of the space condition line and the 15°C constant db-temperature line.

(i) We read off the following values directly from the psychrometric chart (Fig. 4.5); wb-temperature at 1 = 12.8°C, relative humidity at 1 = 78%.

The mass flow rate of dry air is obtained from Eq. (5.35) as

$$\dot{Q}_s = \dot{m}_a c_{am}(t_2 - t_1)$$

Substituting numerical values we have

$$\dot{m}_a = \frac{23}{[1.02(24-15)]} = 2.505 \text{ kgs}^{-1}$$

(ii) The refrigeration capacity of the cooling coil is given by

$$\dot{Q}_{ref} = \dot{m}_a(h_2 - h_1)$$

$$\dot{Q}_{ref} = 2.505(48 - 36.3) = 29.3 \text{ kW}$$

where the enthalpies are obtained directly from the psychrometric chart.

The total heat load on the space is 29.13kW. This value should be equal to $\dot{Q}_{ref}$ by energy conservation.

(iii) The apparatus dew-point temperature or the ideal coil surface temperature is the temperature at the point of intersection 3, of the line 1-2 and the saturation line. We read this temperature as, $t_d = t_3$ =10.5°C. The bypass factor of the coil is given by Eq. (5.21) as

$$b = \frac{t_1 - t_d}{t_2 - t_d} = \frac{15-10.5}{24-10.5} = 0.33$$

Example 5.8 A summer air conditioning system, consisting of a cooling coil and a reheat coil, supplies air to a space maintained at 24°C db-temperature and 18°C wb-temperature. The sensible and latent heat loads on the space are 11 kW and 10 kW respectively. The conditions of the air leaving the cooling coil are 12°C db-temperature and 90% relative humidity. The pressure is constant at 101.3 kPa. Determine (i) the db-temperature and wb-temperature of the supply air, (ii) the dry air mass flow rate of the air supplied, (iii) the refrigeration capacity of the cooling coil, (iv) the rate of heat input by the reheat coil, and (v) the bypass factor and apparatus dew-point of the cooling coil.

Solution A schematic diagram of an air conditioning system with a reheating coil is shown in Fig. 5.9(a). The processes are indicated on the psychrometric chart in Fig. E5.8.1 above.

The state 2 of the return air is located by the intersection of the 24°C constant db-temperature line and the 18°C constant wb-temperature line. The state 3 of the air leaving cooling coil is located by the intersection of the 12°C constant db-temperature line and the 90% constant relative humidity curve. The sensible heat ratio, SHR is given by

$$SHR = \frac{\dot{Q}_s}{\dot{Q}_s + \dot{Q}_l} = \frac{11}{11+10} = 0.524$$

We first draw a line on the SHR-protractor pointing in the direction of the value 0.524. The space condition line is obtained by drawing a line through 2, parallel to the above line on the protractor. During the reheating process, the state of the air follows a horizontal line through 3. The state 1, of the supply air is located by the intersection of the space condition line and the horizontal line through 3 as indicated in Fig. E5.8.1. We read off the following values directly from the psychrometric chart (Fig. 4.5).

(i) The db-temperature and wb-temperature of the supply air are 16.6°C and 13°C respectively.

(ii) The dry air mass flow rate of the air supplied is obtained from Eq. (5.35) as

$$\dot{Q}_s = \dot{m}_a c_{am}(t_2 - t_1)$$

Substituting numerical values we have

$$\dot{m}_a = 11/[1.02(24 - 16.6) = 1.457 \text{ kgs}^{-1}$$

(iii) The refrigeration capacity of the cooling coil is given by

$$\dot{Q}_{ref} = \dot{m}_a(h_2 - h_3)$$

$$\dot{Q}_{ref} = 1.457(51 - 32) = 27.69 \text{ kW}$$

where the enthalpies are obtained directly from the psychrometric chart.

(iv) The heat input to the reheating coil is given by

$$\dot{Q}_{rh} = \dot{m}_a c_{am}(t_1 - t_3)$$

$$\dot{Q}_{rh} = 1.457 \times 1.02(16.6 - 12) = 6.84 \text{ kW}$$

The difference between the energy removed from the system by the cooling coil and the energy supplied by the reheating coil is, (27.69-6.84) = 20.85 kW. The total heat load on the system is 21 kW. For overall energy balance the two quantities above must be equal.

(v) The apparatus dew-point temperature or the ideal coil surface temperature is the temperature at the point of intersection 4, of the line 1-3 and the saturation line. We read this temperature as, $t_d = t_4 = 9.4°C$. The bypass factor of the coil is given by Eq. (5.21) as

$$b = \frac{t_1 - t_d}{t_2 - t_d} = \frac{12 - 9.4}{24 - 9.4} = 0.178$$

Example 5.9 An air conditioning system supplying air to a space with a sensible heat load of 14 kW and a latent heat load of 9 kW has a cooling coil and a bypass path as shown schematically in Fig. E5.9.1(a). The db-temperature of the space is maintained at 26°C. The dry air mass flow rate of supply air is $1.2 kgs^{-1}$. Outdoor ventilation air at 34°C db-temperature and 50% relative humidity is introduced into the system with a dry air mass flow rate of $0.26 kgs^{-1}$. The air leaving the cooling coil is fully saturated at a db-temperature of 6°C. The pressure is constant at 101.3 kPa. Determine (i) db-temperature and relative humidity of the supply air to the space, (ii) the wb-temperature of the space, (iii) the temperature of the air entering the cooling coil, and (iv) the refrigeration capacity of the cooling coil.

Solution

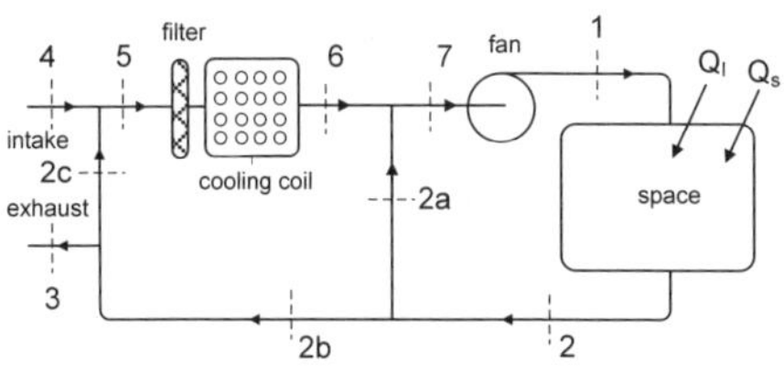

Fig. E5.9.1(a) Schematic diagram

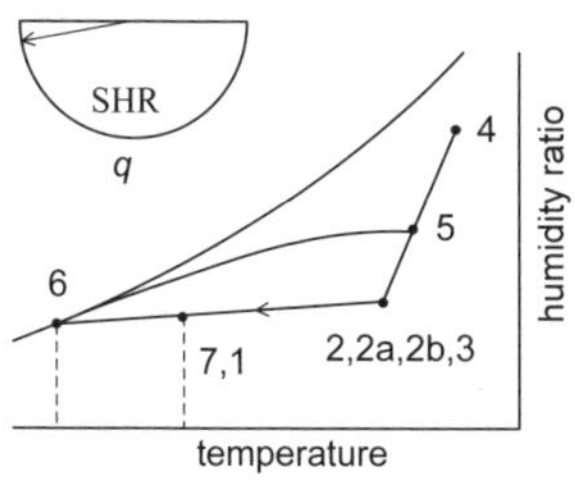

Fig. E5.9.1(b) Psychrometric chart

The state 6 of the air leaving the cooling coil is located by the intersection of the 6°C db-temperature line and the saturation curve. The air at 6 mixes with return air at state 2 to produce supply air at state 1. Now the space condition line is given by 1-2. Therefore it is clear that points 6, 2 and 1 must all lie on the condition line. The sensible heat ratio, SHR is given by

$$SHR = \frac{\dot{Q}_s}{\dot{Q}_s+\dot{Q}_l} = \frac{14}{14+9} = 0.608$$

We first draw a line on the SHR-protractor pointing in the direction of the value of 0.608 as indicated on the psychrometric chart in Fig. E5.9.1(b). The space condition line is obtained by drawing a line through 6, parallel to the above line on the protractor. The intersection of this line and the 26°C db-temperature gives the state 2 of the return air. The sensible heat load is given by

$$\dot{Q}_s = \dot{m}_a c_{am}(t_2 - t_1)$$

$$1.2 \times 1.02(26 - t_1) = 14$$

(i) Hence the supply air temperature is 14.56°C.
The state 1 of the supply air is located by the intersection of the 14.56°C db-temperature line and the space condition line. From the psychrometric chart (Fig. 4.5) we read off the relative humidity at 1 as 77%.

(ii) The wb-temperature of the space is obtained as, t_{w2} = 19°C. The dry air mass flow rates at points 6 and 2a in Fig. 5.9.1(a) are obtained from the lengths of the lines 1-6 and 1-2 on the psychrometric chart. This gives

$$\dot{m}_{a6} = 1.2(L_{12}/L_{61}) = 0.68 \text{ kgs}^{-1}$$

Mass balance at the mixing junctions 2a-6-7 and 2c-4-5 gives

$$\dot{m}_{a2a} = 1.2 - 0.68 = 0.52 \text{ kgs}^{-1}$$

$$\dot{m}_{a2c} = 0.68 - 0.26 = 0.42 \text{ kgs}^{-1}$$

The state 4 of the outdoor fresh air is located by the intersection of the 34°C constant db-temperature line and the 50% constant relative humidity curve. The outdoor air at 4 is mixed with return air at 2 to produce air at 5 that enters the cooling coil. Therefore state 5 is obtained by dividing the line 2-4 such that (L_{45}/L_{25}) = 0.42/0.26.

(iii) We obtain the db-temperature at 5 as 29°C.

(iv) The refrigeration capacity of the cooling coil is given by

$$\dot{Q}_{ref} = \dot{m}_a(h_5 - h_6)$$

$$\dot{Q}_{ref} = 0.68(63 - 20.8) = 28.7 \text{ kW}$$

where the enthalpies are obtained directly from the psychrometric chart.

It is instructive to check the overall energy balance of the system. This can be written in the form

$$\dot{Q}_s + \dot{Q}_l = \dot{Q}_{ref} - \dot{m}_{a4}(h_4 - h_2)$$

Substituting the relevant numerical values in the above equation we obtain the LHS and RHS of the energy balance equation respectively as

$$(14 + 9) = 23 \text{ kW}; \; 28.7 - 0.26(77.5 - 54) = 22.6 \text{ kW}$$

The energy balance is closely satisfied. Note that this example involves several adiabatic mixing processes. For these processes we could use the approximate relation given by Eq. (5.8) to check the accuracy of some of the quantities obtained from the psychrometric chart.

Example 5.10 The winter air conditioning system shown schematically in Fig. E5.10.1(a) has a preheat coil, an air washer and a reheat coil as processing units. It supplies air at 38°C db-temperature to a space maintained at 21°C db-temperature and 13°C wb-temperature. The sensible and latent heat loads on the space are 50 kW and 10 kW respectively.

Of the return air from the space, 40% by dry air mass is discharged to the ambient and replaced with fresh outdoor air at 2°C db-temperature and 95% relative humidity. In the preheater, the air is heated to a db-temperature of 21°C. Assume that the water added in the air washer is at the wb-temperature of the air entering it. The pressure is constant at 101.3 kPa. Determine (i) mass flow rate of supply air, (ii) the rates of heat input in the preheat coil and the reheat coil, (iii) the mass flow rate of water in the air washer, and (iv) the saturation effectiveness of the air washer.

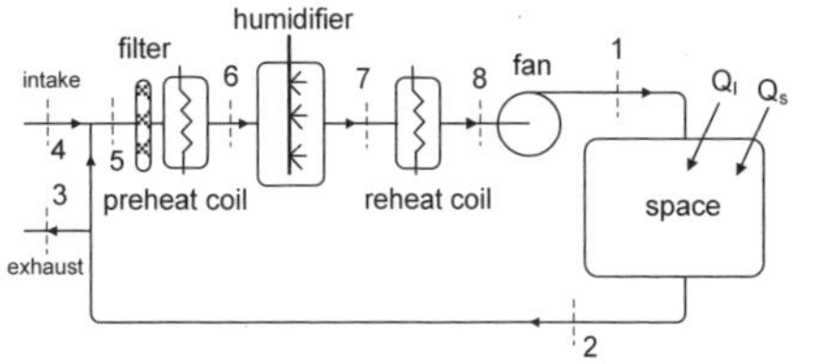

Fig. E5.10.1(a) Schematic diagram

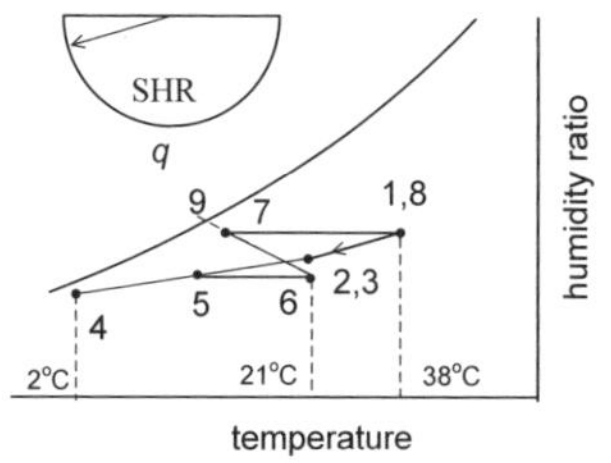

Fig. E5.10.1(b) Psychrometric chart

Solution The state of the air at different points is indicated on the psychrometric chart in Fig. E5.10.1(b). The state 2 of the return air is located by the intersection of the 21°C constant db-temperature line and the 13°C constant wb-temperature line. The state 4 of the outdoor air is located by the intersection of the 2°C db-temperature line and the 95% relative humidity line. To obtain the state of the mixed air at 5, we divide the line 4-2 in the ratio 4:6. In the preheater the air is heated to a state 6 with a constant humidity ratio. Therefore the state 6 is located by the intersection of the horizontal line through 5 and the 21°C constant db-temperature line. The sensible heat ratio, SHR is given by

$$SHR = \frac{\dot{Q}_s}{\dot{Q}_s + \dot{Q}_l} = \frac{50}{50+10} = 0.83$$

In order to locate the state 1 of the supply air we draw a line through 2 in the direction of the line on the SHR-protractor pointing to the value 0.83. The intersection of this line with the 38°C constant db-temperature line locates state 1. The air is heated with a constant humidity ratio in the reheater. Therefore the state 7 of the air entering the reheater must lie on the horizontal line through 1. The state of the air leaving the humidifier lies along the wb-temperature line through 6. Therefore the intersection of the above two lines locates the state 7. Now that we have located all the relevant state points on the psychrometric chart (Fig. 4.5), the following values are read off directly.

$t_{db5} = 13.5°C$, $t_{db7} = 15.2°C$, $t_{wb6} = 12.5°C$, $\omega_7 = 0.0077$, $\omega_6 = 0.0053$

(i) The dry air mass flow rate of the air supplied is obtained from Eq. (5.35) as

$$\dot{Q}_s = \dot{m}_a c_{am}(t_2 - t_1)$$

Substituting numerical values we have

$$\dot{m}_a = 50/[1.02(38 - 21)] = 2.88 \text{ kgs}^{-1}$$

(ii) The heat input to the preheating coil is given by

$$\dot{Q}_{ph} = \dot{m}_a c_{am}(t_6 - t_5)$$

$$\dot{Q}_{ph} = 2.88 \times 1.02(21 - 13.5) = 22.0 \text{ kW}$$

The heat input to the reheating coil is given by

$$\dot{Q}_{rh} = \dot{m}_a c_{am}(t_1 - t_7)$$

$$\dot{Q}_{rh} = 2.88 \times 1.02(38 - 15.2) = 67.0 \text{ kW}$$

(iii) The mass flow rate of water in the air washer is given by

$$\dot{m}_w = \dot{m}_a(\omega_7 - \omega_6)$$

$$\dot{m}_w = 2.88(0.0077 - 0.0053) = 6.9 \times 10^{-3} \text{ kgs}^{-1}$$

(iv) The saturation effectiveness of the air washer is

$$e_c = \frac{t_6 - t_7}{t_6 - t_{w6}} = \frac{21 - 15.2}{21 - 12.5} = 68.2\%$$

It is instructive to check the overall energy balance for the system. This can be written in the form

$$\dot{Q}_s + \dot{Q}_l = \dot{Q}_{ph} + \dot{Q}_{rh} + \dot{m}_{a4}(h_4 - h_2)$$

Substituting the relevant numerical values in the above equation we obtain the LHS and RHS of the energy balance equation respectively as

$$(50 + 10) = 60 \text{ kW}$$

and $$22 + 67.0 + 2.88 \times 0.4(12.7 - 37) = 61.0 \text{ kW}$$

There is discrepancy of about 1.7% in the energy balance. Note that we have neglected the energy of the water entering the air washer which is about 0.4 kW.

Example 5.11 An evaporative cooler is used to cool a space which has a sensible heat load of 15 kW and a latent heat load of 4.8 kW. The dry air mass flow rate of the supply air is 1.4 kgs^{-1}. The air leaves the cooler with a db-temperature of 18°C and a relative humidity of 100%. The outdoor air db-temperature is 35°C. The pressure is constant at 101.3 kPa. Determine (i) the db-temperature and relative humidity of the space, (ii) the outdoor relative humidity, and (iii) the mass flow rate of water.

Solution A schematic diagram of the system and the corresponding psychrometric chart are shown in Fig. 5.12(a) and (b) respectively. The state 1 of the air entering the space is located on the saturation curve at a db-temperature of 18°C. The sensible heat load is given by

$$\dot{Q}_s = \dot{m}_a c_{am}(t_2 - t_1)$$

$$1.4 \times 1.02(t_2 - 18) = 15$$

(i) Hence the db-temperature of the space is 28.5°C.
The sensible heat ratio, SHR is given by

$$SHR = \frac{\dot{Q}_S}{\dot{Q}_S + \dot{Q}_l} = \frac{15}{15+4.8} = 0.758$$

We first draw a line on the SHR-protractor pointing in the direction of the value 0.758. The space condition line is obtained by drawing a line through 1 parallel to the above line on the protractor. This line intersects the 28.5°C constant db-temperature line at 2. The relative humidity at 2 is read off from the psychrometric chart as 59%.

(ii) We assume that in the evaporative cooler the state of the air follows the constant wb-temperature line at 35°C to the state 1. Therefore the state 3 of the outdoor air is located by the intersection of the 35°C db-temperature line and the constant wb-temperature line at state 1. We read off the relative humidity at 3 as 17%.

(iii) The mass flow rate of water to the cooler is given by

$$\dot{m}_w = \dot{m}_a(\omega_1 - \omega_3)$$

$$\dot{m}_w = 1.4(0.013 - 0.006) = 9.8 \times 10^{-3}\ \text{kgs}^{-1}$$

Example 5.12 An air conditioning system with a cooling and dehumidifying coil supplies air to two zones A and B which have individual reheat coils as shown schematically in Fig. 5.13(a). The design data for the zones are summarized in Table E5.12.

Table E5.12. Summary of space conditions

Zone	Q_s (kW)	Q_l(kW)	t_{db} (°C)	ϕ (%)	SHR
A	90	30	22	40	0.75
B	70	35	26	35	0.67

Of the total mass flow of air returning to the cooling coil from the zones, 25% is discharged to the ambient and replaced with an equal quantity of outdoor air at 33°C db-temperature and 60% relative humidity. Air leaves the cooling coil at 5°C db-temperature and 95%

relative humidity. The pressure is constant at 101.3 kPa. Determine (i) the supply air mass flow rates to the two zones, (ii) the heat input by the two reheat coils, and (iii) the refrigeration capacity of the cooling coil.

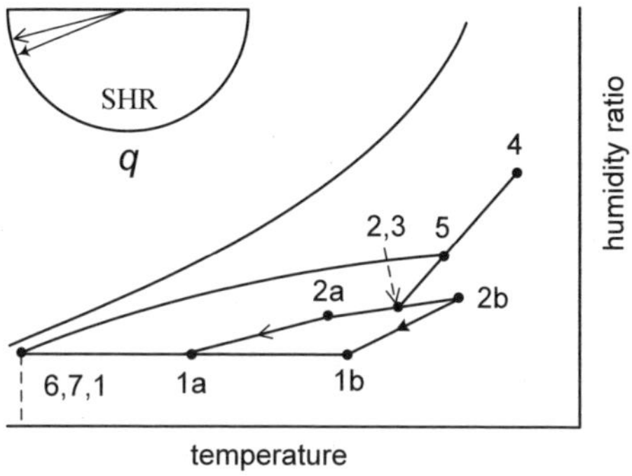

Fig. E5.12.1 Psychrometric chart

Solution We locate the states 2a and 2b of the air in the two spaces using the respective db-temperature and relative humidity. State 6 of the air leaving the cooling coil is located by the intersection of the 5°C db-temperature line and the 95% relative humidity line as indicated on the psychrometric chart in Fig. E5.12.1.

The SHR of each zone, listed in Table E5.12, is calculated using the given heat loads of the zones. With the aid of the SHR-protractor we draw the condition lines of the zones through points 2a and 2b.

During the heating processes in the two reheaters, the humidity ratio of the air streams remain constant. Therefore the supply air states 1a and 1b for the two spaces are located by the points of intersection of the respective condition lines and the horizontal line drawn through 6. This gives the db-temperatures at 1a and 1b as 11.8°C and 16°C respectively.

(i) The dry air mass flow rate of the air supplied to a zone is given by

$$\dot{Q}_s = \dot{m}_a c_{am}(t_2 - t_1)$$

Substituting the numerical values applicable to the two zones A and B in the above equation we obtain the following air flow rates:

$$\dot{m}_{aA} = 90/[1.02(22 - 11.8) = 8.65 \text{ kgs}^{-1}$$

$$\dot{m}_{aB} = 70/[1.02(26 - 16) = 6.86 \text{ kgs}^{-1}$$

(ii) The heat input rate to a reheating coil is given by

$$\dot{Q}_{rh} = \dot{m}_a c_{am}(t_6 - t_1)$$

Substituting the numerical values applicable to the two zones A and B in the above equation we obtain the following heat input rates:

$$\dot{Q}_{rhA} = 8.65 \times 1.02(11.8 - 5) = 60 \text{ kW}$$

$$\dot{Q}_{rhB} = 6.86 \times 1.02(16 - 5) = 77 \text{ kW}$$

The mixing of the air streams at 2a and 2b results in the state 2. We apply the approximate relation given by Eq. (5.8) to find the temperature at 2. Hence we have

$$(8.65 + 6.86)t_2 = 8.65 \times 22 + 6.86 \times 26$$

Therefore, $t_2 = 23.8$°C.

We can now locate state 2 on the psychrometric chart. The ratio of the mass flow rates of outdoor air at state 4 and return air at state 2 is 1:3. These air streams mix before they enter the cooling coil to produce air of state 5. We locate the state 5 of the mixture by dividing the line 2-4 in the inverse ratio of the mass flow rates. This gives, $t_5 = 26.2$°C.

(iv) The refrigeration capacity of the cooling coil is given by

$$\dot{Q}_{ref} = (\dot{m}_{aA} + \dot{m}_{aB})(h_5 - h_6)$$

$$\dot{Q}_{ref} = 15.5(52 - 18) = 527 \text{ kW}$$

The total energy input to the system is

$$\dot{Q}_{in} = Q_{tA} + Q_{tB} + Q_{rhA} + Q_{rhB} + \dot{m}_{ven}(h_4 - h_2)$$

$$\dot{Q}_{in} = 120 + 105 + 60 + 77 + 15.5 \times 0.25 \times (83 - 42) = 521 \text{ kW}$$

For overall energy balance, $\dot{Q}_{in} = \dot{Q}_{ref}$.
The overall energy balance is satisfied within about 1%. Note that the enthalpy of the condensate water has been neglected.

Example 5.13 A variable air volume (VAV) summer air conditioning system, shown schematically in Fig. 5.15(a), supplies air to two zones A and B whose heat loads and db-temperatures are summarized in the Table E5.13.

Table E5.13 Summary of space conditions

Zone	Q_s (kW)	Q_l(kW)	t_{db} (°C)	SHR
A	90	30	22	0.75
B	70	35	26	0.67

Of the total mass flow of air returning to the cooling coil from the two zones, 25% is discharged to the ambient and replaced with an equal mass of outdoor air at 33°C db-temperature and 60% relative humidity. Air leaves the cooling coil at 5°C db-temperature and 95% relative humidity. The pressure is constant at 101.3 kPa. Determine (i) the supply air mass flow rates to the two zones, (ii) the relative humidity of two zones, and (iii) the refrigeration capacity of the cooling coil.

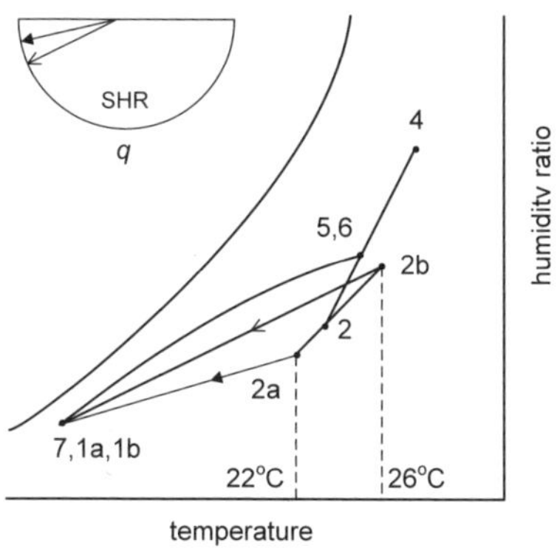

Fig. E5.13.1 Psychrometric chart

Solution The state 7 of the air leaving the cooling coil is located by the intersection of the 5°C db-temperature line and the 95% relative humidity line as indicated on the psychrometric chart in Fig. E5.13.1. The SHR- values of the two zones, listed in the Table E5.13, are calculated using the given heat loads of the zones.

With the aid of the SHR-protractor we draw the condition lines of the two zones through point 7, which is the supply air state for both zones. The points of intersection of the condition lines for zones A and B with the 22°C and 26°C db-temperature lines respectively, give the states of the air in zones A and B.

(i) The dry air mass flow rate of the air supplied to a zone is given by

$$\dot{Q}_s = \dot{m}_a c_{am}(t_2 - t_1)$$

Substituting the numerical values applicable to the two zones A and B in the above equation we obtain the following air flow rates:

$$\dot{m}_{aA} = 90/[1.02(22 - 5) = 5.19 \text{ kgs}^{-1}$$

$$\dot{m}_{aB} = 70/[1.02(26 - 5) = 3.27 \text{ kgs}^{-1}$$

(ii) We read off from the psychrometric chart (Fig. 4.5) the relative humidity of zone A and zone B as 46% and 47% respectively.
The outdoor air is at state 4. We locate the states 2 and 5 by following the steps given in worked example 5.12. Hence we obtain

$$t_2 = 23.5°\text{C} \text{ and } t_5 = 25.9°\text{C}$$

(iii) The refrigeration capacity of the cooling coil is given by

$$\dot{Q}_{ref} = (\dot{m}_{aA} + \dot{m}_{aB})(h_5 - h_7)$$

$$\dot{Q}_{ref} = 8.46(54.5 - 18) = 308.8 \text{ kW}$$

As an exercise the reader is encouraged to check the overall energy balance for the system.

Example 5.14 A variable air volume (VAV) system used for winter air conditioning is shown schematically in Fig. E5.14.1(a). It supplies air to two zones A and B whose heat loads and db-temperatures are summarized in the Table E5.14.

Table E5.14. Summary of space conditions

Zone	Q_s (kW)	Q_l(kW)	t_{db} (°C)	SHR
A	13	9	25	0.59
B	7	1.8	20	0.795

Of the total mass flow of air returning from the zones, 20% is discharged to the ambient and replaced with an equal mass of outdoor air at 5°C db-temperature and 50% relative humidity. The relative humidity of the air leaving the air washer is 95%. Air is supplied to the zones at 35°C db-temperature and 30% relative humidity. The pressure is constant at 101.3 kPa. Determine (i) the mass flow rates of air to the two zones, (ii) the rates of heat input by the preheater and the reheater, and (iii) the rate of moisture addition by the air washer.

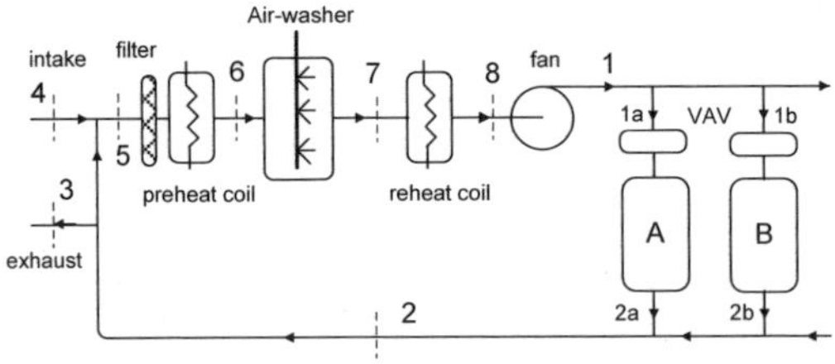

Fig E5.14.1(a) Schematic diagram of VAV heating system

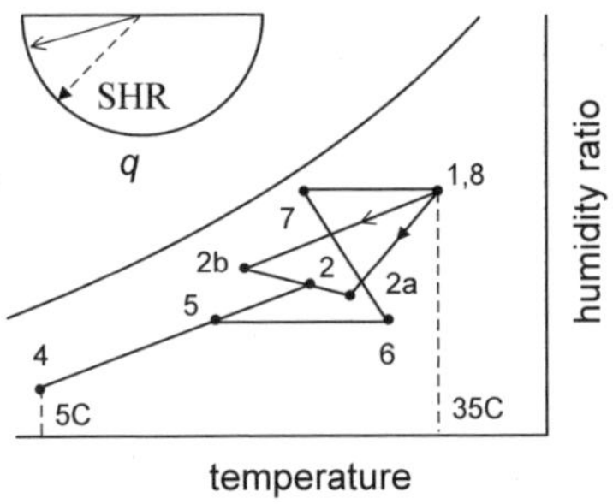

Fig E5.14.1(b) Psychrometric chart

Solution The state 1 of the supply air to the two zones is located by the intersection of the 35°C db-temperature line and the 30% relative humidity line as indicated on the psychrometric chart in Fig. E5.14.1(b). The SHR of the two zones, listed in Table E5.14, are calculated using the given heat loads of the zones. With the aid of the SHR-protractor (Fig. 4.5) we draw the condition lines of the two zones through point 1, which is the supply air state for both zones. The points of intersection of the condition lines for zones A and B with the 25°C and 20°C db-temperature lines, respectively, give the states of air in the two zones as 2a and 2b.

(i) The dry air mass flow rate of the air supplied to a zone is given by

$$\dot{Q}_s = \dot{m}_a c_{am}(t_2 - t_1)$$

Substituting the numerical values applicable to the two zones A and B in the above equation we obtain the following dry air flow rates:

$$\dot{m}_{aA} = 13/[1.02(35 - 25) = 1.27 \text{ kgs}^{-1}$$

$$\dot{m}_{aB} = 7/[1.02(35 - 20) = 0.46 \text{ kgs}^{-1}$$

Hence the total dry air mass flow rate through the heaters is 1.73kgs^{-1}.

We locate the state 2 of the mixed air from the two zones by dividing the line 2a-2b in the inverse ratio of the mass flow rates (0.46:1.27). This air is mixed with outdoor air at state 4 to produce air at state 5. The state 5 is located by dividing the line 2-4 in the ratio of 1:4. The air at state 5 is heated with a constant humidity ratio in the preheater. Therefore state 6 should lie on the horizontal line through 5.

Now the humidity ratio is constant during the heating process in the reheater. We are given that the relative humidity of the air leaving the air washer at 7 is 95%. Therefore state 7 is located by the point of intersection of the horizontal line through 8 and the 95% constant relative humidity line.

The humidification process in the air washer is assumed to occur with a constant wb-temperature. Therefore we locate state 6 by drawing the constant wb-temperature line through 7 to intersect the horizontal line drawn earlier through 5. We have now located all the relevant state points of air on the psychrometric chart (Fig. 4.5). Hence we obtain the following values:

$t_{db5} = 19.86°\text{C}$, $t_{db6} = 24.6°\text{C}$, $t_{db7} = 15.8°\text{C}$, $\omega_7 = 0.0107$, $\omega_6 = 0.0071$

(ii) The heat input rate by the preheater is

$$\dot{Q}_{ph} = \dot{m}_a c_{am}(t_6 - t_5)$$

$$\dot{Q}_{ph} = 1.73 \times 1.02(24.6 - 19.8) = 8.47 \text{ kW}$$

The heat input rate by the reheater is

$$\dot{Q}_{rh} = \dot{m}_a c_{am}(t_8 - t_7)$$

$$\dot{Q}_{rh} = 1.73 \times 1.02(35 - 15.8) = 33.9 \text{ kW}$$

(iii) The mass of moisture added in the air washer is

$$\dot{m}_w = \dot{m}_a(\omega_7 - \omega_6)$$

$$\dot{m}_w = 1.73(0.0107 - 0.0071) = 6.23 \times 10^{-3}\ \text{kgs}^{-1}$$

As an exercise the reader is encouraged to check the overall energy balance for the system.

Example 5.15 A dual-duct air conditioning system, shown schematically in Fig. E5.15.1(a), has a heating coil and a cooling coil. It supplies air to two zones A and B whose design data are summarized in the Table E5.15.

Table E5.15. Summary of space conditions

Zone	Q_s (kW)	Q_l(kW)	t_{db} (°C)	$\dot{m}_a$ (kgs^{-1})	SHR
A	90	30	22	8	0.75
B	70	35	26	14	0.67

Of the total mass flow rate of air returning from the zones, 25% is discharged to the ambient and replaced with an equal amount of outdoor air at 33°C db-temperature and 60% relative humidity. The air leaves the cooling coil at 5°C db-temperature and 95% relative humidity. The air leaving the heating coil is at a db-temperature of 28°C. The pressure is constant at 101.3 kPa. Determine (i) the mass flow rates of air leaving the cooling coil and the heating coil, (ii) the heat input rate by the heating coil, and (iii) refrigeration capacity of the cooling coil.

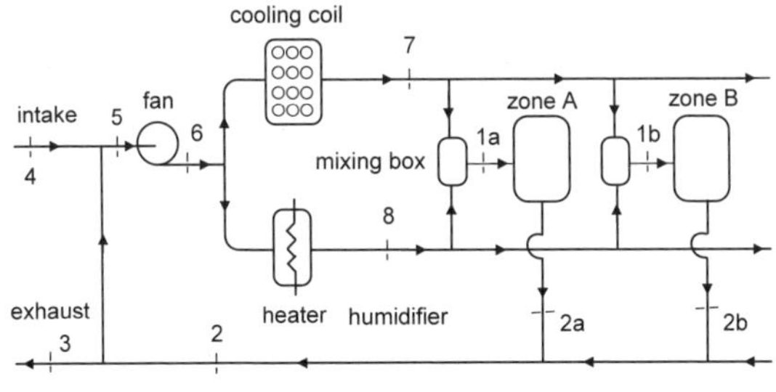

Fig E5.15.1(a) Schematic diagram of dual-duct system

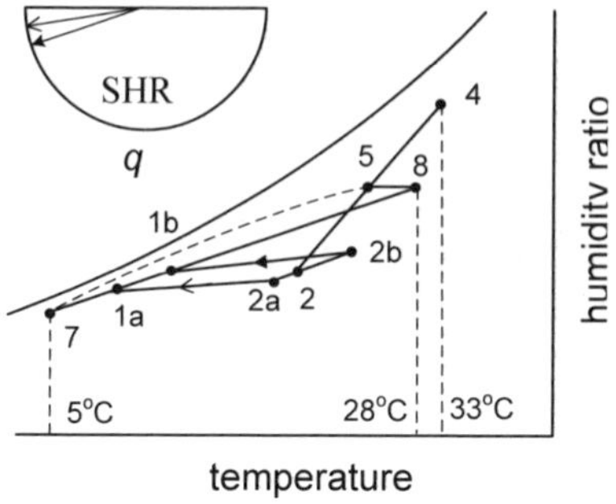

Fig E5.15.1(b) Psychrometric chart

Solution Since the humidity ratios of the air in the two zones are not given it is not possible to locate their states on the psychrometric chart without making an initial guess of one of the humidity ratios. However, we could determine the relevant mass flow rates of air using the given db-temperatures. Now the sensible heat load of a zone is given by

$$\dot{Q}_s = \dot{m}_a c_{am}(t_2 - t_1)$$

Substituting the numerical values applicable to the two zones A and B in the above equation we obtain the following supply air temperatures.

$$t_{a1A} = 22 - \frac{90}{1.02\times8} = 11.0°\text{C}$$

$$t_{a1\text{B}} = 26 - \frac{70}{1.02\times14} = 21.1°\text{C}$$

The above supply air temperatures are produced by mixing cold air at 5°C and hot air at 28°C. For these mixing processes we apply Eq. (5.8) as follows:

$$5\dot{m}_{a7A} + 28(8 - \dot{m}_{a7A}) = 8 \times 11.0$$

$$5\dot{m}_{a7B} + 28(14 - \dot{m}_{a7B}) = 14 \times 21.1$$

From the above equations we obtain the mass flow rates (kgs^{-1}) of cold and hot air to each zone as:

$$\dot{m}_{a7A} = 5.91, \;\; \dot{m}_{a8A} = (8 - 5.91) = 2.09$$

$$\dot{m}_{a7B} = 4.2, \;\;\; \dot{m}_{a8B} = (14 - 4.2) = 9.8$$

(i) Therefore the mass flow rates (kgs^{-1}) of the air leaving the cooling coil and the heating coil are respectively

$$\dot{m}_{ac} = (5.91 + 4.2) = 10.11$$

$$\dot{m}_{ah} = (2.09 + 9.8) = 11.89$$

We apply Eq. (5.8) to the mixing of the return air streams from the two zones to obtain

$$(8 + 14)t_2 = 22 \times 8 + 26 \times 14$$

Therefore $t_2 = 24.5°C$.

Applying Eq. (5.8) to the mixing of return air and outdoor air we have

$$t_5 = 0.25 \times 33 + 0.75 \times 24.5 = 26.63°C$$

As a first guess assume that relative humidity of the air leaving the heating coil at 8 is 50%. We can now locate the state on the psychrometric chart (Fig. 4.5) using the db-temperature of 28°C and the relative humidity of 50%. The states 1a and 1b of the air entering the zones A and B must lie on the line 8-7. Since we know the db-temperatures of the supply air as, $t_{a1A} = 11.0°C$ and $t_{a1B} = 21.1°C$, we can locate the states 1a and 1b on the line 8-7.

The SHR for the two zones are listed in Table E5.15. We draw lines in the direction of the respective SHR-values through the points 1a and 1b to intersect the corresponding zone db-temperatures of 22°C and 26°C. This locates the points 2a and 2b on the psychrometric chart. We draw the line 2a-2b to intersect the constant db-temperature line at 24.5°C, which is the temperature of the mixed air stream at 2. The state 4 of the outdoor air is located by the intersection of the 33°C db-temperature line and the 60% relative humidity line. Mixing of the outdoor air at 4 and the return air at 2 gives the temperature 5 as 26.6°C, as was calculated above.

During the heating process in the coil the state of the air follows the horizontal line through 5 which has to pass through the point 8. We recall that point 8 was located by making an initial guess of the relative humidity at 8 as 50%. After several iterations a relative humidity of 53% was found to satisfy all the above equations.

(ii) The heat input rate by the heater is

$$\dot{Q}_h = 11.89 \times 1.02(28 - 26.5) = 18.2 \text{ kW}$$

(iii) The refrigeration capacity of the cooling coil is given by

$$\dot{Q}_{ref} = \dot{m}_{ac}(h_5 - h_7)$$

$$\dot{Q}_{ref} = 10.11(58.3 - 18) = 407.4 \text{ kW}$$

As an exercise the reader is encouraged to check the overall energy balance for the system. Note that we have neglected the enthalpy of the condensate water leaving the cooling coil.

Problems

P5.1 Saturated ambient air with a db-temperature of 5°C and a mass flow rate of 0.9 kgs^{-1} is divided into two streams. One stream passes through a heating section and leaves it with a relative humidity of 25%. The conditions of the other stream that bypasses the heater remains unchanged. The two streams are then mixed to produce the supply air stream at 24°C. The pressure is constant at 101.3 kPa. Determine (i) heat input by the heating coil, and (ii) the mass flow of air through the bypass section.
[*Answers*: (i) 17.44 kW, (ii) 0.115 kgs^{-1}]

P5.2 Ambient air enters a cooling coil at 24°C db-temperature and 50% relative humidity with a dry air mass flow rate of 0.9 kgs^{-1}. The air leaving the cooling coil at 9°C is reheated to 13°C and 70% relative humidity. The pressure is constant at 101.3 kPa. Determine (i) the dew-point of the ambient air, (ii) the rate of moisture removal in the cooling coil, (iii) the refrigeration capacity of the cooling coil, and (iv) the heat input rate of the heating coil. Compare the results obtained using ideal gas expressions with those obtained using the psychrometric chart.
[*Answers*: (i) 12.8°C, 0.0025 kgs^{-1}, 20.05kW, 3.67 kW]

P5.3 Moist air enters a cooling coil at 33°C db-temperature and 24.5°C wb-temperature, and leaves at 15°C db-temperature and 90% relative humidity. The refrigeration capacity of the coil is 28 kW. The

pressure is constant at 101.3 kPa. The coil is represented using the bypass model. Determine (i) the bypass factor, (ii) the coil surface temperature (apparatus dew-point), (iii) the dry air mass flow rate through the coil, and (iv) the sensible heat ratio (SHR) for the cooling process.
[*Answers*: (i) 0.18, (ii) 11°C, (iii) 0.79 kgs^{-1}, (iv) 0.52]

P5.4 Moist air at 13°C db-temperature and 55% relative humidity enters a winter air conditioning system, consisting a preheater and an air washer. The supply air leaving the air washer is at 32°C db-temperature and 18.3°C wb-temperature. The mass flow rate of dry air is 0.3 kgs^{-1}. The pressure is constant at 101.3 kPa. Determine (i) the db-temperature of the air at the exit of the preheater, (ii) the heat input rate by the preheater, (iii) the rate of moisture addition in the air washer, and (iv) the saturation effectiveness of the air washer.
[*Answers*: (i) 37.8°C, (ii) 7.6 kW, (iii) 7.35×10^{-4} kgs^{-1}, (iv) 30%]

P5.5 The sensible and latent heat losses from a space are 22 kW and 9.5 kW respectively. The space is maintained at 21°C db-temperature and 7°C dew-point temperature. Conditioned air is supplied to the space at 35°C db-temperature. The pressure is constant at 101.3 kPa. Determine (i) the wb-temperature of the supply air, and (ii) the mass flow rate of air.
[*Answers*: (i) 20°C, (ii) 1.54 kgs^{-1}]

P5.6 A winter air conditioning system supplying air to a space at the rate of 2 kgs^{-1} has a heater and a steam humidifier as the processing units. The total heat loss from the space is 28 kW and the SHR is 0.6. The space is maintained at 21°C db-temperature and 14.5°C wb-temperature. Before entering the heater, 40% by mass of the return air is exhausted and replaced with an equal quantity of outdoor air at 7°C and 10% relative humidity. The air enters the humidifier where saturated steam at 120°C is sprayed into the air stream. The pressure is constant at 101.3 kPa (a) Draw a schematic diagram of the system. (b) Calculate (i) the rate of heat input by the heater, and (ii) the rate of moisture addition by the humidifier.

[*Answers*: (i) 26.7 kW, (ii) 0.0104 kgs^{-1}]

P5.7 An air conditioning system, shown schematically in Fig. 5.10(a), supplies air at the rate of 4 kgs^{-1} to a space maintained at a db-temperature of 27°C and relative humidity of 50%. The sensible and latent heat loads on the space are 46 kW and 20 kW respectively. Outdoor air at 35°C db-temperature and 24°C wb-temperature is introduced at the rate of 1.1 kgs^{-1}. The relative humidity of the air leaving the cooling coil is 90%. The pressure is constant at 101.3 kPa. Determine (i) the supply air temperature and relative humidity, (ii) the mass flow rate of the air that bypasses the cooling coil, and (iii) the refrigeration capacity of the cooling coil.
[*Answers*: (i) 15.7°C, 81%, (ii) 0.66 kgs^{-1}, (iii) 82.8 kW]

P5.8 A constant volume, dual-duct, air conditioning system supplies air to a single zone, maintained at 25°C db-temperature. The system has been designed for a maximum heating load of 8 kW, and a maximum sensible cooling load of 7 kW. The heating coil and the cooling coil supply air at 40°C and 14°C db-temperature respectively. The pressure is constant at 101.3 kPa. Determine the following quantities when the space is experiencing a sensible cooling load of 4.5 kW: (i) the supply air temperature, (ii) the mass flow rates of air from the hot and cold air ducts.
[*Answers*: (i) 17.9°C, (ii) 0.094 kgs^{-1}, 0.53 kgs^{-1}]

P5.9 A variable air volume (VAV) air conditioning system supplies air to a single zone maintained at 26°C db-temperature and 20°C wb-temperature. The system is designed to meet a sensible cooling load of 32 kW and a latent cooling load of 21.5 kW, for which the supply air db-temperature is 16°C. When the system operates under part-load conditions the sensible and latent cooling loads are 5.2 kW and 1.4 kW respectively. The mass flow rate of supply air is 25% of the mass flow rate under design conditions. The pressure is constant at 101.3 kPa. Determine (i) the mass flow rate of supply air under design conditions, (ii) the supply air relative humidity under design conditions, and (iii) the

supply air db-temperature and relative humidity under part-load conditions.
[*Answers*: (i) 3.14 kgs^{-1}, (ii) 83%, (iii) 19.5°C, 81%]

P5.10 A dual-duct multi-zone air conditioning system, consisting of a heating coil and a cooling coil, supplies air to 3 identical zones, maintained at 24°C db-temperature and 17°C wb-temperature. For each zone the sensible and latent heat loads are 34 kW and 22 kW respectively. Of the total return air mass flow from the zones, 25% is discharged and replaced with outdoor air at 35°C db-temperature and 40% relative humidity. The air leaving the heating coil is at 40°C db-temperature. At the exit of the cooling coil the air is at 10°C db-temperature and 90% relative humidity. The pressure is constant at 101.3 kPa. (a) Draw a schematic diagram of the system. (b) Determine (i) the mass flow rate of air through the heating coil, (ii) the heat input by the heating coil, (iii) the mass flow rate of air through the cooling coil, and (iv) the refrigeration capacity of the cooling coil. Check the overall energy balance for the system.
[*Answers*: (i) 5.31 kgs^{-1}, (ii) 71.76 kW, (iii) 13.2 kgs^{-1} (iv) 347.2 kW]

P5.11 A dual-duct air conditioning system similar to that shown in Fig. 5.14(a), supplies air to two zones A and B with the design conditions summarized in Table P5.11. Zone A gains sensible and latent heat while zone B looses sensible heat.

Table P5.11. Summary of space conditions

Zone	Q_s (kW)	Q_l(kW)	t_{db} (°C)	RH%
A	5.6	4.5	21	60
B	7	0	22	40

Of the total return air dry mass flow, 50% is discharged and replaced with an equal quantity of outdoor air at 15.5°C db-temperature and 40% relative humidity. The supply air in the cold duct is saturated at 4.5°C. The air in the hot air duct is at 35°C db-temperature and 20% relative humidity. The pressure is constant at 101.3 kPa. Determine (i) the supply air mass flow rates to the two zones, (ii) the mass flow rates of hot air to the two zones, (iii) the heat input by the heater, (iv) the refrigeration

capacity of the cooling coil, and (v) the rate of moisture addition in the air washer. Check the overall energy balance for the system.
[*Answers*: (i) 0.48kgs^{-1}, 1.49kgs^{-1}, (ii) 0.08kgs^{-1}, 1.08kgs^{-1}, (iii) 22.8 kW, (iv) 12.56 kW, (v) 0.00145kgs^{-1}]

References

1. *ASHRAE Handbook* - 2013 *Fundamentals*, American Society of Heating, Refrigeration and Air Conditioning Engineers, Atlanta, 2013.
2. *ASHRAE Handbook* - 2012 *HVAC Systems and Equipment*, American Society of Heating, Refrigeration and Air Conditioning Engineers, Atlanta, 2012.
3. Kuehn, Thomas H., Ramsey, James W. and Threlkeld, James L., *Thermal Environmental Engineering*, 3rd edition, Prentice-Hall, Inc., New Jersey, 1998.
4. Rogers, G. F. C. and Mayhew Y. R., *Thermodynamic and Transport Properties of Fluids.* 5th ed. Blackwell, Oxford, U.K. 1998.
5. Stoecker, Wilbert F. and Jones, Jerold W., *Refrigeration and Air Conditioning*, International Edition, McGraw-Hill Book Company, London, 1982.

Chapter 6

Direct-Contact Transfer Processes and Equipment

6.1 Introduction

In chapter 5 we considered the psychrometric analysis of a number of heating and air conditioning systems consisting of air processing equipment like mixing sections, heaters, cooling coils, air washers, and humidifiers. The latter three devices are called direct-contact devices because the energy and mass transfer between air and water occurs across the interface separating them. The control volume based models developed in chapter 5 to represent these equipment related only the inlet and outlet conditions of the air passing through the device. With the inclusion of semi-empirical design parameters, like the bypass factor, and the saturation effectiveness these models were adequate for purposes of psychrometric analysis.

However, for the design, sizing, and performance evaluation of direct-contact air processing equipment we need to analyze in greater detail the heat and mass transfer processes occurring within the device. The distinguishing features of these processes are: (i) there is direct physical contact between air and water, and (ii) there is simultaneous mass and energy exchange between air and water.

In the present chapter we shall develop heat and mass transfer models for air–water direct-contact equipment like air washers, dehumidifiers, and cooling towers. However, before we embark on the development of these detailed models we shall first review the basic principles of mass diffusion and convection.

6.2 Review of Mass Transfer Principles

In chapter 2 we presented the basic equations governing the transfer of heat by conduction and convection. In a similar manner, the transfer of mass occurs by two processes known as *mass diffusion* and *mass convection*. Although the detailed physical mechanisms of heat transfer and mass transfer are different, these transfer processes are governed by equations having similar mathematical forms.

6.2.1 *Steady mass diffusion through a plane wall*

We shall consider mass diffusion of a single species in a stationary medium, for example, a solid or a stagnant gas. A practical example of such a process, of interest in air conditioning systems, is the diffusion of water vapor through porous building materials.

While the temperature difference is the driving potential for heat conduction, the difference in *mass fraction* (concentration) of a species is the driving potential for mass diffusion. The mass fraction, m_i of species i in a mixture of species is defined as the ratio of the density ρ_i of species i to the total density ρ.

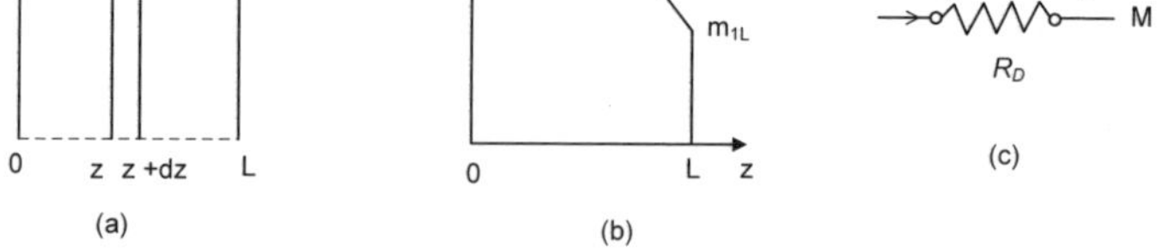

Fig. 6.1 (a) Mass diffusion through wall, (b) Concentration profile, (c) Network element

Consider the plane wall shown schematically in Fig. 6.1(a) through which a species 1 is diffusing steadily. The mass flux of the species 1 is governed by *Fick's law of diffusion* which may expressed as [3,5]

$$j_1 = -\rho D_{12} \frac{dm_1}{dz} \tag{6.1}$$

where j_1 [kgm^{-2} s^{-1}] is the diffusive mass flux of species 1, ρ [kgm^{-3}] is the local density, m_1 is the mass fraction or mass concentration of species 1, and D_{12} [m^2 s^{-1}] is the constant of proportionality, called the *binary diffusion coefficient* or *mass diffusivity*. The negative sign signifies that mass diffusion occurs in the direction of decreasing concentration.

We now apply the *principle of conservation of mass* [5] to the elemental volume of the wall shown in Fig. 6.1(a), assuming steady-state mass diffusion, with no chemical reactions producing or consuming species 1. The temperature of the wall is assumed uniform. Hence we have

$$\frac{d(Aj_1)}{dz} = 0 \tag{6.2}$$

where A is the area of cross section of the wall.
Integrating Eq. (6.2) we obtain

$$Aj_1 = constant = \dot{M}_1 \tag{6.3}$$

where $\dot{M}_1$ is the total mass flow rate of the species 1.
Substituting from Eq. (6.1) in (6.3) we have

$$\frac{\dot{M}_1}{\mathrm{A}} = -\rho D_{12}\frac{dm_1}{dz} \tag{6.4}$$

The solution of Eq. (6.4) is

$$m_1 = -\left(\frac{\dot{M}_1}{A\rho D_{12}}\right)z + c \tag{6.5}$$

Equation (6.5) shows that the mass concentration of species 1 varies linearly with distance, z as depicted in Fig. 6.1(b). The constant of integration c is found by substituting the following boundary conditions: at $z = 0$, $m_1 = m_{10}$ and at $z = L$, $m_L = m_{1L}$.
Hence we have

$$\dot{M}_1 = \frac{m_{10}-m_{1L}}{(L/\rho D_{12}A)} = \frac{m_{10}-m_{1L}}{R_d} \tag{6.6}$$

We notice that Eq. (6.6) is analogues to Ohm's law where the voltage difference is equivalent to the concentration difference, $(m_{10} - m_{1L})$ at the two surfaces of the wall (see Fig. 6.1b) and the current is equivalent to the total mass flow rate $\dot{M}_1$. Therefore we define the *mass diffusion resistance* as

$$R_d = \frac{L}{\rho D_{12} A} \tag{6.7}$$

The resulting equivalent mass diffusion network element is shown in Fig. 6.1(c).

6.2.2 *Steady convection mass transfer*

The analogy between heat transfer and mass transfer can be extended to study *forced convection mass transfer* where a species diffuses through a fluid in motion. A practical example is the drying of a moist material by an air stream flowing over it.

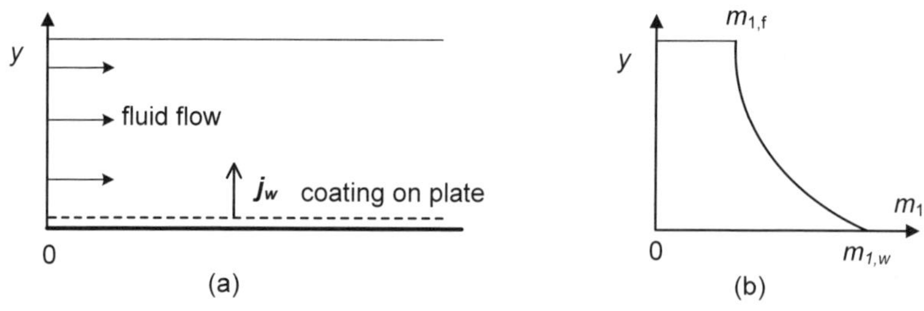

Fig. 6.2 (a) Forced convection mass transfer, (b) Concentration profile

Shown schematically in Fig. 6.2(a) is a fluid stream flowing *steadily*, at constant temperature, over a *thin* coating on a horizontal plate [3]. The coating is made of a substance that dissolves in the fluid as it flows over it. The distribution of the mass fraction (concentration) of the substance in the fluid stream at a section far downstream from the entry section, is depicted in Fig. 6.2(b). The mass fraction of the substance decreases progressively to a lower value in the 'undisturbed' fluid stream, commonly called the *free stream*. The substance is transported across the fluid stream by diffusion and convection as in the case of convective heat transfer, discussed earlier in section 2.7.1.

The concentration profile of the substance shown in Fig. 6.2(b) has the steepest gradient at the plate. Moreover, a *thin layer* of fluid adjacent to the plate is stationary due to the presence of the solid plate. The transfer of substance through this *thin* stationary layer of fluid occurs by mass *diffusion* and therefore the mass flux is given by Fick's law as

$$j_w = -\rho D_{12}\left[\frac{dm_1}{dy}\right]_{y=0} \tag{6.8}$$

where m_1 is the concentration of the substance.

The mass transfer rate from the coating to the bulk fluid, which occurs by *forced convection*, is expressed in terms of a *mass transfer coefficient*, h_m [kgs^{-1}m^{-2}] using the following rate equation

$$j_w = h_m\left(m_{1,w} - m_{1,f}\right) \tag{6.9}$$

where the mass transfer 'driving potential' is the difference in concentration at the plate, $m_{1,w}$ and the free stream, $m_{1,f}$. Notice that Eq. (6.9) is similar in form to 'Newton's law of cooling' expressed by Eq. (2.36). From Eqs. (6.8) and (6.9) we obtain the *mass transfer coefficient* as

$$h_m = \frac{-\rho D_{12}\left[\frac{dm_1}{dy}\right]_{y=0}}{(m_{1,w}-m_{1,f})}$$

The mass transfer coefficient for different fluids and flow geometries are obtained from correlations analogues to the heat transfer correlations discussed section 2.7.

6.3 Simplified Model for Simultaneous Heat and Mass Transfer

The mass diffusion and mass convection processes considered in the preceding section took place under *isothermal* conditions. However, in most practical direct-contact transfer equipment like humidifiers, cooling coils, and cooling towers mass transfer processes driven by concentration differences, are *coupled* to heat transfer processes driven by temperature differences.

In this section we develop a simplified model for simultaneous heat and mass transfer based on the concept of *enthalpy potential*. The physical situation is depicted schematically in Fig. 6.3.

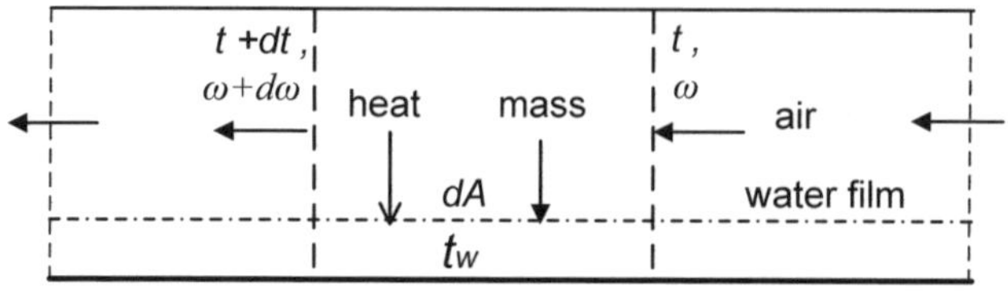

Fig.6.3 Simultaneous heat and mass transfer

Moist air flows steadily over a flat plate maintained at a temperature below the dew-point temperature of the incoming air. A film of water is formed on the plate due to condensation of moisture from the air. The plate surface receives *sensible* heat due convection from the air and *latent* heat due to condensation of water vapor.

The mass flow rate of water vapor, $d\dot{m}_w$, and the sensible heat flow rate, $d\dot{q}_c$, across an elemental area, dA from the air to the water film are given by the respective rate equations for mass transfer and sensible heat transfer as

$$d\dot{m}_w = h_d(\omega - \omega_s)dA \tag{6.10}$$

$$d\dot{q}_c = h_c(t - t_w)dA \tag{6.11}$$

where h_d is the *convective mass transfer coefficient*, and h_c is the *convective heat transfer coefficient.* The average temperature and humidity ratio of the air stream at the section are t and ω respectively. The corresponding properties of the air layer adjacent to the water film are t_w and ω_s respectively.

Notice that in Eq. (6.10) we have approximated the concentrations in the rate equation (6.9) by the respective humidity ratios of water vapor. This approximation is valid for most air conditioning applications due to the relatively low mass concentration of water vapor in ambient air [1].
The rate of latent heat transfer, $d\dot{q}_l$ due to mass transfer of water vapor is given by

$$d\dot{q}_l = h_{fg}d\dot{m}_w \tag{6.12}$$

where h_{fg} is the latent heat of vaporization.
Substituting from Eq. (6.10) in Eq. (6.12) we have

$$d\dot{q}_l = h_{fg}h_d(\omega - \omega_s)dA \tag{6.13}$$

Applying the steady-state sensible and latent energy balance to the elemental control volume we obtain the following equations:

$$d\dot{q}_c = -\dot{m}_a c_{pa} dt \tag{6.14}$$

$$d\dot{q}_l = -\dot{m}_a h_{fg} d\omega \tag{6.15}$$

where $\dot{m}_a$ is the steady mass flow rate of dry air. The changes in temperature and humidity ratio across the elemental control volume *dA* are dt and $d\omega$ respectively.
Manipulating Eqs. (6.11) to (6.15) we obtain

$$\frac{dt}{d\omega} = \frac{h_c(t-t_w)}{c_{pa}h_d(\omega-\omega_s)} = Le\frac{(t-t_w)}{(\omega-\omega_s)} \tag{6.16}$$

In Eq. (6.16) we have introduced a dimensionless quantity called the *Lewis number*, defined as:

$$Le = \frac{h_c}{c_{pa}h_d} \tag{6.17}$$

The Lewis number plays an important role in combined heat and mass transfer processes, occurring in air washers, humidifiers, dehumidifiers and cooling towers.

Now the total rate of energy transfer, $d\dot{q}_t$ to the water film from the air is

$$d\dot{q}_t = d\dot{q}_c + d\dot{q}_l \tag{6.18}$$

Substituting from Eqs. (6.11) and (6.13) in Eq. (6.18) we obtain

$$d\dot{q}_t = h_c(t - t_w)dA + h_{fg}h_d(\omega - \omega_s)dA \tag{6.19}$$

Using Eq. (4. 19), the difference between the mean enthalpy of the air stream, h the saturated air enthalpy at the water temperature, h_s may be expressed as

$$h - h_s = c_{pm}(t - t_w) + h_{go}(\omega - \omega_s) \tag{6.20}$$

Substituting for $(t - t_w)$ from Eq. (6.20) in Eq. (6.19) we obtain

$$d\dot{q}_t = \frac{h_c(h-h_s)dA}{c_{pm}} - \frac{h_c h_{go}(\omega-\omega_s)dA}{c_{pm}} + h_{fg}h_d(\omega - \omega_s)dA \tag{6.21}$$

Assuming that the Lewis number, *Le* = 1, Eq. (6.21) may be written as

$$d\dot{q}_t = \left(\frac{h_c}{c_{pm}}\right)(h - h_s)dA + \left(\frac{h_c}{c_{pm}}\right)(h_{fg} - h_{go})(\omega - \omega_s)dA \qquad (6.22)$$

It is found that for most practical situations involving direct-contact equipment in air conditioning systems, the second term on the RHS of Eq. (6.22) is much smaller than the first term.

For example, when a water film at 36°C looses heat and moisture to ambient air at 24°C and 40% relative humidity, the second term is about 2% of the first term. Moreover, the second term can be interpreted as the enthalpy of the make–up water supplied to maintain a *steady* water film on the plate.

Using the above approximation we express the *total energy transfer rate* from air to water in the form

$$d\dot{q}_t = \left(\frac{h_c}{c_{pm}}\right)(h - h_s)dA \qquad (6.23)$$

where the second term in Eq. (6.22) is neglected.

In Eqs. (6.10) and (6.11), the moisture transfer rate and the sensible heat transfer rate are expressed in terms of the humidity ratio difference and the temperature difference respectively. The latent heat transfer in Eq. (6.13) is proportional to the humidity ratio difference. It is interesting to note that in Eq. (6.23), the *total energy* transfer from the air to the water film is expressed in terms of the enthalpy difference, (h-h_s), usually called the *enthalpy potential.* We shall use Eq. (6.23) to model cooling towers in section 6.5 and cooling coils in chapter 7.

6.4 Air Washers or Humidifiers

An air washer is a device used to humidify air by spraying water as fine droplets into the air stream, as depicted schematically in Fig. 6.4. The water is sprayed with the aid of a series of nozzles, supplied with water from a sump-tank, using an external pump.

Located at the exit of the air washer is a draft eliminator which helps minimize the carry over of water droplets with the air stream. The water that does not evaporate falls to the bottom and is returned to the sump-tank to be recirculated by the pump. Make-up water is supplied from an external source to compensate for the water entering the air stream by evaporation.

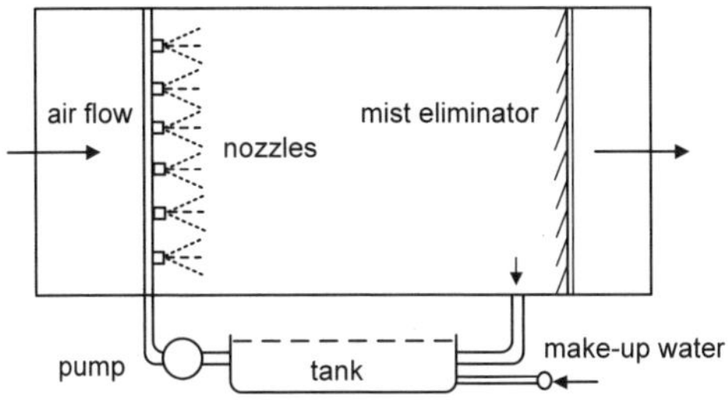

Fig. 6.4 Schematic diagram of an air washer

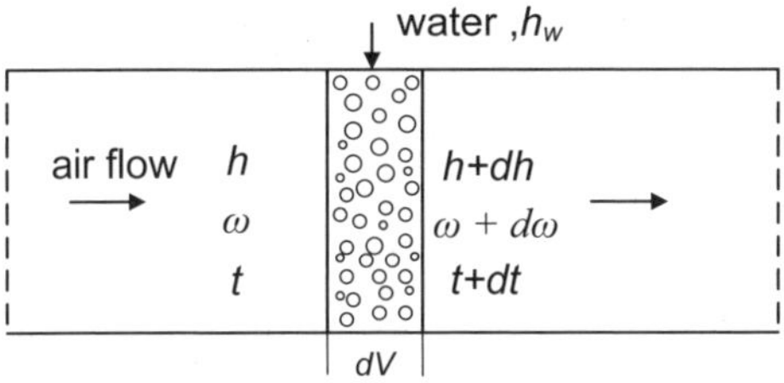

Fig. 6.5 Physical model of the air washer

6.4.1 *Analysis of air washers*

The physical model of the air washer is depicted in Fig. 6.5, where for clarity the water drops are indicated by large circles. The water drops receive heat by convection from the air while the evaporating water releases energy at a rate equal to the enthalpy of evaporation. The mean values of the dry-bulb temperature, the specific enthalpy, and the humidity ratio of the air are denoted by t, h and ω respectively.

Consider the infinitesimal control volume, dV of the flow section in the air washer, shown in Fig. 6.5. Apply the conservation equations of mass and energy to the *overall control volume* dV, neglecting changes in kinetic and potential energy, and any heat interactions with the surroundings. Thus we obtain the following equations:

$$d\dot{m}_{ev} = \dot{m}_a d\omega \tag{6.24}$$

$$\dot{m}_a dh = h_w d\dot{m}_{ev} \tag{6.25}$$

where $\dot{m}_a$ and $\dot{m}_{ev}$ are the dry air flow rate and the evaporation rate of water respectively. The specific enthalpy of water is h_w. From Eqs. (6.24) and (6.25) we have

$$\frac{dh}{d\omega} = h_w \tag{6.26}$$

Using Eq. (4.19) we express the enthalpy of moist air as

$$h = c_{pm}t + h_{go}\omega = (c_{pa} + \omega c_{pw})t + h_{go}\omega \tag{6.27}$$

Differentiate Eq. (6.27) with respect to ω and substitute from Eq. (6.26) to obtain

$$\frac{dt}{d\omega} = -\frac{(h_{go}+c_{pw}t-h_w)}{c_{pm}} = -\frac{h_{fg}}{c_{pm}} \tag{6.28}$$

Now consider the mass and heat transfer between the water drops and the air flowing past them (Fig. 6.5). Let the number water drops per unit volume be n_d and the mean heat and mass transfer area between a drop and air be a_d. The drops receive heat from the air by convection for which the driving potential is the temperature difference between the air and the water. The rate of evaporation from a water drop is proportional to the difference in humidity ratio between the air and saturated air at the water temperature. The water vapor produced by evaporation transfers the net enthalpy of evaporation, $(h_g\text{-}h_f) = h_{fg}$, to the air. The steady-state rate equations for mass and heat transfer from the water drops to the air may be written respectively as

$$d\dot{m}_{ev} = h_d a_d(\omega_s - \omega)n_d dV \tag{6.29}$$

$$d\dot{q}_c = h_c a_d(t - t_w)n_d dV \tag{6.30}$$

where ω_s is the humidity ratio of saturated air at the constant water temperature, t_w and $d\dot{q}_c$ is the rate of convective heat transfer. The heat and mass transfer coefficients are h_c and h_d respectively.

Applying the energy balance equation to the *drops* we have

$$d\dot{q}_c = h_{fg} d\dot{m}_{ev} \tag{6.31}$$

We define the mean mass and heat transfer area per unit volume, A_v as $A_v = n_d a_d$. Substituting in Eq. (6.31) from Eqs. (6.29) and (6.30) we obtain

$$h_d A_v (\omega_s - \omega) h_{fg} dV = h_c A_v (t - t_w) dV \tag{6.32}$$

$h_d A_v$ is called the *volumetric mass transfer coefficient.*
We define the Lewis number as

$$Le = \frac{h_c}{c_{pm} h_d}$$

Hence Eq. (6.32) may be written in the form

$$(\omega_s - \omega) h_{fg} = Le c_{pm} (t - t_w) \tag{6.33}$$

Differentiating Eq. (6.33) with respect to ω we have

$$\frac{dt}{d\omega} = -\frac{h_{fg}}{Le c_{pm}} \tag{6.34}$$

Comparing Eqs. (6.28) and (6.34), we conclude that for an air washer operating under steady-state conditions the Lewis number, $Le = 1$.

Now the wet-bulb temperature, t_{wb} of the incoming air is obtained by applying Eq. (4.39). Hence we have

$$h_{wb}(t_{wb}) = h + [\omega_{wb}(t_{wb}) - \omega] h_{fwb}(t_{wb}) \tag{6.35}$$

Substituting for enthalpy from Eq. (4.19) in Eq. (6.35) and assuming that c_{pm} is constant we obtain

$$c_{pm}(t - t_{wb}) = (\omega_{wb} - \omega)\left(h_{go} - h_{fwb}\right) \cong h_{fg}(\omega_{wb} - \omega) \tag{6.36}$$

We observe that when $Le = 1$, Eq. (6.33) and Eq. (6.36) are equivalent. Therefore we conclude that with *continuous recirculation* of water in an air washer, the water temperature attains the *wet-bulb temperature of the incoming air*. Moreover, comparing Eq. (6.35) with Eq. (4.58) it is clear that the slope of the process line of the air washer on the psychrometric chart follows the constant *wet-bulb temperature* line through the state point of the incoming air.

6.4.2 *Efficiency and number of transfer units* (*NTU*)

The water mass balance equations Eqs. (6.24) and (6.29) can be combined to the form

$$\dot{m}_a d\omega = h_d A_v(\omega_{wb} - \omega)dV$$

$$\frac{d\omega}{dV} = \left(\frac{h_d A_v}{\dot{m}_a}\right)(\omega_{wb} - \omega) \tag{6.37}$$

Solve the first-order differential Eq. (6.37), subject to the boundary conditions: $\omega = \omega_1$ at $V = 0$ and $\omega = \omega_2$ at $V = V_t$, the total volume of the air washer. Hence we have

$$\frac{(\omega_{wb} - \omega_2)}{(\omega_{wb} - \omega_1)} = e^{-NTU} \tag{6.38}$$

where the number of transfer units (NTU) is defined as

$$NTU = \frac{h_d A_v V}{\dot{m}_a} \tag{6.39}$$

The efficiency of the air washer, η_w is defined as the actual rate of evaporation of water to the maximum possible rate of evaporation. The latter condition occurs when the air leaving the air washer is saturated at the wet-bulb temperature of the incoming air. Hence we have

$$\eta_w = \frac{\dot{m}_a(\omega_2 - \omega_1)}{\dot{m}_a(\omega_{wb} - \omega_1)} \tag{6.40}$$

Substituting from Eq. (6.38) in Eq. (6.40) we obtain the following expression for the efficiency

$$\eta_w = 1 - e^{-NTU} \tag{6.41}$$

We can use Eq. (6.33) to express the differences in ω in terms of the differences in t. Thus we transform Eq. (6.40) to the form

$$\eta_w = \frac{(t_1 - t_2)}{(t_1 - t_{wb})} \tag{6.42}$$

Recall that we used Eq. (6.42) in solving some of the worked examples in chapter 5.

6.5 Cooling Towers

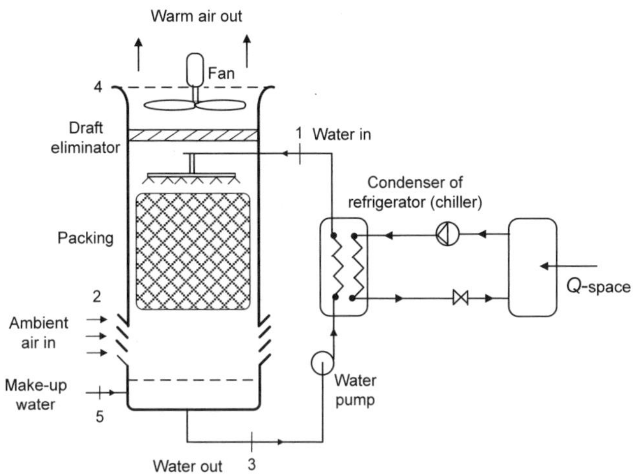

Fig. 6.6 Schematic diagram of cooling tower

A cooling tower, shown schematically in Fig. 6.6, is a direct-contact heat and mass transfer device with wide ranging engineering applications. The main function of cooling towers is to facilitate the rejection of heat from condensers of refrigerators and thermal power plants to the atmosphere.

In the counter-flow arrangement depicted in Fig. 6.6, warm water entering the cooling tower from the condenser of a refrigerating system (chiller plant) at 1 is spayed on to a packing or fill with the aid of a series of nozzles located at the top of the tower. The packing helps distribute the water in the form of a thin film. A fan at the top of the tower draws ambient air through vents located at the bottom of the outer wall. The air moving up through the pores in the packing exchanges heat and mass by direct contact with the water film flowing down.

The water film cools by convective heat transfer to the air, and by supplying the latent heat of vaporization needed to evaporate the water. The cold water collecting at the bottom of the cooling tower is circulated by a pump to the condenser. A constant water level is maintained in the pool at the bottom by a steady supply of make-up water. The draft eliminator located at the top of the tower helps minimize the carryover of water drops with the warm air leaving at the top of the tower.

6.5.1 *Analysis of cooling towers*

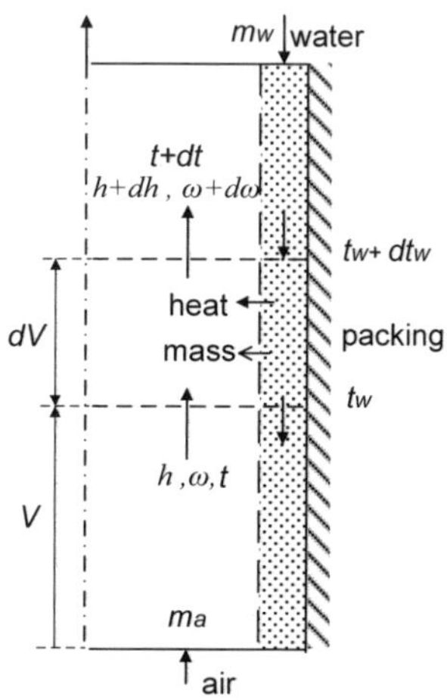

Fig. 6.7 Physical model of a cooling tower

In order to analyze the heat and mass transfer processes in a counter-flow cooling tower we consider the simplified physical model shown in Fig. 6.7. The water flows down by gravity as a thin film over the packing. The air flows up through the open space of the packing.

Consider an elemental volume, dV bounded by two horizontal sections through the tower. The enthalpy and the humidity ratio of the air change due to heat and mass transfer interactions between the water film and the air. Let the contact area for heat and mass transfer per unit volume of the tower be A_v and the heat and mass transfer coefficients be h_c and h_d, respectively.

The air receives heat from the water by convection for which the driving potential is the temperature difference between the water and the air. The rate of evaporation from the water film is proportional to the difference in humidity ratio between the bulk air and saturated air at the water temperature. The water vapor produced by evaporation transfers the enthalpy of evaporation, h_{fg} to the air.

The rate equations for mass transfer, $d\dot{m}_w$, and convective heat transfer, $d\dot{q}_c$, under steady conditions may be written respectively as [4]

$$d\dot{m}_w = h_d A_v (\omega_s - \omega) dV \quad (6.43)$$

$$d\dot{q}_c = h_c A_v (t_w - t) dV \quad (6.44)$$

where ω_s is the humidity ratio of saturated air at the water temperature, t_w.

The heat and mass transfer area per unit volume is A_v.

The conservation equations of mass and energy for the control volume, *dV* under steady conditions are as follows.

Mass balance for water:

$$d\dot{m}_w = \dot{m}_a d\omega \quad (6.45)$$

Sensible heat balance for air:

$$\dot{m}_a c_{pm}(t + dt) = \dot{m}_a c_{pm} t + d\dot{q}_c$$

$$\dot{m}_a c_{pm} dt = d\dot{q}_c \quad (6.46)$$

Energy balance for the water film:

$$(h_w + dh_w)(\dot{m}_w + d\dot{m}_w) = h_w \dot{m}_w + h_g d\dot{m}_w + d\dot{q}_c$$

$$\dot{m}_w dh_w = (h_g - h_w) d\dot{m}_w + d\dot{q}_c$$

$$h_w d\dot{m}_w + \dot{m}_w dh_w = h_g d\dot{m}_w + d\dot{q}_c \quad (6.47)$$

Energy balance for air and water:

$$\dot{m}_a h + (\dot{m}_w + d\dot{m}_w)(h_w + dh_w) = (h + dh)\dot{m}_a + \dot{m}_w h_w$$

$$h_w d\dot{m}_w + \dot{m}_w dh_w = \dot{m}_a dh \quad (6.48)$$

Substituting in Eqs. (6.43) and (6.44) from Eqs. (6.45) and (6.46) respectively we obtain

$$\frac{dt}{d\omega} = Le \frac{(t_w - t)}{(\omega_s - \omega)} \quad (6.49)$$

where *Le* is the Lewis number.

Manipulating Eqs. (6.46), (6.47), and (6.48) we obtain

$$\frac{dh}{d\omega} = h_g + c_{pm} \frac{dt}{d\omega} \quad (6.50)$$

Substituting from Eq. (6.49) in Eq. (6.50) we have

$$\frac{dh}{d\omega} = h_g + c_{pm} Le \frac{(t_w - t)}{(\omega_s - \omega)} \tag{6.51}$$

Express the air enthalpy h, and the enthalpy h_s, of saturated air at the water temperature in terms of Eq. (4.19), assuming c_{pm} is constant. Thus we obtain

$$h_s - h = c_{pm}(t_w - t) + h_{go}(\omega_s - \omega) \tag{6.52}$$

Substituting for $(t_w - t)$ from Eq. (6.52) in Eq. (6.51) we have

$$\frac{dh}{d\omega} = \frac{Le(h_s - h)}{(\omega_s - \omega)} + h_g - Le h_{go} \tag{6.53}$$

Substituting from Eq. (6.45) in Eq. (6.48) we obtain

$$\frac{dt_w}{d\omega} = \left(\frac{dh}{d\omega} - h_w\right)\left(\frac{\dot{m}_a}{\dot{m}_w c_w}\right) \tag{6.54}$$

where the change in enthalpy of water is expressed as, $dh_w = c_w dt_w$.

The distributions of the air enthalpy, h the humidity ratio, ω and the water temperature, t_w along the height of the cooling tower are obtained by solving Eqs. (6.53) and (6.54) simultaneously. The distribution of the dry-bulb temperature of air, t is obtained by solving Eq. (6.49).

The solution of these equations can be carried out graphically using the psychrometric chart [1] or numerically. A MATLAB code to solve the above equations numerically is given in Appendix A6.1. We shall demonstrate the various solution procedures in the worked examples to follow in this chapter.

6.5.2 *Enthalpy potential based model for cooling towers*

In this section we shall develop a simplified model for the cooling tower using the *enthalpy potential* as the driving force for energy transfer [7]. Consider the elemental control volume of the physical model depicted in Fig. 6.7 where the heat and water vapor are transferred from the water to the air across an interfacial area dA. Note that the interfacial area may also be expressed in the form

$$dA = A_v dV$$

where A_v is the heat and mass transfer area per unit volume, and dV is the volume.

The energy balance equations for the air and the water are

$$d\dot{q}_t = \dot{m}_a dh_a = \dot{m}_w c_w dt_w \quad (6.55)$$

The rate equation for the *total energy* transfer between the air and the water is expressed in terms of the *enthalpy potential* following Eq. (6.23). Hence we have

$$d\dot{q}_t = \left(\frac{h_c}{c_{pm}}\right)(h_s - h_a)dA \quad (6.56)$$

In this simplified model we assume the flow rates of air and water are constant. Integration of Eq. (6.55) gives

$$h_a - h_{ai} = \left(\frac{\dot{m}_w c_w}{\dot{m}_a}\right)(t_w - t_{wo}) \quad (6.57)$$

where h_{ai} and t_{wo} are the air enthalpy and water temperature at the bottom of the cooling tower.
From Eqs. (6.55) and (6.56) it follows that

$$\frac{h_c dA}{c_{pm}} = \frac{\dot{m}_w c_w dt_w}{h_s - h_a} \quad (6.58)$$

The required area of the cooling tower to achieve a given change in temperature of the water is obtained by integrating Eq. (6.58). This gives

$$\frac{h_c A}{c_{pm}} = \dot{m}_w c_w \int_{t_{wo}}^{t_{wi}} \frac{dt_w}{h_s - h_a} \quad (6.59)$$

The simultaneous solution of Eqs. (6.57) and (6.59) can be carried out numerically. The procedure will be illustrated in the worked examples.

The quantity $h_c A/c_{pm}$ is called the number of transfer units or *NTU* of the cooling tower. The NTU, which characterizes the performance of the cooling tower, is a function of air and water flow rates and their flow patterns over the packing.

6.5.3 *Approach and range of cooling towers*

The lowest temperature to which water could be cooled in a cooling tower is the wet-bulb temperature of the incoming air. The difference between the actual water outlet temperature and the wet-bulb temperature, commonly called the *approach* of the cooling tower, is an important design parameter of the cooling tower. The higher the NTU of the cooling tower, the lower the approach. The *range* of the cooling tower is the difference between the water inlet and outlet temperatures.

The practical details of air washers, cooling towers, and other direct contact transfer equipment are discussed in the *ASHRAE Handbook - 2012 HVAC Systems and Equipment* [2].

6.6 Property Relations for Moist Air and Water

Design-analysis of cooling towers could be done using the psychrometric chart or numerical procedures. In the former case, all required property data may be obtained directly from the psychrometric chart. However, for numerical computations, property data have to obtained from tabulations available in the literature. Listed below are analytical expressions for some of the more common properties, obtained by fitting polynomials to data tabulated in steam tables [6] for the temperature range from 0°C to 60°C.

(i) Enthalpy of *saturated* moist air, h_s (kJkg^{-1}) at 101.3 kPa is given by [7]:

$$h_s = 9.3625 + 1.786t_s + 1.1135 \times 10^{-2}{t_s}^2 + 9.8855 \times 10^{-4}{t_s}^3 \tag{6.60}$$

where t_s (°C) is thc air temperature.
A quadratic expression for h_s (kJkg^{-1}) that is less accurate but computationally more convenient is given by:

$$h_s = 10.9075 + 1.2205t_s + 5.692 \times 10^{-2}{t_s}^2 \tag{6.61}$$

(ii) Enthalpy of *saturated* water vapor, h_g (kJkg^{-1}) is given by:

$$h_g = 2500.7 + 1.854t - 5.0 \times 10^{-4}t^2 - 6.0 \times 10^{-6}t^3 \tag{6.62}$$

where t (°C) is the vapor temperature.

(iii) Enthalpy of *liquid* water, h_w (kJkg^{-1}) is given by:

$$h_w = 0.002 + 4.198t - 3.0 \times 10^{-4}t^2 \tag{6.63}$$

where t (°C) is the water temperature.

6.7 Worked Examples

Example 6.1 Dry air blows across a wet-bulb thermometer whose bulb is enclosed in a damp cover supplied with water. (i) The thermometer reads 18.3°C. What is the temperature of the *dry* air? (ii) If the air flowing across the bulb is at 32.2°C while the wet-bulb temperature remains at 18.3°C, calculate the relative humidity of the air stream. The ambient pressure is constant at 101.3 kPa. Forced convection heat and mass transfer correlations, give the ratio, $(h_m/h_c) = 1.11$.

Solution In a wet-bulb thermometer, the air flowing over the damp bulb of the thermometer exchanges heat by forced convection with the bulb. This heat supplies the energy needed to evaporate water from the surface of the bulb. The sensible and latent heat rate equations are respectively,

$$Q_{con} = h_c A(t_{db} - t_{wb}) \quad \text{(E6.1.1)}$$

$$Q_{evp} = h_m A(m_{1s} - m_{1e})h_{fg} \quad \text{(E6.1.2)}$$

where A is the area of the bulb. The species 1 is water vapor. The concentration of water vapor at the wet surface, and the air stream are m_{1s} and m_{1e} respectively. The latent heat of vaporization is h_{fg}. Under steady conditions the two energy flow rates given by Eqs. (E6.1.1) and (E6.1.2) are equal. Therefore

$$h_c A(t_{db} - t_{wb}) = h_m A(m_{1s} - m_{1e})h_{fg} \quad \text{(E6.1.3)}$$

(i) Now for the wb-temperature of 18.3°C we obtain the following data from the steam tables [6]: $h_{fg} = 2457$ kJkg^{-1} and $P_g = 2.103$ kPa.

The air at the wet surface is saturated with water vapor. We find the humidity ratio of this air by applying Eq. (4.13) -

$$\omega = \frac{0.622\phi P_g(t)}{P - \phi P_g(t)} = \frac{0.622 \times 2.103}{101.3 - 2.103} = 0.0132$$

The vapor concentration at the surface is

$$m_{1s} = \frac{\omega}{1+\omega} = \frac{0.0132}{1.0132} = 0.013$$

Note that the difference between the humidity ratio, ω and the vapor concentration, m_{1s} is about 1%. Since the air stream is *dry*, $m_{1e} = 0$.

Substituting numerical values in Eq. (E6.1.3) we have

$$(t_{db} - 18.3) = 1.11(0.013 - 0) \times 2457$$

Therefore the dry bulb temperature is 53.75°C.

(ii) Substituting the db–temperature of 32.2°C in Eq. (E6.1.3) we have

$$(32.2 - 18.3) = 1.11(0.013 - m_{1e}) \times 2457$$

Hence the water vapor concentration in the air stream is

$$m_{1e} = 7.903 \times 10^{-3}$$

The humidity ratio is obtained from the relation

$$m_{1e} = \frac{\omega}{1+\omega} = 7.903 \times 10^{-3}$$

The humidity ratio is 7.966×10^{-3}. The saturated vapor pressure at 32.2°C is 4.838 kPa [6].

The relative humidity is obtained by substituting in Eq. (4.13). Hence we have

$$\omega = \frac{0.622\phi P_g(t)}{P-\phi P_g(t)} = \frac{0.622\times 4.838\phi}{101.3-4.838\phi} = 7.966 \times 10^{-3}$$

Therefore relative humidity, ϕ of the air stream is 26.5%

Notc that although the operation of the wet-bulb thermometer is simple, the determination of the relative humidity is somewhat complex.

Example 6.2 Consider the idealized flow arrangement depicted in Fig. 6.3 where air at 30°C and 50% relative humidity flows over a film of water maintained at a constant temperature. The pressure is constant 101.3 kPa. For water temperatures of (i) 28°C, (ii) 20°C and (iii) 13°C, calculate the rates of sensible heat transfer, the latent heat transfer and total energy transfer per unit area of the water film. Assume that the convective heat transfer coefficient is 40 $\mathrm{Wm^{-2}K^{-1}}$.

Solution The rate of sensible heat transfer is given by Eq. (6.11) as

$$d\dot{q}_c = h_c(t_w - t)dA \qquad \text{(E6.2.1)}$$

The rate of latent heat transfer is given by Eq. (6.13) as

$$d\dot{q}_l = \left(\frac{h_c}{c_{pm}}\right)(\omega_s - \omega)h_{fg}dA \tag{E6.2.2}$$

The rate of total heat transfer is given by Eq. (6.23) as

$$d\dot{q}_t = \left(\frac{h_c}{c_{pm}}\right)(h_s - h)dA \tag{E6.2.3}$$

It should be noted that in the above equations we have assumed that the Lewis number, $Le = h_c/h_d c_{pm} = 1$.

We now substitute in Eqs. (E6.2.1)–(E6.2.3) the numerical data pertinent to each of the given sets of conditions to determine the various energy transfer rates per unit area.

For moist air at 30°C and 50% relative humidity we obtain the following properties from the psychrometric chart (Fig. 4.5).

$$t_a = 30°\text{C},\ \omega_a = 0.0133,\ h_a = 64\ \text{kJkg}^{-1},\ c_{pm} = 1.02\ \text{kJkg}^{-1}\text{K}^{-1}$$

(i) For water at 28°C

$$h_{fg} = 2434.8\ \text{kJkg}^{-1}\ [6],\ \omega_s = 0.0241,\ h_s = 90.2\ \text{kJkg}^{-1}$$

$$\frac{d\dot{q}_c}{dA} = h_c(t_w - t) = 40(28 - 30) = -80\ \text{Wm}^{-2}$$

$$\frac{d\dot{q}_l}{dA} = \left(\frac{h_c}{c_{pm}}\right)(\omega_s - \omega)h_{fg} = \frac{40(0.0241-0.0133)2434.8}{1.02} = 1031.2\ \text{Wm}^{-2}$$

$$\frac{d\dot{q}_t}{dA} = \left(\frac{h_c}{c_{pm}}\right)(h_s - h) = \frac{40(90.2-64)}{1.02} = 1027.5\ \text{Wm}^{-2}$$

For the given conditions the sensible heat flow is from air to water while the latent heat flow due to evaporation is from water to air. The net energy flow is from water to air and therefore external heat has to be supplied to maintain the water at a steady temperature.

(ii) For water at 20°C

$$h_{fg} = 2453.7\ \text{kJkg}^{-1}\ [6],\ \omega_s = 0.01468,\ h_s = 57.25\ \text{kJkg}^{-1}$$

$$\frac{d\dot{q}_c}{dA} = h_c(t_w - t) = 40(20 - 30) = -400\ \text{Wm}^{-2}$$

$$\frac{d\dot{q}_l}{dA} = \left(\frac{h_c}{c_{pm}}\right)(\omega_s - \omega)h_{fg} = \frac{40(0.01468-0.0133)2453.7}{1.02} = 132.8\ \text{Wm}^{-2}$$

$$\frac{d\dot{q}_t}{dA} = \left(\frac{h_c}{c_{pm}}\right)(h_s - h) = \frac{40(57.25-64)}{1.02} = -264.7 \text{ Wm}^{-2}$$

For the given conditions the sensible heat flow is from air to water while the latent heat flow due to evaporation is from water to air. However, the net energy flow is from air to water and therefore heat has to be removed to maintain the water at a steady temperature.

(iii) For water at 13°C

$$h_{fg} = 2470.8 \text{ kJkg}^{-1} \text{ [6]}, \omega_s = 0.0094, h_s = 36.8 \text{ kJkg}^{-1}$$

$$\frac{d\dot{q}_c}{dA} = h_c(t_w - t) = 40(13 - 30) = -680 \text{ Wm}^{-2}$$

$$\frac{d\dot{q}_l}{dA} = \left(\frac{h_c}{c_{pm}}\right)(\omega_s - \omega)h_{fg} = \frac{40(0.0094-0.0133)2470.8}{1.02} = -377.9 \text{ Wm}^{-2}$$

$$\frac{d\dot{q}_t}{dA} = \left(\frac{h_c}{c_{pm}}\right)(h_s - h) = \frac{40(36.8-64)}{1.02} = -1066.7 \text{ Wm}^{-2}$$

For the given conditions the sensible heat flow is from air to water. The latent heat flow is negative. This is due to condensation of water vapor from air. The net energy flow is from air to water and therefore heat has to be removed to maintain the water at the steady temperature. This situation corresponds to the energy flows that occur in a cooling and dehumidifying coil.

Example 6.3 Ambient air at 32°C and 50% relative humidity enters an air washer at the rate of 2 kgs^{-1}. The humidity ratio of the air at the exit is 0.0175. The face area of the air washer is 1.1 m^2. For the flow conditions in the air washer, the volumetric mass transfer coefficient, $h_d A_v$ = 1.35 kgs^{-1}m^{-3}. The pressure is 101.3 kPa. Calculate (i) the dry-bulb temperature of the air at the exit, (ii) the efficiency of the air washer, and (iii) the length of the air washer.

Solution The condition of the air through an air washer follows the wet-bulb temperature line through the inlet state. The important requirement is that the water in the air washer is continuously recirculated.

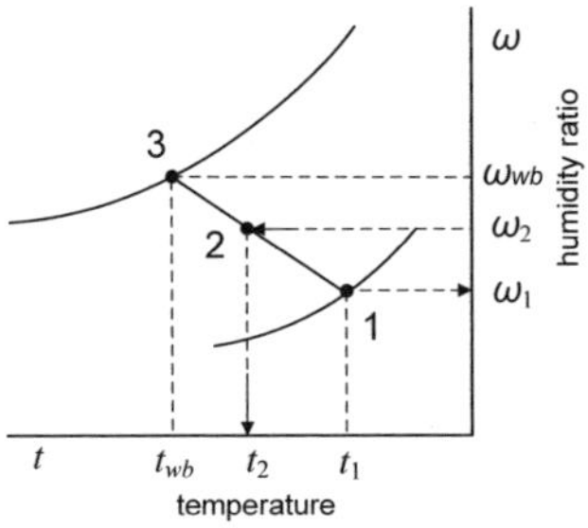

Fig. E6.3.1 Psychrometric chart

From the given values of the temperature and relative humidity we locate the state point 1, of the inlet air, on the psychrometric chart as shown in Fig. E.6.3.1. Draw the constant wet-bulb temperature line 1-3 through 1. At the exit the humidity ratio, $\omega_2 = 0.0175$. Hence we locate state 2 on the line 1-3. The following values obtained directly from the psychrometric chart (Fig. 4.5).

$$\omega_1 = 0.015,\ \omega_{wb} = 0.0185,\ t_2 = 26.0°\text{C}$$

(i) Therefore the dry-bulb temperature at exit is 25.9°C.

(ii) The efficiency of the air washer is given by Eq. (6.40) as

$$\eta_w = \frac{\dot{m}_a(\omega_2 - \omega_1)}{\dot{m}_a(\omega_{wb} - \omega_1)}$$

Substituting numerical values in the above expression we have

$$\eta_w = \frac{(0.0175 - 0.015)}{(0.0185 - 0.015)} = 71.4\%$$

(iii) The efficiency is related to the NTU by the expression

$$\eta_w = 1 - e^{-NTU}$$

Substituting in the above equation, we obtain the NTU = 1.25. Now the NTU is given by Eq. (6.40) as

$$NTU = \frac{h_d A_v V}{\dot{m}_a}$$

Substituting the given numerical data we obtain the volume as V = 1.85m^3. The face area of the air washer is 1.1 m^2. Therefore the length is, 1.85/1.1 = 1.68 m.

Example 6.4 Ambient air at 101.3 kPa, and 38°C enters an air washer whose face area and length are 3.8 m^2 and 1.8 m respectively. The humidity ratio at entry is 0.0115 and the face velocity is 1.8 ms^{-1}. The air exits the air washer with a dry-bulb temperature of 28°C. The temperature of the water in the air washer is 23°C. Determine (i) the efficiency of the air washer using expressions (6.40) and (6.42), and (ii) the value of the volumetric mass transfer coefficient, $h_d A_v$.

Solution The states of the air are depicted in the psychrometric chart in Fig. E6.3.1. We locate the state of air at entry from the given temperature and humidity ratio. The wet-bulb temperature at 1 is 23°C, which is equal to the temperature of the steadily recirculating water in the air washer. State 2, of the exit air is located by the intersection of the 28°C db-temperature line and the 23°C constant wb-temperature line. The following values are obtained directly from the psychrometric chart (Fig. 4.5): $\omega_2 = 0.0157$, $\omega_{wb} = 0.01775$.
The efficiency of the air washer is given by Eq. (6.40) as

$$\eta_w = \frac{\dot{m}_a(\omega_2 - \omega_1)}{\dot{m}_a(\omega_{wb} - \omega_1)}$$

Substituting numerical values in the above expression we have

$$\eta_w = \frac{(0.0157 - 0.0115)}{(0.01775 - 0.0115)} = 67.2\%$$

Substituting numerical data in Eq. (6.42) we obtain

$$\eta_{wt} = \frac{(38-28)}{(38-23)} = 66.7\%$$

The efficiency is related to the NTU by the expression

$$\eta_w = 1 - e^{-NTU}$$

Therefore when the efficiency is 67.2%, the NTU = 1.1147.
The specific volume at 1 is 0.897 m^3kg^{-1}. The mass flow rate of air is given by

$$\dot{m}_a = \frac{A_f v_f}{v_{a1}} = 1.8 \times \frac{3.8}{0.897} = 7.625 \text{ kgs}^{-1}$$

The volume of the air washer is, $V = 3.8 \times 1.8 = 6.84$ m^3
Now the NTU is given by Eq. (6.39) as

$$NTU = \frac{h_d A_v V}{\dot{m}_a}$$

Substituting numerical data in the above expression we obtain the volumetric mass transfer coefficient as, $h_d\, A_v = 1.24$ kgs^{-1}m^{-3}.

Example 6.5 Ambient air at a pressure of 101.3kPa, temperature 30°C and 50% relative humidity enters an air washer with a face velocity of 1.6 ms^{-1}. The face area and NTU of the air washer are 1.6m^2 and 0.6 respectively.
The water in the air washer is continuously recirculated. Calculate (i) the efficiency of the air washer, (ii) the rate of supply of make-up water, and (iii) the temperature and relative humidity of the leaving air.

Solution The states of the air are depicted in the psychrometric chart in Fig. E6.3.1. We locate the state of air at entry from the given temperature and relative humidity. The following values are obtained directly from the psychrometric chart (Fig. 4.5):

$$\omega_1 = 0.0134,\ \omega_{wb} = 0.0167,\ v_{a1} = 0.877 \text{ m}^3\text{kg}^{-1}$$

The efficiency is related to the NTU by the expression

$$\eta_w = 1 - e^{-NTU}$$

The NTU is given as 0.6. Substituting in the above equation we obtain the efficiency = 45.1%.
The efficiency of the air washer is given by Eq. (6.40) as

$$\eta_w = \frac{\dot{m}_a(\omega_2 - \omega_1)}{\dot{m}_a(\omega_{wb} - \omega_1)}$$

Substituting numerical values in the above expression we have

$$\frac{(\omega_2 - 0.0134)}{(0.0167 - 0.0134)} = 0.451$$

Therefore, $\omega_2 = 0.0149$. The state 2 of the outlet air is located by the intersection of the constant wb-temperature line through 1 and the humidity ratio line of 0.0149. We read the temperature and relative humidity at 2 as 26.2°C and 69%.
The mass flow rate of dry air is

$$\dot{m}_a = \frac{A_f v_f}{v_{a1}} = 1.6 \times \frac{1.6}{0.877} = 2.919 \text{ kgs}^{-1}$$

The rate of supply of make-up water is given by

$$\Delta \dot{m}_w = \dot{m}_a(\omega_2 - \omega_1)$$

Substituting numerical values in the above expression we have

$$\Delta \dot{m}_w = 2.191(0.0149 - 0.0134) = 0.00438 \text{ kgs}^{-1}$$

Example 6.6 The dry-bulb and wet-bulb temperatures of air at entry to an air washer are 38°C and 18°C respectively. The mass flow rate is 1.8 kgs^{-1} and the face velocity is 2 ms^{-1}. The volume is 2m^3. The efficiency of the air washer is 70%. Determine (i) the dry bulb temperature and relative humidity of the leaving air, (ii) the volumetric mass transfer coefficient, $h_d A_v$, (iii) the length of the air washer, and (iv) the efficiency using Eq. (6.42).

Solution The states of the air are depicted in the psychrometric chart in Fig. E6.3.1. We locate the state of air at entry from the given dry-bulb and wet-bulb temperatures. The following values are obtained directly from the psychrometric chart (Fig. 4.5)

$$\omega_1 = 0.00475,\ \omega_{wb} = 0.013,\ v_{a1} = 0.8875 \text{ m}^3\text{kg}^{-1}$$

The efficiency of the air washer is given by Eq. (6.40) as

$$\eta_w = \frac{\dot{m}_a(\omega_2 - \omega_1)}{\dot{m}_a(\omega_{wb} - \omega_1)}$$

Substituting numerical values in the above expression we have

$$\frac{(\omega_2 - 0.00475)}{(0.013 - 0.00475)} = 0.7$$

Therefore, $\omega_2 = 0.01053$.

The state 2 of the outlet air is located by the intersection of the constant wb-temperature line through 1 and the humidity ratio line of 0.01053. From the psychrometric chart we obtain the temperature and relative humidity at 2 as 23.7°C and 58% respectively.

The efficiency is related to the NTU by the expression

$$\eta_w = 1 - e^{-NTU}$$

Substituting for efficiency in the above equation we obtain the, NTU = 1.204.

Now the NTU is given by Eq. (6.39) as

$$NTU = \frac{h_d A_v V}{\dot{m}_a}$$

Substituting numerical data in the above expression we obtain the mass transfer coefficient as, $h_d\,A_v = 1.08$ kgs^{-1}m^{-3}.

The specific volume at 1 is 0.8875 m^{3}kg^{-1}. The mass flow rate of air is given by

$$\dot{m}_a = \frac{A_f v_f}{v_{a1}} = A_f \times \frac{2}{0.8875} = 1.8 \text{ kgs}^{-1}$$

Therefore the face area is 0.8 m^2.

The length of the air washer is, $L = 2/0.8 = 2.5$ m.

Substituting numerical data in Eq. (6.42) we obtain

$$\eta_{wt} = \frac{(38-23.7)}{(38-18)} = 71.5\%$$

Example 6.7 Saturated ambient air at 2°C enters a preheater of a winter air conditioning system with a volume flow rate of 4.5 m^{3}s^{-1}. The air then passes through an air washer where water is continuously recirculated. The face velocity is 2.5ms^{-1} and the mass transfer coefficient, $h_d\,A_v = 1.8$ kgs^{-1}m^{-3}. The dry-bulb temperature and degree of saturation of the air leaving the air washer are 20°C and 0.7 respectively. Calculate (i) the wet-bulb and dry-bulb temperatures of the air at entry to the air washer, (ii) the heat input rate of the preheater, (iii) the rate of supply of make-up water for the air washer, and (iii) the volume of the air washer.

Solution The states of the air as it passes through the preheater and the air washer are depicted in the psychrometric chart in Fig. E6.7.1. We locate the initial state 1 of the air on the saturation curve at 2°C. During the heating process in the preheater, the humidity ratio of the air is constant at 0.0044. Therefore the state of the air leaving the preheater should lie on the horizontal line through 1.

Now the temperature of the air leaving the air washer at 3 is 20°C and the degree of saturation is 0.7. The 20°C constant db-temperature line intersects the saturation curve at 4, where the humidity ratio is 0.01475.

Therefore the humidity ratio of the air leaving the washer at state 3 is, 0.7×0.01475 = 0.0103.

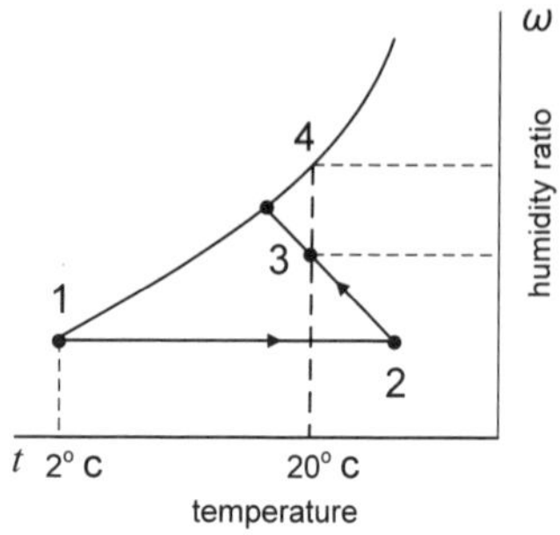

Fig. E6.7.1 Psychrometric chart

State 3 is located by the intersection of the 20°C constant db-temperature line and the 0.0103 humidity ratio line. The process 2-3 in the air washer follows the constant wet-bulb temperature line through 3. The intersection of this line and the horizontal line through 1 locates the state 2 of the air entering the air washer.

From the psychrometric chart (Fig. 4.5) we obtain the dry-bulb and wet-bulb temperatures at 2 as 34°C and 16.4°C respectively. The mass flow rate of air is (4.5/0.784) = 5.74 kgs^{-1}. The heat supply rate in the preheater is given by

$$\dot{Q}_{ph} = \dot{m}_a(h_2 - h_1)$$

We obtain the air enthalpies at 1 and 2 from the psychrometric chart. Substituting these values in the above equation we have

$$\dot{Q}_{ph} = 5.74(45 - 13) = 186.5 \text{ kW}$$

The rate of supply of make-up water is given by

$$\dot{m}_w = \dot{m}_a(\omega_3 - \omega_2)$$

Substituting numerical values we have

$$\dot{m}_w = 5.74(0.0103 - 0.0044) = 0.0338 \text{ kgs}^{-1}$$

The efficiency of the air washer is given by Eq. (6.40) as

$$\eta_w = \frac{\dot{m}_a(\omega_3 - \omega_2)}{\dot{m}_a(\omega_{wb} - \omega_2)}$$

Substituting numerical values in the above expression we have

$$\eta_w = \frac{(0.0103-0.0044)}{(0.0117-0.0044)} = 0.808$$

The efficiency is related to the NTU by the expression

$$\eta_w = 1 - e^{-NTU}$$

Substituting for efficiency in the above equation we obtain the, NTU = 1.65.

Now the NTU is given by Eq. (6.39) as

$$NTU = h_d A_v V / \dot{m}_a$$

Substituting numerical data in the above expression we obtain the volume of the air washer as 5.26 m^3.

Example 6.8 A water-cooled condenser of a central air conditioning system rejects heat to the ambient through a counter-flow cooling tower. Water enters the cooling tower at 35°C and leaves at 22°C. The mass flow rate of water is 19 kgs^{-1}. Ambient air enters the cooling tower at 20°C and 60% relative humidity. The air leaves at 30°C and 100% relative humidity. The temperature of the make-up water is 25°C. The pressure is constant at 101.3 kPa. Calculate (i) the mass flow rate of air, (ii) the rate of supply of make-up water, and (iii) the heat rejection rate in the condenser.

Solution In this example we consider the overall performance of a cooling tower using the control volume approach. We refer to the schematic diagram depicted in Fig. 6.6 where the various entry and exit ports are indicated. The following properties of moist air are obtained from the psychrometric chart (Fig. 4.5).

$$h_{a2}=42.5 \text{ kJkg}^{-1},\ \omega_{a2}=0.00875,\ h_{a4}=101 \text{ kJkg}^{-1},\ \omega_{a4}=0.02725$$

The water mass balance equation gives

$$\dot{m}_{w5} = \dot{m}_a(\omega_4 - \omega_2) = 0.0185\dot{m}_a \qquad \text{(E6.8.1)}$$

Apply the steady-flow energy equation to the cooling tower, neglecting changes in kinetic and potential energy of the streams, and any heat exchange with the surroundings. Thus we have

$$\dot{m}_a(101 - 42.5) = \dot{m}_w \times 4.19(35 - 22) + \dot{m}_{w5} \times 4.19 \times 25 \quad \text{(E6.8.2)}$$

Solving Eqs. (E6.8.1) and (E6.8.2) simultaneously we obtain

$$56.56\dot{m}_a = 1034.93$$

Therefore the mass flow rate of air is 18.3 kgs^{-1}.
Substituting in Eq. (E6.8.1) we obtain the mass flow rate of make-up water as 0.345 kgs^{-1}.

The heat rejection rate of the condenser is given by

$$\dot{Q}_c = 19 \times 4.19(53 - 22) = 1034.9 \text{ kW}$$

Example 6.9 Hot water with a mass flow rate of 19 kgs^{-1} enters a counter-flow cooling tower at 35°C and leaves at 22°C. Ambient air with a mass flow rate of 18.6 kgs^{-1} enters the cooling tower at 20°C and 60% relative humidity. The air leaves at 30°C and 100% relative humidity. The pressure is constant at 101.3 kPa. Outline a graphical procedure using the psychrometric chart to obtain the condition line of the air passing through the cooling tower.

Solution

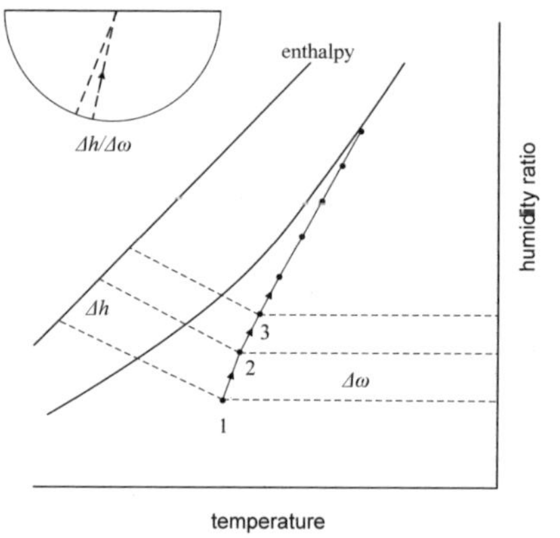

Fig. E6.9.1 Graphical method for cooling tower design

The graphical procedure to solve the equations governing the properties of air and water in a counter-flow cooling tower is illustrated in the psychrometric chart in Fig. E6.9.1. The enthalpy of moist air and the temperature of water are given by Eqs. (6.53) and (6.54) as follows:

$$\frac{dh}{d\omega} = \frac{Le(h_s - h)}{(\omega_s - \omega)} + h_g - Leh_{go} \quad \text{(E6.9.1)}$$

$$\frac{dt_w}{d\omega} = \left(\frac{dh}{d\omega} - h_w\right)\left(\frac{\dot{m}_a}{\dot{m}_w c_w}\right) \quad \text{(E6.9.2)}$$

These equations maybe written in the following numerical form in terms of finite changes in enthalpy and temperature:

$$\frac{\Delta h}{\Delta\omega} = \frac{Le(h_s - h)}{(\omega_s - \omega)} + h_g - Leh_{go} \quad \text{(E6.9.3)}$$

$$\Delta t_w = \left(\frac{dh}{d\omega} - h_w\right)\left(\frac{\dot{m}_a}{\dot{m}_w c_w}\right)\Delta\omega \quad \text{(E6.9.4)}$$

For forced convection of air the Lewis number is given by, $Le = (\alpha/D)^{2/3}$ where α and D are the thermal diffusivity, and the moisture diffusivity respectively [4]. We use the data in Ref. [4] to obtain the Lewis number at the mean air temperature of 25°C as 0.897.

The state of air at the entry section 1 to the cooling tower is located by the given temperature and relative humidity. The following properties of air at 1, are obtained from the psychrometric chart: $h_{a1} = 42.2$ kJkg^{-1}, $\omega_1 = 0.00873$.

The water leaving at 1 is at 22°C. The following properties of saturated air at 22°C are obtained directly from the psychrometric chart. $h_{s1} = 64.7$ kJkg^{-1}, $\omega_{s1} = 0.0168$.

From the steam tables [6] we obtain the following properties of water. At 22°C, $h_{g1} = 2541.2$ kJkg^{-1} and $h_{w1} = 92.2$ kJkg^{-1}. The enthalpy of water vapor at 0°C is given by $h_{go} = 2467.8$ kJkg^{-1}.

We now substitute the above numerical values in Eq. (E6.9.1) to obtain the enthalpy-humidity ratio, $(dh/d\omega)$ at 1 as

$$\frac{dh}{d\omega} = \frac{0.897(64.7 - 42.2)}{(0.0168 - 0.00873)} + 2541.2 - 0.897 \times 2467.8 = 2830$$

Applying the overall energy balance equation to the cooling tower we obtain the change of air enthalpy as

$$h_{out} - h_{in} = \dot{m}_w c_w (t_{wi} - t_{wo})/\dot{m}_a = 55.6 \text{ kJkg}^{-1}$$

For the graphical construction consider 10 equal increments of air enthalpy of Δh = 5.56 kJkg^{-1}. We start by drawing a straight line through point 1with a slope of 2830 kJkg^{-1}, using the enthalpy-humidity ratio protractor. This line intersects the constant enthalpy line for 47.76 (42.2+5.56) kJkg^{-1} at point 2 on the psychrometric chart. The humidity ratio at 2, ω_2 = 0.0107. The change in water temperature is obtained by substituting in Eq. (E6.9.4). This gives

$$\Delta t_w = (2830 - 92.2)\left[\frac{18.6}{4.19\times 19}\right] \times (0.0107 - 0.00873) = 1.26$$

Therefore the water temperature at 2 is t_w = 1.26+22 = 23.26°C. The following properties of saturated air at 23.26°C are obtained directly from the psychrometric chart. h_{s2} = 69.5 kJkg^{-1}, ω_{s2} = 0.0182.

From the steam tables [6] we obtain the following properties of water h_{g2} = 2543.6 kJkg^{-1} and h_{w2} = 97.5 kJkg^{-1}.

We substitute the above numerical values in Eq. (E6.9.1) to obtain the enthalpy-humidity ratio, $(dh/d\omega)$ at 2 as

$$\frac{dh}{d\omega} = \frac{0.897(69.5 - 47.76)}{(0.0182 - 0.0107)} + 2543.6 - 0.897 \times 2467.8 = 2930$$

We now draw a straight line through 2 with a slope of 2930 kJkg^{-1} using the protractor. The intersection of this line with the constant enthalpy line for 53.3 = (47.76+5.56) kJkg^{-1} gives point 3 on the psychrometric chart.

The humidity ratio at 3, ω_3 = 0.0126. The change in water temperature is obtained by substituting in Eq. (E6.9.4). This gives

$$\Delta t_w = (2930 - 97.5)\left[\frac{18.6}{4.19\times 19}\right] \times (0.0126 - 0.0107) = 1.257$$

Therefore the water temperature at 3 is t_w= 1.257+23.26 = 24.5°C.

The following properties of saturated air at 24.5°C are obtained directly from the psychrometric chart. h_{s2} = 74.53 kJkg^{-1}, ω_{s2} = 0.0196.

From the steam tables [6] we obtain the following properties of water. h_{g2} = 2545.9 kJkg^{-1} and h_{w2} = 102.7 kJkg^{-1}.

The procedure outlined above is carried out to locate point 4 on the chart. At the last point, 11 on the *condition line* the air enthalpy is 97.8 kJkg^{-1}. The numerical data obtained in the graphical construction are summarized in Table E6.9.1. The difference in water temperature at point 11 is due to inaccuracies in the graphical procedure.

Table E6.9.1 Numerical data from graphical construction

Point	h_{air}	ω_{air}	ω_{sat}	$dh/d\omega$	t_{water}
1	42.2	0.00873	0.0168	2833	22
2	47.7	0.0107	0.0182	2940	23.26
3	53.3	0.0126	0.0196	3032	24.5
4	58.9	0.0144	0.0212	3105	25.77
5	64.4	0.0162	0.0229	3156	27.02
6	70.0	0.0180	0.0247	3189	28.28
7	75.5	0.0197	0.0266	3205	29.53
8	81.1	0.0215	0.0286	3208	30.78
9	86.7	0.0232	0.0307	3203	32.03
10	92.2	0.0249	0.0330	3192	33.27
11	97.8	0.0267	0.0354	-	34.54

Example 6.10 Consider the cooling tower described above in worked example 6.9 with the same conditions of operation. Assume that the volumetric mass transfer coefficient, $h_d A_v$ = 0.536 kgs^{-1}m^{-3}. Calculate the volume of the cooling tower.

Solution In worked example 6.9 we obtained the condition line for air which is a plot of air temperature versus humidity ratio. The data are tabulated in Table E6.9.1. In order to calculate the volume of the cooling tower we use the following procedure.

The rate equation for water mass transfer is

$$d\dot{m}_w = h_d A_v(\omega_s - \omega)dV \qquad \text{(E6.10.1)}$$

where ω_s is the humidity ratio of saturated air at the water temperature.

The mass conservation equation for water for the control volume, dV under steady conditions is

$$d\dot{m}_w = \dot{m}_a d\omega \qquad \text{(E6.10.2)}$$

From Eqs. (E6.10.1) and (E6.10.2) we have

$$dV = \frac{\dot{m}_a d\omega}{h_d A_v(\omega_s - \omega)} \qquad \text{(E6.10.3)}$$

The volume of the cooling tower is obtained by integrating Eq. (E6.10.3). Hence we have

$$V = \int_{in}^{out} \frac{\dot{m}_a d\omega}{h_d A_v(\omega_s - \omega)} \tag{E6.10.4}$$

The integral maybe evaluated using the trapizoidal rule as illustrated in the Fig. E6.10.1 below. The integral is expressed as a sum of the areas of the trapeziums. This gives

$$V = \left(\frac{\dot{m}_a}{h_d A_v}\right) \sum_1^N \left[\left(\frac{1}{\omega_{s,i+1} - \omega_{i+1}}\right) - \left(\frac{1}{\omega_{s,i} - \omega_i}\right)\right] \frac{(\omega_{i+1} - \omega_i)}{2}$$

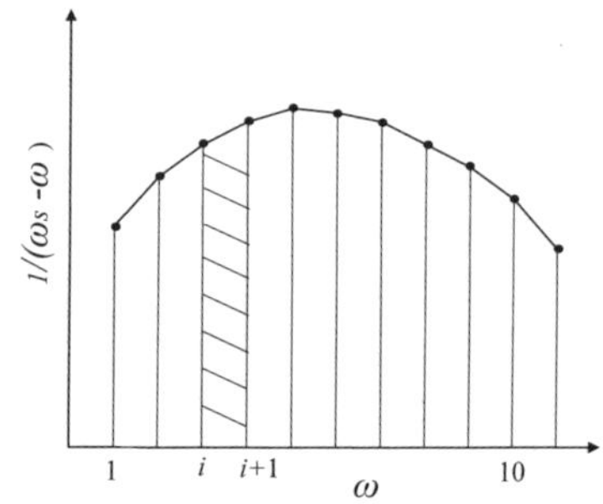

Fig. E6.10.1 Evaluation of integral

The numerical data required for evaluating the above summation are available in Table E6.9.1. Hence we obtain the volume of the cooling tower as 86.1m^3.

Example 6.11 The water and air mass flow rates of a counter-flow cooling tower are 18.5 kgs^{-1} and 15.5 kgs^{-1} respectively. Water enters the cooling tower at 35°C and leaves at 28°C. Ambient air enters the cooling tower at 32°C dry-bulb temperature, and 55% relative humidity. The mass volumetric transfer coefficient, $h_d A_v = 0.54$ kgs^{-1}m^{3}.

(i) Write a numerical algorithm to analyze the counter-flow cooling tower using the detailed model.

(ii) Obtain the condition line for air.

(iii) Obtain the dry-bulb temperature of air.

(iv) Calculate the volume of the cooling tower.

Solution In the detailed model of the counter-flow cooling tower, the enthalpy of moist air, the temperature of water, and the dry-bulb temperature of air are given by Eqs. (6.53), (6.54) and (6.49) as follows:

$$\frac{dh}{d\omega} = \frac{Le(h_s - h)}{(\omega_s - \omega)} + h_g - Leh_{go} \quad \text{(E6.11.1)}$$

$$\frac{dt_w}{d\omega} = \left(\frac{dh}{d\omega} - h_w\right)\left(\frac{\dot{m}_a}{\dot{m}_w c_w}\right) \quad \text{(E6.11.2)}$$

$$\frac{dt}{d\omega} = Le\frac{(t_w - t)}{(\omega_s - \omega)} \quad \text{(E6.11.3)}$$

Equations (E6.11.1) and (E6.11.2) are two coupled first-order differential equations in h and t_w. These equation may be solved by adopting the Euler–Cauchy(E–C) method. Although there are several improvements to the E–C method, we shall develop our computer algorithm based on the E–C method. The important steps are given below.

(i) Divide the difference in water temperature at the top and bottom of the tower into N equal intervals, Δt_w. These divisions represent the nodal points 1 to (N+1). In general, $[t_{w,(i+1)} - t_{w,i}] = \Delta t_w$.

(ii) Equations (E6.11.1) and (E6.11.2) are now expressed in finite-difference form using the nodal values of the variables in the equations. This gives

$$\frac{h_{(i+1)} - h_i}{\omega_{(i+1)} - \omega_i} = \frac{Le(h_{s,i} - h_i)}{(\omega_{s,i} - \omega_i)} + h_{g,i} - Leh_{go} \quad \text{(E6.11.4)}$$

$$t_{w,(i+1)} - t_{w,i} = \left(\frac{h_{(i+1)} - h_i}{\omega_{(i+1)} - \omega_i} - h_{w,i}\right)\left(\frac{\dot{m}_a}{\dot{m}_w c_w}\right)\left[\omega_{(i+1)} - \omega_i\right] \quad \text{(E6.11.5)}$$

$$\frac{t_{(i+1)} - t_i}{\omega_{(i+1)} - \omega_i} = \frac{Le(t_{w,i} - t_i)}{(\omega_{s,i} - \omega_i)} \quad \text{(E6.11.6)}$$

(iii) To evaluate the properties of saturated air and water we use the polynomial expressions given in section 6.6.

The enthalpy of *saturated* moist air, $h_{s,i}$ (kJkg^{-1}) at the nodal water temperature, $t_{w,i}$ (°C) is given by Eq. (6.60).

The enthalpy of *saturated* water vapor, $h_{g,i}$ (kJkg^{-1}) at the nodal water temperature, $t_{w,i}$ (°C) given by Eq. (6.62).

The humidity ratio of saturated air at water temperature, $t_{w,i}$ (°C) is

$$\omega_{s,i} = \frac{(h_{s,i} - c_{pa} t_{w,i})}{h_{g,i}}$$

The enthalpy of *liquid* water, $h_{w,i}$ (kJkg^{-1}) is given by Eq. (6.63).

(iv) The design data for the given problem are: $\dot{m}_a = 15.5$ kgs^{-1}, $\dot{m}_w = 18.5$ kgs^{-1}, $c_w = 4.19$ kJkg^{-1}K^{-1}, $Le = 0.897$, $h_{go} = 2467.8$ kJkg^{-1}.

The boundary conditions for node 1, at the bottom of the tower are: t_{w1}=28°C, t_1 =32.0°C, $\phi_1 = 55\%$, $h_1 = 74.12$ kJkg^{-1}, $\omega_1 = 0.01648$, $\omega_{s1} = 0.02428$, $h_{s1} = 89.97$ kJkg^{-1}, $h_{g1} = 2552.1$ kJkg^{-1}.

For node (N+1) at the top of the tower, $t_{w,(N+1)} = 35$°C. Choose the number of tower sections as, N = 10. Therefore $\Delta t_w = (35 - 28)/10 = 0.7$°C.

(v) The computation begins with node, i =1. Substitute the relevant numerical values for node 1, listed above, in the RHS of Eq. (E6.11.4) to evaluate the LHS of the equation. Substitute the resulting value of the slope ($\Delta h/\Delta\omega$) in Eq. (E6.11.5) to calculate ω_2 at node 2. Substitute in Eq. (E6.11.4) to calculate h_2. Substitute in Eq. (E6.11.6) to calculate the dry-bulb temperature t_2. Use the polynomial expressions listed in (iii) above to calculate all the other relevant properties at node 2.

(vi) Repeat the above procedure to calculate the properties at all the nodes up to node 11 at the top of the tower.

(vii) The computational procedure was carried using the MATLAB software package. The code in listed in Appendix A6.1 at the end of this chapter. The computed results are summarized in Table E6.11.1.

Table E6.11.1 Numerical data from computation

Node	t_{water}	ω_{air}	ω_{sat}	$t_{dry\ bulb}$
1	28	0.01648	0.02428	32
2	28.7	0.0182	0.0253	31.2
3	29.4	0.0198	0.0264	30.7
4	30.1	0.0213	0.0275	30.4
5	30.8	0.0228	0.0286	30.36
6	31.5	0.0242	0.0298	30.45
7	32.2	0.0255	0.0310	30.68
8	32.9	0.0268	0.0323	31
9	33.6	0.0281	0.0336	31.4
10	34.3	0.0293	0.0349	31.8
11	35	0.0305	0.0363	32.3

(viii) The volume of the cooling tower is obtained using the following expression derived in worked example 6.10:

$$V = \left(\frac{\dot{m}_a}{h_d A_v}\right) \sum_1^N \left[\left(\frac{1}{\omega_{s,i+1} - \omega_{i+1}}\right) - \left(\frac{1}{\omega_{s,i} - \omega_i}\right)\right] \frac{(\omega_{i+1} - \omega_i)}{2}$$

This gives the volume of the cooling tower as 66.9m^3.

Example 6.12 Consider the cooling tower described in worked example 6.11 with the same operating conditions. Use the computer program developed to study the effect of, (i) the number of volume sections, (ii) the mass transfer coefficient, and (iii) the assumed Lewis number, on the computed tower volume.

Solution The results presented in worked example 6.11 are based on several assumed parameters. In design simulations it is important to examine the sensitivity of the predictions of the simulation to values of these parameters. The sensitivity study is carried out by varying each parameter in turn over a range. Tabulate below are the results of this sensitivity study.

(i) Effect of the number of volume sections or nodes on the volume of the cooling tower:

N	5	10	15	20	25	50
V(m^3)	68.3	66.96	66.55	66.35	66.23	66.0

We notice that when the number of nodes exceeds about 10, the predicted volume of the cooling tower is nearly constant.

(ii) Effect of Lewis number (*Le*) on the volume:

Le	0.86	0.88	0.89	0.91	0.92	0.93
V(m^3)	67.0	66.98	66.98	66.94	66.92	66.91

It is clear from the data in the table above that, the tower volume is relatively insensitive to the value of Lewis number. It is therefore reasonable to choose the value corresponding to the average air temperature.

(iii) Effect of the mass transfer coefficient (h_dA_v) on the volume:

h_dA_v	0.5	0.52	0.54	0.56	0.58	0.6
V(m^3)	72.3	69.5	66.9	64.5	62.35	60.27

We notice from the data tabulated above that the tower volume is strongly dependent on the value of the volumetric mass transfer coefficient. This is evident from the expression in Eq. (E6.10.4), in which the volume is inversely proportional to the volumetric mass transfer coefficient.

Example 6.13 Consider the cooling tower described in worked example 6.11 with the same mass flow rates of air and water but different operating conditions for water and air. Water enters the cooling tower at 38°C, and ambient air enters the cooling tower at 34°C dry-bulb temperature and 50% relative humidity. The volume of the cooling tower is 66.9 m^3 and the volumetric mass transfer coefficient, h_dA_v = 0.54 kgs^{-1}m^{-3}. Use the computer program developed in worked example 6.11 to determine the outlet temperature of water.

Solution In this example we are considering the same cooling tower as in worked example 6.11, whose volume is specified. However, the water and air inlet conditions have changed due to changes in operating conditions. We use the same computer code, given in

Appendix A6.1, in an iterative manner to simulate the given *off-design* conditions.

In addition to the new air and water inlet conditions given, we make an initial guess of the water outlet temperature. We then run the program to determine the cooling tower volume. The water outlet temperature is varied in small steps until the tower volume is equal to the design volume of 66.9 m^3, obtained in worked example 6.11. The results of the iterations are summarized below in Table E6.13.1.

E6.13.1 Effect of water temperature (t_{wo}) on the tower volume

t_{wo}(°C)	29	30	29.5	29.1	29.145
V(m^3)	70.9	48.6	55.6	68.1	66.9

The conditions of air and water under the given off-design conditions are summarized in Table E6.13.2.

Table E6.13.2 Results for off-design operating conditions

Node	t_{water}	ω_{air}	ω_{sat}	$t_{dry\ bulb}$
1	29.145	0.01677	0.0260	34
2	30.04	0.01895	0.02739	32.97
3	30.92	0.0210	0.0288	32.33
4	31.81	0.0229	0.03034	32.02
5	32.69	0.0247	0.03196	31.97
6	33.58	0.0265	0.03355	32.13
7	34.46	0.0281	0.03525	32.43
8	35.35	0.0298	0.0370	32.86
9	36.23	0.0314	0.0336	33.35
10	37.1	0.0329	0.0408	33.9
11	38	0.0345	0.0427	34.48

Example 6.14 Water entering a counter-flow cooling tower with a mass flow rate of 18.5 kgs^{-1} is cooled from 35°C to 28°C. The conditions of air at entry are 32°C and 55% relative humidity. The air flow rate is 15.5 kgs^{-1}. Assume that the volumetric mass transfer coefficient, $h_d A_v$ = 0.54 $kgs^{-1}m^{-3}$. Use the simplified model to calculate the volume of the cooling tower.

Solution Using the approximate model, the required volume is obtained by evaluating the integral in Eq. (6.59). We assume that the

Lewis number is one. Also, the heat transfer area, $dA = A_v\, dV$. Hence we have

$$h_d A_v V = \dot{m}_w c_w \int_{t_{wo}}^{t_{wi}} \frac{dt_w}{h_s - h_a} \qquad \text{(E6.14.1)}$$

We divide the total water temperature drop across the tower into N equal intervals. The approximate form of the energy balance for water and air gives the enthalpy change of air as

$$\dot{m}_a \Delta h_a = \dot{m}_w c_w \Delta t_w$$

We express Eq. (E6.14.1) in the finite difference form

$$h_d A_v V = \dot{m}_w c_w \Delta t_w \sum_{i=1}^{N} \frac{1}{(h_{sm,i} - h_{am,i})} \qquad \text{(E6.14.2)}$$

where the mean value of the enthalpy for the temperature interval, i is expressed as, $h_{m,i} = 0.5(h_{i+1} + h_i)$. This applies both to the air enthalpy and the saturation enthalpy, which are denoted by the subscripts a and s respectively. For the numerical solution we consider 10 equal temperature intervals for water. The pertinent numerical data from the computation are tabulated below.

Table E6.14.1 Numerical data from computation

Node	t_{water}	h_{air}	h_{sat}	h_{sm}	h_{am}	$1/(h_{sm}-h_{am})$
1	28	74.18	89.97	91.65	75.93	0.06358
2	28.7	77.68	93.34	95.07	79.43	0.06392
3	29.4	81.18	96.80	98.59	82.93	0.06387
4	30.1	84.68	100.37	102.2	86.43	0.06343
5	30.8	88.18	104.0	105.9	89.93	0.06260
6	31.5	91.68	107.78	109.7	93.43	0.06141
7	32.2	95.18	111.65	113.6	96.93	0.05988
8	32.9	98.68	115.6	117.6	100.4	0.05808
9	33.6	102.18	119.7	121.8	103.9	0.05604
10	34.3	105.68	123.87	126.0	107.4	0.05381
11	35	109.18	128.16	-	-	

Substituting numerical values in Eq. (E6.14.2) we have

$$0.54V = 18.5 \times 4.19 \times 0.7 \times 0.6066$$

Hence the volume of the cooling tower is 60.96 m^3. It should be noted that the design and operating conditions in this example are the same as those in worked example 6.11. The exact model used in example 6.11 gave a cooling tower volume of 66.9 m^3 while the approximate model used in this example predicted a volume of 60.96 m^3.

Example 6.15 Outline a numerical solution procedure based on the Euler-Heun method to solve the governing equations of a counter-flow cooling tower. Develop a MATLAB code to implement the procedure. Use the program to compute the volume of the cooling tower described in worked example 6.11.

Solution In the detailed model of the counter-flow cooling tower, the enthalpy of moist air and the temperature of water are given by Eqs. (6.53) and (6.54) as follows:

$$\frac{dh}{d\omega} = \frac{Le(h_s - h)}{(\omega_s - \omega)} + h_g - Leh_{go} \qquad \text{(E6.15.1)}$$

$$\frac{dt_w}{d\omega} = \left(\frac{dh}{d\omega} - h_w\right)\left(\frac{\dot{m}_a}{\dot{m}_w c_w}\right) \qquad \text{(E6.15.2)}$$

The numerical accuracy of the solution may be improved using the Euler-Heun method. We first compute the slope $(dh/d\omega)_1$ using Eq. (E6.15.1) as was done in worked example 6.11. Substituting in Eq. (E6.15.2) we obtain $\Delta\omega$ for the assumed change Δt_w of the water temperature. Substitution in Eq. (E6.15.1) gives the change Δh of the air enthalpy.

Hence we obtain the first set of approximate values of the enthalpy and humidity ratio at the next node as $(h+\Delta h)$ and $(\omega+\Delta\omega)$. These values together with the saturated air properties at the next node are then re-substituted in Eq. (E6.15.1) to obtain the slope $(dh/d\omega)_2$ at the next node. We now compute an improved value of the slope as $0.5[(dh/d\omega)_1 + (dh/d\omega)_2]$. This procedure is usually called a predictor-corrector method because at each step of the computation we first predict a value by Eq. (E6.15.1) and then correct by using the mean slope. The MATLAB code given in Appendix A6.1 can be easily modified to implement the above numerical procedure. For the sake of brevity, this is left as an exercise to the reader.

For the design and operating conditions in worked example 6.11, the Euler-Heun method predicts the cooling tower volume as 65.5 m^3.

Problems

P6.1 Ambient air at 101.3 kPa flows over a wet and dry bulb psychrometer. The db-temperature is 31.8°C and the wb-temperature measures 26.8°C. The Lewis number is 0.92. Determine the relative humidity of the air.
[*Answer*: 68.5%]

P6.2 Consider the idealized flow arrangement depicted in Fig. 6.3 where air at 32°C and 40% relative humidity flows over a film of water maintained at a constant temperature of 24°C. The pressure is constant at 101.3 kPa. The area of the exposed surface of the water film is 2.3 m^2. (i) Calculate the rates of sensible heat transfer, the latent heat transfer, and total energy transfer from the water to the air. (ii) Calculate the external energy input required to maintain the water film under steady conditions. Assume that the convective heat transfer coefficient is 45 $Wm^{-2}K^{-1}$.
[*Answers*: (i) −828W, 1724.8W, 969W, (ii) 898W]

P6.3 An air washer has a face area of 4.1 m^2 and a length of 1.8 m. Ambient air at 36°C and 30% relative humidity enters the air washer with a face velocity of 2.0 ms^{-1}. The dry-bulb temperature of the air at exit is 29°C. The temperature of the water in the air washer is 24°C. The pressure is constant at 101.3 kPa. Calculate (i) the efficiency of the air washer, (ii) the NTU, and (iii) the volumetric mass transfer coefficient $h_d A_v$.
[*Answers*: (i) 51.7%, (ii) 0.728, (iii) 0.906 $kgs^{-1}m^{-3}$]

P6.4 The face area and NTU of an air washer are 1.6 m^2 and 0.65 respectively. Ambient air at 32°C db-temperature and 21°C wb-temperature enters the air washer with a face velocity of 1.8 ms^{-1}. The water in the air washer is continuously recirculated. The pressure is constant at 101.3 kPa. Calculate (i) the efficiency of the air washer, (ii) the rate of supply of make-up water, and (iii) the dry bulb temperature and relative humidity of the leaving air.
[*Answers*: (i) 47.8%, (ii) 7.1 gms^{-1}, (iii) 26.7°C, 60%]

P6.5 Ambient air at 34°C db-temperature and 21°C wb-temperature enters an air washer whose face area is 1.2 m^2. The mass flow rate of air is 2.1 kgs^{-1}. The relative humidity of the air at the exit is 57%. The pressure is constant at 101.3 kPa. For the flow situation in the air washer, the volumetric mass transfer coefficient, $h_d A_v = 1.4$ $kgs^{-1}m^{-3}$. Calculate (i) the efficiency of the air washer, and (ii) the length of the air washer. [*Answers*: (i) 50.5%, (ii) 0.878m]

P6.6 Ambient air at 20°C db-temperature and 60% relative humidity enters a counter-flow cooling tower. The air leaves at 30°C and 100% relative humidity. Water enters the cooling tower at 34°C and leaves at 21°C. The mass flow rate of water is 18 kgs^{-1}. The temperature of the make-up water is 24°C. The pressure is constant at 101.3 kPa. Calculate (i) the mass flow rate of air, and (ii) the rate of supply of make-up water. [*Answers*: (i) 17.3 kgs^{-1}, (ii) 0.32 kgs^{-1}]

P6.7 A counter-flow cooling tower is used to cool water from 36°C to 28°C. Ambient air enters the cooling tower at 30°C db-temperature and 23°C wb-temperature. The mass flow rates of water and air are 19 kgs^{-1} and 16 kgs^{-1} respectively. The volumetric mass transfer coefficient, h_dA_v = 0.56 $kgs^{-1}m^{-3}$. The pressure is constant at 101.3 kPa.

(i) Obtain the condition line for the air.

(ii) Calculate the db-temperature and relative humidity of air at the exit.

(iii) Calculate the volume of the cooling tower.

Use the detailed model of the cooling tower.

[*Answers*: (ii) 32.2°C, 98% (iii) 51.9 m^3]

P6.8 Water enters a counter-flow cooling tower at 38°C with a mass flow rate of 19 kgs^{-1}. The conditions of ambient air at entry to the cooling tower are 34°C dry-bulb temperature and 55% relative humidity. The mass flow rate of air is 16 kgs^{-1}. The pressure is constant at 101.3 kPa. The volume of the cooling tower is 75.6m^3. For the prevailing conditions the volumetric mass transfer coefficient, $h_dA_v = 0.55$ $kgs^{-1}m^{-3}$. Using the detailed model of the cooling tower, calculate (i) the outlet

temperature of water, and (ii) the dry-bulb temperature and relative humidity of air at the outlet.
[*Answers*: (i) 29.45°C, (ii) 34.7°C, 98.7%]

P6.9 Water entering a counter-flow cooling tower with a mass flow rate of 18 kgs^{-1} is cooled from 36°C to 28°C. The conditions of the air at entry to the cooling tower are 33°C db-temperature and 24°C wb-temperature. The mass flow rate of air is 16 kgs^{-1}. Assume that the volumetric mass transfer coefficient, $h_d A_v$ = 0.56 kgs^{-1}m^{-3}. Calculate the volume of the cooling tower using the simplified model of the cooling tower.
[*Answer*: 52.4 m^3]

P6.10 Use the numerical solution procedure based on the Euler-Heun method to obtain the volume of the counter-flow cooling tower described in problem P6.9, for the same operating conditions. Use the detailed model of the cooling tower.
[*Answer*: 55.8 m^3]

P6.11 A counter-flow cooling tower has a volume of 70.2 m^3. Warm water at 34°C enters the cooling tower with a flow rate of 20 kgs^{-1}. The conditions of ambient air entering the tower are 29°C dry-bulb temperature and 24°C wet-bulb temperature. The mass flow rate of air is 18 kgs^{-1}. Assume that the volumetric mass transfer coefficient, $h_d A_v$ = 0.56 kgs^{-1}m^{-3}. Use the simplified model of the cooling tower to obtain the outlet temperature of water.
[*Answer*: 27.0°C]

References

1. *ASHRAE Handbook* - 2013 *Fundamentals*, American Society of Heating, Refrigeration and Air Conditioning Engineers, Atlanta, 2013.
2. *ASHRAE Handbook* - 2012 *HVAC Systems and Equipment*, American Society of Heating, Refrigeration and Air Conditioning Engineers, Atlanta, 2012.

3. Bejan, Adrian, *Heat Transfer*, John Wiley & Sons, Inc., New York, 1993.
4. Kuehn, Thomas H., Ramsey, James W. and Threlkeld, James L., *Thermal Environmental Engineering*, 3rd edition, Prentice-Hall, Inc., New Jersey, 1998.
5. Mills, Anthony F., *Mass transfer*, Prentice Hall, New Jersey, 2001.
6. Rogers G. F. C. and Mayhew Y. R., *Thermodynamic and Transport Properties of Fluids*. 5th ed. Blackwell, Oxford, U.K. 1998.
7. Stoecker, Wilbert F. and Jones, Jerold W., *Refrigeration and Air Conditioning*, International Edition, McGraw-Hill Book Company, London, 1982.

Appendix A6.1 - MATLAB Code for Cooling Tower Design

```
% cooling tower design using exact model
% data from worked example 6.11
clear all
nv=10;                  % number of sections of tower = nv
taii=32;                % inlet temperature of air, °C
rhai=0.55;              % relative humidity of air at inlet
pamb=101.3;             % ambient air pressure, kPa
paii=4.754;    % saturation vapor pressure at air inlet temperature, kPa
% humidity ratio of air at inlet
waii=0.622*rhai*paii/(pamb-rhai*paii);
twoo=35;                % inlet water temperature, °C
twii=28;                % outlet water temperature, °C
tw(1)=twii;             % water temperature at node 1, °C
ta(1)=taii  ;           % air temperature at node 1, °C
amair=15.5;             % mass flow rate of air, kgs^-1
amwat=18.5;             % mass flow rate of water, kgs^-1
amr=amair/amwat;        % ratio of mass flow rates
ca=1;                   % specific heat capacity of air, kJkg^-1K^-1
hdav=0.54;              % volumetric transfer coefficient; kgs^-1m^-3
% coefficients for vapor enthalpy cubic expression
a0=2500.7;
a1=1.854;
```

```
a2=-0.0005;
a3=-6.0e-06;
A=[a3 a2 a1 a0];
hgii = polyval(A,taii);            %saturation vapor enthalpy at air inlet
                                   % temperature, kJkg^-1
haii=ca*taii+waii*hgii;            % inlet air enthalpy, kJkg^-1
ha(1)=haii;                        % air enthalpy at node 1, kJkg^-1
wa(1)=waii;                        % air humidity ratio at node 1
% coefficients for saturated air enthalpy- cubic expression
b0=9.3625;
b1=1.7861;
b2=0.01135;
b3=9.8855e-04;
B=[b3 b2 b1 b0];
% coefficients for liquid enthalpy- quadratic expression
c0=0.002;
c1=4.198;
c2=-0.0003;
C=[c2 c1 c0];
dtw=(twoo-twii)/nv;        % water temperature  increment, °C
cw=4.19;                   % specific  heat capacity of water,  kJkg^-1K^-1
alew=0.897;                %  Lewis number , dimensionless
hole=2467.8*alew;          % parameter in Eq.(E6.11.1)
nv1=nv+1;
% solve the governing equations by integration
for i=1:nv;
   hw(i)=polyval(C,tw(i));
   hg(i)=polyval(A,tw(i));
   hs(i)=polyval(B,tw(i));
   ws(i)=(hs(i)-ca*tw(i))/hg(i);
% calculate the gradient (dh/dw) at node i
   dhdw(i)=alew*(hs(i)-ha(i))/(ws(i)-wa(i)) +hg(i)-hole;
% calculate the change in humidity ratio at next node (i+1)
   dwa(i)=dtw*cw/(amr*(dhdw(i)-hw(i)));
% calculate the change in air enthalpy at next node (i+1)
   dha(i)=dhdw(i)*dwa(i);
```

```
% calculate all variables at next node, (i+1)
   tw(i+1)=tw(i)+dtw
   ha(i+1)=ha(i)+dha(i)
   wa(i+1)=wa(i)+dwa(i)
   ta(i+1)=ta(i)+alew*(tw(i)-ta(i))*(wa(i+1)-wa(i))/(ws(i)-wa(i))
 % check computed air enthalpies by using the air temperature and
 % humidity ratio
   hgfa = polyval(A,ta(i+1));
   hach(i+1)=ca*ta(i+1)+wa(i+1)*hgfa;
 end
% values of variables at last node , (nv+1)
hw(nv1)=polyval(C,tw(nv1));
hg(nv1)=a0+a1*tw(nv1)+a2*tw(nv1)^2+a3*tw(nv1)^3 ;
hg(nv1)=polyval(A,tw(nv1));
hs(nv1)=polyval(B,tw(nv1));
ws(nv1)=(hs(nv1)-ca*tw(nv1))/hg(nv1);
% compute cooling tower volume
sum=0;
for i=1:nv
   i1=i+1;
   fun1=1/(ws(i)-wa(i));
   fun2=1/(ws(i1)-wa(i1));
   fun3=wa(i1)-wa(i);
   sum=sum+0.5*(fun1+fun2)*fun3;
 end
  vol=amair*sum/hdav              % volume of cooling tower, m^3
```

Chapter 7

Heat Exchangers and Cooling Coils

7.1 Introduction

Heat exchangers are devices where heat is transferred from a hot fluid to a cold fluid across a separating solid wall. They are used widely in heating and air conditioning systems as subcomponents. For example, gas furnaces that provide hot air for heating homes, are equipped with heat exchangers that transfer heat from the combustion gases to the air supplied to the building. In these heating equipment, hot combustion gases flow through tubes and air flows over the tubes. An alternative home heating arrangement is where air flowing over a bank of tubes is heated by steam or hot water flowing through the tubes. To overcome the relatively poor heat transfer characteristics of air, fins are usually attached on the outside of the tubes to increase the heat transfer area.

Heat exchangers are also used widely for cooling applications in air conditioning systems. The condensers of domestic air conditioners or window units, are cooled by blowing ambient air across a coiled tube bundle carrying refrigerant. In central air conditioning systems used for cooling large buildings, the condensers of the refrigeration plants are cooled by water flowing through banks of horizontal tubes over which the refrigerant condenses. The cooling water is then pumped to a cooling tower where heat is rejected to the atmosphere.

In small room air conditioners, the return air from the room is blown across a coiled tube bundle inside which cold refrigerant evaporates. The main function of this evaporator is to cool and dehumidify air. In larger central air conditioning systems, chilled water produced in the refrigerant plant (chiller unit) flows through finned tubes in a heat exchanger, commonly called an air handling unit (AHU). Return air from the conditioned space is blown over the finned tubes of the AHU.

The energy transfer processes in evaporators and AHUs, differ from those in other types of heat exchangers mentioned above due to condensation of water vapor on the outside of the finned tubes. Therefore evaporators and AHUs are called *wet-coil* heat exchanges because both sensible heat and latent heat are transferred from the air to the coolant flowing through the tubes.

The published literature on the analysis and design of *dry-coil* heat exchangers, where there is no condensation of water vapor, is vast because of their widespread industrial applications. This is evident from the numerous text books, handbooks, and research journals dedicated to the subject of heat transfer. Therefore in the present chapter we shall only consider dry-coil heat exchangers that are of special relevance to heating and air conditioning applications, and that too, in a concise fashion. However, wet-coil heat exchangers will be analyzed in greater detail because of their importance in the design of cooling and dehumidifying equipment.

7.2 Design–Analysis of Dry-Coil Heat Exchangers

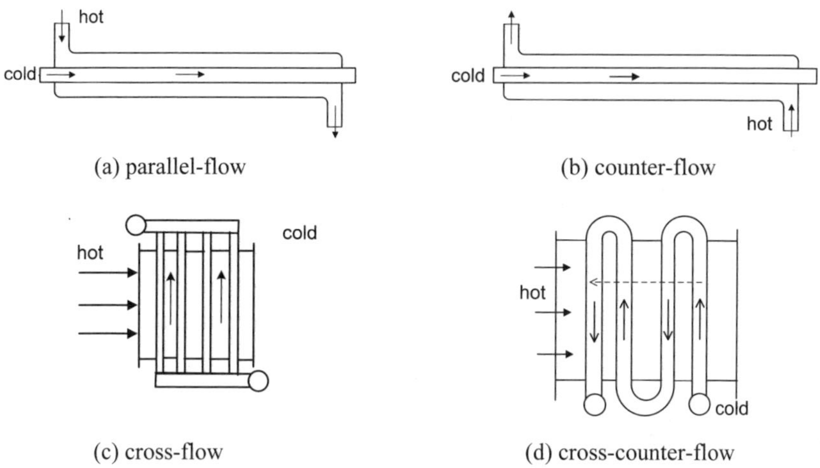

Fig. 7.1 Flow configurations of heat exchangers

7.2.1 *Some common types of heat exchangers*

Heat exchangers are usually classified according to the flow arrangement of the hot and cold fluids. Four common flow arrangements are depicted in Fig. 7.1.

(i) In the *parallel-flow* tubular heat exchanger shown schematically Fig. 7.1(a) the cold fluid flows through the inner tube while the hot fluid flows in the same direction through the annulus between the tube and the outer cylinder.

(ii) In the *counter-flow* tubular heat exchanger shown in Fig. 7.1(b) the two fluids flow in opposite directions.

(iii) Shown schematically in Fig. 7.1(c) is a *cross-flow* heat exchanger where the flow directions of the hot and cold fluids are at right angles. The cold fluid entering the inlet header is divided into four parallel streams and each stream flows through a horizontal tube. The fluid streams then enter the outlet header where they mix. The hot fluid, flows in the large duct, over the horizontal tubes, in a direction perpendicular to the cold fluid.

(iv) A *mixed-flow* heat exchanger which involves a combination of cross-flow and counter-flow is depicted in Fig. 7.1(d). Here the cold fluid flows inside the horizontal tubes, connected in series. The hot fluid flows at right angles over the tubes. For each horizontal tube the hot and cold fluid flow paths are perpendicular as in a cross-flow heat exchanger. However, due to the flow of the same hot fluid through the different horizontal tubes, its overall flow direction also includes a component opposite (counter) to the hot fluid. Such a flow arrangement is called *cross-counter* flow. Heating and cooling coils with pure cross-flow and mixed-flow (cross-counter flow) arrangements are commonly used in heating and air conditioning applications.

7.2.2 *Analysis of counter-flow heat exchangers*

The performance characteristics of a *counter-flow* heat exchanger is often used as a base line to develop mathematical expressions and design charts for the performance of other types of heat exchangers. There are two well-established procedures called the *log mean temperature difference* (*LMTD*) method and the *effectiveness-NTU* method [2] that are used to design heat exchangers. We shall first develop the physical model and the governing energy equations for a typical counter-flow heat exchanger which forms the basis of the above design methods.

Consider the tubular counter-flow heat exchanger depicted schematically in Fig. 7.2. Assume that: (i) the mass flow rates of the hot and cold fluids are constant, (ii) the thermophysical properties like the specific heat capacities are constant, (iii) the overall heat transfer coefficient is constant, (iv) heat exchange with the surroundings is negligible, and (v) the operating conditions are steady.

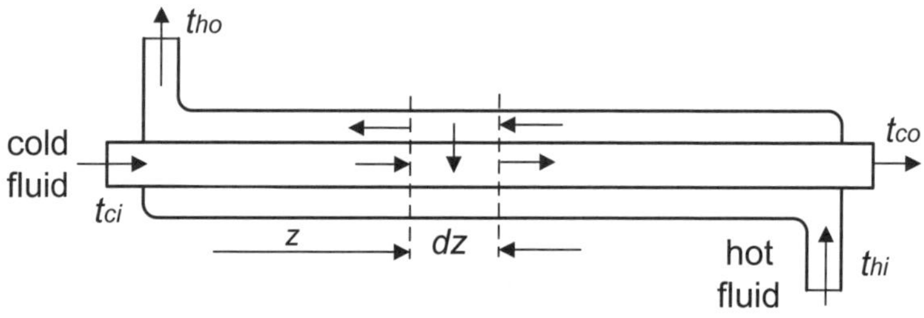

Fig. 7.2 Physical model of counter-flow heat exchanger

Let the mass flow rates of the hot and cold fluids and their specific heat capacities be $\dot{m}_h$, $\dot{m}_c$, c_h and c_c respectively. Let the inner and outer radii of the inner tube and its length be r_i, r_o and L respectively. The forced convection heat transfer coefficient for flow in the inner tube is h_i, and for flow in the annulus it is h_o. The thermal conductivity of the material of the tube is k. The overall heat transfer coefficient from the hot fluid to the cold fluid is obtained from the thermal network shown in Fig. 7.3 (see section 2.5.1).

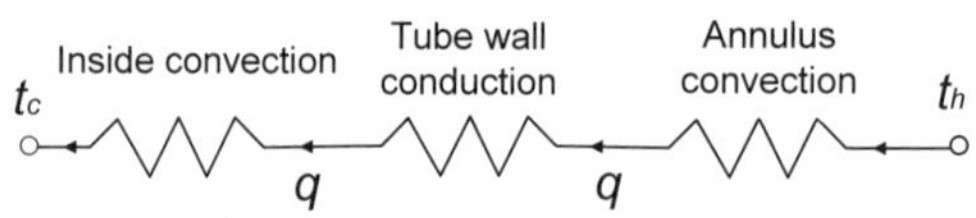

Fig. 7.3 Thermal network for heat exchanger

This gives the overall heat transfer coefficient U based on the inner tube area as

$$\frac{1}{2\pi r_i LU} = \frac{1}{2\pi r_i L h_i} + \frac{ln(\frac{r_o}{r_i})}{2\pi kL} + \frac{1}{2\pi r_o L h_o} \tag{7.1}$$

Consider a control volume of width dz at a distance z from the end of the inner tube as shown in Fig. 7.2. The heat transfer rate from the hot fluid to the cold fluid across the inner wall is given by

$$dq = 2\pi r_i U(t_h - t_c)dz \tag{7.2}$$

where t_h and t_c are the temperatures of hot and cold fluids respectively.

Apply the energy balance equation to the hot and cold fluids to obtain the following equations:

$$dq = \dot{m}_c c_c dt_c = \dot{m}_h c_h dt_h \tag{7.3}$$

For overall energy balance of the heat exchanger we have

$$\dot{Q} = \dot{m}_c c_c(t_{co} - t_{ci}) = \dot{m}_h c_h(t_{hi} - t_{ho}) \tag{7.4}$$

where $\dot{Q}$ is the *total* heat transfer rate. The inlet and outlet temperatures of the fluids are denoted by the subscripts i and o respectively.

Express the fluid temperature difference at any section as

$$\theta = t_h - t_c \tag{7.5}$$

Manipulating Eqs. (7.2), (7.3) and (7.5) we obtain the following differential equation for θ:

$$\frac{d\theta}{\theta} = U\left(\frac{1}{\dot{m}_h c_h} - \frac{1}{\dot{m}_c c_c}\right) dA \tag{7.6}$$

where the elemental heat transfer area, $dA = 2\pi r_i dz$.

Solving Eq. (7.6) and applying the boundary conditions at the entrance and the exit sections of the heat exchanger we obtain

$$ln\left(\frac{t_{hi}-t_{co}}{t_{ho}-t_{ci}}\right) = UA\left(\frac{1}{\dot{m}_h c_h} - \frac{1}{\dot{m}_c c_c}\right) \tag{7.7}$$

where the *total* heat transfer area is, $A = 2\pi r_i L$.

The two design methods mentioned earlier, namely the LMTD method and the effectiveness-NTU method are based on Eqs. (7.4) and (7.7). We shall now outline the main steps of these design methods.

7.2.3 *The LMTD method*

The most common heat exchanger design problem is to determine the heat transfer area required to meet a given design heat load or duty. The fluid temperatures and the flow rates are usually specified. We eliminate the heat capacity rates ($\dot{m}_h c_h$) and ($\dot{m}_c c_c$) between Eqs. (7.4) and (7.7) to obtain the following expression for the total heat transfer rate:

$$\dot{Q} = \frac{UA[(t_{ho}-t_{ci})-(t_{hi}-t_{co})]}{ln[(t_{ho}-t_{ci})/(t_{hi}-t_{co})]} \qquad (7.8)$$

Equation (7.8) is now expressed in terms of the *log mean temperature difference* (LMTD) as

$$\dot{Q} = UA\,(LMTD) \qquad (7.9)$$

where the *LMTD* is defined by the expression

$$LMTD = \frac{[(t_{ho}-t_{ci})-(t_{hi}-t_{co})]}{ln[(t_{ho}-t_{ci})/(t_{hi}-t_{co})]} \qquad (7.10)$$

Notice that Eq. (7.9) has the general form of the expression for convective heat transfer rate which is the product of *UA* and the temperature difference. However, the temperature difference now is the LMTD instead of the simple temperature difference as in Newton's law of cooling.

For the design of other types of heat exchangers, such as cross flow heat exchangers, Eq. (7.9) is modified to include an *LMTD-correction factor F*. Hence we have

$$\dot{Q} = UA\,F(LMTD) \qquad (7.11)$$

The correction factor in available in the form of charts in standard texts on heat transfer and more extensively in heat exchanger design handbooks [4].

7.2.4 *The effectiveness–NTU method*

The effectiveness of a heat exchanger, ε is defined as the ratio of the actual heat transfer rate to the maximum possible heat transfer rate. The maximum heat transfer rate, compatible with the laws of thermodynamics is given by

$$\dot{Q}_{max} = (\dot{m}c)_{min}(t_{hi} - t_{ci}) \tag{7.12}$$

where $(\dot{m}c)_{min}$ represents the *smaller* of the capacity rates $(\dot{m}_h c_h)$ and $(\dot{m}_c c_c)$ of the two fluids. Therefore the effectiveness is given by

$$\epsilon = \frac{\dot{Q}_{act}}{\dot{Q}_{max}} = \frac{\dot{m}_h c_h(t_{hi}-t_{ho})}{(\dot{m}c)_{min}(t_{hi}-t_{ci})} = \frac{\dot{m}_c c_c(t_{co}-t_{ci})}{(\dot{m}c)_{min}(t_{hi}-t_{ci})} \tag{7.13}$$

From Eq. (7.7) it follows that

$$\frac{t_{ho}-t_{ci}}{t_{hi}-t_{co}} = exp\left[UA\left(\frac{1}{\dot{m}_c c_c} - \frac{1}{\dot{m}_h c_h}\right)\right] = \alpha \text{ (say)} \tag{7.14}$$

Eliminating the ratio of the temperature differences between Eqs. (7.13) and (7.14) it is possible to show that

$$\varepsilon = \frac{1-\alpha}{\left[\frac{(\dot{m}c)_{min}}{\dot{m}_h c_h} - \frac{\alpha(\dot{m}c)_{min}}{\dot{m}_c c_c}\right]} \tag{7.15}$$

Substituting for α from Eq. (7.14) in Eq. (7.15) we have

$$\varepsilon = \frac{1-exp\left[UA\left(\frac{1}{\dot{m}_c c_c}-\frac{1}{\dot{m}_h c_h}\right)\right]}{\frac{(\dot{m}c)_{min}}{\dot{m}_h c_h}-\left[\frac{(\dot{m}c)_{min}}{\dot{m}_c c_c}\right]exp\left[UA\left(\frac{1}{\dot{m}_c c_c}-\frac{1}{\dot{m}_h c_h}\right)\right]} \tag{7.16}$$

Design charts for effectiveness may be developed using Eq. (7.16) by choosing the following dimensionless parameters:

$$c_r = \frac{(\dot{m}c)_{min}}{(\dot{m}c)_{max}} \quad \text{and} \quad NTU = \frac{UA}{(\dot{m}c)_{min}} \tag{7.17}$$

Equation (7.17) may be written in terms of the above dimensionless parameters in the form

$$\varepsilon = \frac{1-exp\left[NTU\left(\frac{(\dot{m}c)_{min}}{\dot{m}_c c_c}-\frac{(\dot{m}c)_{min}}{\dot{m}_h c_h}\right)\right]}{\frac{(\dot{m}c)_{min}}{\dot{m}_h c_h}-\left[\frac{(\dot{m}c)_{min}}{\dot{m}_c c_c}\right]exp\left[NTU\left(\frac{(\dot{m}c)_{min}}{\dot{m}_c c_c}-\frac{(\dot{m}c)_{min}}{\dot{m}_h c_h}\right)\right]} \tag{7.18}$$

By considering the two cases: $(\dot{m}c)_{min} = \dot{m}_c c_c$ and $(\dot{m}c)_{min} = \dot{m}_h c_h$, it is possible to show that Eq. (7.18) has the following general form:

$$\varepsilon = \frac{1-exp[NTU(1-c_r)]}{c_r-exp[NTU(1-c_r)]} \tag{7.19}$$

We note that if $c_r = 1$, then the effectiveness given by Eq. (7.19) becomes indeterminate. This singularity occurs when the two capacity rates $\dot{m}_h c_h$ and $\dot{m}_c c_c$ are equal. This difficulty is resolved by performing the foregoing analysis assuming equal capacity rates for the two fluids, which gives the limiting value of the effectiveness as

$$\varepsilon = \frac{NTU}{1+NTU} \tag{7.20}$$

We have used Eqs. (7.19) and (7.20) to develop the design curves, shown in Fig. 7.4, for a counter flow heat exchanger.

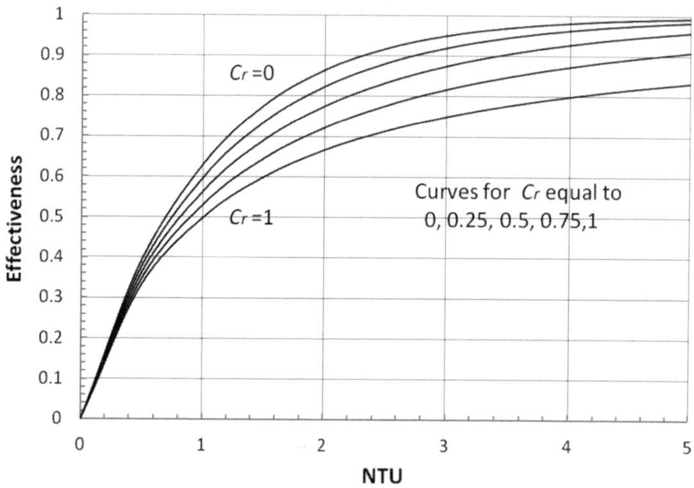

Fig. 7.4 Effectiveness versus NTU for a counter-flow heat exchanger

We shall illustrate the application of Eq. (7.19) in the worked examples to follow in this chapter.

7.2.5 *Evaporators and condensers*

Evaporators and condensers differ from the general counter flow heat exchanger considered thus far because one of the fluids undergoes phase change, that is condensation or evaporation. If there is no subcooling or superheating of the fluid undergoing phase change then its temperature remains constant during the flow through the heat exchanger. This simplifies the analysis of evaporators and condensers significantly.

In evaporators and condensers of small domestic air conditioning units, the refrigerant undergoes phase change inside coiled tube bundles.

Air flows outside the tubes in a cross-flow arrangement. In central air conditioning systems used in large buildings the refrigerant undergoes phase change over a bank of horizontal tubes through which water flows.

For the sake of brevity, we shall illustrate the analysis of condensers and evaporators by considering the general arrangement shown schematically in Fig. 7.5(a).

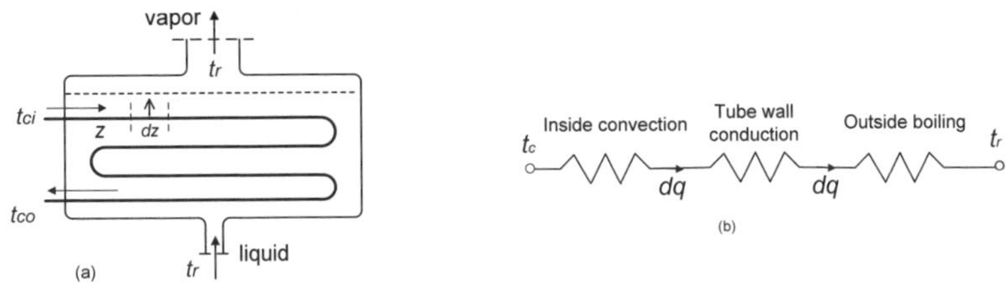

Fig. 7.5 (a) Schematic diagram of evaporator, (b) thermal network

Here heat is transferred from the water flowing inside the tubes to the pool of refrigerant in the outer shell. The refrigerant undergoes phase change by a process commonly called *pool boiling*, while the water is chilled due to sensible heat transfer. The modes of heat transfer at a typical location along the tube, includes forced convection inside the tube, conduction through the tube wall, and pool boiling outside the tube. The thermal network for the heat transfer from the water to the refrigerant is shown in Fig. 7.5(b). The overall heat transfer coefficient U, based on the inner tube area, is obtained from the network (see section 2.5.1) as

$$\frac{1}{2\pi r_i L U} = \frac{1}{2\pi r_i L h_i} + \frac{ln(\frac{r_o}{r_i})}{2\pi k L} + \frac{1}{2\pi r_o L h_b} \tag{7.21}$$

where the inner and outer radii of the tube and its length are r_i , r_o and L respectively. The boiling heat transfer coefficient is h_b, the forced convection heat transfer coefficient is h_i, and the thermal conductivity of the tube wall is k. Heat transfer correlations for boiling heat transfer are available in Refs. [2,3].

Consider a control volume of width dz at a distance z from the end of the inner tube as shown in Fig. 7.5(a). The heat transfer rate from the water to the refrigerant across the tube wall is given by

$$dq = 2\pi r_i U(t_c - t_r)dz \tag{7.22}$$

where t_c and t_r are the temperatures of water and refrigerant respectively.

Applying the energy balance equation to the control volume we have

$$dq = -\dot{m}_c c_c dt_c \tag{7.23}$$

From Eqs. (7.22) and (7.23) it follows that

$$\dot{m}_c c_c dt_c = -2\pi r_i U(t_c - t_r)dz \tag{7.24}$$

Since the refrigerant temperature, t_r is constant, Eq. (7.24) can be solved in a straightforward manner. With the substitution of the inlet and exit water temperatures t_{ci} and t_{co} in the solution we obtain

$$\frac{t_{co}-t_r}{t_{ci}-t_r} = exp\left(\frac{-UA}{\dot{m}_c c_c}\right) \tag{7.25}$$

where the total heat transfer area is, $A = 2\pi r_i L$.

Now the maximum possible heat transfer from the water is given by

$$\dot{Q}_{max} = \dot{m}_c c_c (t_{ci} - t_r) \tag{7.26}$$

We define the *effectiveness* of the evaporator as

$$\epsilon = \frac{\dot{Q}_{act}}{\dot{Q}_{max}} = \frac{\dot{m}_c c_c (t_{ci}-t_{co})}{\dot{m}_c c_c (t_{ci}-t_r)} \tag{7.27}$$

Manipulating Eqs. (7.25), (7.26) and (7.27) we obtain the following expression for the effectiveness

$$\epsilon = 1 - exp(-NTU) \tag{7.28}$$

where *NTU*, the number of transfer units, is defined as

$$NTU = \frac{UA}{\dot{m}_c c_c} \tag{7.28a}$$

We note that for heat exchangers where one of the fluids undergoes phase change, the expression for the effectiveness is much simpler than that for a standard counter-flow heat exchanger. Note that it is possible to obtain Eq. (7.28) from Eq. (7.18) by assuming that the capacity rate of the fluid undergoing phase change is *infinity* because its temperature does not change. We shall consider the design of condensers and evaporators in the worked examples.

7.2.6 *Cross-flow heat exchangers*

In air conditioning systems, cross-flow heat exchangers are commonly used to heat air using hot water or steam. The air flows over the outside of the tubes which usually have fins attached to them. The fins increase the heat transfer area and thereby enhance the heat transfer process.

We shall illustrate the analysis of a simple cross-flow heat exchanger by considering a heating coil with a single vertical row of tubes. The expressions obtained for the fluid temperatures could be extended in a recursive manner to coils with multiple rows of tubes [4].

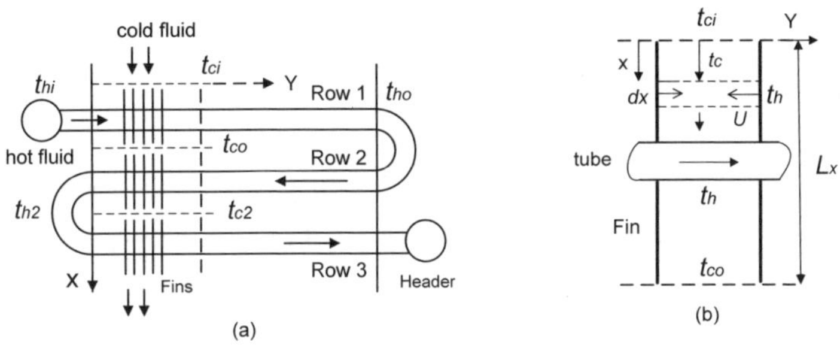

Fig. 7.6 (a) Cross flow heat exchanger (plan view), (b) Heat flow from fins to air

A general arrangement of a cross-flow heat exchanger with three tube passes is depicted in Fig. 7.6(a). Fins are attached on the outside of the tubes. At any section of the tube, the hot fluid flowing through it is well mixed. However, the cold fluid approaching the tube at right angles is unmixed due to the presence of the fins which form separate flow channels for the cold fluid. The details of a typical flow channel between two fins is shown enlarged in Fig. 7.6(b).

The x, y and z directions are along the fin, along the tube and the vertical respectively. Consider a control volume of thickness dx in the x-direction. Let the number of fins and therefore the number of flow channels be n. Due to the close spacing of fins in a typical heat exchanger, any end effects may be neglected.

Let the mass flow rates of the hot and cold fluids and their specific heat capacities be $\dot{m}_h$, $\dot{m}_c$, c_h and c_c respectively. Let the overall heat transfer coefficient for the cold fluid be U. Now the flow rate per channel is $\dot{m}_c/n$. Assume that the temperatures of the fin surfaces are uniform and equal to the local hot fluid temperature t_h, which is constant at the location (y), but varies along the tube.

Applying the energy balance equation to the control volume we obtain

$$(\dot{m}_c/n)c_c dt_c = (2L_z dx)U(t_h - t_c) \tag{7.29}$$

where L_z is the height of the channel. Integrating Eq. (7.29) we have

$$\ln\left(\frac{t_h - t_{co}}{t_h - t_{ci}}\right) = \frac{-2L_z U n L_x}{\dot{m}_c c_c} = \frac{-AU}{\dot{m}_c c_c} \tag{7.30}$$

where t_{ci} and t_{co} are the cold fluid temperatures at the inlet and outlet of the flow channel. The total external heat transfer area, $A = 2nL_xL_z$. From Eq. (7.30) it follows that:

$$\frac{t_h - t_{co}}{t_h - t_{ci}} = \exp(-\alpha_1) \tag{7.31}$$

where, $\alpha_1 = \frac{AU}{\dot{m}_c c_c}$. Rearranging Eq. (7.31) we obtain

$$t_{co} - t_{ci} = (t_h - t_{ci})(1 - e^{-\alpha_1}) \tag{7.32}$$

Apply the energy balance equation to the hot fluid in a small control volume of the tube between the two fins in Fig. 7.6(b). This gives

$$\dot{m}_h c_h \Delta t_h = -(\dot{m}_c c_c/n)(t_{co} - t_{ci}) = \left(\frac{\dot{m}_c c_c}{L_y}\right)(t_{co} - t_{ci})\Delta y \tag{7.33}$$

where the distance between two fins is, $\Delta y = L_y/n$.

Substituting in Eq. (7.33) from Eq. (7.32) we have

$$\dot{m}_h c_h \Delta t_h = -\left(\frac{\dot{m}_c c_c}{L_y}\right)(1 - e^{-\alpha_1})(t_h - t_{ci})\Delta y \tag{7.34}$$

Since the fin spacing Δy is very small, we may treat Eq. (7.34) as equivalent to the following differential equation:

$$\dot{m}_h c_h \left(\frac{dt_h}{dy}\right) = -\left(\frac{\dot{m}_c c_c}{L_y}\right)(1 - e^{-\alpha_1})(t_h - t_{ci}) \tag{7.35}$$

The solution of Eq. (7.35) gives

$$ln\left(\frac{t_{ho} - t_{ci}}{t_{hi} - t_{ci}}\right) = -\left(\frac{\dot{m}_c c_c}{\dot{m}_h c_h}\right)(1 - e^{-\alpha_1}) \tag{7.36}$$

where t_{hi} and t_{ho} are the hot fluid temperatures at the inlet and outlet of the tube. Equation (7.36) can be rearranged to the form

$$t_{hi} - t_{ho} = (1 - e^{-\alpha_2})(t_{hi} - t_{ci}) \quad (7.37)$$

where

$$\alpha_2 = \left(\frac{\dot{m}_c c_c}{\dot{m}_h c_h}\right)(1 - e^{-\alpha_1}) = \left(\frac{\dot{m}_c c_c}{\dot{m}_h c_h}\right)\left[1 - exp\left(-\frac{AU}{\dot{m}_c c_c}\right)\right] \quad (7.38)$$

The outlet temperature of the hot fluid is obtained from Eq. (7.37) as

$$t_{ho} = t_{hi} - (1 - e^{-\alpha_2})(t_{hi} - t_{ci}) \quad (7.39)$$

Now the cold fluid streams leaving the different channels between the fins may be assumed to mix completely. We obtain the mixed-mean outlet temperature, t_{com} by writing the overall energy balance equation. Hence we have

$$\dot{m}_c c_c(t_{com} - t_{ci}) = \dot{m}_h c_h(t_{ho} - t_{ci})$$

Substituting in the above equation from Eq. (7.39) we obtain

$$t_{com} = t_{ci} + \left(\frac{\dot{m}_h c_h}{\dot{m}_c c_c}\right)(1 - e^{-\alpha_2})(t_{hi} - t_{ci}) \quad (7.40)$$

The effectiveness of the heat exchanger is defined as the ratio of the actual heat transfer to the maximum possible heat transfer.

Now the maximum heat transfer rate is given by

$$\dot{Q}_{max} = (\dot{m}c)_{min}(t_{hi} - t_{ci}) \quad (7.41)$$

where $(\dot{m}c)_{min}$ represents the *smaller* of the two capacity rates $(\dot{m}_h c_h)$ and $(\dot{m}_c c_c)$ of the two fluids. Hence the effectiveness is

$$\epsilon = \frac{\dot{Q}_{act}}{\dot{Q}_{max}} = \frac{\dot{m}_h c_h(t_{hi} - t_{ho})}{(\dot{m}c)_{min}(t_{hi} - t_{ci})} \quad (7.42)$$

Substituting from Eq. (7.39) in Eq. (7.42) we obtain the effectiveness as

$$\epsilon = \frac{\dot{m}_h c_c}{(\dot{m}c)_{min}}(1 - e^{-\alpha_2}) \quad (7.43)$$

where α_2 is defined in Eq. (7.38).

It is interesting to note that the outlet fluid temperatures t_{ho} and t_{com} given by Eqs. (7.39) and (7.40), respectively are the fluid inlet temperatures for the second row of tubes. We could easily modify these expressions to obtain the fluid outlet temperatures t_{h2} and t_{c2} for the second tube as indicated in Fig. 7.6(a). Hence the effectiveness of a two-pass coil with *cross-parallel* flow arrangement could be deduced.

The same procedure could be applied to a *cross-counter* situation where the hot fluid flows through the tube passes in the opposite direction. In this case we need to make an initial guess of the fluid temperature at the outlet and adjust it in an iterative manner until the given inlet temperature is obtained.

7.2.7 *Efficiency of extended surfaces*

Extended surfaces, or finned surfaces, are used extensively in heat exchangers for heating and cooling air. The addition of fins to a surface greatly increases the heat transfer area, which lowers the average temperature difference between the hot and cold fluids significantly.

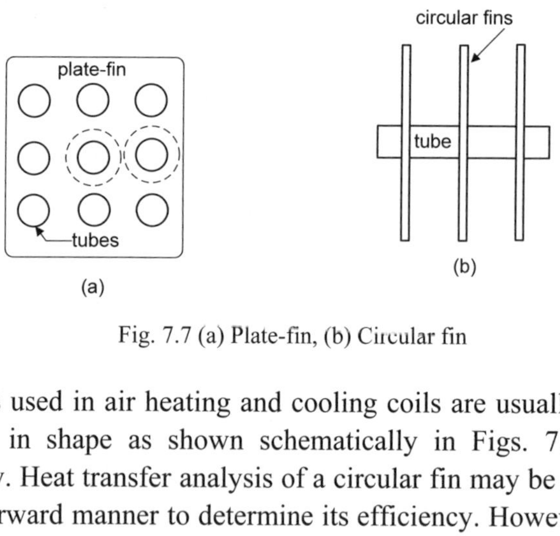

Fig. 7.7 (a) Plate-fin, (b) Circular fin

The fins used in air heating and cooling coils are usually rectangular or circular in shape as shown schematically in Figs. 7.7(a) and (b) respectively. Heat transfer analysis of a circular fin may be carried out in a straightforward manner to determine its efficiency. However, plate fins are difficult to model accurately due to the complex interaction of the heat flows from the different tubes attached to the same fin.

Nevertheless, an approximate analysis of plate fins may be developed by identifying a circular area on the plate around each tube whose boundary may be treated as *adiabatic* [2]. This concept of equivalence is indicated in Fig. 7.7(a).

In view of the practical usefulness of circular fins in heating and cooling coils we shall now develop a heat transfer model for it with the aim of obtaining the temperature distribution and the fin efficiency.

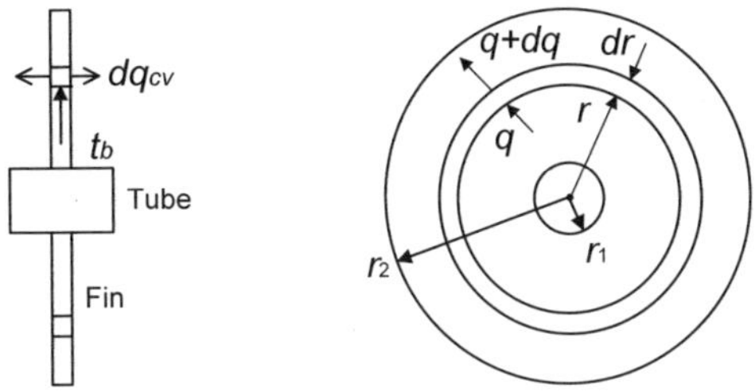

Fig. 7.8 Physical model of circular fin

Consider the physical model of a circular fin of constant thickness 2δ whose inner and outer radii are r_1 and r_2 respectively. Two views of the fin are shown schematically in Fig. 7.8. Assume that heat is conducted radially along the fin in a symmetrical manner so that the temperature distribution in the fin is a function solely of the radius, r. Due to the small thickness of the fin, the temperature across the fin may be assumed uniform at any radius. There is convective heat transfer from the fin surface to the surrounding fluid whose temperature, t_a may be assumed constant. Moreover, the average convective heat transfer coefficient, h_c for this process is assumed constant.

Applying the steady-state energy balance equation to a small control of thickness dr we have

$$q = (q + dq) + 2 \times 2\pi r dr \times h_c(t - t_a)$$

$$\frac{dq}{dr} = -4\pi r h_c(t - t_a) \tag{7.44}$$

Note that the term on the RHS of Eq. (7.44) is the heat transfer from the *two sides* of the fin by convection. Applying Fourier's law of heat conduction we obtain the conduction heat transfer rate as

$$q = -(2\pi r \times 2\delta)k\left(\frac{dt}{dr}\right) \tag{7.45}$$

Substituting from Eq. (7.45) in Eq. (7.44) we have

$$4\pi\delta k \frac{d}{dr}\left(r\frac{dt}{dr}\right) = 4\pi r h_c (t - t_a) \tag{7.46}$$

$$\frac{d}{dr}\left(r\frac{dt}{dr}\right) = r\beta^2 (t - t_a) \tag{7.47}$$

where $\beta^2 = h_c/\delta k$.

We now express Eq. (7.47) in terms of the following dimensionless variables

$$x = r\beta \quad \text{and} \quad \theta = \frac{(t-t_a)}{(t_b-t_a)}$$

This gives the following differential equation for θ:

$$x^2 \frac{d^2\theta}{dx^2} + x\frac{d\theta}{dx} - x^2\theta = 0 \tag{7.48}$$

We assume that the fin-base temperature is specified and that the heat transfer rate at the outer edge or 'fin tip' is zero. These boundary conditions may be expressed in the form:

$$\text{(i) } r = r_1 : t = t_b \quad \text{and} \quad \text{(ii) } r = r_2 : \frac{dt}{dr} = 0 \tag{7.49a}$$

$$\text{or (i) } x = \beta r_1 : \theta = 1 \quad \text{and} \quad \text{(ii) } x = \beta r_2 : \frac{d\theta}{dx} = 0 \tag{7.49b}$$

The differential equation (7.48) is a modified Bessel's equation whose general solution may be expressed as

$$\theta = c_1 I_0(x) + c_2 K_0(x) \tag{7.50}$$

where I_0 and K_0 are zero-order modified Bessel functions of the first and second kinds respectively. The constants c_1 and c_2 are obtained by applying the boundary conditions given by Eq. 7.49(b). Hence we have

$$c_1 = \frac{K_1(x_2)}{G(x_1, x_2)} \quad \text{and} \quad c_2 = \frac{I_1(x_2)}{G(x_1, x_2)} \tag{7.51}$$

where I_1 and K_1 are first-order modified Bessel functions, and the function G is given by

$$G = I_0(x_1)K_1(x_2) + I_1(x_2)K_0(x_1) \tag{7.52}$$

The heat transferred from the fin to the surrounding fluid by convection is equal to the heat entering through the base of the fin by conduction. This is given by Fourier's law as

$$\dot{Q}_{cb} = -(2\pi r_1 \times 2\delta)k\left(\frac{dt}{dr}\right)_{r=r_1} = -4\pi\delta k(t_b - t_a)\beta\left(\frac{d\theta}{dz}\right)_{z=z_1} \quad (7.53)$$

Now the maximum possible convective heat transfer to the surrounding fluid occurs if the *entire fin* is at the fin-base temperature. Hence we have

$$\dot{Q}_{max} = 2\pi({r_2}^2 - {r_1}^2)h_c(t_b - t_a) \quad (7.54)$$

The efficiency of the fin is defined as the ratio of the actual heat transfer to the maximum possible heat transfer. Manipulating Eqs. (7.50) to (7.54), we obtain the following expression for the fin efficiency:

$$\eta_f = \frac{\dot{Q}_{cb}}{\dot{Q}_{max}} = \frac{(2r_1/\beta)[K_1(\beta r_1)I_1(\beta r_2) - I_1(\beta r_1)K_1(\beta r_2)]}{({r_2}^2 - {r_1}^2)[K_0(\beta r_1)I_1(\beta r_2) + I_0(\beta r_1)K_1(\beta r_2)]} \quad (7.55)$$

The values of the various Bessel functions are available in tabular form in Ref. [5].

The heat transfer from the fin to the surrounding fluid may be expressed in terms of the fin-base temperature and the fin efficiency as

$$\dot{Q}_{fin} = A_{fin}\eta_f h_c(t_b - t_a) \quad (7.56)$$

We note from Eq. (7.56) that the fin may be represented by a thermal network element as shown in Fig. 7.10(b). The fin thermal resistance is given by

$$R_{fin} = \frac{1}{A_{fin}\eta_f h_c}$$

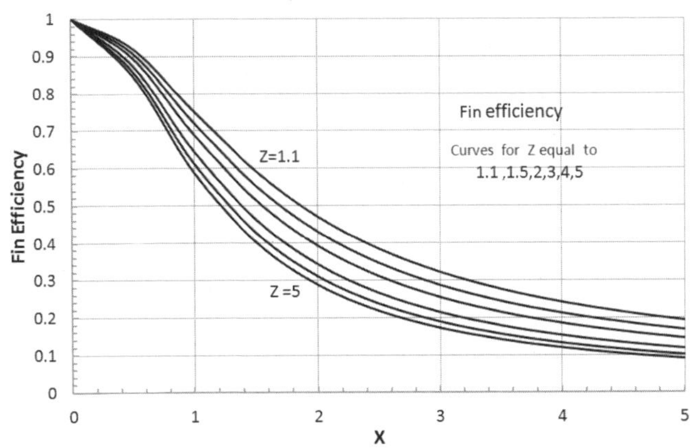

Fig.7.9 Efficiency of circular fins. $z = r_o/r_i$ and $x = (r_o - r_i)\sqrt{h_c/k\delta}$

The family of curves for the efficiency of circular fins was produced by evaluating Eq. (7.55). The required Bessel functions were obtained directly from the toolkit in the MATLAB software package.

7.2.8 Overall heat transfer coefficient for finned tubes

The heat flow paths through a single fin and two equal sections of the tube on either side of a typical finned tube are depicted in Fig. 7.10(a). Heat is first transferred by convection from the fluid in the tube to the tube wall, which is followed by conduction through the tube wall. A fraction of this heat is then transferred to the surrounding fluid by convection from the tube surface and the rest by convection from the fin surface. The latter two heat flow paths are parallel.

The representative section in Fig. 7.10(a), consisting of a single fin and two short equal lengths of the tube, may be thought of as a 'unit-cell' which characterizes the finned tube, when end effects are ignored. The thermal network for the heat flow in the unit-cell is shown in Fig. 7.10(b).

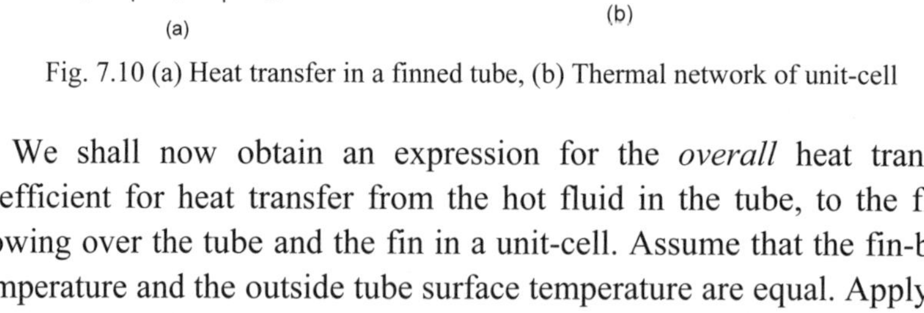

Fig. 7.10 (a) Heat transfer in a finned tube, (b) Thermal network of unit-cell

We shall now obtain an expression for the *overall* heat transfer coefficient for heat transfer from the hot fluid in the tube, to the fluid flowing over the tube and the fin in a unit-cell. Assume that the fin-base temperature and the outside tube surface temperature are equal. Applying Ohm's law to the equivalent thermal network in Fig. 7.10(b) we have

$$t_h - t_c = Q\left[R_{ic} + R_{tw} + \left(\frac{1}{R_f} + \frac{1}{R_{ts}}\right)^{-1}\right] \tag{7.57}$$

The individual thermal resistances given by

$$R_{ic} = \frac{1}{(A_{ti}h_{ti})} \quad , \quad R_{tw} = \frac{\delta_t}{(A_{ti}k_t)}$$

$$R_f = \frac{1}{(A_f\eta_f h_c)} \quad \text{and} \quad R_{ts} = \frac{1}{(A_{to}h_{to})}$$

The inside and outside convective heat transfer coefficients for the tube are h_{ti} and h_{to} respectively. The tube wall thickness is δ_t, and the thermal conductivity is k_t. The efficiency of the fin is η_f, and its area is A_f.

The overall heat transfer coefficient U_o based on the inner tube area A_{ti} of the unit-cell is defined by Eq. (7.58) as

$$Q = A_{ti}U_o(t_h - t_c) \tag{7.58}$$

From Eqs. (7.57) and (7.58) it follows that

$$\frac{1}{U_o A_{ti}} = R_{ic} + R_{tw} + \left(\frac{1}{R_f} + \frac{1}{R_{ts}}\right)^{-1} \tag{7.59}$$

We shall illustrate the application of the various models developed in the preceding sections to the design of heat exchangers in the worked examples to follow in this chapter.

7.3 Wet-Coil Heat Exchangers or Cooling Coils

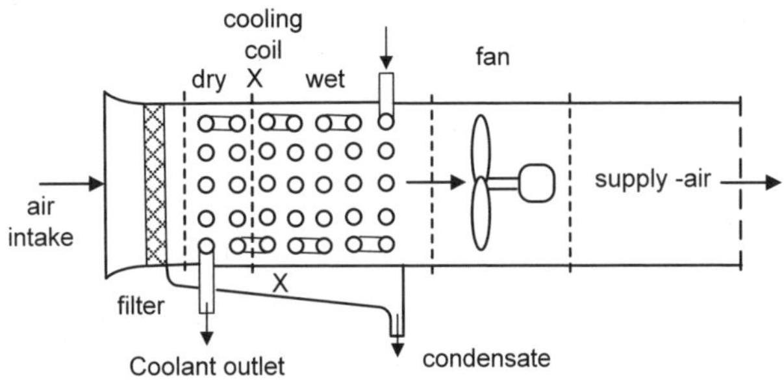

Fig. 7.11 Cooling and dehumidifying coil

In this section we shall extend our studies on heat exchangers to cooling and dehumidifying coils which are used widely in air conditioning systems. These coils are called *wet-coil* heat exchangers because of the presence of a water film on the outside of the tubes due to condensation of water vapor from the air flowing over the tube bundle.

A typical wet-coil heat exchanger is shown schematically in Fig. 7.11. The cooling fluid flowing through the bank of tubes could either be a refrigerant, or chilled water produced in a separate refrigeration plant. The coolant flows in series through different rows of vertical tubes. Therefore the overall direction of flow of the coolant is counter to the air flowing over the tubes.

Due to heat transfer from the air to the coolant, the average temperature of the air decreases in the flow direction. In the arrangement shown in Fig. 7.11, the air experiences only sensible cooling over the first two rows of tubes. However, at the third row of tubes condensation of moisture commences because the tube surface temperature is at the *dew-point temperature of the incoming air*. As the air passes over the rest of the tube rows condensation continues and the water produced forms a thin film over the tubes in the row. The water drips down due to gravity and eventually falls to the condensation pan from which it is discharged to the atmosphere.

7.3.1 *Physical processes in wet-coils*

The *dry-coil* section of the heat exchanger in Fig. 7.11 can be analyzed using the models developed in the preceding sections of this chapter. However, to analyze the *wet-coil* section, we need to consider the details of the simultaneous heat and mass transfer processes that occur during condensation.

But first it is instructive to develop a simplified model of the cooling coil where it is easier to identify the physical processes involved. For this purpose we consider the idealized heat exchanger, shown schematically in Fig. 7.12. Here air flows over the flat upper surface of a rectangular duct through which a cooling fluid flows in the opposite direction.

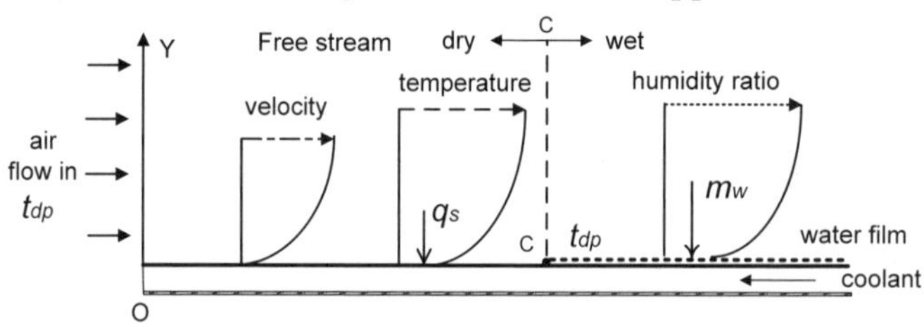

Fig. 7.12 Flow of moist air over a flat coolant duct

Assume that the velocity, temperature and humidity ratio distributions of the air stream at the entrance section OY are uniform across the flow cross section. As air flows over the plate, its velocity at the surface reduces to zero because of friction. A typical velocity profile at a downstream section is shown in Fig. 7.12. The layer of air adjacent to the cold surface looses heat to the coolant and its temperature becomes equal to the plate temperature, as indicated in the temperature profile in Fig. 7.12.

The relatively thin layers of air in which the above velocity and temperature changes occur are called the *velocity boundary layer* and *thermal boundary layer* respectively. Detailed analysis of the properties of these boundary layers are available in numerous text books on heat transfer including Refs. [2,5].

The humidity ratio of air remains uniform as long as the surface temperature is above the dew-point temperature of the air at the entrance section. In Fig. 7.12, condensation of water vapor commences at section, CC where the plate temperature is equal to the dew point temperature, t_{dp}. Since the coolant is approaching section CC from the right, the surface temperature to the right of section CC is below t_{dp}, and therefore condensation continues along this part of the plate. Due to condensation, the humidity ratio of air decreases from the free-stream to the surface as shown in Fig. 7.12. This in turn produces a *vapor concentration boundary layer* with its origin at section CC.

In chapter 6 we developed several models to analyze simultaneous heat and mass transfer processes associated with the evaporation from a film of water to air flowing over it. We showed that for most applications involving ambient air, the *total energy flow* due to sensible and latent heat transfer is proportional to the difference in enthalpy between the ambient air and saturated air at the water film temperature. The latter difference in enthalpy was called the *enthalpy potential* and we shall use this concept to analyze wet-coil heat exchangers in the next section.

7.3.2 *Analysis of wet-coil heat exchangers*

Shown in Fig. 7.13(a) is a section of a finned tube with a thin film of water on its surface due to condensation of water vapor from the air flowing over it. The tube and the fin are at a temperature below the dew point of the incoming air. The corresponding thermal network is depicted in Fig. 7.13(b).

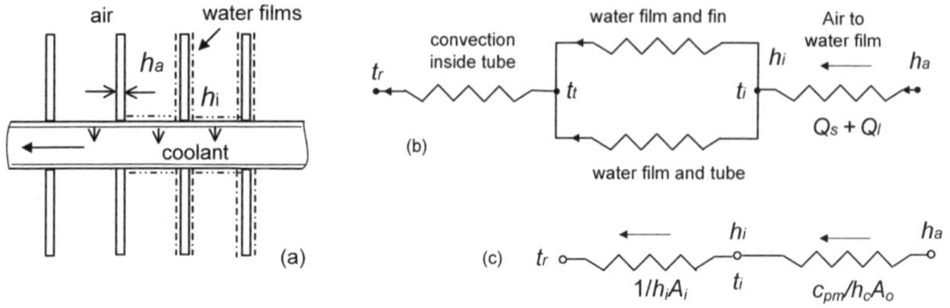

Fig. 7.13 (a) Finned tube with water film, (b) Overall network, (c) Equivalent network

The sensible heat flux, Q_s and the latent heat flux due to condensation, Q_l enter the coolant flowing through tube via the water film, the fin, and the tube wall. A detailed analysis of these heat and mass transfer processes in the finned tube heat exchanger is beyond the scope of this book. However, for the interested reader, the heat and mass transfer analysis of a finned tube with a water film is available in Ref. [4]. The analysis is similar to that in section 7.2.7 except the inclusion of a thin water film on the surfaces of the fin and the tube.

We shall now develop a simplified model where the thermal resistances of the water film, the fins, the tube wall are included in a suitably averaged overall heat transfer coefficient h_i based on the inner tube area, as indicated in Fig. 7.13(c). The rows of coils in the real cooling coil in Fig. 7.11 are represented by a series of control volumes through which moist air and coolant flow in opposite directions as shown schematically in Fig. 7.14.

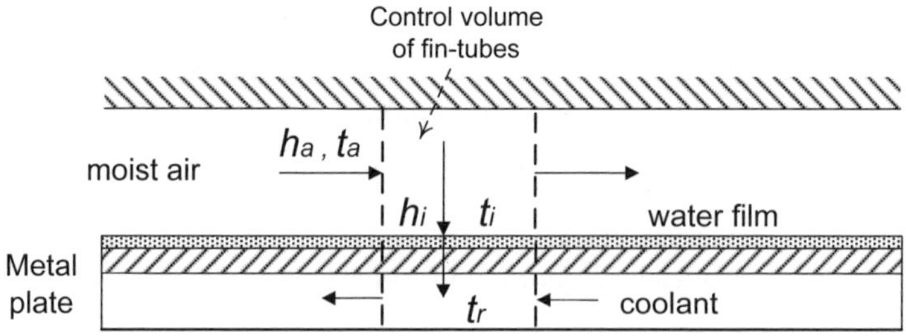

Fig. 7.14 Simplified model of a cooling coil

For the typical control volume shown in Fig. 7.14, let the coolant-side heat transfer area be, δA_i and the outside area of the coil, including the fins, over which the air flows be, δA_o. The total energy transfer rate from the *air to the water film* may be written in terms of the *enthalpy potential* [see Eq. (6.23)] as

$$\delta q_t = (h_c/c_{am})(h_a - h_i)\delta A_o \qquad (7.60)$$

where h_a is the enthalpy of moist air at dry-bulb temperature t_a, and humidity ratio ω_a. The enthalpy and humidity ratio of saturated air at the temperature of the water film, t_i are h_i and ω_i respectively.

The rate of *sensible* heat transfer from the *water film to the coolant* is given by

$$\delta q_t = h_i(t_i - t_r)\delta A_i \qquad (7.61)$$

where t_r is the temperature of the coolant.

From Eqs. (7.60) and (7.61) it follows that:

$$\frac{t_i - t_r}{h_a - h_i} = \frac{h_c \delta A_o}{c_{am} h_i \delta A_i} = \alpha \text{ (say)} \qquad (7.62)$$

We note that Eq. (7.62) is applicable at any section of the heat exchanger where there is *condensation* of water vapor. Furthermore, for a given cooling coil the RHS of Eq. (7.62) is equal to a constant, say α.

In section 6.6 we presented the following cubic relationship for the enthalpy of saturated air, h_i (kJkg^{-1}) at temperature t_i and pressure 101.3 kPa:

$$f_3(t_i) = 9.3625 + 1.786t_i + 1.1135 \times 10^{-2} t_i^2 + 9.8855 \times 10^{-4} t_i^3$$

(7.63)

A quadratic expression for the saturation enthalpy that is less accurate, but computationally more convenient, is given by

$$f_2(t_i) = 10.9075 + 1.2205t_i + 5.692 \times 10^{-2} t_i^2 \qquad (7.64)$$

We can find the water film temperature, t_i at any section of the heat exchanger by eliminating the enthalpy h_i between Eq. (7.62) and one of the Eqs. (7.63) or (7.64). This would require the solution of a cubic equation or a quadratic equation in t_i, depending on the polynomial expression chosen.

7.3.3 Numerical model for wet-coils

In order to illustrate the numerical design procedure for a wet-coil heat exchanger we consider an air handling unit (AHU) where the coolant is chilled water.

Assume that the following quantities are specified: (i) the wet and dry bulb temperatures of the air at the inlet and exit of the heat exchanger, (ii) the inlet temperature of chilled water, which is flowing in the opposite direction to air, and (iii) the mass flow rates of air and water. Assume that condensation occurs at the air entrance section, 1 of the heat exchanger.

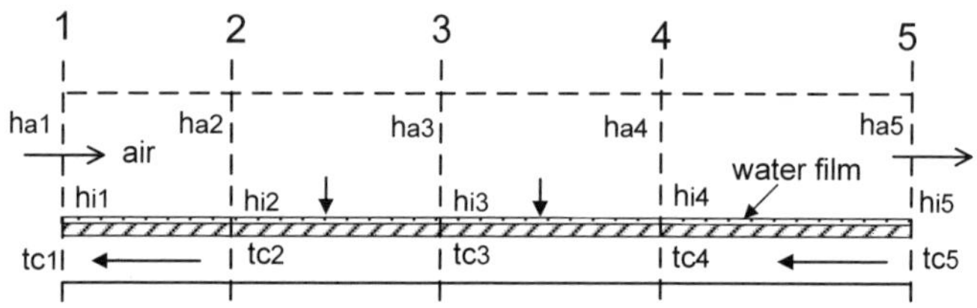

Fig. 7.15 Numerical model for wet-coil heat exchanger

For purposes of illustration of the numerical design procedure we divide the heat exchanger into 4 *unequal* control volumes as shown in Fig. 7.15. It is computationally more convenient to assume that the changes in air enthalpy across the different control volumes are equal. Therefore we have

$$h_{a1} - h_{a2} = h_{a2} - h_{a3} = h_{a3} - h_{a4} = h_{a4} - h_{a5} = \Delta h_a \qquad (7.65)$$

where $$\Delta h_a = \frac{(h_{a1} - h_{a5})}{4}$$

Note that the air enthalpies at the inlet and outlet h_{a1} and h_{a5} are obtained from the psychrometric chart using the specified conditions of the air at these sections.

Now if we neglect the enthalpy of the condensate water leaving each control volume, then for overall energy balance,

$$\dot{m}_c c_w \Delta t_c = \dot{m}_a \Delta h_a$$

where $$\Delta t_c = t_{c1} - t_{c2} = t_{c2} - t_{c3} = t_{c3} - t_{c4} = t_{c4} - t_{c5} \qquad (7.66)$$

Using Eqs. (7.65) and (7.66) we calculate the air enthalpies and the chilled water temperatures at all the sections from 1 to 5.

By eliminating h_i between Eqs. (7.62) and (7.63) we obtain the following cubic equation for t_i:

$$f_3(t_i) = h_a - \frac{(t_i - t_c)}{\alpha} \tag{7.67}$$

We could also select the relation in Eq. (7.64) instead of Eq. (7.63), in which case Eq. (7.67) is a quadratic equation. We solve Eq. (7.67) at the boundaries of the control volumes to obtain the water film temperatures t_{i1} to t_{i5} at the 5 sections. The saturation air enthalpies at the sections are then computed by substituting these temperatures in Eq. (7.63).

The heat transfer areas of the four control volumes are determined by applying the energy balance equation to each control volume. Hence for the nth control volume we have

$$q_{t,n} = \dot{m}_a\left[h_{a,n} - h_{a,(n+1)}\right]$$

The total energy transfer rate, $q_{t,n}$ is now expressed in terms of the *enthalpy potential* using Eq. (7.60). Hence we obtain

$$A_n(h_c/c_{am})\left[\frac{(h_{a,n}+h_{a,(n+1)})}{2} - \frac{(h_{i,n}+h_{i,(n+1)})}{2}\right] = \dot{m}_a\left[h_{a,n} - h_{a,(n+1)}\right] \tag{7.68}$$

Note that in Eq. (7.68) we use the *mean* enthalpies of air, and saturated air to determine the enthalpy potential in Eq. (7.60). The solution of Eq. (7.68) for $n = 1$ to 4 gives the heat transfer areas A_1 to A_4.

Since the heat transfer areas of the control volumes have been determined, the dry bulb temperatures of air, $t_{a,n}$ at the boundaries of the control volumes can be obtained by applying the *sensible* heat balance equation to each control volume. Hence for the nth control volume we have

$$A_n h_c\left[\frac{(t_{a,n}+t_{a,(n+1)})}{2} - \frac{(t_{i,n}+t_{i,(n+1)})}{2}\right] = \dot{m}_a c_{am}\left[t_{a,n} - t_{a,(n+1)}\right] \tag{7.69}$$

Similarly, the humidity ratio of the air, $\omega_{a,n}$ at the control volume boundaries are obtained by applying the water mass balance equation to each control volume. Hence for the nth control volume we have

$$A_n(h_c/c_{am})\left[\frac{(\omega_{a,n}+\omega_{a,(n+1)})}{2}-\frac{(\omega_{i,n}+\omega_{i,(n+1)})}{2}\right]=\dot{m}_a\left[\omega_{a,n}-\omega_{a,(n+1)}\right] \tag{7.70}$$

We have now obtained three properties of the air stream, namely, the enthalpy, the dry-bulb temperature, and the humidity ratio at the boundaries of the different control volumes. Using any two of these properties we are able to plot the *condition line* for the air passing through the cooling coil on the psychrometric chart. A MATLAB code to analyze a cooling and dehumidifying coil using the above numerical procedure is given in Appendix A7.1 at the end of this chapter.

The analysis of *direct expansion* cooling coils (DX coils), which use *refrigerants* instead chilled water as the coolant, is much simpler because the refrigerant temperature remains constant across the control volumes. We shall illustrate the analysis of such a coil in worked example 7.14.

If the heat exchanger has a dry section as shown in Fig. 7.12, then we determine the area of the dry control volume by noting that t_i at the boundary of this control volume is equal to the *dew-point* temperature of the air at the entrance. Detailed analysis of such a situation will be illustrated in worked examples in 7.17 and 7.18.

7.4 Worked Examples

Example 7.1 A multi-tube counter flow heat exchanger has 25 steel tubes of inner diameter 26 mm and outer diameter 30 mm. The inside and outside convective heat transfer coefficients are 520 $\mathrm{Wm^{-2}K^{-1}}$ and 220 $\mathrm{Wm^{-2}K^{-1}}$ respectively. The thermal conductivity of steel is 18 $\mathrm{Wm^{-1}K^{-1}}$. Calculate the overall heat transfer coefficient from the inner fluid to the outer fluid flowing through of heat exchanger.

Solution Consider the length *L* of the heat exchanger. Let the steady heat flow rate from the inner fluid to the outer fluid be *Q*. The convective heat flow rate from the inner fluid to the inner tube surface is

$$Q = 2\pi r_i L h_i (t_{fi} - t_{wi}) \tag{E7.1.1}$$

The rate of conduction heat flow through the tube wall is

$$Q = \frac{2\pi k_w L(t_{wi} - t_{wo})}{ln(r_o/r_i)} \tag{E7.1.2}$$

The rate of convective heat flow from the outer tube surface to the outer fluid is

$$Q = 2\pi r_o L h_o (t_{wo} - t_{fo}) \tag{E7.1.3}$$

If the overall heat transfer coefficient from the inner fluid to the outer fluid based on the *inner tube surface* area is *U*, then the heat transfer rate from the inner fluid to the outer fluid may be written as

$$Q = 2\pi r_i L U (t_{fi} - t_{fo}) \tag{E7.1.4}$$

From Eqs. (E7.1.1) to (E7.1.4) it follows that

$$\frac{1}{U} = \frac{1}{h_i} + \frac{r_i ln(r_o/r_i)}{k_w} + \frac{r_i}{r_o h_o} \tag{E7.1.5}$$

Substituting the given numerical values in Eq. (E7.1.5) we have

$$\frac{1}{U} = \frac{1}{520} + \frac{13\times10^{-3} ln(15/13)}{18} + \frac{13}{15\times220}$$

$$\frac{1}{U} = 1.923 \times 10^{-3} + 0.1033 \times 10^{-3} + 3.939 \times 10^{-3}$$

Hence the overall heat transfer coefficient, *U* is 167.6 $Wm^{-2}K^{-1}$. Note that the thermal resistance of the tube wall is relatively small.

Example 7.2 A single-pass counter-flow heat exchanger is used to heat water from 8°C to 15°C using hot oil at 130°C. The mass flow rates of water and oil are 2.2 kgs^{-1} and 1.9 kgs^{-1} respectively. Their respective specific heat capacities are 4.2 $kJkg^{-1}K^{-1}$ and 1.9 $kJkg^{-1}K^{-1}$. Calculate (i) the temperature of the oil leaving the heat exchanger, (ii) the log-mean temperature difference (LMTD), (iii) the heat transfer area, if the overall heat transfer coefficient is 240 $Wm^{-2}K^{-1}$, and (iv) the effectiveness.

Solution Assume that the operating conditions of the heat exchanger are steady and that there is no heat exchange with the surroundings.

(i) Applying the overall energy balance equation we have

$$\dot{m}_c c_c (t_{co} - t_{ci}) = \dot{m}_h c_h (t_{hi} - t_{ho}) \tag{E7.2.1}$$

Substituting numerical values in Eq. (E7.2.1) we obtain

$$2.2 \times 4.2(15 - 8) = 1.9 \times 1.9(85 - t_{ho})$$

Hence the outlet temperature of the oil is, $t_{ho} = 67.08°C$.

(ii) The LMTD is given by Eq. (7.10) as

$$LMTD = \frac{\Delta t_1 - \Delta t_2}{ln(\Delta t_1/\Delta t_2)} \quad \text{(E7.2.2)}$$

Substituting numerical values in Eq.(E7.2.2) we have

$$LMTD = \frac{(85-15)-(67.08-8)}{ln[(85-15)/(67.08-8)]} = 64.38°C$$

(iii) The total heat transfer rate is given by

$$Q = \dot{m}_c c_c(t_{co} - t_{ci}) = 2.2 \times 4.2 \times 7 = 64.68 \text{ kW}$$

The total heat transfer rate for a counter-flow heat exchanger is given by

$$Q = UA(LMTD)$$

Substituting numerical values in the above equation we have

$$64.68 = 0.24A(64.38)$$

Therefore the heat transfer area is, $A = 4.19$ m^2.

(iv) The maximum possible heat transfer rate is given by

$$Q_{max} = (\dot{m}c)_{min}(t_{hi} - t_{ci})$$

$$Q_{max} = 1.9 \times 1.9(85 - 8) = 278 \text{ kW}$$

The effectiveness of the heat exchanger is given by Eq. (7.13) as

$$\epsilon = \frac{Q}{Q_{max}} = \frac{64.68}{278} = 0.23$$

Example 7.3 Hot water flows in parallel through a single vertical row of tubes as shown in Fig. 7.6. The air flowing over the tubes, with a mass flow rate of 2.3 kgs^{-1}, is unmixed, while the water flowing in the tubes is fully mixed. The water enters at 70°C and leaves at 62°C. The air is heated from 20°C to 40°C. Calculate (i) the mass flow rate of water,

(ii) the product UA of the heat exchanger, and (iii) the effectiveness. Assume that the specific heat capacities of air and water are 1.03 $kJkg^{-1}K^{-1}$ and 4.2 $kJkg^{-1}K^{-1}$.

Solution A cross-flow heat exchanger with a single row of vertical tubes was analyzed in section 7.2.6. We shall apply the expressions obtained directly to solve the present problem.

(i) Applying the overall energy balance equation we have

$$\dot{m}_c c_c(t_{co} - t_{ci}) = \dot{m}_h c_h(t_{hi} - t_{ho}) \quad \text{(E7.3.1)}$$

Substituting numerical values in Eq. (E7.3.1)

$$2.3 \times 1.03 \times (40 - 20) = \dot{m}_h \times 4.2 \times (70 - 62)$$

Hence the mass flow rate of water is 1.41 kgs^{-1}.

(ii) The outlet water temperature is given by Eq. (7.37) as

$$t_{ho} = t_{hi} - (1 - e^{-\alpha_2})(t_{hi} - t_{ci}) \quad \text{(E7.3.2)}$$

Substituting the given numerical values in Eq. (E7.3.2) we have

$$62 = 70 - (1 - e^{-\alpha_2})(70 - 20)$$

Hence $\alpha_2 = 0.174$. From Eq. (7.38) we have

$$\alpha_2 = \left(\frac{\dot{m}_c c_c}{\dot{m}_h c_h}\right)\left[1 - exp\left(-\frac{AU}{\dot{m}_c c_c}\right)\right] \quad \text{(E7.3.3)}$$

Substituting numerical values in Eq. (E7.3.3) we obtain

$$0.174 = \left(\frac{2.3\times1.03}{1.41\times4.2}\right)\left[1 - exp\left(-\frac{AU}{2.3\times1.03}\right)\right]$$

Hence we have the product, $UA = 1.35$ kWK^{-1}.

(iii) The maximum possible heat transfer rate is given by

$$Q_{max} = (\dot{m}c)_{min}(t_{hi} - t_{ci})$$

$$Q_{max} = 2.3 \times 1.03(70 - 20) = 118.45 \text{ kW}$$

The actual heat transfer rate is

$$Q_{act} = 2.3 \times 1.03(40 - 20) = 47.38 \text{ kW}$$

The effectiveness of the heat exchanger is

$$\epsilon = \frac{Q_{act}}{Q_{max}} = \frac{47.38}{118.45} = 0.4$$

Example 7.4 The tubes of an air heater have circular fins of thickness 0.3 mm, with an inner diameter of 40mm and an outer diameter of 80 mm fitted to them. The thermal conductivity of the material of the fins is 160 $Wm^{-1}K^{-1}$. The fin base temperature is 80°C and the temperature of the air flowing over the fins is 15°C. The convective heat transfer coefficient from the fin surface to the air is 24 $Wm^{-2}K^{-1}$. Calculate (i) the fin efficiency, (ii) the maximum possible heat transfer from a fin, and (iii) the actual heat transfer rate from a fin.

Solution (i) The detailed analysis of the heat transfer in a circular fin of constant thickness is given section 7.2.7. The efficiency of the circular fin is given by Eq. (7.55) as

$$\eta_f = \frac{(2r_1/\beta)[K_1(\beta r_1)I_1(\beta r_2)-I_1(\beta r_1)K_1(\beta r_2)]}{(r_2^2-r_1^2)[K_0(\beta r_1)I_1(\beta r_2)+I_0(\beta r_1)K_1(\beta r_2)]} \quad \text{(E7.4.1)}$$

where

$$\beta^2 = \frac{h_c}{\delta k} = \frac{24}{(160\times 0.15\times 10^{-3})} = 1000$$

Therefore $\beta = 31.62$ m^{-1} and $\beta r_1 = 0.6325$, $\beta r_2 = 1.265$.

The values of the required Bessel functions are obtained directly using the besseli(n,x) and besselk(n,x) functions in the MATLAB software package. Here n is the order of the Bessel function, 0 or 1, and x is the argument. Hence we have the following values:

$$I_1(\beta r_1) = 0.3323,\ I_1(\beta r_2) = 0.7677,\ I_0(\beta r_1) = 1.1025$$

$$K_1(\beta r_1) = 1.212,\ K_1(\beta r_2) = 0.393,\ K_0(\beta r_1) = 0.7367$$

Substitute these numerical values in Eq. (E7.4.1). Hence we obtain the fin efficiency as 0.844. Note that we could also obtain the fin efficiency of the circular fin directly from the curves in Fig. 7.9.

(ii) The maximum possible heat transfer occurs if the entire fin surface is at the temperature of the fin base. Therefore

$$\dot{Q}_{max} = 2\pi(r_2^2 - r_1^2)h_c(t_b - t_a)$$

Substituting numerical values in the above equation we have

$$\dot{Q}_{max} = 2\pi(40^2 - 20^2) \times 10^{-6} \times 24(80 - 15) = 11.76 \text{ W}$$

(iii) The actual heat transfer from the fin is $0.844 \times 11.76 = 9.93$ W.

Example 7.5 An air heater using hot water as the heating medium has staggered rows of vertical tubes and plate fins as shown schematically in Fig. E7.5. The following design data on the air heater are available from the manufacturer: outer radius of a tube = 6.3 mm, thickness of a tube = 0.7 mm, horizontal spacing between vertical rows of tubes = 43 mm, vertical spacing between tubes = 38 mm, fin thickness = 0.25 mm, fin spacing = 300 fins per meter, thermal conductivities of the materials of the fins and the tubes are 160 $Wm^{-1}K^{-1}$ and 150 $Wm^{-1}K^{-1}$ respectively. The convective heat transfer coefficients on the outside and the inside are 62 $Wm^{-2}K^{-1}$ and 3350 $Wm^{-2}K^{-1}$ respectively. Calculate the overall heat transfer coefficient from the water to the air.

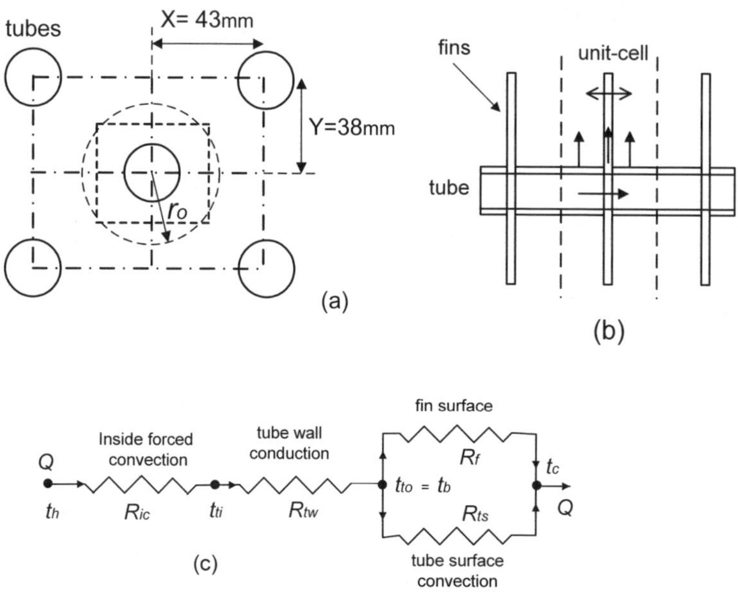

Fig. E7.5 (a) Staggered array of tubes with plate fins, (b) the unit-cell and (c) the thermal network

Solution An exact analysis of heat transfer through the plates is difficult due to the interaction of heat flows from the different tubes. We therefore consider an equivalent circular fin that has the same area as a rectangle of area *xy*, surrounding a tube, as shown in Fig. E7.5(a). Hence the radius, r_o of the equivalent circular fin given by

$$r_o = \sqrt{xy/\pi} = \sqrt{43 \times 38/\pi} = 22.8 \text{ mm}$$

The dimensionless parameters for this circular fin are (see Fig. 7.9):

$$z = \frac{r_o}{r_i} = \frac{22.8}{6.3} = 3.62 \quad \text{and}$$

$$x = (r_o - r_i)\sqrt{\frac{h_c}{k\delta}} = (22.8 - 6.3)10^{-3}\sqrt{\frac{62}{160 \times 0.125 \times 10^{-3}}} = 0.918$$

The efficiency of the fin may be obtained from the curves in Fig. 7.9 or by following the procedure outlined in worked example 7.4. Using the latter we obtain, $\eta_f = 0.663$.

Consider the heat flow in a *unit-cell* consisting of a single equivalent circular fin and two equal lengths of tube on either side as depicted in Fig. E7.5(b). The width of the unit cell is 1000/300 =3.33 mm. The following heat transfer areas can now be calculated.

Inner tube area of the unit cell is

$$A_{ti} = 2\pi \times 5.6 \times 10^{-3} \times 3.33 \times 10^{-3} = 117.17 \times 10^{-6} \text{ m}^2$$

Total fin area of the unit cell is

$$A_f = 2\pi(22.8^2 - 6.3^2) \times 10^{-6} = 3016.9 \times 10^{-6} \text{ m}^2$$

Total area of the outer surface of the tube sections in the unit cell is

$$A_{ts} = 2\pi \times 6.3 \times 10^{-3} \times 3.083 \times 10^{-3} = 122.04 \times 10^{-6} \text{ m}^2$$

The thermal network for the heat flow in the unit cell is shown in Fig. E7.5 (c). The thermal resistances indicated in Fig. E7.5(c) are given by:

$$R_{ic} = \frac{1}{(A_{ti}h_{ti})} = \frac{1}{117.17 \times 3350 \times 10^{-6}} = 2.5476 \text{ K.W}^{-1}$$

$$R_{tw} = \frac{\delta_t}{A_{ti}k_t} = \frac{0.7 \times 10^{-3}}{117.17 \times 150 \times 10^{-6}} = 3.98 \times 10^{-2} \text{ K.W}^{-1}$$

$$R_f = \frac{1}{A_f \eta_f h_c} = \frac{1}{3016.9 \times 10^{-6} \times 0.663 \times 62} = 8.06 \text{ K.W}^{-1}$$

$$R_{ts} = \frac{1}{(A_{ts}h_{to})} = \frac{1}{122.04\times10^{-6}\times62} = 132.2\ \text{K.W}^{-1}$$

The overall thermal resistance is given by Eq. (7.59) as

$$R_{tot} = \frac{1}{U_oA_{ti}} = R_{ic} + R_{tw} + \left(\frac{1}{R_f} + \frac{1}{R_{ts}}\right)^{-1}$$

$$R_{tot} = 2.5476 + 3.98\times10^{-2} + \left(\frac{1}{8.06} + \frac{1}{132.2}\right)^{-1} = 10.18\ \text{K.W}^{-1}$$

Hence the overall heat transfer coefficient, U_o based on the inner tube area is

$$U_0 = \frac{1}{R_{tot}A_{ti}} = \frac{1}{(10.18\times117.17\times10^{-6})} = 838.4\ \text{Wm}^{-2}\text{K}^{-1}$$

Example 7.6 In a heating system, ambient air at 15°C enters the annulus between a cylindrical resistance heater of diameter 4cm, and an outer coaxial metal cylinder of inner diameter 10 cm. The axial lengths of the heater and the cylinder are 2 m. The mass flow rate of air is 0.2 kgs^{-1}. The electrical resistance of the heater is 20 ohms and it produces a uniform heat flux at its outer surface.

A direct current of 10 amps flows through the resistance heater. The convective heat transfer coefficient between the air and the heater is 520 $\text{Wm}^{-2}\text{K}^{-1}$. The cylinder is well insulated on the outside. (i) Obtain expressions for the axial temperature distribution of the air, and the heater surface. (ii) Calculate the temperature of the air leaving the cylinder. (iii) Calculate the temperature of the heater surface at the exit.

Solution

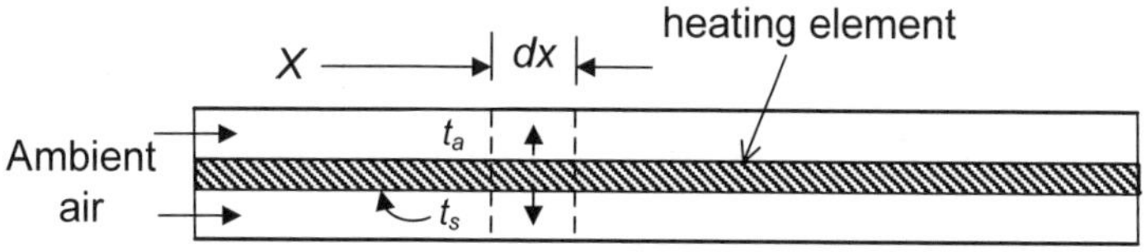

Fig. E7.6 Air heater with electrical heating element

The total electrical energy input rate to the cylindrical heater is given by

$$E = i^2R = 20\times100 = 2000\ \text{W}$$

The heating rate per unit length is Q_e = 1 kW per m. Since the cylindrical heating element is closed, when conditions are steady, all the internal energy generated will be supplied uniformly to the air, neglecting any end losses. Assume that the pipe is well insulated on the outside so that the heat loss to the surroundings is negligible.

Consider the small control volume of length *dx* of the heater shown in Fig. E7.6. Applying the steady-flow energy equation to the control volume we have

$$\dot{m}_a c_a dt_a = 2\pi r_o h_o (t_s - t_a) dx = Q_e dx \qquad \text{(E7.6.1)}$$

where h_o is the convective heat transfer coefficient. From Eq. (E7.6.1) it follows that

$$t_a(x) = \left(\frac{Q_e}{\dot{m}_a c_a}\right) x + t_{ai} \qquad \text{(E7.6.2)}$$

where t_{ai} is the inlet temperature of the air. Substituting numerical values in Eq. (E7.6.2) we have

$$t_a(x) = \left(\frac{1}{0.2 \times 1.02}\right) x + 15 = 4.902x + 15 \qquad \text{(E7.6.3)}$$

which is the expression for the air temperature distribution.

We also have from Eq. (E7.6.1)

$$t_s(x) = \left(\frac{Q_e}{2\pi r_o h_o}\right) + t_a(x) \qquad \text{(E7.6.4)}$$

Substituting from Eq. (E7.6.3) in Eq. (E7.6.4) we obtain

$$t_s(x) = \left(\frac{1}{2 \times \pi \times 0.02 \times 0.52}\right) x + 4.902x + 15$$

$$t_s(x) = 4.902x + 30.3$$

which is the heater surface temperature distribution.

(ii) The air temperature at the exit where, $x = 2$ m is $t_{ao} = 24.8$°C.

(iii) The temperature of the heater surface at the exit is $t_{so} = 40.1$°C

Example 7.7 Chilled water, leaving the chiller of an air conditioning system at 4°C, flows at the rate of 0.8 kgs^{-1} to an air handling unit (AHU) through a metal pipe of inner diameter 10 cm, wall thickness 4

mm and length 50 m. The thermal conductivity of the material of the pipe is 60 $Wm^{-1}K^{-1}$. The pipe has thermal insulation of thickness of 1.5 cm and thermal conductivity 0.06 $Wm^{-1}K^{-1}$ on the outside. The ambient air surrounding the insulation has a db-temperature of 30°C. The outside and inside convective heat transfer coefficients are 30 $Wm^{-2}K^{-1}$ and 560 $Wm^{-2}K^{-1}$ respectively. (i) Obtain an expression for the temperature distribution of the water in the pipe. (ii) Calculate the temperature of the chilled water reaching the AHU.

Solution

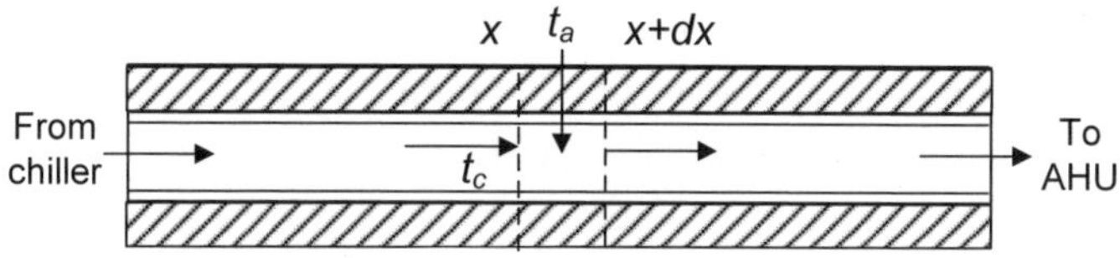

Fig. E7.7 Insulated chilled water pipe

Consider the small section *dx* of the pipe at a distance *x* from the chiller as shown in Fig. E7.7. The overall heat transfer coefficient *U* from the ambient air to the chilled water is given by

$$\frac{1}{U} = \frac{1}{h_i} + \frac{r_i ln(r_o/r_i)}{k_w} + \frac{r_i ln(r_{io}/r_o)}{k_i} + \frac{r_i}{r_{io}h_o} \tag{E7.7.1}$$

where r_i and r_o are the inner and outer radii of the pipe, r_{io} is the outer radius of the insulation. The thermal conductivities of the pipe and insulation are k_w and k_i respectively. The inner and outer heat transfer coefficients are h_i and h_o respectively.

Substituting numerical values in Eq. (E7.7.1) we obtain

$$\frac{1}{U} = \frac{1}{560} + \frac{0.05ln(0.054/0.05)}{60} + \frac{0.05ln(0.069/0.054)}{0.06} + \frac{0.05}{0.069x30}$$

Hence we have, $U = 4.34\ Wm^{-2}K^{-1}$

Applying the energy balance equation to a small control volume of width *dx* we have

$$\dot{m}_c c_c dt_c = 2\pi r_i U(t_a - t_c)dx \tag{E7.7.2}$$

The solution of Eq. (E7.7.2) gives the chilled water temperature distribution $t_c(x)$ as

$$\frac{t_a - t_c(x)}{t_a - t_{ci}} = exp\left(-\frac{2\pi r_i U x}{\dot{m}_c c_c}\right) \quad \text{(E7.7.3)}$$

where t_{ci} is the inlet chilled water temperature.

The water temperature t_{co}, at the AHU (x = 50m) is obtained by substituting numerical values in Eq. (E7.7.3). Hence we have

$$\frac{30 - t_{co}}{30 - 4} = \exp\left(\frac{-2\pi \times 0.05 \times 4.34 \times 50}{0.8 \times 4200}\right) = 0.98$$

Therefore the chilled water temperature at the AHU is 4.5°C.

Example 7.8 Shown schematically in Fig. E7.8 is a heat recovery system, commonly called as a run-around coil, where warm air discharged from a room is used to preheat ventilation air from the ambient. A pump circulates water at the rate of 1.8 kgs^{-1} through two heat exchangers, each of which has an effectiveness of 0.6. Air discharged from the room, maintained at 28°C, flows through one of the heat exchangers at the rate of 4.5 kgs^{-1}. Ambient air at 2°C flows through the other heat exchanger at the rate of 4.5kgs^{-1}. Calculate (i) the rate of heat transfer from room air to water, and (ii) the temperature of ambient air leaving the heat exchanger. Assume that the specific heat capacities of air and water are 1.02 kJkg^{-1}K^{-1} and 4.2 kJkg^{-1}K^{-1}.

Solution

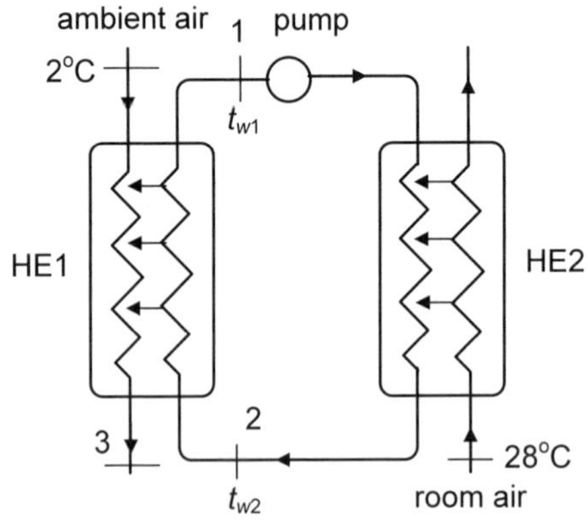

Fig.E7.8 Heat recovery system for building

Now for both counter flow heat exchangers HE1 and HE2 depicted in Fig. E7.8,

$$(\dot{m}c)_{min} = \dot{m}_a c_a = 4.5 \times 1.02 = 4.59 \text{ kWK}^{-1}$$

Applying the energy balance equation to the two heat exchangers we obtain the heat transfer rates as

$$Q_1 = 4.59(t_{w2} - 2)\varepsilon_1 = 2.754(t_{w2} - 2) \qquad \text{(E7.8.1)}$$

$$Q_2 = 4.59(28 - t_{w2})\varepsilon_2 = 2.754(28 - t_{w1}) \qquad \text{(E7.8.2)}$$

Applying the energy equation to the water in the two heat exchangers we have

$$Q_w = 1.8 \times 4.2(t_{w2} - t_{w1}) \qquad \text{(E7.8.3)}$$

From the energy balance of the two heat exchangers it follows that

$$Q_1 = Q_2 = Q_w \qquad \text{(E7.8.4)}$$

Manipulating Eqs. (7.8.1) to (7.8.4) we obtain

$$\frac{Q_w}{2.754} + \frac{Q_w}{2.754} - \frac{Q_w}{1.8 \times 4.2} = 28 - 2 = 26$$

Hence the heat transfer rate, $Q_w = 43.8$ kW.

Applying the energy equation to air in HE1 we have

$$Q_1 = Q_w = 43.8 = 4.5 \times 1.02(t_3 - 2)$$

Therefore the temperature of air leaving HE1 is, $t_{a3} = 11.54°\text{C}$.

Example 7.9 A solar collector producing hot air for heating a house is shown schematically in Fig. E7.9. Ambient air at 12°C enters the rectangular channel of width 1 m and length 2m of the collector with a mass flow rate of 0.018 kgs^{-1}. The upper surface of the channel is a transparent glass sheet while the bottom surface is a metal sheet coated with solar radiation absorbing paint.

On a sunny day the metal surface absorbs solar radiation at the rate of 0.6 kWm^{-2}. All this energy flows to the air by convection. The convective heat transfer coefficient from the inner glass surface to air flowing in the channel is 40 Wm^{-2}K^{-1} and the heat transfer coefficient from the outer glass surface to the ambient is 10 Wm^{-2}K^{-1}. (i) Obtain an expression for the axial temperature distribution of the air. (ii) Calculate the temperature of the air leaving the collector.

Solution A schematic view of the solar air heater is shown in Fig. E7.9. The solar radiation passing through the transparent glass cover is absorbed completely by the metal absorber plate. Since the absorber plate is well insulated at the back, all the solar radiation absorbed by the plate is transferred to the air stream by convection when conditions are steady. Thus the air stream gets heated. However, the air stream looses a fraction of its enthalpy to the glass sheet by convection. The glass, in turn, transfers some of its energy to the ambient.

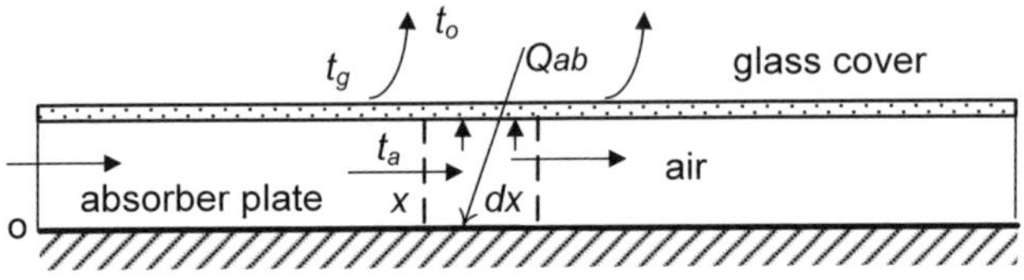

Fig. E7.9 Solar air heater

Consider a small control volume of length *dx* at a distance *x* from the air entrance section. Applying the steady-flow energy equation to this control volume we have

$$\dot{m}_a c_a dt_a = Q_{ab} B dx + h_{ag}(t_a - t_g) B dx \qquad \text{(E7.9.1)}$$

where h_{ag} is the heat transfer coefficient from the air to the glass and B is the width of the channel. The glass and air temperatures are t_g and t_a respectively.

Applying the energy balance equation to the glass element we obtain

$$h_{ag}(t_a - t_g) B dx = h_{go}(t_g - t_o) B dx \qquad \text{(E7.9.2)}$$

where h_{go} is the heat transfer coefficient from the outer glass surface to the ambient. The ambient temperature is t_o.

Eliminating t_g between Eqs. (E7.9.1) and (E7.9.2) we have

$$\dot{m}_a c_a dt_a = Q_{ab} B dx - U(t_a - t_o) B dx \qquad \text{(E7.9.3)}$$

where $$\frac{1}{U} = \frac{1}{h_{go}} + \frac{1}{h_{ag}} = \frac{1}{40} + \frac{1}{10} = 0.125$$

Rearranging Eq. (E7.9.3) we obtain

$$dt_a = \left(\frac{BU}{\dot{m}_a c_a}\right)\left[\left(t_o + \frac{Q_{ab}}{U}\right) - t_a\right]dx \qquad \text{(E7.9.4)}$$

Substituting numerical values in Eq. (E7.9.4) we have

$$dt_a = \left(\frac{1\times 8}{0.018\times 1020}\right)\left[\left(12 + \frac{600}{8}\right) - t_a\right]dx$$

$$dt_a = 0.436(87 - t_a)dx \qquad \text{(E7.9.5)}$$

The solution of Eq. (E7.9.5) gives the air temperature distribution as

$$\frac{87 - t_a(x)}{87 - t_{ai}} = exp(-0.436x)$$

where t_{ai} = 12°C is the inlet air temperature at x = 0. Hence we have

$$t_a(x) = 87 - 75exp(-0.436x)$$

Therefore the exit air temperature at, x = 2m is 55.6°C.

Example 7.10 The shell and tube evaporator of a refrigeration plant producing chilled water has 15 stainless steel tubes of outer diameter 20 mm. The total water flow rate through the tubes is 0.09 kgs^{-1} and the temperature at the entrance is 12°C. Refrigerant R134a at a saturation temperature of 6°C undergoes *pool boiling* on the outer surface of the tubes. The forced convection heat transfer coefficient for water flowing through the tubes is 950 $Wm^{-2}K^{-1}$.

(i) Calculate the *pool boiling* heat transfer coefficient assuming that the tube surface temperature is 9°C.

(ii) Calculate the overall heat transfer coefficient between the water and the refrigerant. Neglect the effects of the tube wall thickness.

(iii) If the effectiveness of the evaporator is 0.6, calculate the length of a tube and the water outlet temperature.

Solution We calculate the heat transfer coefficient h_b, for pool boiling on the surface of the tube using the correlation developed by W.M. Rohsenow [2,5]. This may be expressed in the form

$$Nu = \frac{h_b L_c}{k_l} = \frac{Ja^2}{C^3(Pr_l)^m} \qquad \text{(E7.10.1)}$$

where the Jakob number, Ja is given by

$$Ja = c_l(t_s - t_{sat})/h_{fg} \tag{E7.10.2}$$

and the characteristic length, L_c is given by

$$L_c = \left[\frac{\sigma}{g(\rho_l - \rho_v)}\right]^{1/2} \tag{E7.10.3}$$

where Pr_l is the liquid Prandtl number, t_s is the surface temperature, t_{sat} is the saturation temperature, and σ is the surface tension. The constants C and m are specific to the surface and the fluid undergoing pool boiling. The following properties of refrigerant R134a at 6°C are obtained from data tables in Ref. [5].

$$\rho_l = 1273.8 \text{ kgm}^{-3}, \rho_v = 17.7 \text{ kgm}^{-3}, h_{fg} = 104.06 \text{ kJkg}^{-1},$$

$$c_l = 1.353 \text{ kJkg}^{-1}\text{K}^{-1}, \sigma = 10.86 \times 10^{-3} \text{ Nm}^{-1},$$

$$k_l = 90.7 \times 10^{-3} \text{ Wm}^{-1}\text{K}^{-1}, Pr_l = 3.98.$$

Acceleration due to gravity, $g = 9.81\text{ms}^{-2}$. For the stainless steel surface [5], C= 0.015 and m = 2.
We substitute the above numerical values in Eqs. (E7.10.1) to (E7.10.3). Hence we obtain

$$L_c = \left[\frac{10.86\times10^{-3}}{9.81(1273.8-17.7)}\right]^{1/2} = 0.938 \times 10^{-3} \text{ m}$$

$$Ja = \frac{1.353(9-6)}{194.06} = 0.0209$$

$$Nu = \frac{0.938\times10^{-3}h_b}{90.7\times10^{-3}} = \frac{0.0209^2}{0.015^3(3.98)^2} = 8.17$$

(i) Therefore the pool boiling heat transfer coefficient, h_b is 790 $\text{Wm}^{-2}\text{K}^{-1}$.

(ii) The overall heat transfer coefficient is given by

$$\frac{1}{U} = \frac{1}{h_b} + \frac{1}{h_c} = \frac{1}{790} + \frac{1}{950} = 2.318 \times 10^{-3}$$

Hence the overall heat transfer coefficient, U is 431 $\text{Wm}^{-2}\text{K}^{-1}$.

(iii) The flow rate of water per tube is 0.09/15 = 0.006 kgs^{-1}.
Now the effectiveness of the evaporator is given by Eq. (7.28) as

$$\epsilon = 1 - exp(-NTU) = 0.6$$

Therefore $$NTU = 0.916 = \frac{UA_t}{\dot{m}_w c_w} = \frac{0.431 A_t}{0.006 \times 4.2}$$

Hence the area of heat transfer, $A_t = 0.0535\ \text{m}^2$.

Now $$A_t = \pi DL = \pi \times 20 \times 10^{-3} L = 0.0535\ \text{m}^2$$

Therefore the length of the evaporator is, $L = 0.85$ m.

(iv) Applying Eq. (7.27) to the evaporator we have

$$Q_a = \dot{m}_w c_w (t_{wi} - t_{wo}) = \dot{m}_w c_w (t_{wi} - t_{ref})\varepsilon$$

Substituting numerical values in the above equation we obtain

$$(12 - t_{wo}) = (12 - 6) \times 0.6$$

Hence the water outlet temperature is 8.4°C

Example 7.11 Refrigerant 134a flows at the rate of 0.09 kgs^{-1} through the inner tube of a double-pipe evaporator while water flows in the opposite direction at the rate of 0.1 kgs^{-1}. The inner diameter of the tube is 24 mm. The qualities (dryness) of the refrigerant at the entrance and exit are 0.05 and 0.45 respectively. The respective entry temperatures of the refrigerant and water are 5°C and 25°C. The convective-boiling heat transfer coefficient for the refrigerant in the tube has been estimated as 8.0 $\text{kWm}^{-2}\text{K}^{-1}$. The convective heat transfer coefficient for water in the annulus is 3.6 $\text{kWm}^{-2}\text{K}^{-1}$. Calculate (i) the temperature of the water leaving the evaporator, (ii) the effectiveness of the evaporator, (iii) the length of the evaporator, and (iv) the water temperature at the middle of the evaporator. Neglect the effects due to the tube wall thickness.

Solution (i) Applying the overall energy balance equation to the evaporator we have

$$\dot{m}_w c_w (t_{wi} - t_{wo}) = \dot{m}_r h_{fg} (x_o - x_i) \tag{E7.11.1}$$

where h_{fg} is the latent heat of vaporization of the refrigerant which is obtained from the tables in Ref. [7]. x_i and x_o are qualities of the refrigerant at the inlet and outlet of the evaporator respectively. Substituting numerical values in Eq. (E7.11.1) we obtain

$$0.1 \times 4.18(25 - t_{wo}) = 0.09 \times 194.58(0.45 - 0.05)$$

Hence we have the outlet water temperature as, $t_{wo} = 8.24$°C.

(ii) The overall heat transfer coefficient, U from water to refrigerant is given by

$$\frac{1}{U} = \frac{1}{h_{ref}} + \frac{1}{h_{water}} = \frac{1}{8.0} + \frac{1}{3.6} = 0.4028$$

Therefore U= 2.483 $\text{kWm}^{-2}\text{K}^{-1}$

Applying Eq. (7.27) to the evaporator we have

$$Q_a = \dot{m}_w c_w(t_{wi} - t_{wo}) = \dot{m}_w c_w(t_{wi} - t_{ref})\varepsilon$$

Substituting numerical values in the above equation we obtain

$$(25 - 8.24) = (25 - 5)\varepsilon$$

Hence the effectiveness is 0.838.

(iii) Applying Eq. (7.28) to the evaporator we have

$$\varepsilon = 1 - exp(-NTU) = 0.838$$

Therefore, $$NTU = 1.82 = \frac{UA_t}{\dot{m}_w c_w} = \frac{2.483A_t}{0.1 \times 4.18}$$

Hence the area of heat transfer of the evaporator, $A_t = 0.3064$ m^2.

Now $$A_t = \pi DL = \pi \times 24 \times 10^{-3}L = 0.3064 \text{ m}^2$$

Therefore the length of the evaporator is 4.06 m.

(iv) We obtain the water temperature at the mid-point of the evaporator, t_{mp} by applying Eq. (7.25) with the area equal to 0.5A. Hence we have

$$\frac{t_{mp} - t_r}{t_{ci} - t_r} = exp\left(\frac{-0.5UA}{\dot{m}_w c_w}\right)$$

Substituting numerical values in the above equation we have

$$\frac{t_{mp} - 5}{25 - 5} = exp\left(\frac{-0.5 \times 2.483 \times 0.3064}{0.1 \times 4.18}\right) = 0.4025$$

Hence the water temperature at the mid-point is, $t_{mp} = 13.05$°C.

Example 7.12 An air conditioning system has an air-cooled condenser that rejects 60 kW of heat to the air flowing over the finned tubes at the rate of 7.4 kgs^{-1}. The outside heat transfer area of the condenser is 200 m^2 and the overall heat transfer coefficient from refrigerant to air is 35 $Wm^{-2}K^{-1}$. The temperature of the condensing refrigerant is 50°C. Calculate (i) the temperature of the air at the inlet to the condenser, (ii) the effectiveness of the condenser, and (iii) the LMTD.

Solution Assume that the refrigerant temperature in the condenser is constant. The *NTU* of a condenser was defined in Eq. (7.28a) as

$$NTU = \frac{UA}{\dot{m}_a c_a} = \frac{200\times35\times10^{-3}}{7.4\times1.02} = 0.927$$

The effectiveness of the condenser is given by Eq. (7.28) as

$$\varepsilon = 1 - exp(-NTU) = 0.604$$

The maximum possible heat transfer rate from the refrigerant to the air is

$$\dot{Q}_{max} = \dot{m}_a c_a (t_{ref} - t_{ai})$$

Therefore the actual heat transfer rate is given by

$$\dot{Q} = \varepsilon \dot{Q}_{max} = \dot{m}_a c_a (t_{ref} - t_{ai})\varepsilon$$

$$60 = 7.4 \times 1.02 \times (50 - t_{ai}) \times 0.604$$

Hence temperature of air at the inlet is, t_{ai} = 36.8°C.
For a counter-flow evaporator, the LMTD is given by Eq. (7.11) as

$$\dot{Q} = UA\,(LMTD)$$

$$60 = 35 \times 10^{-3} \times 200 \times (LMTD)$$

Hence the LMTD is 8.57°C.

Example 7.13 The width, breadth and height of a vertical rectangular-duct condenser are 0.2 m, 0.5 m and 1.1 m respectively. Air entering the duct at 34°C flows up through the duct at the rate of 0.5 kgs^{-1}. Saturated refrigerant R113 at 47.7°C condenses on the two sides of the duct with dimensions of 1.1 m by 0.5 m as shown in Fig. E7.13.1. The two narrow sides of the duct are well insulated. Calculate (i) the average laminar condensation heat transfer coefficient, assuming that the two surfaces where condensation occurs are at a uniform temperature of 43°C, (ii) the total rate of heat transfer to the air, (iii) the total rate of condensation of refrigerant, (iv) the outlet temperature of the air, and (v) the effectiveness of the condenser.

Solution (i) We use Nusselt's theory to calculate the laminar condensation heat transfer coefficient on a vertical plate. The analysis and the final form of the correlation is available in most standard text books on heat transfer, including Refs. [2,5]. The average heat transfer coefficient is given by

$$h_{ave} = 0.943\left[\frac{h_{fg}g(\rho_l-\rho_v)k_l^{3}}{L(t_{sat}-t_{wall})\nu_l}\right]^{1/4} \quad \text{(E7.13.1)}$$

where t_{wall} is the temperature of the plate which is assumed uniform and t_{sat} is the saturation temperature of refrigerant vapor surrounding the plate. The kinematic viscosity of the liquid refrigerant is ν_l and its thermal conductivity is k_l. The following properties of refrigerant R113 at 47.7°C are obtained from the tables in Ref. [5]:

$$h_{fg} = 144\times10^{3}\ \text{Jkg}^{-1},\ k_l = 0.07\ \text{Wm}^{-1}\text{K}^{-1},\ \rho_l = 1507\ \text{kgm}^{-3},$$

$$\rho_l = 7.1\ \text{kgm}^{-3},\ \nu_l = 0.34\times10^{-6}\ \text{m}^2\text{s}^{-1}.$$

Acceleration due to gravity, $g = 9.81\text{ms}^{-2}$.

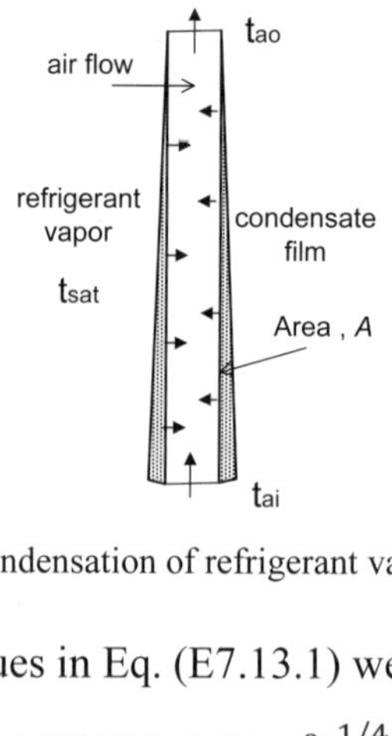

Fig. E7.13.1 Condensation of refrigerant vapor on plate

Substituting numerical values in Eq. (E7.13.1) we have

$$h_{ave} = 0.943\left[\frac{144\times10^{3}\times9.81(1507-7.1)0.07^{3}}{1.1(47.7-43)0.34\times10^{-6}}\right]^{1/4} = 0.756\ \text{kWm}^{-2}\text{K}^{-1}$$

(ii) The total heat transfer rate from the condensing refrigerant to the air flowing through the rectangular duct is given by (see Fig. E7.13.1)

$$\dot{Q}_{tot} = 2Ah_{ave}(t_{vap} - t_{wall})$$

$$\dot{Q}_{tot} = 2 \times 1.1 \times 0.5 \times 0.756(47.7 - 43) = 3.908 \text{ kW}$$

(iii) Applying the energy balance equation to the refrigerant we have

$$\dot{Q}_{tot} = \dot{m}_{con}h_{fg} = 144\dot{m}_{con} = 3.908 \text{ kW}$$

Therefore the rate of condensation of refrigerant, $\dot{m}_{con}$ is 0.027 kgs^{-1}.

(iv) Applying the energy balance equation to the air we have

$$\dot{Q}_{tot} = \dot{m}_a c_a(t_{ao} - t_{ai})$$

$$3.908 = 0.5 \times 1.1(t_{ao} - 34)$$

Therefore the outlet air temperature, t_{ao} = 41.1°C.

(v) Applying Eq. (7.27) to the condenser we obtain

$$\dot{m}_a c_a(t_{ao} - t_{ai}) = \dot{m}_a c_a(t_{ref} - t_{ai})\varepsilon$$

$$(41.1 - 34) = (47.7 - 34)\varepsilon$$

Therefore the effectiveness of the condenser is 0.52.

Example 7.14 Ambient air at 29°C db–temperature and 67% relative humidity enters a cooling and dehumidifying coil (DX coil) with a mass flow rate of 2.75 kgs^{-1}. The pressure is 101.3 kPa. Refrigerant 134a at 7°C flows through the vertical rows of horizontal tubes as shown schematically in Fig. E7.14.1. The total mass flow rate of the refrigerant is 0.4 kgs^{-1}. The face area of the coil is 0.96 m^2.

The air-side heat transfer area of the coil is 15 m^2 per m^2 of face area per row of tubes. The ratio of the air-side to the refrigerant-side heat transfer areas is 14. The cooling capacity of the coil is 52 kW. The air-side and refrigerant-side convective heat transfer coefficients are 0.065 kWm^{-2}K^{-1} and 2.05 kWm^{-2}K^{-1} respectively. The latent heat of vaporization of the refrigerant is 192.7 kJkg^{-1}. Calculate (i) the face velocity, (ii) the enthalpy of air at the outlet, (iii) the total air-side surface area of the coil, (iv) the number of rows of tubes, and (v) the change in quality of the refrigerant.

Solution As indicated schematically in Fig. E7.14.1(a) the refrigerant flows through rows of horizontal tubes. The overall flow direction of the refrigerant is counter to that of moist air. The condensate film formed on the tube surfaces drains by gravity to the condensate pan at the bottom from which it is discharged.

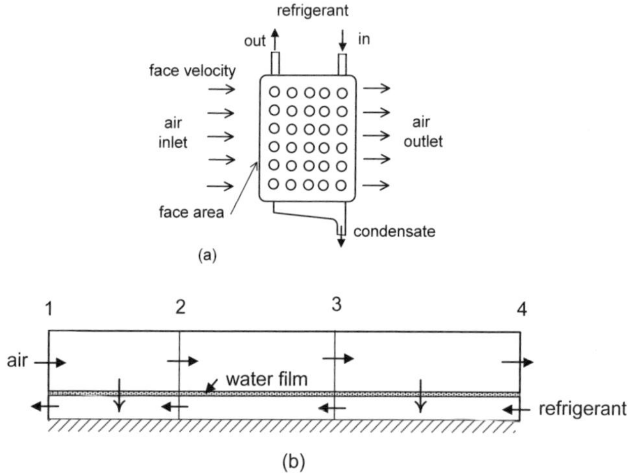

Fig. E7.14.1 (a) Direct-expansion (DX) cooling coil, (b) Physical model

(i) The specific volume of the inlet air is obtained from the psychrometric chart (Fig. 4.5) as 0.877 m^3kg^{-1}. The face velocity is given by

$$V_{face} = \frac{\dot{m}_a}{(\rho_{air}A_{face})} = \frac{2.75\times 0.877}{0.96} = 2.5 \text{ ms}^{-1}$$

(ii) The idealized physical model of the cooling coil is shown in Fig. E7.14.1(b). Here the air flows over the top surface of a flat duct through which refrigerant flows in the opposite direction. We shall assume the duct surface temperature at the entrance section 1, to be below the dew-point temperature of air. Therefore condensation of water vapor would begin at the entrance section 1 of the coil. However, if the calculation gives a surface temperature above the dew-point, we shall revise the analysis assuming the presence of a dry section in the coil.

From Eqs. (7.60) and (7.61) we obtained the following relation.

$$\frac{t_i - t_r}{h_a - h_i} = \frac{h_c \delta A_o}{c_{am} h_r \delta A_i} = \alpha \text{ (say)} \qquad \text{(E7.14.1)}$$

Substituting the given numerical data we have

$$\frac{t_i - t_r}{h_a - h_i} = \frac{65 \times 14}{1.02 \times 2050} = 0.435 \qquad \text{(E7.14.2)}$$

Equation (7.14.2) is applicable to any section of the coil where condensation occurs. From the given inlet conditions we obtain the air enthalpy at section 1 as, $h_{a1} = 71.8$ kJkg^{-1}. Applying the overall energy balance to the air we have

$$\dot{Q}_c = \dot{m}_a \left(h_{a,in} - h_{a.out}\right)$$

Substituting numerical values in the above equation we obtain

$$52 = 2.75\left(71.8 - h_{a,out}\right)$$

Therefore the exit enthalpy of the air is 53 kJkg^{-1}.

In order to illustrate the details of the computational procedure we divide the coil into 3 control volumes as shown in Fig. E7.14.1(b). The heat transfer areas of the control volumes are such that the enthalpy drop of the air is the same for all of them. This enthalpy drop is given by

$$\Delta h_a = \left(h_{a,in} - h_{a,out}\right)/3 = (71.8 - 53)/3 = 6.27$$

Therefore the air enthalpies at the boundaries of the control volumes are:

$$h_{a1} = 71.8\ , \quad h_{a2} = 65.53, \quad h_{a3} = 59.26, \quad h_{a4} = 53 \text{ kJkg}^{-1}$$

We now calculate the water film or condensate temperatures at the boundaries of the control volumes by using Eq. (E7.14.2). For ease of computation we shall use the quadratic relationship between the saturation air enthalpy and the temperature [Eq. (7.64)].

$$h_i = f_2(t_i) = 10.9075 + 1.2205 t_i + 5.692 \times 10^{-2} {t_i}^2 \qquad \text{(E7.14.3)}$$

We now substitute this expression for h_i in Eq. (E7.14.2) to obtain the quadratic equation

$$10.9075 + 1.2205t_i + 5.692 \times 10^{-2}t_i^2 - h_a + \frac{(t_i - t_r)}{0.435} = 0 \quad \text{(E7.14.4)}$$

The value of the air enthalpy, h_a at the boundary of each control volume is substituted in Eq. (E7.14.4) and the resulting quadratic equation is solved to obtain the water film temperature at the location. For boundary 1 have

$$10.9075 + 1.2205t_{i1} + 5.692 \times 10^{-2}t_{i1}^2 - 71.8 + \frac{(t_{i1} - 7)}{0.435} = 0$$

The roots of the above quadratic equation are: 17.1 and −78.9. We ignore the negative root and take the physically meaningful positive root as the water film temperature at 1. Therefore t_{i1}= 17.1°C.

The dew-point temperature of the air at the inlet is obtained from the psychrometric chart as 22.2°C. Since the computed coil surface temperature of 17.1°C at the inlet is below the dew-point, condensation commences at the inlet section as was initially assumed.

The saturation air enthalpy at the water film temperature is obtained by substituting the value of t_{i1}= 17.1°C, in Eq. (E7.14.3). This gives h_{i1} = 48.46 kJkg^{-1}. The above procedure is repeated for the boundaries of the three control volumes to obtain the data summarized in Table E7.14.1.

Table E7.14.1 Summary of computed values

Boundary	Air enthalpy (kJkg^{-1})	Water film temperature (°C)	Saturation air enthalpy (kJkg^{-1})
1	71.8	17.1	48.46
2	65.5	15.97	44.9
3	59.3	14.78	41.38
4	53	13.56	37.92

We now apply Eq. (7.68) to each control volume to obtain the heat transfer area on the air-side. For the first control volume we have

$$A_1(h_c/c_{am})\left[\frac{(h_{a,1}+h_{a,2})}{2} - \frac{(h_{i,1}+h_{i,2})}{2}\right] = \dot{m}_a\left[h_{a,1} - h_{a,2}\right]$$

Substituting numerical values in the above equation we obtain

$$A_1(0.065/1.02)\left[\frac{(71.8+65.5)}{2} - \frac{(48.86+44.9)}{2}\right] = 2.75[71.8 - 65.5]$$

Hence the air-side heat transfer area of control volume 1, A_1 = 12.2 m^2.

Applying Eq. (7.68) to the next 2 control volumes we obtain the following heat transfer areas: $A_2 = 14.05$ m^2, $A_3 = 16.39$ m^2.
The total air-side heat transfer area is, $A_{tot} = A_1 + A_2 + A_3 = 42.5$ m^2.
Now the total air-side heat transfer may be expressed as

$$A_{tot} = 15 A_{face} N_{rows}$$

Therefore the number rows of tubes is

$$N_{rows} = \frac{42.5}{(15 \times 0.96)} = 2.95$$

The number of vertical rows of tubes is 3.

Applying the energy balance equation to the refrigerant we have

$$\dot{Q}_c = \dot{m}_{ref}(x_{out} - x_{in}) h_{fg}$$

where x is the quality and h_{fg} is the enthalpy of evaporation.
Substituting numerical values in the above equation we obtain

$$52 = 0.4 \times (x_{out} - x_{in}) \times 192.7$$

Therefore the change in quality of the refrigerant is 0.67.

Example 7.15 Consider the direct expansion cooling coil described in worked example 7.14. For the same operating conditions, calculate the following: (i) the distribution of the dry-bulb temperature of air, (ii) the distribution of the humidity ratio of air, and (iii) the total condensation rate. Plot the coil condition line on the psychrometric chart.

Solution We consider the same 3 control volumes shown in Fig. E7.14.1 (b) for which we determined the air-side heat transfer areas. Applying Eq. (7.69) to control volume 1 we have

$$A_1 h_c \left[\frac{(t_{a,1} + t_{a,2})}{2} - \frac{(t_{i,1} + t_{i,2})}{2}\right] = \dot{m}_a c_{am} \left[t_{a,1} - t_{a,2}\right] \quad \text{(E7.15.1)}$$

Substituting numerical values in Eq. (E7.15.1) we obtain

$$12.2 \times 0.065 \left[\frac{(29 + t_{a,2})}{2} - \frac{(17.1 + 15.97)}{2}\right] = 2.75 \times 1.02\left[29 - t_{a,2}\right]$$

Therefore the air temperature at section 2, $t_{a,2}$ =25.9°C.

Applying Eq. (7.69) to each of the other two control volumes we obtain: $t_{a,3}$ =22.97°C and $t_{a,4}$ =20.16°C.

Knowing the water film temperatures, t_i at the different control volume boundaries we are now able to obtain the saturation humidity ratios, ω_i using the psychrometric chart. Hence we have the following values of the humidity ratio of saturated air:

$$\omega_{i,1}= 0.0122,\ \omega_{i,2}= 0.0113,\ \omega_{i,3}= 0.0104,\ \omega_{i,4}= 0.0095.$$

Applying Eq. (7.70) to control volume 1 we have

$$A_1(h_c/c_{am})\left[\frac{(\omega_{a,1}+\omega_{a,2})}{2}-\frac{(\omega_{i,1}+\omega_{i,2})}{2}\right]=\dot{m}_a\left[\omega_{a,1}-\omega_{a,2}\right] \qquad \text{(E7.15.2)}$$

Substituting numerical values in Eq. (E7.15.2) we obtain

$$\left(12.2\times\frac{0.065}{1.02}\right)\left[\frac{(0.0167+\omega_{a,2})}{2}-\frac{(0.0113+0.0122)}{2}\right]=2.75\left[0.0167-\omega_{a,2}\right]$$

Therefore $\omega_{a,2} = 0.0155$.

Applying Eq. (7.70) to each of the other two control volumes we obtain: $\omega_{a,3} = 0.0142$ and $\omega_{a,4} = 0.0128$.

The total rate of condensation of water is given by

$$\dot{m}_{con}=\dot{m}_a\left(\omega_{a,1}-\omega_{a,4}\right) \qquad \text{(E7.15.3)}$$

Substituting numerical values in Eq. (E7.15.3) we obtain the total condensation rate as

$$\dot{m}_{con}=2.75(0.0167-0.0128)=0.0107\ \text{kgs}^{-1}$$

Since we now have the dry-bulb temperature and the humidity ratio of the air as it passes through the cooling coil, we can plot these values on the psychrometric chart (Fig. 4.5) to obtain the coil condition line as shown in Fig. E7.15.1.

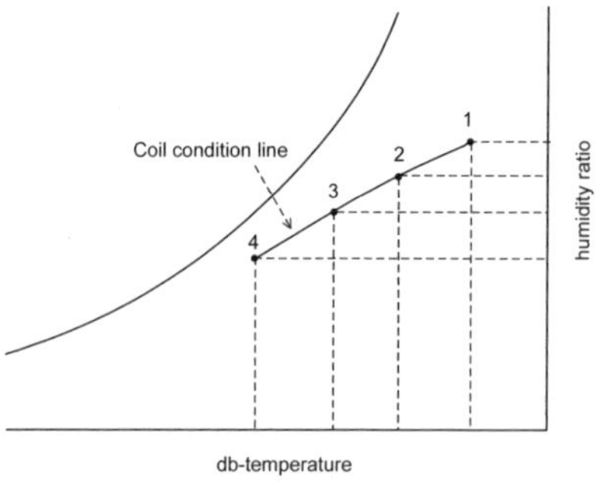

Fig. E7.15.1 Coil condition line 1-2-3-4

Example 7.16 Chilled water enters a counter-flow cooling and dehumidifying coil at 6°C and leaves at 12°C. The mass flow rate of water is 3.2 kgs^{-1}. Moist air entering the coil at 26°C db–temperature and 21°C wb–temperature flows in the opposite direction with a mass flow rate of 2.5 kgs^{-1}. The air-side and water-side convective heat transfer coefficients are 0.055 $kWm^{-2}K^{-1}$ and 3.0 $kWm^{-2}K^{-1}$ respectively. The ratio of air-side area to the water-side area is 16. Determine (i) the total air-side area of the cooling coil, (ii) the air temperature distribution, (iii) the air humidity ratio distribution, and (iv) the total rate of condensation.

Solution The idealized physical model of the cooling coil is shown in Fig. E7.14.1(b). Here air flows over the top surface of a flat duct through which chilled water flows in the opposite direction. From the given conditions of air we obtain the dew-point temperature at the entrance section 1 from the psychrometric chart (Fig. 4.5) as 18.9°C. We shall assume that the duct surface temperature at 1, is below the dew-point temperature of air. Therefore condensation of water vapor would begin at the entrance section of the coil. However, if subsequent computations indicate that the surface temperature at 1 is above the dew-point, we shall revise the analysis assuming the presence of a dry section in the coil.

From Eqs. (7.60) and (7.61) we obtain the following relation:

$$\frac{t_i - t_w}{h_a - h_i} = \frac{h_c \delta A_o}{c_{am} h_r \delta A_i} = \alpha \text{ (say)} \tag{E7.16.1}$$

t_w is the chilled water temperature.
Substituting the given numerical data in Eq. (E7.16.1) we have

$$\frac{t_i - t_w}{h_a - h_i} = \frac{55 \times 16}{1.02 \times 3000} = 0.2876 \tag{E7.16.2}$$

Equation (E7.16.2) is applicable to any section of the cooling coil where condensation occurs. From the given conditions we obtain the air enthalpy at section 1 as h_{a1} = 60.5 $kJkg^{-1}$. Applying the overall energy balance to air and water we have

$$\dot{m}_a (h_{a,in} - h_{a.out}) = \dot{m}_w c_w (t_{w,out} - t_{w,in})$$

Substituting numerical values in the above equation we have

$$2.5(60.5 - h_{a.out}) = 3.2 \times 4.2(12 - 6)$$

Therefore the enthalpy of air at the exit section is 28.24 $kJkg^{-1}$.

In order to illustrate the details of the computational procedure we divide the coil into 3 control volumes as shown in Fig. E7.14.1(b). The heat transfer areas of the control volumes are such that the enthalpy drop of air across each control volume is the same. This enthalpy drop is given by

$$\Delta h_a = \frac{(h_{a,in} - h_{a,out})}{3} = \frac{(60.5 - 28.24)}{3} = 10.75$$

Therefore the air enthalpies at the boundaries of the control volumes are

$$h_{a1} = 60.5, \quad h_{a2} = 49.75, \quad h_{a3} = 39, \quad h_{a4} = 28.25 \text{ kJkg}^{-1}$$

The chilled water temperature change across the control volumes is given by

$$\Delta t_w = \frac{(t_{w,out} - t_{w,in})}{3} = \frac{(12-6)}{3} = 2°C$$

Therefore the water temperatures at the boundaries of the control volumes are:

$$t_{w1} = 12, \quad t_{w2} = 10, \quad t_{w3} = 8, \quad t_{w4} = 6°C$$

We now calculate the water film or condensate temperatures at the boundaries of the control volumes using Eq. (E7.16.2). For ease of computation we shall use the quadratic relationship between the saturation air enthalpy and the temperature given by Eq. (7.64).

$$h_i = f_2(t_i) = 10.9075 + 1.2205t_i + 5.692 \times 10^{-2}{t_i}^2 \qquad \text{(E7.16.3)}$$

Substituting this expression for h_i in Eq. (E7.16.2) we obtain the following quadratic equation:

$$10.9075 + 1.2205t_i + 5.692 \times 10^{-2}{t_i}^2 - h_a + \frac{(t_i - t_w)}{0.2876} = 0 \quad \text{(E7.16.4)}$$

The value of the air enthalpy h_a at the boundary of each control volume is substituted in Eq. (E7.16.4) and the resulting quadratic equation is solved to obtain the water film temperature at the location. For boundary 1 we have

$$10.9075 + 1.2205t_{i1} + 5.692 \times 10^{-2}{t_{i1}}^2 - 60.5 + \frac{(t_{i1} - 12)}{0.2876} = 0$$

The roots of the above quadratic equation are: 16.24 and −99.0. We ignore the negative root and take the physically meaningful positive root as the water film temperature at 1. Therefore t_{i1}= 16.24°C. The dew-point temperature of the air at the inlet section 1 is 18.9°C. Since the computed coil surface temperature of 16.24°C at the inlet is below the dew-point, condensation commences at the inlet section, as was initially assumed.

The saturation air enthalpy at the water film temperature is obtained by substituting, t_{i1}= 16.24°C, in Eq. (E7.16.3). This gives, h_{i1} = 45.74 $kJkg^{-1}$. The above procedure is repeated for the other three boundaries of the control volumes to obtain the data summarized in Table E7.16.1.

Table E7.16.1 Summary of computed values

Boundary	Air enthalpy ($kJkg^{-1}$)	Water film temperature (°C)	Saturation air enthalpy ($kJkg^{-1}$)
1	60.5	16.24	45.74
2	49.75	13.47	37.68
3	39	10.55	30.12
4	28.24	7.45	23.17

We now apply Eq. (7.68) to each control volume to obtain the heat transfer area on the air-side. For control volume 1 we have

$$A_1(h_c/c_{am})\left[\frac{(h_{a,1}+h_{a,2})}{2}-\frac{(h_{i,1}+h_{i,2})}{2}\right]=\dot{m}_a\left[h_{a,1}-h_{a,2}\right] \qquad \text{(E7.16.5)}$$

Substituting numerical values in Eq. (E7.16.5) we obtain

$$A_1(0.055/1.02)\left[\frac{(60.5+49.75)}{2}-\frac{(45.75+37.68)}{2}\right]=2.5[60.5-49.75]$$

Hence the air-side heat transfer area of control volume 1, A_1 = 37.17 m^2. Applying Eq. (7.68) to the next two control volumes we obtain the following air-side areas: A_2 = 47.6 m^2, A_3 = 71.51 m^2.
The total air-side heat transfer area is,

$$A_{tot}=A_{23}+A_{34}+A_{45}=156.3 \text{ m}^2$$

Applying Eq. (7.69) to control volume 1 we have

$$A_1h_c\left[\frac{(t_{a,1}+t_{a,2})}{2}-\frac{(t_{i,1}+t_{i,2})}{2}\right]=\dot{m}_a c_{am}\left[t_{a,1}-t_{a,2}\right] \qquad \text{(E7.16.6)}$$

Substituting numerical values in Eq. (E7.16.6) we obtain

$$37.17 \times 0.055\left[\frac{(26+t_{a,2})}{2}-\frac{(16.24+13.47)}{2}\right]=2.5\times 1.02\left[26-t_{a,2}\right]$$

Therefore $t_{a,2}$ =19.62°C.

We apply Eq. (7.69) to each of the other two control volumes to obtain the air temperatures as: $t_{a,3}$ =14.46°C and $t_{a,4}$ =9.71°C.

Knowing the water film temperatures, t_i we are now able to obtain the saturation humidity ratios, ω_i at the different control volume boundaries using the psychrometric chart (Fig. 4.5). Hence we obtain the following values of the humidity ratio of saturated air:

$$\omega_{i,1}=0.0115,\ \omega_{i,2}=0.0095,\ \omega_{i,3}=0.0077,\ \omega_{i,4}=0.0062.$$

Applying Eq. (7.70) to control volume 1 we have

$$A_1(h_c/c_{am})\left[\frac{(\omega_{a,1}+\omega_{a,2})}{2}-\frac{(\omega_{i,1}+\omega_{i,2})}{2}\right]=\dot{m}_a\left[\omega_{a,1}-\omega_{a,2}\right] \quad \text{(E7.16.7)}$$

Substituting numerical values in Eq. (E7.16.7) we obtain

$$\left(37.17\times\frac{0.055}{1.02}\right)\left[\frac{(0.0136+\omega_{a,2})}{2}-\frac{(0.0115+0.0095)}{2}\right]=2.5\left[0.0136-\omega_{a,2}\right]$$

Therefore $\omega_{a,2}$ =0.0118.

We apply Eq. (7.70) to each of the other two control volumes. Hence we obtain: $\omega_{a,3}$ = 0.0096 and $\omega_{a,4}$ = 0.0073.

The total rate of condensation of water is given by

$$\dot{m}_{con}=\dot{m}_a\left(\omega_{a,1}-\omega_{a,4}\right) \quad \text{(E7.16.8)}$$

Substituting numerical values in Eq. (E7.16.8) we obtain the condensation rate as

$$\dot{m}_{con}=2.5(0.0136-0.0073)=0.01575\ \text{kgs}^{-1}$$

Since we now have the dry-bulb temperature and the humidity ratio of the air as it passes through the cooling coil, we can plot these values on the psychrometric chart to obtain the coil condition line shown in Fig. E7.15.1.

Example 7.17 Moist air flows at the rate of 0.32 kgs^{-1} through a direct-expansion (DX) type cooling coil. The inlet conditions of the air are 32°C db–temperature and 50% relative humidity. The air leaving the coil is saturated at a temperature of 14°C. The refrigerant flowing in the opposite direction is at 12°C. The following data are applicable to the coil: the ratio of the air-side to refrigerant-side heat transfer areas is 18; the air-side and refrigerant-side heat transfer coefficients are 0.11 $kWm^{-2}K^{-1}$ and 2.4 $kWm^{-2}K^{-1}$ respectively. Calculate (i) the dew-point temperature of the air at the entrance, (ii) the db–temperature and enthalpy of the air when condensation begins, and (iii) the total area of the cooling coil.

Solution

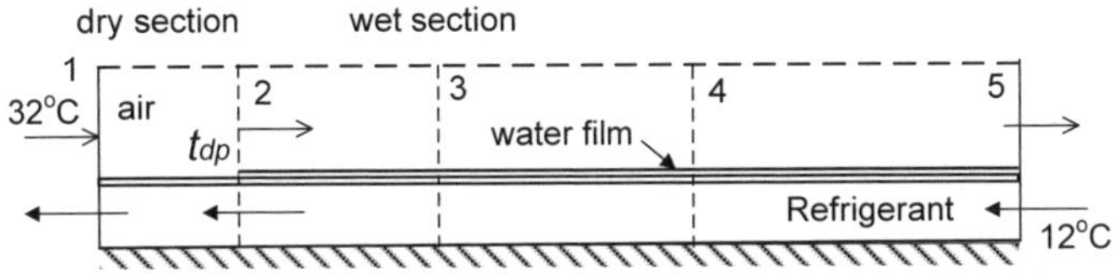

Fig. E7.17.1 Direct-expansion cooling coil with dry section

The simplified physical model of the cooling coil is shown in Fig. E7.17.1. Apply the heat balance equation to a *small* area at the entrance section. Hence we have

$$(32 - t_1)\delta A_o h_o = (t_1 - 12)\delta A_i h_i \qquad \text{(E7.17.1)}$$

where t_1 is the plate temperature at section 1, which is assumed to be dry.

Substituting numerical values in Eq. (E7.17.1) we obtain

$$(32 - t_1)18 \times 0.11 = (t_1 - 12)2.4$$

Therefore $t_1 = 21.04$°C.

We use the psychrometric chart to obtain the dew-point temperature of air at the entrance section as 20.4°C. Since the plate temperature is above the dew-point, condensation will *not commence* at the entrance section 1. Let 2 in Fig. E7.17.1 denote the section where condensation begins. Therefore the *plate temperature* at 2, $t_2 = 20.4$°C, the *dew-point temperature of incoming air*. Since condensation *just* begins at 2, we apply the following heat balance equation to a *small* area at 2.

$$(t_{a,2} - t_2)\delta A_o h_o = (t_2 - 12)\delta A_i h_i \qquad \text{(E7.17.2)}$$

Substituting numerical values in Eq.(E7.17.2) we obtain

$$(t_{a,2} - 20.4)18 \times 0.11 = (20.4 - 12)2.4$$

Hence we have the air temperature at 2, as $t_{a,2} = 30.58°C$.

In order to calculate the area of the dry section 1-2 we treat the control volume 1-2 as a heat exchanger with known inlet and outlet fluid temperatures. The log mean temperature difference for this heat exchanger is obtained by applying Eq. (7.10). Hence we have

$$LMTD = \frac{(32-12)-(30.58-12)}{\ln\left(\frac{32-12}{30.58-12}\right)} = 19.28$$

The overall heat transfer coefficient based on the air-side area is given by

$$\frac{1}{U_o} = \frac{18}{h_i} + \frac{1}{h_o} = \frac{1}{0.11} + \frac{18}{2.4} = 16.6$$

The total heat transfer rate in the dry section 1-2 is

$$\dot{Q}_{12} = \dot{m}_a c_a(t_{a,1} - t_{a,2}) = 0.32 \times 1.02 \times (32 - 30.58) = 0.463$$

Also, $$\dot{Q}_{12} = U_o A_{12}\, F(\,LMTD)$$

where $F = 1$ for counter-flow. Substituting numerical values in the above equation we obtain the dry section area as

$$A_{12} = \frac{0.463 \times 16.6}{19.28} = 0.4 \text{ m}^2$$

Then we now proceed to analyze the wet section of the coil using a procedure similar to that used in worked example 7.14. The *wet section* of the coil is divided into 3 control volumes as shown in Fig. E7.17.1. Across each control volume the enthalpy change of the air is assumed to be the same.

The enthalpy of air at 2 is obtained from the psychrometric chart, knowing the humidity ratio and the db-temperature at 2. Note that $\omega_1 = \omega_2 = 0.015$, because there is no condensation in section 1-2. Hence we have $h_{a,2} = 70.6$ kJkg^{-1}. The enthalpy of saturated air leaving the coil at 5 is $h_{a,5} = 39.5$ kJkg^{-1}.

From Eqs. (7.60) and (7.61) we obtained the following relation:

$$\frac{t_i - t_r}{h_a - h_i} = \frac{h_c \delta A_o}{c_{am} h_r \delta A_i} \qquad \text{(E7.17.3)}$$

Substituting the given numerical data we have

$$\frac{t_i - t_r}{h_a - h_i} = \frac{110 \times 18}{1.02 \times 2400} = 0.809 \qquad \text{(E7.17.4)}$$

Equation (7.17.4) is applicable to all sections of the coil from 2 to 5.

Now the heat transfer areas of the three control volumes are such that the enthalpy drop of the air is the same for all of them. This enthalpy drop is given by

$$\Delta h_a = \frac{(h_{a,2} - h_{a,5})}{3} = \frac{(70.6 - 39.5)}{3} = 10.37$$

Therefore the air enthalpies at the boundaries of the control volumes are:

$$h_{a1} = 70.6, \quad h_{a2} = 60.23, \quad h_{a3} = 49.86, \quad h_{a4} = 39.5$$

We now calculate the water film or condensate temperatures at the boundaries of the control volumes by using Eq. (E7.17.4). For ease of computation we shall use the quadratic relationship between the saturation air enthalpy and the temperature, given by Eq. (7.64).

$$h_i = f_2(t_i) = 10.9075 + 1.2205 t_i + 5.692 \times 10^{-2} t_i^2 \qquad \text{(E7.17.5)}$$

We substitute the expression for h_i in Eq. (E7.17.5) to obtain the equation

$$10.9075 + 1.2205 t_i + 5.692 \times 10^{-2} t_i^2 - h_a + \frac{(t_i - 12)}{0.809} = 0 \qquad \text{(E7.17.6)}$$

The air enthalpy, h_a at the boundary of each control volume is substituted in Eq. (E7.17.6) and the resulting quadratic equation is solved to obtain the water film temperature at the section. For boundary 3 we have

$$10.9075 + 1.2205 t_{i3} + 5.692 \times 10^{-2} t_{i3}^2 - 70.6 + \frac{(t_{i3} - 12)}{0.8088} = 0$$

The solution gives, t_{i3} = 18.33°C. The saturation air enthalpy at the water film temperature is obtained by substituting, t_{i3} = 18.33°C, in Eq. (E7.14.2). This gives h_{i3} = 52.4 kJkg^{-1}. The above procedure is repeated for the boundaries of the other two control volumes to obtain the data summarized in Table E7.17.1. Note that at boundary 2 condensation *just* begins.

Table E7.17.1 Summary of computed values

Boundary	Air enthalpy ($kJkg^{-1}$)	Water film temperature (°C)	Saturation air enthalpy ($kJkg^{-1}$)
2	70.6	20.5	60.0
3	60.2	18.3	52.4
4	49.9	16	45.0
5	39.5	13.5	37.7

We now apply Eq. (7.68) to each 'wet' control volume to obtain the heat transfer area on the air-side. For the control volume 2-3 we have

$$A_{23}(h_c/c_{am})\left[\frac{(h_{a,2}+h_{a,3})}{2}-\frac{(h_{i,2}+h_{i,3})}{2}\right]=\dot{m}_a\left[h_{a,2}-h_{a,3}\right]$$

Substituting numerical values in the above equation we obtain

$$A_{23}(0.11/1.02)\left[\frac{(70.6+60.2)}{2}-\frac{(60.0+52.4)}{2}\right]=0.32[70.6-60.2]$$

Hence the air-side heat transfer area of control volume 2-3, A_{23} = 3.34 m^2. Applying Eq. (7.68) to the other two control volumes we obtain the following areas: A_{34} = 4.83 m^2, A_{45} = 9.13 m^2.
The total air-side heat transfer area of the *wet section* 2-5 is

$$A_{wet}=A_{23}+A_{34}+A_{45}=17.3\ \text{m}^2$$

The area of the *dry section* 1-2 was calculated to be 0.4 m^2. Therefore the *total* air-side area of the coil is 17.7 m^2.

Example 7.18 Moist air enters a cooling and dehumidifying coil at 28°C db–temperature and 21°C wb–temperature with a mass flow rate 2.5 kgs^{-1}. Chilled water flows through the coil in the opposite direction at the rate 3 kgs^{-1}. The entering and leaving temperatures of water are 10.5°C and 16°C respectively. The air-side and water-side heat transfer coefficients are 0.06 $kWm^{-2}K^{-1}$ and 3.0 $kWm^{-2}K^{-1}$ respectively. The ratio of the air-side area to the water-side area is 16. Determine (i) the dry area of the coil, (ii) the wet area of the coil, (iii) the air temperature and relative humidity distribution of the air, (iv) the water temperature distribution, and (v) the total the condensation rate. Plot the air condition line on the psychrometric chart.

Solution

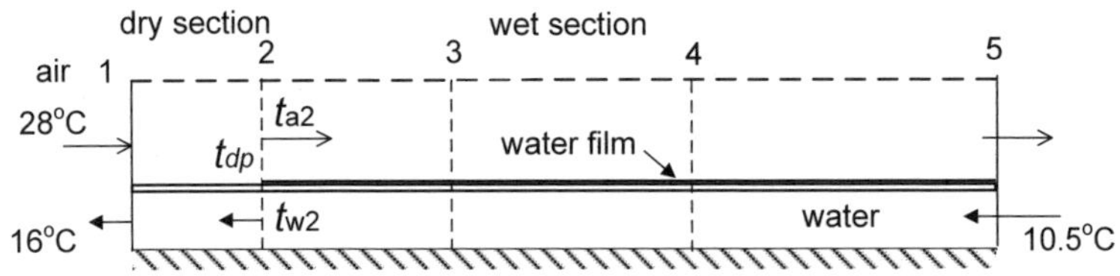

Fig. E7.18.1 Physical model of the chilled water cooling coil

The physical model of the cooling coil is shown in Fig. E7.18.1. Apply the heat balance equation to a *small* area at the entrance section 1. This gives

$$(28 - t_1)\delta A_o h_o = (t_1 - 16)\delta A_i h_i \qquad \text{(E7.18.1)}$$

where t_1 is the plate temperature at section 1, which is assumed to be dry.

Substituting numerical values in Eq. (E7.18.1) we have

$$(28 - t_1) \times 16 \times 0.06 = (t_1 - 16) \times 3$$

Therefore $t_1 = 18.91$°C.

We use the psychrometric chart (Fig. 4.5) to obtain the dew-point temperature of air at the entrance section as 17.8°C. Since the plate temperature at 1 is above the dew-point, condensation will not commence at the entrance section. Let 2 in Fig. E7.18.1 denote the section where condensation just begins. Therefore the *plate temperature* at 2, $t_2 = 17.8$°C, the dew-point temperature. Since condensation *just* begins at 2, we apply the following heat balance equation to a *small* area at 2.

$$(t_{a,2} - t_2)\delta A_o h_o = (t_2 - t_{w,2})\delta A_i h_i \qquad \text{(E7.18.2)}$$

Substituting numerical values in Eq. (E7.18.2) we obtain

$$(t_{a,2} - 17.8) \times 16 \times 0.06 = (17.8 - t_{w,2}) \times 3$$

$$0.32 t_{a,2} = 23.5 - t_{w,2} \qquad \text{(E7.18.3)}$$

Apply the overall energy balance equation to the '*dry*' control 1-2.

$$\dot{m}_a c_a (t_{a,1} - t_{a,2}) = \dot{m}_w c_w (t_{w,1} - t_{w,2}) \qquad \text{(E7.18.4)}$$

Substituting numerical values in Eq. (E7.18.4) we obtain

$$2.5 \times 1.02 \times (28 - t_{a,2}) = 3 \times 4.2 \times (16 - t_{w,2})$$

$$10.33 + 0.2023t_{a,2} = t_{w,2} \quad \text{(E7.18.5)}$$

Solving Eqs. (E7.18.3) and (E7.18.5) simultaneously we have

$$t_{a,2} = 25.2°\text{C} \quad \text{and} \quad t_{w,2} = 15.43°\text{C}.$$

In order to calculate the area of the *dry section* 1-2 we treat the control volume 1-2 as a counter-flow heat exchanger with known inlet and outlet fluid temperatures. The log mean temperature difference for this heat exchanger is obtained by applying Eq. (7.10). Hence we have

$$LMTD = \frac{(28-16)-(25.2-15.43)}{\ln\left(\frac{28-16}{25.2-15.43}\right)} = 10.85$$

The overall heat transfer coefficient based on the air-side area is given by

$$\frac{1}{U_o} = \frac{1}{h_o} + \frac{16}{h_i} = \frac{1}{0.06} + \frac{16}{3} = 22$$

The total heat transfer rate in the dry section 1-2 is

$$\dot{Q}_{12} = \dot{m}_a c_a (t_{a,1} - t_{a,2}) = 2.5 \times 1.02 \times (28 - 25.2) = 7.14 \text{ kW}$$

Also $$\dot{Q}_{12} = U_o A_{12} F(LMTD) = 7.14$$

where the correction factor, $F = 1$ for counter-flow. Substituting numerical values in the above equation we obtain the dry section area as

$$A_{12} = \frac{7.14 \times 22}{10.85} = 14.47 \text{ m}^2$$

Then we now proceed to analyze the *wet section* of the coil using a procedure similar to that used in worked example 7.16. The wet section of the coil is divided into 3 control volumes as shown in Fig. E7.18.1. Across each control volume the enthalpy change of the air is assumed to be the same.

The enthalpy of the air at 2 is obtained from the psychrometric chart, by knowing the humidity ratio and the db–temperature at 2. Note that $\omega_1 = \omega_2 = 0.0128$, because there is no condensation in the dry section 1-2. Hence we have $h_{a,2} = 58 \text{ kJkg}^{-1}$.

Applying the overall energy balance equation to the wet section 2-5 of the coil we have

$$\dot{m}_a\left(h_{a,2} - h_{a,5}\right) = \dot{m}_w c_w\left(t_{w,2} - t_{w,5}\right) \qquad \text{(E7.18.6)}$$

Substituting numerical values in Eq. (E7.18.6) we obtain

$$2.5 \times \left(58 - h_{a,5}\right) = 3 \times 4.2 \times (15.43 - 10.5) \qquad \text{(E7.18.7)}$$

Therefore the enthalpy of the air leaving the coil at 5 is $h_{a,5} = 33.15$ kJkg^{-1}.

From Eqs. (7.60) and (7.61) we obtained the following relation:

$$\frac{t_i - t_w}{h_a - h_i} = \frac{h_c \delta A_o}{c_{am} h_r \delta A_i} \qquad \text{(E7.18.8)}$$

t_w is the chilled water temperature.
Substituting the given numerical data we have

$$\frac{t_i - t_w}{h_a - h_i} = \frac{60 \times 16}{1.02 \times 3000} = 0.3137 \qquad \text{(E7.18.9)}$$

Equation (E7.18.9) is applicable to all the *wet sections* of the coil from 2 to 5.

The heat transfer areas of the three wet control volumes are such that the enthalpy drop of the air is the same for all of them. This enthalpy drop is given by

$$\Delta h_a = \frac{\left(h_{a,2} - h_{a,5}\right)}{3} = \frac{(58 - 33.15)}{3} = 8.28$$

Therefore the air enthalpies at the boundaries of the control volumes are:

$$h_{a2} = 58, \quad h_{a3} = 49.72, \quad h_{a4} = 41.44, \quad h_{a5} = 33.15$$

The chilled water temperature change across the sections is given by

$$\Delta t_w = \frac{\left(t_{w,2} - t_{w,5}\right)}{3} = \frac{(15.43 - 10.5)}{3} = 1.64$$

Therefore the chilled water temperatures at the boundaries of the control volumes are:

$$t_{w2} = 15.43, \quad t_{w3} = 13.8, \quad t_{w4} = 12.15, \quad t_{w5} = 10.5$$

We now calculate the water film or condensate temperatures at the boundaries of the control volumes by using Eq. (E7.18.9). For ease of computation we shall use the quadratic relationship between the saturation air enthalpy and temperature.

$$h_i = f_2(t_i) = 10.9075 + 1.2205t_i + 5.692 \times 10^{-2}t_i^2 \qquad \text{(E7.18.10)}$$

Substituting the above expression for h_i in Eq. (E7.18.9) we obtain the quadratic equation

$$10.9075 + 1.2205t_i + 5.692 \times 10^{-2}t_i^2 - h_a + \frac{(t_i - t_w)}{0.3137} = 0$$

The value of the air enthalpy h_a at the boundary of each control volume is substituted in the above equation and the resulting quadratic equation is solved to obtain the water film temperature at the section. For boundary 3 have

$$10.9075 + 1.2205t_{i3} + 5.692 \times 10^{-2}t_{i3}^2 - 49.72 + \frac{(t_{i3} - 13.8)}{0.3137} = 0$$

The positive root of the above quadratic equation gives, t_{i3}= 15.62°C. The saturation air enthalpy at the water film temperature is obtained by substituting the value of t_{i3}= 15.62°C, in Eq. (E7.18.10). This gives h_{i3} = 43.86 kJkg^{-1}. The above procedure is repeated for the other two boundaries of the control volumes to obtain the data summarized in Table E7.18.1.

Table E7.18.1 Summary of computed values

Boundary	Air enthalpy (kJkg^{-1})	Water film temperature (°C)	Saturation air enthalpy (kJkg^{-1})
2	58	17.8	50.55
3	49.72	15.62	43.87
4	41.43	13.4	37.46
5	33.15	11.06	31.37

We now apply Eq. (7.68) to each of the wet control volumes to obtain the heat transfer area on the air side. For control volume 2-3 we have

$$A_{23}(h_c/c_{am})\left[\frac{(h_{a,2}+h_{a,3})}{2} - \frac{(h_{i,2}+h_{i,3})}{2}\right] = \dot{m}_a\left[h_{a,2} - h_{a,3}\right]$$

Substituting numerical values in the above equation we obtain

$$A_{23}(0.06/1.02)\left[\frac{(58+49.72)}{2} - \frac{(50.55+43.86)}{2}\right] = 2.5[58 - 49.72]$$

Hence the air-side heat transfer area of control volume 2-3, A_{23} = 52.94 m^2.

Applying Eq. (7.68) to the other two 'wet' control volumes we obtain the following areas: $A_{34} = 71.62$ m^2, $A_{45} = 122.17$ m^2.
The total air-side heat transfer area of the wet section 2-5 is,

$$A_{wet} = A_{23} + A_{34} + A_{45} = 246.7 \text{ m}^2$$

The area of the dry section 1-2 is, $A_{dry} = 14.47$ m^2. Therefore the total air-side area is 261 m^2.

Applying Eq. (7.69) to control volume 2-3 we have

$$A_{23}h_c\left[\frac{(t_{a,2}+t_{a,3})}{2} - \frac{(t_{i,2}+t_{i,3})}{2}\right] = \dot{m}_a c_{am}\left[t_{a,2} - t_{a,3}\right]$$

Substituting numerical values in the above equation we obtain

$$52.94 \times 0.06\left[\frac{(25.2+t_{a,3})}{2} - \frac{(17.77+15.62)}{2}\right] = 2.5 \times 1.02\left[25.2 - t_{a,3}\right]$$

Therefore $t_{a,3} = 18.67$°C.

We apply Eq. (7.69) to each of the other two wet control volumes. Hence we have: $t_{a,4} = 14.86$°C and $t_{a,5} = 11.75$°C.

Knowing the water film temperature, t_i we are now able to obtain the saturation humidity ratio, ω_i at the different control volume boundaries using the psychrometric chart. Hence we obtain the following values of the humidity ratio of saturated air:

$$\omega_{i,2} = 0.0128,\ \omega_{i,3} = 0.011,\ \omega_{i,4} = 0.0094,\ \omega_{i,5} = 0.008.$$

Applying Eq. (7.70) to control volume 2-3 we have

$$A_{23}(h_c/c_{am})\left[\frac{(\omega_{a,2}+\omega_{a,3})}{2} - \frac{(\omega_{i,2}+\omega_{i,3})}{2}\right] = \dot{m}_a\left[\omega_{a,2} - \omega_{a,3}\right]$$

Substituting numerical values in the above equation we obtain

$$(52.94 \times \frac{0.06}{1.02})\left[\frac{(0.0128+\omega_{a,3})}{2} - \frac{(0.0128+0.011)}{2}\right] = 2.5\left[0.0128 - \omega_{a,3}\right]$$

Therefore $\omega_{a,3} = 0.0121$.

We apply Eq. (7.70) to each of the other two wet control volumes. Hence we obtain: $\omega_{a,4} = 0.0104$ and $\omega_{a,5} = 0.0084$.

The total rate of condensation of water is given by

$$\dot{m}_{con} = \dot{m}_a\left(\omega_{a,2} - \omega_{a,5}\right)$$

Substituting numerical values in the above equation we obtain the condensation rate as

$$\dot{m}_{con} = 2.5 \times (0.0128 - 0.0084) = 0.011\ \mathrm{kgs}^{-1}$$

Since we now have the variation of the dry-bulb temperature and the humidity ratio of air we are able to plot the condition line on the psychrometric chart. This is shown in Fig. E7.18.2.

Section 1-2 of the condition line represents the properties of air before condensation begins at 2. Note that the *average* air temperature at 2 is above the dew-point temperature and it is only the air layer *adjacent* to the plate that reaches the dew-point temperature at 2. However, as the air flows through the cooling coil the average air temperature will also reach the dew-point.

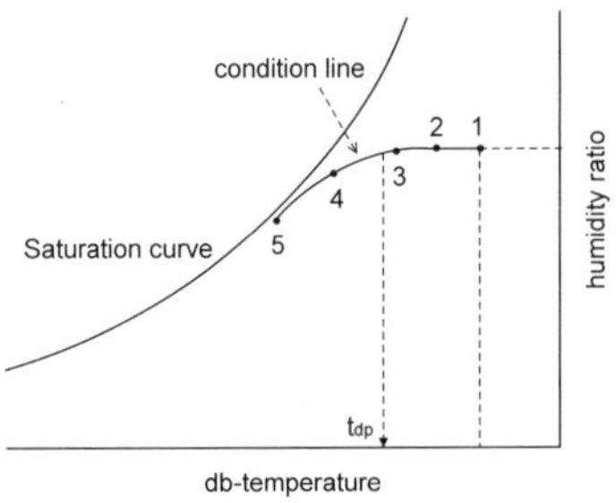

Fig. E7.18.2 Condition line of air

Example 7.19 Develop a MATLAB software program to analyze the performance of wet cooling coils using chilled water as the coolant. Use the data in worked example 7.16 to study the effect of the number of coil sections, and the polynomial function used on the computed heat transfer area.

Solution The main steps of the MATLAB code, given in Appendix A7.1, are summarized below.

(i) Input the design parameters of the cooling coil. These include: the air-side to water-side area ratio, the convective heat transfer coefficients for the air-side and the water-side, and the specific heat capacities of moist air and water.

(ii) Input the operating conditions: the mass flow rates of air and water, the inlet and outlet temperatures of water, or the inlet and outlet enthalpies of air.

(iii) Calculate the parameter α in Eq. (7.62).

(iv) Select the number of control volumes of the wet-coil. Divide the enthalpy change of air and the temperature change of water equally between the control volumes. Hence calculate the air enthalpies and water temperatures at the boundaries of the control volumes.

(v) As the cubic-expression for the saturation air enthalpy, given by Eq. (7.63), is more accurate, we include it in the computer code. Substitute this expression in Eq. (7.62) to obtain the following cubic equation for the water film temperature:

$$c_0 + c_1 t_i + 1.1135 \times 10^{-2} t_i^{\,2} + 9.8855 \times 10^{-4} t_i^{\,3} = 0$$

where $\quad c_0 = 9.3625 - \frac{t_w}{\alpha} - \frac{h_a}{\alpha} \quad$ and $\quad c_1 = 1.786 + \frac{1}{\alpha}$

(vi) Solve the above cubic equation at all the boundaries of the control volumes to determine the respective water film temperatures. MATLB software package includes a convenient expression, *roots* (*c*), that gives the roots of a polynomial equation. Ignore the two imaginary roots and select the physically meaningful positive root. Calculate the saturation air enthalpies, h_i at the boundaries by substituting the water film temperatures in Eq. (7.63).

(vii) Substitute the boundary values of the air enthalpies in Eq. (7.68) to calculate the air-side heat transfer areas of the control volumes.

(viii) Substitute the areas obtained in (vii), and the values of the db–temperature of the upstream boundary in Eqs. (7.69) to compute the db–temperatures at the downstream boundaries.

(ix) Compute the saturation humidity ratios, ω_i at the boundaries using the expression

$$\omega_i h_g(t_i) + c_{pm} t_i = h_i$$

where the saturated vapor enthalpy, h_g is given by Eq. (6.62).

(x) Substitute the areas obtained in (vii), and the values of the humidity ratio of the upstream boundary in Eqs. (7.70) to compute the air humidity ratios at the downstream boundaries.

(a) We performed a parametric study to determine the effect of the polynomial expression used on the computed heat transfer area. The quadratic expression in Eq. (7.64) gives an area of 156.3m^2 while the cubic expression in Eq. (7.63) gives an area of 160.8 m^2.

(b) We performed a parametric study to determine the effect of the number of control volumes on the computed heat transfer area. The results are summarized in Table E7.19.1.

Table E7.19.1 - Effect of number of control volumes on total area

Number	3	4	5	6	7
Area, m^2	160.8	161.65	162.06	162.3	162.44

Example 7.20 Consider the *off-design* operation of the cooling coil described in worked example 7.16. Use the MATLAB computer code developed in example 7.19 to determine the cooling capacity of the coil for chilled water inlet temperatures of 6°C, 5°C, 4°C and 3°C. Assume that all the other conditions remain unchanged.

Solution We have used the computer code developed in example 7.19 to study the performance of the cooling coil described in example 7.16 under *off-design* operating conditions. This has to be done using a trial and error procedure where for each value of the inlet chilled water temperature we guess an initial value for the outlet water temperature.

We use the computer code to calculate the total area of the coil. The outlet temperature is adjusted until the computed total area is 161.65 m^2, which is the *design area* obtained with 4 control volumes as listed in Table E7.19.1. The results obtained are summarized in Table E7.20.1.

Table E7.20.1 Variation of cooling capacity

Water inlet temperature °C	6	5	4	3
Cooling capacity, kW	80.6	85.3	89.8	94.0

The computer code could be used to investigate various other issues concerning the off-design performance of a cooling coil. The code could be modified to include the analysis of a dry section before condensation commences, using the procedure described in worked example 7.18. Also, note that the computer code could be easily modified to solve design problems involving refrigeration coils where the coolant temperature is constant.

Problems

P7.1 Water enters a counter-flow heat exchanger with a flow rate of 0.2 kgs^{-1}. The inlet and outlet temperatures of the water are 20°C and 50°C respectively. A hot heating fluid enters the heat exchanger at 80°C and leaves at 40°C. The specific heat capacity of water is 4.2 kJ $kg^{-1}K^{-1}$. Calculate (i) the LMTD, (ii) the effectiveness, and (iii) the NTU.
[*Answers*: (i) 24.7°C , (ii) 0.67, (iii) 1.62]

P7.2 Refrigerant R134a flows at the rate of 0.09 kgs^{-1} through the inner tube of a double-pipe heat exchanger while water flows in the opposite direction through the annulus at the rate of 0.1 kgs^{-1}. The inner diameter of the tube is 24mm. The inlet temperature of water is 25°C, and the refrigerant is 5°C. The refrigerant quality at the inlet is 0.2 and at the outlet it is 0.6. The convective transfer coefficients on the refrigerant-side and water-side are 8 $kWm^{-2}K^{-1}$ and 3.6 $kWm^{-2}K^{-1}$ respectively. Calculate (i) outlet temperature of water, (ii) the length of the heat exchanger, and (iii) the temperature of the water at the mid-point along the length.
[*Answers*: (i) 8.25°C, (ii) 4.06 m, (iii) 13.05°C]

P7.3 Moist ambient air at 21°C db–temperature and 12°C wb–temperature enters a single-pass cross-flow heat exchanger with a volume flow rate of 2.1 m^3s^{-1}. The air is heated to 42°C. Water enters the

heater at 72°C and leaves at 62°C. The pressure 101.3 kPa. Calculate (i) the mass flow rate of water, (ii) the product (*UA*) of the heat exchanger, (iii) the LMTD, and (iv) the effectiveness.
[*Answers*: (i) 1.28kgs^{-1}, (ii) 1.56 kWK^{-1}, (iii) 35.2°C, (iv) 0.41]

P7.4 Moist ambient air entering a cross-flow heat exchanger at 21°C db-temperature and 32% relative humidity flows over rows of horizontal copper tubes inside which steam condenses at 100°C. The air temperature at the exit is 62°C. The volume flow rate of air at entry is 2.5 m^{3}s^{-1}. The inner and outer diameters of the tubes are 12.2 mm and 13.5 mm respectively.

Circular plate aluminum fins of thickness 0.25 mm and outer diameter 50 mm are fitted on the outside of the tubes. There are 250 fins per meter of tube. The air-side and steam-side heat transfer coefficients are 50 Wm^{-2}K^{-1} and 1200 Wm^{-2}K^{-1} respectively. The thermal conductivities of copper and aluminum are 370 Wm^{-1}K^{-1} and 160 Wm^{-1}K^{-1} respectively. Calculate (i) the efficiency of a fin, (ii) the overall heat transfer coefficient based on the inner tube area, (iii) the effectiveness, and (iv) the total length of the tubes.
[*Answers*: (i) 0.66, (ii) 496 Wm^{-2}K^{-1}, (iii) 0.519, (iii) 117m]

P7.5 A single-pass evaporator of a water chiller has 100 horizontal copper tubes of inner diameter 15mm and wall thickness 1.5mm through which water flows with a total mass flow rate of 1.8 kgs^{-1}. Refrigerant 134a undergoes pool boiling at 2°C on the outer surface of the tubes. The water is cooled from 14°C to 6°C. The water-side and refrigerant-side heat transfer coefficients are 1100 Wm^{-2}K^{-1} and 980 Wm^{-2}K^{-1} respectively. Calculate (i) the overall heat transfer coefficient based on the inner tube area, (ii) the LMTD (iii) the effectiveness of the evaporator, and (iv) the length of a tube.
[*Answers*: (i) 567 Wm^{-2}K^{-1}, (ii) 7.28°C, (iii) 0.67, (iv) 3.13m]

P7.6 The evaporator of a water chiller has 15 stainless tubes of outer diameter 20mm and length 0.5 m through which water flows at a total flow rate of 0.05 kgs^{-1}. Water enters the evaporator at 18°C. Refrigerant 134a at 6°C undergoes pool boiling on the outer surface of the tubes.

Assume that the outer surface of the tubes is at a uniform temperature of 9°C. Calculate (i) the pool boiling heat transfer coefficient, and (ii) the outlet temperature of the water. Outline a procedure to check whether the assumed tube surface temperature is accurate.
[*Answers*: (i) 790 $Wm^{-2}K^{-1}$, (ii) 12.6°C]

P7.7 Saturated refrigerant R134a at 0°C enters a tube of outer diameter 50 mm and height 0.8 m at the bottom, with a quality of 0.2. The mass flow rate of the refrigerant is 0.025 kgs^{-1}. Saturated steam at 8°C condenses on the outside of the tube. (i) Assuming the tube wall to be at a uniform temperature of 4°C, calculate the average laminar condensation heat transfer coefficient. (ii) Calculate the total steam condensation rate. (iii) Calculate the quality of the refrigerant leaving the tube at the top.
[*Answers*:(i)5.25 $kWm^{-2}K^{-1}$, (ii) 1.06gs^{-1}, (iii) 0.73]

P7.8 Refrigerant flowing through a direct-expansion cooling coil enters at 8°C with a quality of 0.25. Moist air flowing in the opposite direction enters the coil at 30°C, db–temperature and 21°C wb–temperature. The mass flow rates of refrigerant and air are 2.6 kgs^{-1} and 3 kgs^{-1} respectively. The outside and inside heat transfer coefficients are 0.06 $kWm^{-2}K^{-1}$ and 2 $kWm^{-2}K^{-1}$ respectively. The ratio of the outside to inside heat transfer areas is 15. The latent heat of vaporization of the refrigerant is 198 $kJkg^{-1}$. The air enthalpy at the exit of the coil is 32 $kJkg^{-1}$. Calculate (i) the outside area of the coil, (ii) temperature and humidity ratio of air at the exit, (iii) the quality of the refrigerant leaving the coil, and (iv) the cooling capacity of the coil. Plot the condition line of the air on the psychrometric chart.
[*Answers*: (i) 157.5m^2, (ii) 11.45°C, 0.0081, (iii) 0.686, (iv) 86.4kW]

P7.9 Air flowing through a direct-expansion cooling coil at the rate of 3.2 kgs^{-1} enters with a db–temperature of 31°C and a wb–temperature of 21°C. At the exit the air is at 13°C db-temperature and 90% relative humidity. The temperature of the refrigerant is 12°C. The outside and inside heat transfer coefficients are 0.065 $kWm^{-2}K^{-1}$ and 2.6 $kWm^{-2}K^{-1}$ respectively. The ratio of the outside to inside heat transfer areas is 15.

Calculate (i) the area of the dry section of the coil, (ii) the total area of the coil, (iii) the total cooling capacity, and (iv) the temperature and humidity ratio of the air at the exit. Plot the condition line of the air on the psychrometric chart.
[*Answers*: (i) 11.3m^2, (ii) 268 m^2, (iii) 83.8 kW, (iv) 12.24°C, 0.0088]

P7.10 Air at 28°C db-temperature and 55% relative humidity enters a chilled water cooling coil with a mass flow rate of 3 kgs^{-1}. Chilled water entering at 10°C flows in the opposite direction with a mass flow rate of 2.8 kgs^{-1}. The leaving water temperature is 17°C. The outside and inside heat transfer coefficients are 0.06 kWm^{-2}K^{-1} and 3 kWm^{-2}K^{-1} respectively. The ratio of the outside to inside heat transfer areas is 16. Calculate (i) the temperature of the air when condensation just begins, (ii) the total area of the coil, and (iii) the cooling capacity.
[*Answers*: (i) 24.2°C, (ii) 274m^2, (iii) 82.3 kW]

P7.11 Refrigerant at 9°C flows through a direct-expansion cooling coil with a total air-side area of 120m^2. Air enters the coil at 31°C db–temperature and 60% relative humidity, with a mass flow rate of 3.5 kgs^{-1}. The outside and inside heat transfer coefficients are 0.075 kWm^{-2}K^{-1} and 3 kWm^{-2}K^{-1} respectively. The ratio of the outside to inside heat transfer areas is 16. Show that vapor condensation will occur at the entrance section of the coil. Calculate (i) the cooling capacity, and (ii) the db–temperature and humidity ratio of the air at the exit. Plot the condition line on the psychrometric chart.
[*Answers*: (i) 122.7 kW, (ii) 14.4°C, 0.011]

P7.12 Consider the off-design operation of the direct-expansion coil described in problem P7.8. The air inlet conditions remain unchanged but the refrigerant temperature decreases to 5°C. Calculate (i) the cooling capacity, and (ii) the temperature and humidity ratio of air at the exit of the coil.
[*Answers*: (i) 103.2 kW, (ii) 9.0°C, 0.0068]

References

1. *ASHRAE Handbook* - 2013 *Fundamentals*, American Society of Heating, Refrigeration and Air Conditioning Engineers, Atlanta, 2013.
2. Bejan, Adrian, *Heat Transfer*, John Wiley & Sons, Inc. New York, 1993.
3. Bejan, Adrian and Kraus, Allen D., *Heat Transfer Handbook*, John Wiley & Sons, Inc. New York, 2003.
4. Kuehn, Thomas H., Ramsey, James W. and Threlkeld, James L., *Thermal Environmental Engineering*, 3rd edition, Prentice-Hall, Inc., New Jersey, 1998.
5. Mills, Anthony F., *Heat Transfer*, Irwin, Richard D., Inc., Boston, MA, 1992.
6. Mills, Anthony F., *Mass Transfer*, Prentice Hall, New Jersey, 2001.
7. Rogers G. F. C. and Mayhew Y. R., *Thermodynamic and Transport Properties of Fluids*. 5th ed. Blackwell, Oxford, U.K. 1998
8. Stoecker, Wilbert F. and Jones, Jerold W., *Refrigeration and Air Conditioning*, International Edition, McGraw-Hill Book Company, London, 1982.

Appendix A7.1 - MATLAB Code for Design of Chilled Water Coils

```
% uses cubic expression for saturation air enthalpy
% numerical values from worked example 7.16
% condensation occurs at the air entrance section 1
 cpm=1.02          % specific heat capacity of air, kJkg^-1K^-1
 ma=2.5            % mass flow rate of air, kgs^-1
tain=26            % dry-bulb temperature of air at inlet, °C
wain=0.0136        % humidity ratio of air at inlet
hain=60.5          % enthalpy of air at inlet, kJkg^-1
twin=6             % inlet water temperature, °C
twout=12           % outlet water temperature, °C
maw=3.2            % mass flow rate of water, kgs^-1
cw=4.2             % specific heat capacity of water, kJkg^-1K^-1
hcc=0.055          % air-side heat transfer coefficient, kWm^-2K^-1
```

```
hcw=3                        % water-side heat transfer coefficient, kWm-2K-1
arer=16                      % ratio of air-side area to water-side area
nc=3                         % number of control volumes or sections
qc=maw*cw*(twout-twin)    % cooling capacity of the coil, kW
alfa=arer*hcc/(hcw*cpm)   % parameter alpha for the coil
tw(1)=twout
ha(1)=hain
% compute water temperatures and air enthalpies at the CV boundaries
for i=1:nc
ii=i+1
tw(ii)=tw(i)-(twout-twin)/nc
ha(ii)=ha(i)-qc/(nc*ma)
end
% coefficients for water vapor enthalpy cubic polynomial
A0=2500.7;
A1=1.854;
A2=-0.0005;
A3=-6.0e-06;
ac=[A3,A2,A1,A0];
% coefficients for the saturated air enthalpy cubic polynomial
b0=9.3625;
b1=1.7861;
b2=0.01135;
b3=9.8855e-04;
B=[b3 b2 b1 b0];
% compute coefficients of the cubic equation for water film temperature
c(1)=b3;
c(2)=b2
c(3)=b1 +1/alfa;
for i= 1:(nc+1);
c(4)=b0 -tw(i)/alfa -ha(i);
% compute the roots of the cubic equation
r=roots(c);
ti1=r(1);
ti2=r(2);
ti3=r(3);     % physically meaningful root
```

```
% compute saturation air enthalpy and humidity ratio
ti(i)=ti3
hi(i) = polyval(B,ti3);
hidat=[ha(i), hi, ti]
hgi(i)=polyval(ac,ti(i));
wi(i)=(hi(i)-cpm*ti(i))/hgi(i);
end
% compute the heat transfer areas and total area of coil
aht=0 ;                    % total heat transfer area of coil on air-side
for k = 1:nc;
xxa=0.5*(ha(k)+ha(k+1));
xxi=0.5*(hi(k)+hi(k+1));
ACV(k)=ma*cpm*(ha(k)-ha(k+1))/(hcc*(xxa-xxi))
aht=aht+ACV(k)
end
% compute dry-bulb temperature of air
ta(1)=tain
for j= 1:nc;
fac=2*ma*cpm/(hcc*ACV(j));
xxa=ti(j)+ti(j+1);
jj=j+1;
ta(jj)=(xxa+fac*ta(j)-ta(j))/(1+fac);
end
% compute humidity ratio of air
wa(1)=wain;
for j= 1:nc;
fac=2*ma*cpm/(hcc*ACV(j));
xxa=wi(j)+wi(j+1);
jj=j+1;
wa(jj)=(xxa+fac*wa(j)-wa(j))/(1+fac);
end
% compute air enthalpy for checking computational accuracy
nc1=nc+1;
ACV(nc1)=0 ;
datin=[ma,hain,tain,wain,maw,twin,twout,hcc,hcw,arer]
qcw=ma*(hain-ha(nc1));
```

```
qcww=maw*cw*(twout-twin);
for i= 1:nc1
hga(i)=polyval(ac,ta(i));
hacl(i)=cpm*ta(i)+hga(i)*wa(i);
 errha(i)=(ha(i)-hacl(i))/ha(i);    % fractional difference in enthalpy
 twdat=[ha(i),hacl(i),errha(i),wa(i),ta(i),ti(i),wi(i),hi(i), aht,qcw,qcww]
 end
```

Chapter 8

Steady Heat and Moisture Transfer Processes in Buildings

8.1 Introduction

In chapter 5 we considered psychrometric design aspects of winter heating systems and summer cooling systems, applicable to single-zone and multi-zone buildings. The detailed physical modeling of several subcomponents of these air conditioning systems like heaters, heat exchangers, humidifiers, cooling coils, and cooling towers was presented in chapters 6 and 7. The sizes of these subcomponents depend critically on the heating and cooling loads experienced by the conditioned spaces in the building.

The winter *heating load* of a building is the rate at which heat has to be supplied by the heating system of the building, to maintain its indoor temperature and humidity within specified limits. The main contributor to the heating load of a space is the heat loss to the outside ambient across the building envelope. Moreover, the energy input required to heat any cold ambient air, entering the building through cracks or openings in the envelope, to the temperature of the space will contribute indirectly to the heating load of the building.

In general, the heating load of a building is time-varying due to several factors. The heat loss through the building envelope, consisting of the walls, roofs and other structural components, is transient due to the thermal mass of the structural materials. Furthermore, the external weather conditions such as the ambient temperature, the wind speed and the solar radiation intensity that drive the energy flows vary with time.

For winter heating load estimation , a conservative design approach, recommended in the *ASHRAE Handbook* - 2013 *Fundamentals* [1], is to ignore the heat inputs due to solar radiation entering through fenestrations and the energy input from lights, occupants and equipment within the building because these energy flows help reduce the heat input required from the heating system. Similarly, the energy stored in the structural elements of the building envelop tends to reduce the required heat input. Therefore, for estimating the winter heating load we need to consider, in detail, only *steady heat losses* through building envelope components and the heating *load due to infiltration* of cold ambient air. We shall analyze these energy transfer processes in the following sections.

8.2 Steady Heat Transfer through Multi-Layered Structures

The most common designs of walls and roofs of buildings are multilayered structures. A typical arrangement of such a structure is depicted in Fig. 8.1.

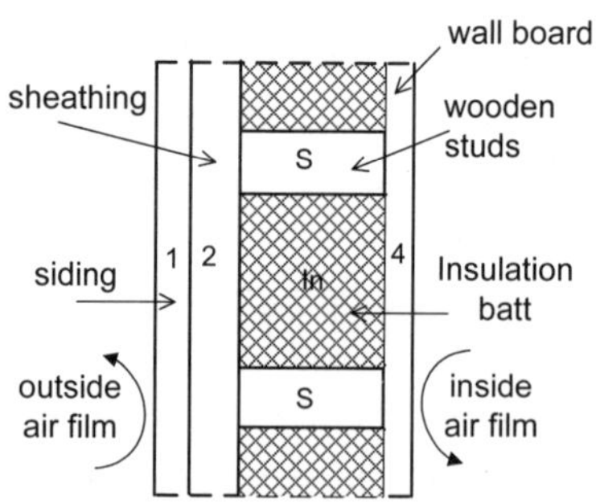

Fig. 8.1 Typical arrangement of a multilayered wall structure

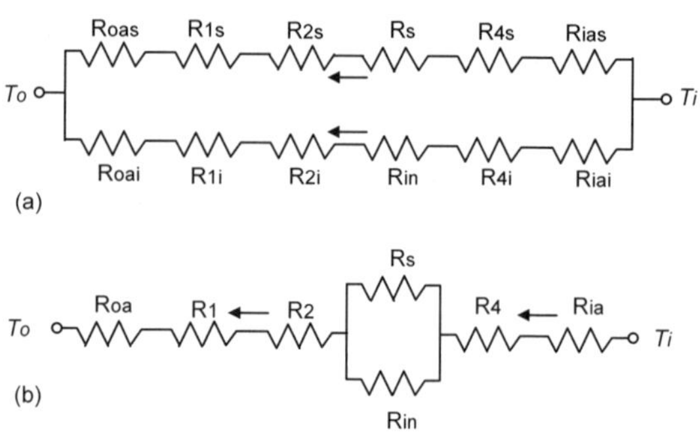

Fig. 8.2 Thermal networks: (a) Parallel path, (b) Isothermal plane

The wall consists of an outer siding layer, 1 followed by a sheathing layer, 2 that bears against a wooden frame made of vertical beams called studs, (s). The space between these studs is filled with insulation material (in). The inside of the wall consists of a layer of wall board, 4. Convective heat transfer occurs between the inner and outer surfaces of the wall and the adjacent air layers. These wall surfaces also exchange thermal radiation with other surfaces in the surroundings. It should be noted that we have considered a wall consisting of the layers of materials described above only as a representative design for developing the heat flow analysis. Many other wall configurations are listed in Ref. [1].

In section 2.4 we presented the analysis of one-dimensional heat conduction problems using equivalent thermal networks. We shall now extend the network approach to conduction situations involving both series and parallel heat flow paths as in the case of the wall shown in Fig. 8.1.

There are two possible representations of the heat flow paths through the wall that lead to the two thermal networks depicted in Fig. 8.2. The corresponding analytical methods are called the *parallel path method* and the *isothermal plane method.* For complicated geometrical shapes involving metal sections, a third approach called the *zone method* is recommended in Ref. [1].

8.2.1 *Parallel path method*

In the parallel path method we assume that heat flow from the inside air at temperature, T_i to the outside air at temperature T_o, or vice versa, occurs through two parallel paths. The first path, through the studs, (s) has a heat flow area equal to the total cross sectional area, A_s of the studs while the second path, through the insulation, (in) has a heat flow area equal to the total cross sectional area of the insulation, A_{in}. The thermal resistances of the various sections of the wall along these two heat flow paths are as follows [see Fig. 8.2(a)]:

$$R_{oas} = \frac{1}{A_s h_o}, \quad R_{1s} = \frac{L_1}{A_s k_1}, \quad R_{2s} = \frac{L_2}{A_s k_2}, \quad R_s = \frac{L_s}{A_s k_s}, \quad R_{4s} = \frac{L_4}{A_s k_4},$$

$$R_{ias} = \frac{1}{A_s h_i}, \quad R_{oai} = \frac{1}{A_{in} h_o}, \quad R_{1i} = \frac{L_1}{A_{in} k_1}, \quad R_{2i} = \frac{L_2}{A_{in} k_2},$$

$$R_{in} = \frac{L_{in}}{A_{in}k_{in}}, \quad R_{4i} = \frac{L_4}{A_{in}k_4}, \quad R_{iai} = \frac{1}{A_{in}h_i}$$

where, L_n and k_n are respectively the thickness and thermal conductivity of the nth layer in Fig. 8.1. The inside and outside heat transfer coefficients, which include both convection and radiation heat transfer, are h_i and h_o respectively. The total resistance of each path is obtained by adding the individual resistances that are in series. Hence we have

$$R_{to,s} = R_{oas} + R_{1s} + R_{2s} + R_s + R_{4s} + R_{ias}$$

$$R_{to,in} = R_{oai} + R_{1i} + R_{2i} + R_{in} + R_{4i} + R_{iai}$$

The two resistances, $R_{to,s}$ and $R_{to,in}$ above are in parallel, as seen from the thermal network in Fig. 8.2(a). Therefore the overall resistance for the complete wall is given by Eq. (2.13) as

$$R_{wall} = \left(\frac{1}{R_{to,s}} + \frac{1}{R_{to,in}}\right)^{-1} \tag{8.1}$$

8.2.2 *Isothermal plane method*

In the isothermal plane method we assume that the heat flow paths are parallel only through the insulation and the studs. For the other layers, the temperature along any lateral plane normal to the direction of heat flow is assumed uniform. This implies excellent heat flow in the lateral direction and therefore the heat flow paths through these layers are in series as seen in the equivalent thermal network in Fig. 8.2(b). The individual thermal resistances of the wall sections are as follows:

$$R_{oa} = \frac{1}{A_w h_o}, \quad R_1 = \frac{L_1}{A_w k_1}, \quad R_2 = \frac{L_2}{A_w k_2}, \quad R_s = \frac{L_s}{A_s k_s}, \quad R_4 = \frac{L_4}{A_w k_4},$$

$$R_{in} = \frac{L_{in}}{A_{in}k_{in}}, \quad R_{ia} = \frac{1}{A_w h_i}$$

where the total wall area, $A_w = A_s + A_{in}$.

The overall thermal resistance of the wall is given by

$$R_{wall} = R_{oa} + R_1 + R_2 + \left(\frac{1}{R_s} + \frac{1}{R_{in}}\right)^{-1} + R_4 + R_{ia} \tag{8.2}$$

For complex thermal networks it is often convenient to define a *unit thermal resistance* for each element i as

$$\bar{R}_i = \frac{L_i}{k_i} = A_i R_i$$

Hence we can express Eq. (8.2) in the form

$$\bar{R}_{wall} = \bar{R}_{oa} + \bar{R}_1 + \bar{R}_2 + \left(\frac{a_s}{\bar{R}_s} + \frac{a_i}{\bar{R}_i}\right)^{-1} + \bar{R}_4 + \bar{R}_{ia} \qquad (8.2a)$$

where $a_s = A_s/A_w$ and $a_i = A_i/A_w$ are the fractional areas of the studs and the insulation respectively.

Once the overall thermal resistance, R_{wall} is obtained from Eqs. (8.1) or (8.2) we can calculate the *overall heat transfer coefficient*, U_o, which is given by the heat transfer rate equation

$$Q = A_w U_o \Delta T = \frac{\Delta T}{R_{wall}} \qquad (8.3)$$

where, $\Delta T = T_i - T_o$, is the overall temperature difference across the wall. From Eq. (8.3) it follows that

$$U_o = \frac{1}{A_w R_{wall}} = \frac{1}{\bar{R}_{wall}} \qquad (8.4)$$

8.2.3 *Zone method*

The zone method is recommended in the *ASHRAE Handbook* - 2013 *Fundamentals* [1] as a simplified procedure to estimate the heat flow through structures with widely spaced metal members of complex cross sectional areas like I-sections and T-sections. For computational purposes, the structure is divided into two zones A and B. Zone A contains the metal elements with the complex cross sectional shape. The width of zone A is computed using empirical formulae obtained from detailed two-dimensional computer simulations of the structure.

Zone B includes the section of the structure without the metal components where the heat flow is one-dimensional. The individual equivalent thermal resistance of zones A and B are determined using the isothermal plane method. These equivalent thermal resistances of the two zones A and B are then combined using the parallel flow method to determine the overall thermal resistance of the structure. We shall illustrate the application of the zone method to a flat roof section in worked example 8.5.

The *ASHRAE Handbook* - 2013 *Fundamentals* [1] gives a table of values of thermal conductivities and thermal resistances of different building envelope materials. The thermal conductivities of a few building materials, extracted from Ref. [1], are listed in Table 8.1 for purposes of illustration.

Table 8.1 Thermal conductivity of some building materials*

Material	Density, $kg.m^{-3}$	Thermal Conductivity, $Wm^{-1}K^{-1}$
Brick	2400 to 1120	1.21-1.47 to 0.36-0.45
Concrete	2400 to 960	1.4-2.9 to 0.30-0.36
Gypsum	640	0.16
Plywood	450 to 540	0.09 to 0.10
Sheathing	290 to 400	0.055 to 0.067
Glass-fiber	10 to 14	0.045 to 0.048
Mineral wool	16 to 130	0.040 to 0.035
Polystyrene	25 to 40	0.022 to 0.03
Softwoods	500 to 660	0.13 to 0.16
Roofing	1600 to 2300	0.43 to 1.15

*Values extracted from Table 1, Page 26.7-*ASHRAE Handbook* - 2013 *Fundamentals* [1]

8.2.4 *Radiation heat transfer coefficient*

In chapter 2 we considered several heat transfer situations involving simultaneous convection and radiation. The analysis of these problems are somewhat tedious because of the fourth-power law governing radiation heat transfer. Fortunately, in most air conditioning applications, the temperature difference between the radiating surfaces is relatively small, and this allows us to develop a simplified approach based on the *radiation heat transfer coefficient.*

Consider a flat surface like a roof losing heat by thermal radiation and convection to the surrounding ambient. The sky, with which the surface exchanges radiation, may be treated as a *large* hemispherical surface at a temperature T_o. Let the emissivity, the area, and the absolute temperature of the surface be ε, A and T_s respectively. Then the net rate of radiation heat transfer between the surface and the sky is given by Eq. (2.76) as

$$Q_r = A\sigma\epsilon\left(T_s{}^4 - T_o{}^4\right) \tag{8.5}$$

We can rewrite Eq. (8.5) in terms of the temperature difference, $\Delta T = (T_s - T_o)$ and the average temperature, $T_{av} = (T_s + T_o)/2$ (K) as [3]

$$Q_r = A\sigma\varepsilon(T_s - T_o)\left[4T_{av}^{\ 3}\left(1 + \frac{0.25\Delta T^2}{T_{av}^{\ 2}}\right)\right] \tag{8.6}$$

For most practical air conditioning applications, the difference in temperature, ΔT is small compared to the average temperature, T_{av}, expressed in degrees Kelvin. Therefore the error caused by neglecting the second term within the square brackets in Eq. (8.6) is relatively small. Hence we can express the radiation heat transfer rate as

$$Q_r \cong A\left(4T_{av}^{\ 3}\sigma\varepsilon\right)(T_s - T_o) = Ah_r(T_s - T_o) \tag{8.7}$$

where $h_r = 4\varepsilon\sigma T_{av}^{\ 3}$, is the *radiation heat transfer coefficient.*

The convective heat transfer rate is given by

$$Q_c = Ah_c(T_s - T_a) \tag{8.8}$$

where T_a is the ambient air temperature.
The total heat transfer rate is

$$Q_t = Q_r + Q_c \tag{8.9}$$

Substituting in Eq. (8.9) from Eqs. (8.7) and (8.8) we have

$$Q_t = Ah_r(T_s - T_o) + Ah_c(T_s - T_a) \tag{8.10}$$

Now the sky temperature, T_o can be closely approximated by the ambient air temperature, T_a. With this assumption we obtain a convenient expression for the total heat transfer rate in the form

$$Q_t = Ah_{cr}(T_s - T_a) \tag{8.11}$$

where the *combined convection–radiation heat transfer coefficient* is given by

$$h_{cr} = h_c + h_r = h_c + 4\varepsilon\sigma T_{av}^{\ 3} \tag{8.12}$$

The *ASHRAE Handbook* - 2013 *Fundamentals* [1] gives a table of values for h_{cr} for different orientations of surfaces with different surface emissivities. We have listed a few representative values, extracted from Ref. [1], in Table 8.2.

Table 8.2 Surface heat transfer coefficients - still air*

Orientation of Surface	Heat flow Direction	Emittance = 0.9 h_{cr} (Wm^{-2}K^{-1})	Emittance = 0.2 h_{cr} (Wm^{-2}K^{-1})	Emittance = 0.05 h_{cr} (Wm^{-2}K^{-1})
Horizontal	Upward	9.26	5.17	4.32
Slope 45°	Upward	9.09	5.0	4.15
Vertical	Horizontal	8.29	4.2	3.35
Slope 45°	Downward	7.5	3.41	2.56
Horizontal	Downward	6.13	2.1	1.25

For moving air (i) at speed 6.7 ms^{-1}, h_{cr} = 34 Wm^{-2}K^{-1}
(ii) at speed 3.4 ms^{-1}, h_{cr} = 22.7 Wm^{-2}K^{-1}

*Values extracted from Table 10, Page 26.20 - *ASHRAE Handbook* - 2013 *Fundamentals* [1]

8.2.5 *Heat transfer in gas filled cavities*

Gas filled spaces or cavities are present in several building envelop components including, wall sections, roof sections and multi-glazed fenestrations like windows, doors and sky lights. In wall sections the space between solid layers contains air whereas in some windows the space between the glass panes is filled with a gas like argon or krypton. Heat transfer across the gas space occurs due to radiation and convection.

Most surfaces encountered in building envelope components may be treated as gray surfaces (see section 2.8.8) for which the radiation transfer depends mainly on the emissivity of the surfaces. The convective heat transfer, on the other hand, is affected by several factors including the properties of the gas filling the space, the orientation of gas layer, direction of heat flow, and the thickness of the gas space.

Consider a gas filled space bounded by two gray surfaces 1 and 2 with emissivities ε_1 and ε_2 at absolute temperatures T_1 and T_2 respectively. The expression for radiation heat transfer rate per unit area was obtained in section 2.8.9 [Eq. (2.71)] as

$$Q_r = \sigma\varepsilon_{eff}\left({T_1}^4 - {T_2}^4\right) \tag{8.13}$$

where the effective emissivity is

$$\varepsilon_{eff} = \left(\frac{1}{\varepsilon_1} + \frac{1}{\varepsilon_2} - 1\right)^{-1} \tag{8.14}$$

Now for most practical situations encountered in building systems the difference in temperature between the surfaces is much smaller than the average temperature of the surfaces, expressed in degrees Kelvin. Therefore we can write the radiation heat transfer rate in terms of the *radiation heat transfer coefficient*. Hence from Eq. (8.7) we have

$$Q_r = 4\sigma\varepsilon_{eff}{T_{av}}^3(T_1 - T_2) = h_r(T_1 - T_2) \tag{8.15}$$

where $h_r = 4\varepsilon_{eff}\sigma{T_{av}}^3$, is the *radiation heat transfer coefficient*.

The convective heat transfer rate in the cavity per unit surface area is given by Eq. (8.8) as

$$Q_c = h_c(T_1 - T_2) \tag{8.16}$$

Now the total heat transfer rate is

$$Q_t = Q_r + Q_c \tag{8.17}$$

Substituting in Eq. (8.17) from Eqs. (8.15) and (8.16) we have

$$Q_t = h_r(T_1 - T_2) + h_c(T_1 - T_2) = h_{ef}(T_1 - T_2) \tag{8.18}$$

where the *convection–radiation heat transfer coefficient* is

$$h_{ef} = h_c + h_r = h_c + 4\varepsilon_{eff}\sigma{T_{av}}^3 \tag{8.19}$$

The *ASHRAE Handbook* - 2013 *Fundamentals* [1] gives a table of values of thermal resistances, $R_{ef} = 1/h_{ef}$, for plane air spaces of different orientations. We have included, in Table 8.3, a few representative values of h_{ef}, calculated using the data in Ref. [1], for 13 mm vertical air spaces for different values of the effective emissivity, ε_{eff}.

Table 8.3 Heat transfer coefficients ($Wm^{-2}K^{-1}$) for 13 mm vertical air spaces*

T_{mean}, C°	ΔT, C°	$\varepsilon_{eff} = 0.03$	$\varepsilon_{eff} = 0.05$	$\varepsilon_{eff} = 0.2$	$\varepsilon_{eff} = 0.5$	$\varepsilon_{eff} = 0.82$
32.2	5.6	2.33	2.44	3.45	5.26	7.14
10.0	16.7	2.22	2.33	3.13	4.55	6.25
10.0	5.6	2.13	2.22	3.03	4.55	6.25
−17.8	11.1	2.0	2.08	2.63	3.85	5.0
−17.8	5.6	1.92	2.0	2.56	3.7	5.0
−45.5	11.1	1.96	2.0	2.44	3.23	4.17
−45.6	5.6	1.79	1.82	2.22	3.03	3.85

* Values extracted from Table 3, Page 26.13-*ASHRAE Handbook* - 2013 *Fundamentals* [1]

8.3 Steady Heat Transfer through Fenestrations

The heat flow through windows and doors consisting of transparent sections, commonly called *fenestrations*, involve several coupled physical processes. Since the main focus of this chapter is the estimation of winter heating loads, we shall only consider convective, conductive and radiative heat flows through the fenestration, due to the difference in temperatures between the indoor and outdoor air. Moreover, for estimating winter heating loads we shall ignore all effects due to the absorption of solar radiation in the transparent material, which is the conservative design approach, recommended in Ref. [1]. However, for estimating the summer cooling load the effects of transmission and absorption of solar radiation in fenestrations are important and, therefore, these effects will be considered in detail in chapter 9.

8.3.1 *Windows and doors*

We shall now consider the heat flow through complete windows and doors consisting of transparent sections, located in opaque frames.

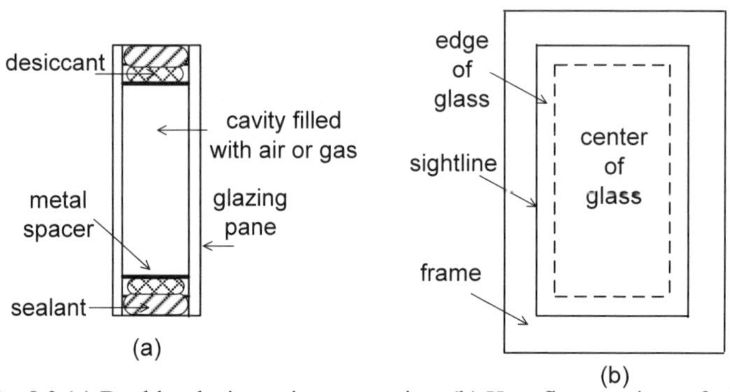

Fig. 8.3 (a) Double-glazing unit construction, (b) Heat flow sections of window

Some important constructional details of a double-glazed window are shown schematically in Fig. 8.3(a). Glass is the most common material used for the two panes but plastic sheets are used in some instances. Low-emissivity coatings may be applied on the inner surfaces of the two panes to reduce the radiation heat transfer across the window. The sealed cavity between the panes is usually filled with air. However, convective heat transfer rate in the cavity could be reduced by substituting a gas such as argon and krypton in lieu of air.

The spacer separating the two panes of glass provides a surface for sealant adhesion. Typical spacers, made of metals like aluminum, usually result in excessive heat flow across them. To minimize this heat flow, warm-edge spaces have been developed using materials such as stainless steel and polymers that have lower thermal conductivities.

Sealants are applied at the edges of windows to minimize moisture and hydrocarbon transmission into the gas filled cavity. In some designs, desiccants such as molecular sieve or silica gel are located at the edges, as shown in Fig. 8.3(a). These absorb the moisture that is initially trapped in the glazing unit during construction, or diffuses through the sealant over time.

The transparent unit of the window is most often placed in a frame made of wood, metal or polymer. The thermal performance of wood and polymer frames are far superior to highly conducing metal frames, typically made of aluminum. Polymer and wood frames have similar thermal and structural performance but wood has low resistance to weather, moisture and organic degradation. The general characteristics of doors are similar to those of windows discussed above. For a more complete description of the characteristics of different window and door types, the reader is referred to the *ASHRAE Handbook - 2013 Fundamentals* [1].

8.3.2 *Overall heat transfer coefficient*

Heat transfer across windows and doors occur through the transparent section as well as the frame. For multi-pane fenestrations where the gaps between the panes are small compared to the area of the panes, heat transfer across the central area may be treated as one-dimensional. Therefore the analysis developed in section 8.2.5 is applicable to this section.

However, the heat flow across the area close to the edge of the window is two-dimensional, and is further complicated by the presence of spacers and sealant sections. The heat transfer across the frame is also two-dimensional.

For design calculations, the window or door is divided into three sections with different average heat transfer coefficients, as shown in Fig. 8.3(b). These sections are called the center of the glass, the edge of the glass, and the frame, and their respective areas are A_{cg}, A_{eg} and A_{fr}. The total heat transfer across the window may be expressed in terms of an average heat transfer coefficient, U_o, which satisfies the overall energy balance equation,

$$(A_{cg} + A_{eg} + A_{fr})U_o\Delta T = A_{cg}U_{cg}\Delta T + A_{eg}U_{eg}\Delta T + A_{fr}U_{fr}\Delta T$$

where the heat transfer coefficients for the center of glass, the edge of glass, and the frame are respectively, U_{cg}, U_{eg}, and U_{fr}. The temperature difference between the inside and outside air is ΔT.
From the above equation we have

$$U_0 = \frac{A_{cg}U_{cg}+A_{eg}U_{eg}+A_{fr}U_{fr}}{A_{cg}+A_{eg}+A_{fr}} \tag{8.20}$$

The heat transfer coefficient for the center of the glass of a double-glazed window is given by

$$\frac{1}{U_{cg}} = \frac{1}{h_{ai}} + \frac{L_1}{k_1} + \frac{1}{h_{ef}} + \frac{L_2}{k_2} + \frac{1}{h_{ao}} \tag{8.21}$$

where the inside and outside *convection–radiation heat transfer coefficients*, defined by Eq. (8.12), are h_{ai} and h_{ao} respectively. The respective thicknesses and thermal conductivities of the two panes are L_1, k_1 and L_2, k_2. The effective heat transfer coefficient for the gas filled cavity, defined by Eq. (8.19), is h_{ef}. A few representative values of h_{ai} and h_{ef} are listed in Tables 8.2 and 8.3 respectively.

The width of the edge-of-glass area, A_{eg} where two-dimensional heat transfer effects are dominant, has been determined using computer models based on conduction-only analysis [1]. The edge-of-glass area is typically taken to be a 63.5 mm wide band around the sightline (see Fig. 8.3b). The edge-of-glass U-factors for a large number of fenestration products have been obtained by computer simulation and their values are tabulated in Ref. [1].

Estimation of heat flow rates through frames of fenestrations is complicated due to the variety of frame configurations, sizes, materials and spacer types used. However, average U-factors for frames made of wood, vinyl and aluminum with different spacer types have been obtained by computer simulation [1].

The *ASHRAE Handbook* - 2013 *Fundamentals* [1] gives a table of values for the center-of-glass U-factor, the edge-of-glass U-factor, and the frame U-factor for a wide variety of fenestration products with single, double, triple and quadruple glazings. We have included a few representative values for *fixed double-glazed* vertical windows of different designs in Table 8.4. We shall use some of these values in the worked examples to follow in this chapter.

Table 8.4 U-values[1] of fixed double-glazed windows ($Wm^{-2}K^{-1}$)*

Gap, Gas in space	Center of glass, U_{cg}	Edge of glass, U_{eg}	Aluminum frame without break, U_{fr}	Aluminum frame with break, U_{fr}
6 mm ,air	3.12	3.63	3.88	3.52
13 mm, air	2.73	3.36	3.54	3.18
6 mm, argon	2.90	3.48	3.68	3.33
13 mm, argon	2.56	3.24	3.39	3.04
6 mm, air[2]	2.95	3.52	3.73	3.38
13 mm, air[2]	2.50	3.20	3.34	2.99
6 mm, argon[2]	2.67	3.32	3.49	3.13
13 mm, argon[2]	2.33	3.08	3.20	2.84

[1] U-values are based on -18°C outdoor and 21°C indoor temperatures, wind speed of $6.7ms^{-1}$.

[2] For the two surfaces facing the gas space, the emissivity, $\varepsilon = 0.60$.

*Values extracted from Table 4, Page 15.8 - *ASHRAE Handbook* - 2013 *Fundamentals* [1]

8.4 Below Grade Heat Transfer in Buildings

Heat losses through basement walls and floors contribute to winter heating loads of residential buildings. A detailed analysis of these heat transfer processes would require the use of three-dimensional transient models of heat conduction. However, simplified one-dimensional conduction models for estimating these heat transfer rates yield results that agree satisfactorily with the predictions of detailed models [1,3]. Moreover, as a conservative design approach we shall ignore transient effects and use the steady heat transfer rates to estimate the below-grade heat losses through walls and floors.

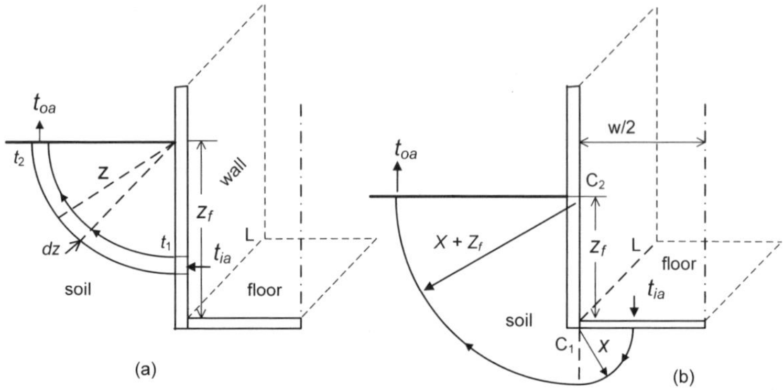

Fig. 8.4 Parameters for below-grade heat transfer models: (a) wall, (b) floor

8.4.1 *Heat transfer through basement walls*

The parameters used in the steady one-dimensional conduction models for heat transfer through the soil surrounding a basement is shown in Fig. 8.4. We shall first consider the heat flow across the wall depicted in Fig. 8.4(a). We assume that the heat flow lines are circular and therefore the constant temperature lines are radial.

Let the length and height of the wall be L and z_f respectively. The inside air temperature is t_{ia} and the outside ambient air temperature is t_{oa}. The length of a typical circular heat flow path of radius z through the soil is

$$L_Z = \frac{\pi z}{2} \tag{8.22}$$

The area of cross section of the heat flow path is $L\delta z$. Hence the thermal resistance (Eq. 2.5) of this heat flow path is

$$R_Z = \frac{L_Z}{k_S L\delta z} = \frac{\pi z}{2k_S L\delta z} \tag{8.23}$$

where k_s is the soil thermal conductivity, assumed uniform and constant.

There are several additional thermal resistances in the above heat flow path that are in series with R_Z. These include the convection–

radiation resistances of the inside air and the outside air, and the thermal resistance of the wall. Let the sum of these *additional resistances*, in series with R_Z, be R_a per *unit area* of the wall. Therefore the total thermal resistance to heat flow from the inside air to the outside ambient air is

$$R_{to} = \frac{\pi z}{2k_s L\delta z} + \frac{R_a}{L\delta z} \tag{8.24}$$

The rate of heat flow through the above path is given by

$$\delta Q_z = \frac{(t_{ia} - t_{oa})}{R_{to}} = \frac{(t_{ia} - t_{oa})L\delta z}{R_a + \pi z/2k_s} \tag{8.25}$$

We determine the total heat flow rate through the wall by adding the contributions from all the small heat flow paths of thickness δz. Thus we obtain the total heat flow, Q_{to}, by integrating Eq. (8.25) as

$$Q_{to} = (t_{ia} - t_{oa})L \int_o^{z_f} \frac{dz}{R_a + \pi z/2k_s} \tag{8.26}$$

Now the overall heat transfer coefficient, U_{bw} for heat flow through the soil is defined by the equation

$$Q_{to} = U_{bw}(Lz_f)(t_{ia} - t_{oa}) \tag{8.27}$$

From Eqs. (8.26) and (8.27) it follows that

$$U_{bw} = \left(\frac{2k_s}{\pi z_f}\right) ln\left(\frac{z_f + 2k_s R_a/\pi}{2k_s R_a/\pi}\right) \tag{8.28}$$

The below-grade U-factors for heat flow from basement walls have been computed for several conditions using Eq. (8.28) and the data are tabulated in Ref. [1].

8.4.2 *Heat transfer through basement floors*

A typical heat flow path through the soil, from the basement floor to the outside air, is shown in Fig. 8.4(b). It consists of two circular arcs, the first with its center, C_1 at the corner of the floor and the wall, and the second with its center, C_2 at the point of intersection of the grade and the wall.

The length of a typical circular heat flow path of radius x, through the soil, is given by

$$L_x = \frac{\pi x}{2} + \left(x + z_f\right)\frac{\pi}{2} = \pi\left(x + \frac{z_f}{2}\right) \quad (8.29)$$

The area of cross section of the heat flow path is $L\delta x$.
Hence the thermal resistance of this path (Eq. 2.5) is

$$R_z = \frac{L_x}{k_s L\delta x} = \frac{\pi(x+z_f/2)}{k_s L\delta x} \quad (8.30)$$

where k_s is the soil thermal conductivity, assumed uniform and constant.

There are several additional thermal resistances in the above heat flow path that are in series with R_Z. These include the convection-radiation resistances of the inside air and the outside air, and the thermal resistance of the floor. Let the sum of these *additional resistances*, in series with R_Z, be R_a per *unit area* of the floor. Therefore the total thermal resistance to heat flow from the inside air to the outside ambient is

$$R_{to} = \frac{\pi(x+z_f/2)}{k_s L\delta x} + \frac{R_a}{L\delta x} \quad (8.31)$$

The rate of heat flow through the above path is given by

$$\delta Q_z = \frac{(t_{ia}-t_{oa})}{R_{to}} = \frac{(t_{ia}-t_{oa})L\delta x}{R_a+\pi(x+z_f/2)/k_s} \quad (8.32)$$

We determine the total heat flow rate through the floor by adding the contributions from all the small heat flow paths of thickness δx. Thus we obtain the total heat flow, Q_{to}, by integrating Eq. (8.32) as

$$Q_{to} = (t_{ia} - t_{oa})L\int_o^{w_b/2} \frac{dx}{R_a+\pi(x+z_f/2)/k_s} \quad (8.33)$$

Now the overall heat transfer coefficient, U_{bf} for heat flow through the soil across the floor is defined by the equation

$$Q_{to} = \frac{U_{bf}(t_{ia}-t_{oa})Lw_b}{2} \quad (8.34)$$

From Eqs. (8.33) and (8.34) it follows that

$$U_{bf} = \left(\frac{2k_s}{\pi w_b}\right) ln\left(\frac{\frac{w_b}{2}+\frac{z_f}{2}+\frac{k_s R_a}{\pi}}{\frac{z_f}{2}+\frac{k_s R_a}{\pi}}\right) \quad (8.35)$$

The below-grade *U*-factors for heat flow from basement floors have been computed for several conditions using Eq. (8.35), and the data are tabulated in Ref. [1].

8.4.3 *Heat transfer through surfaces at grade level*

Floors made of concrete slabs may be heated by: (i) the room heating medium like the hot air delivered to the room, or (ii) a hot fluid flowing in pipes buried within the concrete slab. The steady-state heat loss from the slab, q_{sl} is proportional to its perimeter, p and a heat transfer coefficient based on the perimeter, F_p (Wm^{-1}K^{-1}). Hence we can express the heat transfer rate through the slab as [1]

$$q_{sl} = pF_p\Delta t \tag{8.36}$$

where Δt is the indoor–outdoor temperature difference.

Values of the heat loss coefficient, F_p for different floor constructions are given in Table 24 on page 18.31 of the *ASHRAE Handbook - 2013 Fundamentals* [1].

8.5 Infiltration in Buildings

Infiltration is the unintended air flows into a conditioned space through cracks and openings in the building envelope. These air flows contribute to the winter heating load because the cold air entering the space has to be heated to the indoor air temperature and its humidity has to be increased to the indoor humidity.

8.5.1 *Heating load due to infiltration*

Once the mass flow rate of infiltration air, $\dot{m}_a$ is estimated the heating load, Q_{in} can be calculated in a straightforward manner using the energy balance equation. Hence we have

$$Q_{in} = \dot{m}_a(h_i - h_o) \tag{8.37}$$

where the inside and outside air enthalpies are h_i and h_o respectively. The total heating load can be expressed as a sensible component, Q_s and a latent component, Q_l such that

$$Q_{in} = Q_s + Q_l \tag{8.38}$$

The sensible and latent heat loads are given by

$$Q_s = \dot{m}_a c_{am}(t_i - t_o) \tag{8.39}$$

and

$$Q_l = \dot{m}_a(\omega_i - \omega_o) \tag{8.40}$$

respectively, where c_{am} is the mean specific heat capacity of air. The temperatures and humidity ratios of inside and outside air are respectively t_i, ω_i and t_o, ω_o.

Substituting the density and specific heat capacity of air at *standard conditions* of 1 bar and 15°C, in Eq. (8.39) the following expression is obtained for the *sensible* heat load [1]:

$$Q_s(kW) = 1.23\dot{V}_{in}(t_i - t_o) \tag{8.41}$$

where $\dot{V}_{in}$ (m^3s^{-1}) is the volumetric flow rate of infiltration air.

Similarly, for *standard air* and nominal comfort conditions, the *latent* heat load given by Eq. (8.40) may be expressed as [1]

$$Q_l(kW) = 3010\dot{V}_{in}(\omega_i - \omega_o) \tag{8.42}$$

The infiltration air flow rate is sometimes specified as *air exchanges per hour* (*ACH*). This is given by the expression

$$ACH = 3600\dot{V}_{in}/V \tag{8.43}$$

where V is the building volume.

8.5.2 *Infiltration air flow rates*

Infiltration of air through openings in a building envelope occurs due to the outdoor–indoor air pressure difference across the openings. This pressure difference is the net result of the interaction of three different physical processes called: (i) the stack or buoyancy effect, (ii) the wind effect, and (iii) the mechanical effect.

(i) *The stack effect*

The stack effect is the pressure difference across an opening in the building envelope due to the difference in density between the outside and inside air, resulting from the outside–inside temperature difference. The variation of the pressure, P in a vertical column of quiescent air of height z is governed by the equilibrium equation

$$\frac{dP}{dz} = -\rho g \tag{8.44}$$

where ρ is the density of air and g is the acceleration due to gravity.

If we assume the density, ρ to be constant, then it follows from Eq. (8.44) that the graph of pressure versus height is a straight line with a slope proportional to the air density.

If air is assumed to be an *ideal gas*, then the density is given by

$$\rho = \frac{m}{V} = \frac{P}{R_a T} \tag{8.45}$$

where R_a is the gas constant of air.

Now from Eq. (8.45) we observe that when the air temperature outside is lower than the air temperature inside the building, as in typical winter conditions, the outside air density is higher than the inside air density. This results in the inside and outside pressure distributions shown graphically in Fig. 8.5(a)

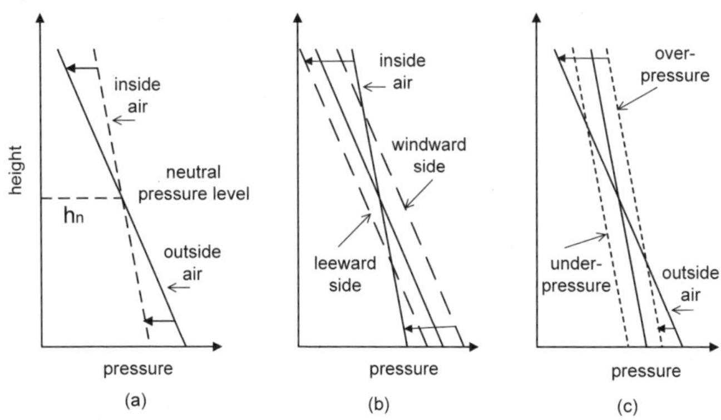

Fig. 8.5 (a) Stack pressure, (b) Stack and wind pressure, (c) Stack and mechanical pressure

If the areas of the cracks or openings in the building envelope are evenly distributed vertically, and the inflow and outflow of air are balanced, then the pressure difference between the inside and outside will be zero at mid-height as seen in Fig. 8.5(a). Below this *neutral plane* there is inflow of air and above it there is outflow of air due to the inside–outside pressure difference.

We now integrate Eq. (8.44), assuming the density to be constant, to obtain the following expression for the pressure distribution

$$P(z) = -g\rho z + C \tag{8.46}$$

where C is the constant of integration.

Applying Eq. (8.46) to air inside and air outside, we obtain the following expressions for the respective pressures at a height z:

$$P_i(z) = -g\rho_i z + P_{inp} + g\rho_i h_n \tag{8.47}$$

$$P_o(z) = -g\rho_o z + P_{onp} + g\rho_o h_n \tag{8.48}$$

where P_{inp} and P_{onp} are the inside and outside pressures at the neutral plane at height, h_n. Subtracting Eq. (8.47) from Eq. (8.48) and noting that at the neutral plane, $P_{inp} = P_{onp}$, we have

$$\Delta P_{st} = P_o - P_i = g(h_n - z)(\rho_o - \rho_i) \tag{8.49}$$

We substitute for the density, ρ from Eq. (8.45) in Eq. (8.49), assuming that the *absolute pressures* inside and outside are approximately equal to P_o. This gives the stack pressure difference as

$$\Delta P_{st}(z) = \left(\frac{gP_o}{R_a}\right)(h_n - z)\left(\frac{1}{T_o} - \frac{1}{T_i}\right) \tag{8.50}$$

The idealized pressure distribution, given by Eq. (8.50), is valid in the absence of vertical separations like floors. In typical buildings, the doors and stairways between floors offer resistance to vertical air flow due to the stack effect. This resistance is accounted for by introducing a correction factor, C_{td} called the *thermal draft coefficient*, such that the actual stack pressure difference is

$$\Delta P_s = C_{td}\Delta P_{st} \tag{8.51}$$

A detailed discussion on the thermal draft coefficient is available in the *ASHRAE Handbook* - 2013 *Fundamentals* [1].

(ii) *The wind effect*

Another cause of air infiltration into buildings is wind impinging on the exterior surfaces of the envelope. On striking the surfaces a fraction of the kinetic energy of the wind is converted to a static pressure head, the magnitude of which depends on the wind direction, air density, surface orientation, and the surrounding conditions. On the windward side of the building the static pressure is positive with respect to the surrounding static pressure whereas on the leeward side it is negative.

Applying Bernoulli's equation, neglecting any energy losses, the static pressure due to the wind effect is

$$\Delta P_w = \frac{C_p \rho U^2}{2} \tag{8.52}$$

where ρ is the air density and U is the wind speed.

The wind pressure coefficient, C_p in Eq. (8.52) is a function of the location on the building envelope and the wind direction. Four values of C_p for wind directions of 0°, 90°, 180°, 270° and a harmonic function to obtain C_p at other angles by interpolation, are available in Ref. [1].

The graphs in Fig. 8.5(b) show the effect of wind on the outside pressure of the building envelope. Here the inside pressure distribution is the same as in Fig. 8.5(a). On the windward side of the building the outside pressure distribution in Fig. 8.5(a) is shifted to the right due the positive wind effect, whereas on the leeward it is shifted to the left due to negative wind effect. Hence on the windward side and the leeward side the neutral pressure planes are shifted up and down respectively, with respect to the mid-height plane. Air infiltration now occurs over a larger height on the windward side whereas the outflow is larger on the leeward side.

(iii) *The mechanical effect*

Another possible cause of air infiltration through building envelopes is the unbalance of the delivery and exhaust flow rates induced by the mechanical system of the building. The resulting pressure difference between the inside and outside air, ΔP_p is called the *mechanical effect.* Over pressurization of a building can cause problems such as doors that do not close properly. Under pressurization, on the other hand, may lead to problems like drawing contaminants into the building from regions outside where their levels are high.

The graphs in Fig. 8.5(c) show the effect of ΔP_p on the inside pressure of the envelope. The outside pressure distribution is the same as that in Fig. 8.5(a). When the building is over pressurized the inside pressure distribution in Fig. 8.5(a) is shifted to the right due the positive ΔP_p, whereas when the building is under pressurized it is shifted to the left due to the negative ΔP_p. As a result, the neutral pressure plane is shifted down and up respectively. Infiltration now occurs over a larger height for negative ΔP_p while the outflow is larger for positive ΔP_p.

The net static pressure difference that drives the infiltration flow through openings in the building envelope is the sum of the stack pressure, the wind pressure, and the mechanical pressure. This net pressure distribution may be obtained by combining the graphs in Figs. 8.5(b) and (c) appropriately.

8.5.3 *Estimation of infiltration flow rates*

In order to quantify the air leakage sites and their magnitudes it is necessary to carry out pressurization tests of the building envelope. Using data from such tests it is possible to determine the *leakage function* for an opening by curve fitting. This may be expressed as

$$Q = c(\Delta p)^n \tag{8.53}$$

where Q is the air flow rate, Δp is the pressure difference, c is a flow coefficient, and n is the pressure exponent.

Procedures to calculate air infiltration flow rate, ranging from simple estimation methods to complex physical models are presented in Ref. [1]. For purposes of illustration, we shall describe a basic model, recommended in Ref. [1] for residential infiltration load calculations. This model is based on the effective air leakage area, A_L (cm^2) at 4 Pa, obtained from whole-building pressurization tests. The infiltration flow rate, Q_{in} (m^{3}s^{-1}) is expressed as

$$Q_{in} = \left(\frac{A_L}{1000}\right)\sqrt{C_s|\Delta t| + C_w U^2} \tag{8.54}$$

where C_s = stack coefficient, (Ls^{-1})^2 (cm^{-4}K^{-1})

Δt = average indoor-outdoor temperature difference, K
C_w = wind coefficient, $(Ls^{-1})^2$ $[cm^{-4} (ms^{-1})^{-2}]$
U = average wind speed at local weather station, ms^{-1}.

The stack coefficient and the wind coefficient for one-, two- and three-story houses for several shelter classes are tabulated in Ref. [1]. We have included a few representative values in Table 8.5. Applications of Eq. (8.54) are given in the worked examples to follow in this chapter.

Table 8.5 Wind and Stack Coefficients for use in Eq. (8.54)*

Wind Coefficients C_W $[L^2s^{-2}cm^{-4}(ms^{-1})^{-2}]$	Class	One-story	Two-story	Three-story
House without shielding	1	3.19×10^{-4}	4.2×10^{-4}	4.94×10^{-4}
Rural isolated house	2	2.46×10^{-4}	3.25×10^{-4}	3.82×10^{-4}
House with building across street	3	1.74×10^{-4}	2.31×10^{-4}	2.71×10^{-4}
House on large lot	4	1.04×10^{-4}	1.37×10^{-4}	1.61×10^{-4}
Building with houses close by	5	0.32×10^{-4}	0.42×10^{-4}	0.49×10^{-4}
Stack coefficient C_s $[L^2s^{-2}cm^{-4}K^{-1}]$	All	1.45×10^{-4}	2.9×10^{-4}	4.35×10^{-4}

*Values extracted from Tables 5 and 6, Page 16.23 - *ASHRAE Handbook* - 2013 *Fundamentals* [1]

8.6 Moisture Transport in Building Structures

Both under winter and summer weather conditions, a building envelope is subjected to temperature and relative humidity gradients. In the winter the air inside has a higher temperature and relative humidity than the ambient air outside. These conditions are reversed when the building is cooled during the summer, and the outside ambient is hot and humid.

Most of the commonly used building envelope materials such as, brick, concrete, gypsum board, and fiberglass insulation are porous materials. Moisture is transported through these porous building materials due to the vapor pressure gradient resulting from temperature and relative humidity differences across the building envelope.

If the vapor pressure at any location within the material becomes higher than the *saturation vapor pressure* corresponding to the temperature at the location, then condensation of vapor occurs. The presence of liquid condensate in the pores of the material decreases its thermal resistance. Over time, the accumulation of condensed moisture within the material results in the growth of mold or mildew which eventually destroys the material.

Common insulation materials like glass fiber are highly porous and therefore offer relative low resistance to moisture diffusion compared to less porous materials like brick or gypsum board. Moreover, the largest temperature drop in multi-layered building structures (Fig. 8.1) such as walls and roofs occurs across the insulation layer. Therefore the insulation layer is particularly vulnerable to moisture condensation problems. As a precaution, a vapor retarder, often a thin sheet of plastic, is placed against the warm side of the insulation layer in most building structures.

Porous thermal insulations are also applied on the outside of chilled water pipes and cold air ducts in summer air conditioning systems. Ambient moisture entering these insulations through cracks and opening on the outer cover or the vapor barrier, condenses within the material. However, a larger fraction of the water vapor, after diffusing through the insulation, condenses on the outer surface of the pipe or the duct. Eventually, this leads to the corrosion of the metal. Experimental and analytical studies on water vapor diffusion and condensation in porous thermal insulations are described in Refs. [6] and [7].

8.6.1 *Fick's Law*

We introduced Fick's law in section 6.1 to study one-dimensional mass diffusion in a wall. For multi-layered building structures, the driving potential for moisture transfer is the vapor pressure difference. Therefore Fick's law may be written in the form [1,2]

$$\dot{m}_w = \frac{A\mu\Delta P_w}{L} \tag{8.55}$$

where $\dot{m}_w$ is the mass flow rate of water vapor, A is the area of vapor flow, ΔP_w is the vapor pressure drop across the thickness L of the material, and μ is the permeability of the material.

Fick's law is applicable to highly porous insulation materials like fiber glass where simple vapor diffusion occurs through the air filled pores of the material. However, for low-porosity materials like brick and gypsum board, the diffusion process is very complex, and involves several mass transport mechanisms like surface diffusion and capillary conduction. For these materials the permeability is a function of the local values of the relative humidity and the temperature within the material.

It is interesting to note that Fick's law of diffusion, given by Eq. (8.55), has the same mathematical form as Fourier's law of heat conduction (Eq. 2.2), the permeability being the material property analogous to the thermal conductivity. Hence we shall apply the methods developed in chapter 2 to analyze one-dimensional heat conduction problems to solve moisture diffusion problems. The application of Eq. (8.55) to condensation problems in building structures will be illustrated in worked example 8.15.

The *ASHRAE Handbook* - 2013 *Fundamentals* [1] gives a table of values of the water vapor permeability of different building envelop materials. For purposes of illustration, the permeability of a few building materials are listed in Table 8.6.

Table 8.6 Permeability of some building materials* [1]

Material	Thickness mm	Average Permeability, μ $ngPa^{-1}s^{-1}m^{-1}$
Concrete (1:2:4 mix)	-	4.6
Concrete block	200	28
Brick	102	4.6
Gypsum	9.5	27
Plywood	6.4	0.7
Sheathing	-	30-70
Glass-fiber batt	-	172
Mineral wool	-	245
Polystyrene	-	1.2
Softwoods	-	0.6-7.8

*Values extracted from Table 6, Page 26.17 - *ASHRAE Handbook* - 2013 *Fundamentals* [1]

8.7 Worked Examples

Example 8.1 An exterior wall of a building has a 10 cm thick layer of face brick on the outside, followed by a layer of 20 cm thick concrete. A 15 cm thick layer of mineral wool insulation is sandwiched between the concrete and a layer of plywood of thickness 10 mm. The wall is 10 m long and 3 m high. The outside and inside heat transfer coefficients are 30 $Wm^{-2}K^{-1}$ and 9 $Wm^{-2}K^{-1}$ respectively. The outside and inside air temperatures are 22°C and −15°C respectively. Calculate (i) the total thermal resistance, (ii) the overall heat transfer coefficient, and (iii) the total heat transfer rate through the wall.

Solution Consider unit area (A = 1 m^2) of the wall where the constituent layers are in series. The thermal resistance of a layer is

$$R_i = \frac{L_i}{A_i k_i}$$

The thermal resistances of the different layers computed using the above equation are tabulated below.

Table E8.1.1 Thermal resistances

Layer, i	*k_i, $Wm^{-1}K^{-1}$	L_i, mm	R_i, m^2KW^{-1}
Brick*	0.81	100	0.123
Concrete	1.8	200	0.11
Mineral wool	0.035	150	4.29
Plywood	0.095	10	0.105

*k-values extracted from Table 1, Page 26.7-*ASHRAE Handbook* - 2013 *Fundamentals* [1]

The thermal resistances of the indoor and outdoor air films, including the contribution due to thermal radiation are

$$R_i = \frac{1}{h_i} = 0.11 \text{ m}^2\text{KW}^{-1} \quad \text{and} \quad R_o = \frac{1}{h_o} = 0.033 \text{ m}^2\text{KW}^{-1}$$

Hence the total thermal resistance is

$$R_{tot} = 0.033 + 0.123 + 0.11 + 4.29 + 0.105 + 0.11 = 4.77 \text{ m}^2\text{KW}^{-1}$$

The overall heat transfer coefficient is given by

$$U_o = \frac{1}{R_{tot}} = 0.21 \text{ Wm}^{-2}\text{K}^{-1}$$

The total heat transfer rate through the wall is

$$Q_{tot} = AU_o(T_i - T_o) = 10 \times 3 \times 0.21 \times (22 + 15) = 233 \text{ W}$$

Example 8.2 A building has an exterior wall with 3.8 cm × 14 cm framing that makes 20 percent of its area. The wall consists of the following layers of materials: 9.5 cm thick gypsum board on the inside of the framing, fiber glass in the spaces between the framing, 12.7 mm thick sheathing next to the insulation and framing, 38 cm thick layer of expanded polystyrene, 100 mm thick layer of brick on the outside. The air temperature inside is 20°C and the outside ambient temperature is −10°C. The wind speed is 12 km h^{-1}. The inside heat transfer coefficient is 8 Wm^{-2}K^{-1}. Calculate (i) the total thermal resistance, (ii) the average heat transfer coefficient, and (iii) the average heat transfer rate through the wall using: (a) the parallel path method, and (b) the isothermal plane method.

Solution The equivalent thermal networks for the parallel flow method and the isothermal plane method are shown in Fig. 8.2(a) and (b) respectively. It is convenient to first compute the *unit thermal resistances*, ($\bar{R}_i = L_i/k_i$), considering unit cross sectional area of each heat transfer path. The results obtained are listed in the table below.

Table E8.2.1 Thermal resistances of wall layers

Layer, i	Actual area, m^2	*k_i, Wm^{-1}K^{-1}	L_i, mm	$\bar{R}_i$, m^{2}KW^{-1}
Brick*	1.0	0.9	100	0.11
Polystyrene	1.0	0.036	38	1.06
Sheathing	1.0	0.055	12.7	0.23
Wood frame	0.2	0.15	140	0.93
Fiber glass	0.8	0.04	140	3.5
Gypsum	1.0	0.16	9.5	0.059

*k-values extracted from Table 1, Page 26.7-*ASHRAE Handbook* - 2013 *Fundamentals* [1]

For a wind speed of 13 km h^{-1} (3.6 ms^{-1}) the heat transfer coefficient, h_o is 25 Wm^{-2}K^{-1} [1]. The unit thermal resistances of the indoor and outdoor air films, including contributions due to thermal radiation are:

$$\bar{R}_i = \frac{1}{h_i} = 0.125 \text{ m}^2\text{KW}^{-1} \quad \text{and} \quad \bar{R}_o = \frac{1}{h_o} = 0.04 \text{ m}^2\text{KW}^{-1}$$

(a) *Parallel flow method*

Consider two parallel heat flow paths from the outside air to the inside air, the first through the wood framing and the second through the fiber glass insulation. The total thermal resistances of these two paths are as follows:

$$R_{fr} = \frac{(0.04+0.11+1.06+0.23+0.93+0.059+0.125)}{A_{fr}}$$

$$R_{fr} = \frac{2.554}{0.2} = 12.77 \text{ KW}^{-1}$$

$$R_{in} = \frac{(0.04+0.11+1.06+0.23+3.5+0.059+0.125)}{A_{in}}$$

$$R_{in} = \frac{5.124}{0.8} = 6.405 \text{ KW}^{-1}$$

The two resistances above are in parallel. Therefore the overall thermal resistance is given by Eq. (8.1) as

$$R_{pr} = \left(\frac{1}{R_{fr}} + \frac{1}{R_{in}}\right)^{-1} = \left(\frac{1}{12.77} + \frac{1}{6.405}\right)^{-1} = 4.27 \text{ KW}^{-1}$$

The overall heat transfer coefficient is

$$U_{pr} = \frac{1}{A_{tot}R_{pr}} = \frac{1}{4.27\times 1} = 0.234 \text{ Wm}^{-2}\text{K}^{-1}$$

The average heat transfer rate through the wall is given by

$$Q_{pr} = AU_{pr}(T_i - T_o) = 1 \times 0.234 \times [20 - (-10)] = 7 \text{ Wm}^{-2}$$

(b) *Isothermal plane method*

In the isothermal plane method we assume that heat flows uniformly through all the layers except the framing and the insulation where the heat flow paths are parallel. The thermal resistances of the latter two paths are given by

$$R_{fr} = \frac{0.93}{0.2} = 4.65 \quad \text{and} \quad R_{ins} = \frac{3.5}{0.8} = 4.375$$

The equivalent resistance of these two parallel paths is

$$R_{eq} = \left(\frac{1}{R_{fr}} + \frac{1}{R_{ins}}\right)^{-1} = \left(\frac{1}{4.65} + \frac{1}{4.375}\right)^{-1} = 2.254 \text{ KW}^{-1}$$

Hence the total thermal resistance is

$$R_{tot} = (0.04 + 0.11 + 1.06 + 0.23 + 2.254 + 0.059 + 0.125) = 3.88$$

The total thermal resistance is 3.88 KW^{-1}

The overall heat transfer coefficient is

$$U_{ip} = \frac{1}{A_{tot}R_{tot}} = 0.258 \text{ Wm}^{-2}\text{K}^{-1}$$

The average heat transfer rate through the wall is given by

$$Q_{ip} = AU_{ip}(T_i - T_o) = 1 \times 0.258[20 - (-10)] = 7.74 \text{ Wm}^{-2}$$

Example 8.3 The flat roof of a building has a 10 mm layer of gypsum wall board on the inside. Next to it are 3.8 cm × 19 cm studs with their centers 30 cm apart. The 19 cm high spacing between the studs is filled with loose cellulose insulation, with a thermal conductivity of 0.043 $\text{Wm}^{-1}\text{K}^{-1}$. The layer of plywood next to it has a thickness of 2 cm and the built up roofing on the outside has a thickness of 10mm. The outside and inside air temperatures are 35°C and 25°C respectively. The wind speed is 12 km h^{-1}. The inside heat transfer coefficient is 10 $\text{Wm}^{-2}\text{K}^{-1}$. Calculate (i) the total thermal resistance, (ii) the average heat transfer coefficient, and (iii) average heat transfer rate through the roof, using: (a) the parallel path method, and (b) the isothermal plane method.

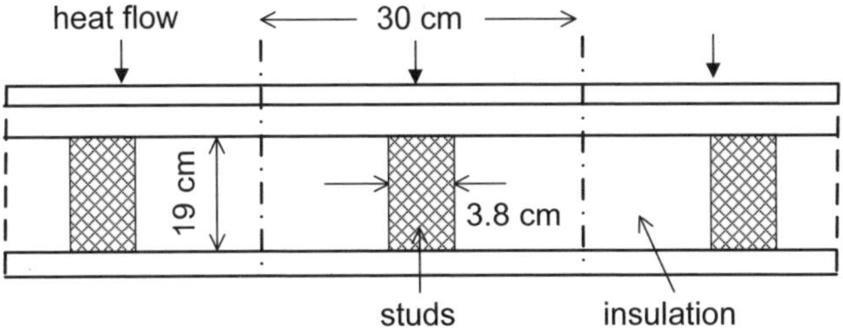

Fig. E 8.3.1 Roof section

Solution Consider a 30 cm wide representative section of the roof as shown schematically in Fig. E8.3.1. The fractional cross sectional areas of the studs and the insulation are:

$$a_{st} = \frac{3.8}{30} = 0.127 \quad \text{and} \quad a_{in} = 1 - 0.127 = 0.873$$

The equivalent thermal networks for the parallel flow method and the isothermal plane method are shown in Fig. 8.2(a) and (b) respectively. It is convenient to first compute the *unit thermal resistances*, ($\bar{R}_i = L_i/k_i$), considering unit cross sectional area of each heat transfer path. The results obtained are listed in the table below.

Table E8.3.1 Thermal resistances of roof layers

Layer, i	Actual area, m^2	k_i, Wm^{-1}K^{-1}	L_i, mm	$\bar{R}_i$, m^{2}KW^{-1}
Roofing*	1.0	0.17	10	0.059
Plywood	1.0	0.095	19	0.2
Insulation	0.873	0.045	190	4.22
Studs	0.127	0.16	190	1.19
Gypsum	1.0	0.16	10	0.0625

**k*-values extracted from Table 1, Page 26.7 - *ASHRAE Handbook* - 2013 *Fundamentals* [1]

The unit thermal resistance of the indoor air film, including the contribution due to thermal radiation is

$$R_i = \frac{1}{h_i} = \frac{1}{10} = 0.1 \ \text{m}^2\text{KW}^{-1}$$

The external convective heat transfer coefficient at a wind speed of 12 kmh^{-1} is obtained from the tabulated data in Ref. [1] as 22.7 Wm^{-2}K^{-1}. Hence the unit thermal resistance of the air film outside is

$$R_o = \frac{1}{h_o} = \frac{1}{22.7} = 0.044 \ \text{m}^2\text{KW}^{-1}$$

(a) *Parallel flow method*

Consider two parallel heat flow paths from the outside air to the inside air, the first through the wooden studs and the second through the cellulose insulation. The total thermal resistances of these two paths are as follows:

$$R_{st} = \frac{(0.1+0.0625+1.19+0.2+0.059+0.043)}{a_{st}}$$

$$R_{st} = \frac{1.655}{0.127} = 13.03 \ \text{KW}^{-1}$$

$$R_{in} = \frac{(0.1+0.0625+4.22+0.2+0.059+0.043)}{a_{in}}$$

$$R_{in} = \frac{4.685}{0.873} = 5.37 \ \text{KW}^{-1}$$

The two resistances above are in parallel. Therefore the overall thermal resistance is

$$R_{pr} = \left(\frac{1}{R_{st}} + \frac{1}{R_{in}}\right)^{-1} = \left(\frac{1}{13.03} + \frac{1}{5.37}\right)^{-1} = 3.8\ \mathrm{KW^{-1}}$$

The overall heat transfer coefficient is

$$U_{pr} = \frac{1}{A_{tot}R_{pr}} = \frac{1}{3.8 \times 1} = 0.262\ \mathrm{Wm^{-2}K^{-1}}$$

The average heat transfer rate through the wall is given by

$$Q_{pr} = AU_{pr}(T_i - T_o) = 1 \times 0.262 \times (35 - 25) = 2.62\ \mathrm{Wm^{-2}}$$

(b) *Isothermal plane method*

In the isothermal plane method we assume that heat flows uniformly through all the layers except the studs and the insulation where the heat flow paths are parallel. The thermal resistances of the latter two paths are given by

$$R_{st} = \frac{1.19}{0.127} = 9.37 \quad \text{and} \quad R_{ins} = \frac{4.22}{0.873} = 4.83$$

The equivalent resistance of these two parallel paths is

$$R_{eq} = \left(\frac{1}{R_{st}} + \frac{1}{R_{ins}}\right)^{-1} = \left(\frac{1}{9.37} + \frac{1}{4.83}\right)^{-1} = 3.19\ \mathrm{KW^{-1}}$$

The total thermal resistance is given by

$$R_{tot} = (0.1 + 0.0625 + 3.19 + 0.2 + 0.059 + 0.043)$$

Hence the total thermal resistance is 3.65 $\mathrm{KW^{-1}}$

The overall heat transfer coefficient is

$$U_{ip} = \frac{1}{A_{tot}R_{tot}} = \frac{1}{3.65 \times 1} = 0.274\ \mathrm{Wm^{-2}K^{-1}}$$

The average heat transfer rate through the wall is given by

$$Q_{ip} = AU_{ip}(T_o - T_i) = 1 \times 0.274 \times (35 - 25) = 2.74\ \mathrm{Wm^{-2}}$$

Example 8.4 A horizontal section of a vertical wall made of three-core blocks of concrete is shown in Fig. E8.4.1. The length, breadth and height of a block are 40 cm, 20 cm and 22 cm respectively. The two faces of a block are 3.5 cm thick and the thickness of a web is 24 mm.

The cores are filled with perlite insulation of thermal conductivity, 0.055 $Wm^{-1}K^{-1}$. The thermal conductivity of concrete is 1.5 $Wm^{-1}K^{-1}$. The outside and inside heat transfer coefficients are 28 $Wm^{-2}K^{-1}$ and 8 $Wm^{-2}K^{-1}$ respectively. Calculate (i) the overall thermal resistance of the wall and (ii) the overall heat transfer coefficient, using the isothermal plane method.

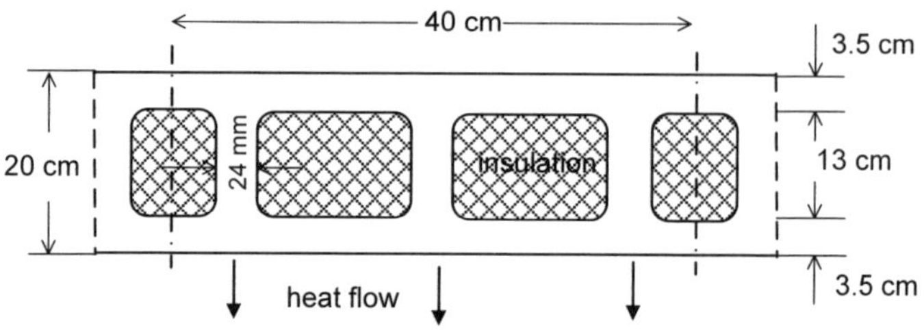

Fig. E8.4.1 Horizontal section of a concrete block wall

Solution A representative section of the concrete block wall of width 40 cm is shown schematically in Fig. E8.4.1. The heat flow paths through the insulation and the concrete webs are parallel. The equivalent thermal network for the isothermal plane method is shown in Fig. 8.2(b). We shall first compute the heat flow areas, the lengths of the heat flow paths, L_i and the *unit thermal resistances*, ($R_i = L_i/k_i$), of each heat transfer path. The results are listed in the table below.

Table E8.4.1 Thermal resistances

Layer, i	Area, cm^2	k_i, $Wm^{-1}K^{-1}$	L_i, cm	$\bar{R}_i$, m^2KW^{-1}
Outer face	40×22 = 880	1.5	3.5	0.023
Concrete webs	(3×2.4×22) = 158	1.5	20-7 = 13	0.087
Insulation cores	22×(40-3×2.4) = 721.6	0.055	20-7 = 13	2.36
Inner face	40×22 = 880	1.5	3.5	0.023

The unit thermal resistances of the indoor and outdoor air films, including contributions due to thermal radiation are:

$$R_i = \frac{1}{h_i} = \frac{1}{8} = 0.125 \text{ m}^2\text{KW}^{-1}$$

and $$R_o = \frac{1}{h_o} = \frac{1}{28} = 0.036 \text{ m}^2\text{KW}^{-1}$$

In the isothermal plane method we assume that heat flows uniformly through the outer and inner concrete faces. Through the webs and the insulation the heat flow paths are parallel. The fractional cross sectional areas of the webs and the insulation cores are

$$a_w = \frac{158.5}{880} = 0.18 \quad \text{and} \quad a_{ins} = 1 - 0.18 = 0.82$$

The thermal resistances of the webs and the cores are

$$R_w = \frac{0.087}{0.18} = 0.48 \quad \text{and} \quad R_{ins} = \frac{2.36}{0.82} = 2.88$$

The equivalent resistance of these two parallel paths is

$$R_{eq} = \left(\frac{1}{R_w} + \frac{1}{R_{ins}}\right)^{-1} = \left(\frac{1}{0.48} + \frac{1}{2.88}\right)^{-1} = 0.41 \text{ KW}^{-1}$$

The total thermal resistance of the wall is given by

$$R_{tot} = (0.125 + 0.023 + 0.41 + 0.023 + 0.036) = 0.62 \text{ KW}^{-1}$$

Hence the total thermal resistance is 0.62 KW^{-1}.
The overall heat transfer coefficient is given by

$$U_{ip} = \frac{1}{A_{tot}R_{tot}} = \frac{1}{0.62 \times 1} = 1.6 \text{ Wm}^{-2}\text{K}^{-1}$$

Example 8.5 The vertical section through a roof deck is shown in Fig. E8.5.1. It has steel beams (k = 48 $\text{Wm}^{-1}\text{K}^{-1}$) supporting a layer of concrete (k = 0.25 $\text{Wm}^{-1}\text{K}^{-1}$) and the roofing ($R$ = 0.059 m^2KW^{-1}). The spacing between the beams is 65 cm. A layer of fiber glass insulation (k = 0.038 $\text{Wm}^{-1}\text{K}^{-1}$) is located below the concrete. The outside and inside heat transfer coefficients are 25 $\text{Wm}^{-2}\text{K}^{-1}$ and 7 $\text{Wm}^{-2}\text{K}^{-1}$ respectively. Calculate (i) the overall thermal resistance of the roof, and (ii) the overall heat transfer coefficient.

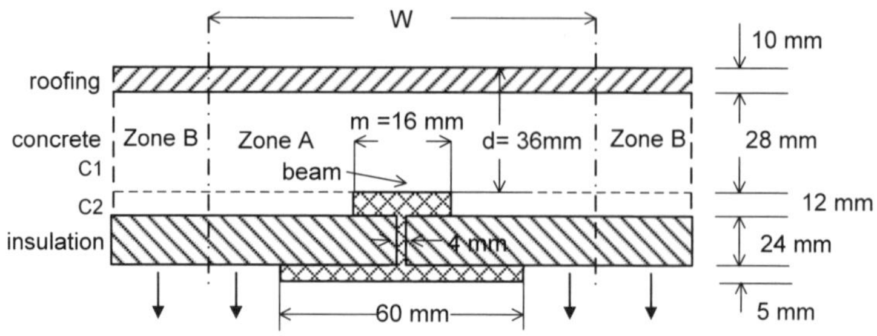

Fig. E8.5.1 Roof deck with I-beams

Solution Figure E8.5.1 shows a section of the roof deck with I-beams spaced 65 cm apart. We shall use the two-zone approach recommended in the *ASHRAE Handbook* - 2013 *Fundamentals* [1] to analyze the heat flow through the roof. A representative section of the roof of width 65 cm is divided into two zones A and B. Zone A is centered on an I- beam and includes a width W of the roof where the heat flow lines are significantly affected by the presence of the highly conducting metal beam. The zone B is the rest of the representative area that is assumed to be unaffected by the presence metal beam.

The width of zone A is calculated using an empirical equation which has been obtained from two-dimensional heat flow simulations. This is given in the *ASHRAE Handbook* - 2009 *Fundamentals* [2] as

$$W = m + 2d$$

where *m* and *d* are the dimensions indicated in Fig. E8.5.1. For a metal surface in contact with still air *d* should not be less than 13 mm. Applying the above equation to the top and bottom of the I-beam we obtain the following:

$$W_t = 16 + 2 \times 36 = 88 \text{ mm} \quad \text{and} \quad W_b = 60 + 2 \times 13 = 86 \text{ mm}$$

We select the larger of the two values as the width of zone A. Hence W = 88 mm. The width of zone B is therefore (650 - 88) = 562 mm.

Consider a meter length of the roof along the beam which has a representative area of 0.65 m^2. The corresponding areas of zones A and B are 0.088 m^2 and 0.562 m^2 respectively. We shall analyze the heat flow in zones A and B using the isothermal plane method.

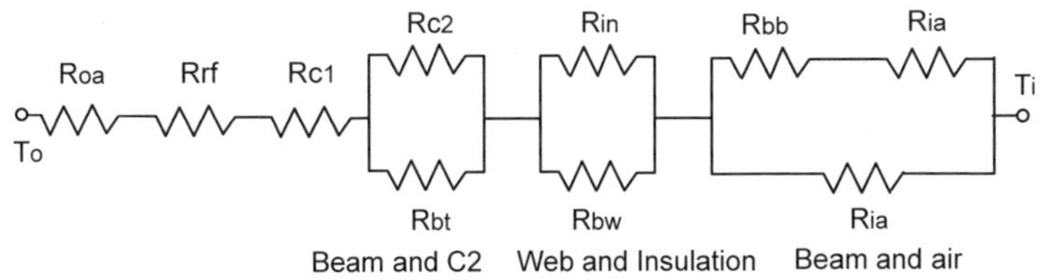

Fig. E8.5.2 Isothermal plane thermal network for zone A

The thermal network for zone A is depicted in Fig. E8.5.2. The heat flow areas and the *unit thermal resistances*, (L_i/k_i) for the different layers are summarized in the Table E8.5.1. There are three separate parallel heat flow paths for which the equivalent unit resistances are given by the general expression in Eq. (8.2a),

$$\bar{R}_{eq} = \left(\frac{a_1}{\bar{R}_1} + \frac{a_2}{\bar{R}_2}\right)^{-1} \quad \text{(E8.5.1)}$$

where $\bar{R}_1$ and $\bar{R}_2$ are the individual *unit thermal resistances* and a_1 and a_2 are the respective area ratios.

Table E8.5.1 Zone A - Unit thermal resistances of sections

Layer, i	Area, cm^2	k_i, Wm^{-1}K^{-1}	L_i, cm	$\bar{R}_i$ m^{2}KW^{-1}
Roofing	880	-	1.0	$R_{rf} = 0.059$
Concrete, C1	880	0.25	2.8	$R_{c1} = 0.112$
Concrete, C2	720	0.25	1.2	$R_{c2} = 0.048$
Beam-top	160	48	1.2	$R_{bt} = 2.5\times10^{-4}$
Insulation	840	0.038	2.4	$R_{in} = 0.632$
Beam-web	40	48	2.4	$R_{bw} = 5\times10^{-4}$
Beam-bottom	600	48	0.5	$R_{bb} = 1.04\times10^{-4}$

The thermal resistances of the indoor and outdoor air films, including contributions due to thermal radiation are

$$\bar{R}_i = \frac{1}{h_i} = \frac{1}{7} = 0.143 \text{ m}^2\text{KW}^{-1}$$

$$\bar{R}_o = \frac{1}{h_o} = \frac{1}{25} = 0.04 \text{ m}^2\text{KW}^{-1}$$

Applying Eq. (E8.5.1) to the parallel paths in Fig. E8.5.2 we obtain the following equivalent unit resistances:

$$\bar{R}_{cb} = \left(\frac{720/880}{0.048} + \frac{160/880}{2.5\times10^{-4}}\right)^{-1} = 1.34 \times 10^{-3}\ \text{m}^2\text{KW}^{-1}$$

$$\bar{R}_{ib} = \left(\frac{840/880}{0.632} + \frac{40/880}{5\times10^{-4}}\right)^{-1} = 0.0109\ \text{m}^2\text{KW}^{-1}$$

$$\bar{R}_{ia} = \left(\frac{280/880}{0.143} + \frac{600/880}{0.143+1.04\times10^{-4}}\right)^{-1} = 0.143\ \text{m}^2\text{KW}^{-1}$$

The total unit thermal resistance is

$$\bar{R}_{tot,A} = (0.04 + 0.059 + 0.112 + 1.34 \times 10^{-3} + 0.0109 + 0.143)$$

The total *unit thermal resistance* of zone A is 0.366 m^2KW^{-1}.

For zone B all the resistances are in series. Hence the total *unit resistance* is

$$\bar{R}_{tot,B} = (0.04 + 0.059 + 0.112 + 0.048 + 0.632 + 0.143)$$

The total *unit thermal resistance* of zone B is 1.034 m^2KW^{-1}.

Now the heat flow through zones A and B are in parallel and therefore the equivalent unit resistance is given by Eq. (E8.5.1). Hence we obtain the average *unit resistance* of the representative roof section as

$$\bar{R}_{roof} = \left(\frac{0.088/0.65}{0.366} + \frac{0.562/0.65}{1.034}\right)^{-1} = 0.83\ \text{m}^2\text{KW}^{-1}$$

The overall heat transfer coefficient is given by

$$U_{roof} = \frac{1}{\bar{R}_{roof}} = \frac{1}{0.83} = 1.2\ \text{Wm}^{-2}\text{K}^{-1}$$

Example 8.6 A building has a double-glazed vertical window with a 13 mm air space between the two glasses. The glass thickness is 6 mm and the thermal conductivity is 0.8 $\text{Wm}^{-1}\text{K}^{-1}$. The emissivity of glass is 0.9. The inside and outside heat transfer coefficients are 8.3 $\text{Wm}^{-2}\text{K}^{-1}$ and 34 $\text{Wm}^{-2}\text{K}^{-1}$ respectively. The indoor conditions are 20°C and 40 percent relative humidity. The outdoor air temperature is −10°C. The ambient pressure is 101.3 kPa. (a) Calculate the rate of heat loss through center of the window. (b) Will condensation occur on the inner surface of the window?

Solution Consider unit area (A=1m^2) of the window. Thermal resistance of each glass pane is

$$R_g = \frac{L_g}{k_g} = \frac{6\times10^{-3}}{0.8} = 7.5\times10^{-3}\ \text{m}^2\text{KW}^{-1}$$

The effective emissivity for the air-filled cavity between the glass panes is given by Eq. (8.14). Hence we have

$$\frac{1}{\varepsilon_{eff}} = \frac{1}{\varepsilon_g} + \frac{1}{\varepsilon_g} - 1 = \frac{2}{0.9} - 1 = 1.22$$

Therefore the effective emissivity is 0.82.

As an initial guess assume the inner and outer glass surface temperatures to be 14°C and −8°C respectively. Therefore the mean air space temperature is 3°C and the temperature difference is 22°C. Table 3, page 26.13 of the *ASHRAE Handbook* - 2013 *Fundamentals* [1] gives the air space heat transfer coefficient for different values of the effective emissivity, the mean air space temperature, and the temperature difference. A few representative values from the above source are included in Table 8.3. For the parameters pertinent to the present example we obtain, by interpolation, the air space heat transfer coefficient, h_c as 6.02 $\text{Wm}^{-2}\text{K}^{-1}$. We shall now use the above value of h_c to estimate of the heat transfer rate through the window.

The overall thermal resistance of the center of the window is

$$\frac{1}{U_o} = \frac{1}{8.3} + 7.5\times10^{-3} + \frac{1}{6.02} + 7.5\times10^{-3} + \frac{1}{34} = 0.33$$

Therefore the first estimate of the overall heat transfer coefficient, U_o is 3.02 $\text{Wm}^{-2}\text{K}^{-1}$. The total heat transfer rate per unit area through the center of the window is given by

$$Q = U_o(T_i - T_o) = 3.02\times30 = 90.6\ \ \text{Wm}^{-2}$$

We now obtain an improved estimate of the inner glass surface temperatures by applying network analogy. Hence we have

$$T_{g1} = 20 - 90.6\left(\frac{1}{8.3} + 7.5\times10^{-3}\right) = 8.4°\text{C}$$

$$T_{g2} = -10 + 90.6\left(\frac{1}{34} + 7.5\times10^{-3}\right) = -6.65°\text{C}$$

Therefore the new mean air space temperature and the temperature difference are 0.9°C and 15°C respectively. From Table 3, page 26.13 in Ref. [1] we obtain the new air space heat transfer coefficient, by interpolation, as 5.7 $\mathrm{Wm^{-2}K^{-1}}$. Therefore the overall heat transfer coefficient for the center of the window is

$$\frac{1}{U_o} = \frac{1}{8.3} + 7.5 \times 10^{-3} + \frac{1}{5.7} + 7.5 \times 10^{-3} + \frac{1}{34} = 0.34$$

Hence the improved estimate of the overall heat transfer coefficient is 2.94 $\mathrm{Wm^{-2}K^{-1}}$. The total heat transfer rate per unit area through the center of the window is given by

$$Q = U_o(T_i - T_o) = 2.94 \times 30 = 88.2 \ \mathrm{Wm^{-2}}$$

The temperature of the inner surface of the window is

$$T_{gi} = 20 - \frac{88.2}{8.3} = 9.4°\mathrm{C}$$

(b) Now when the indoor air is at 20°C and 40% relative humidity, the dew point temperature obtained from the psychrometric chart is 6°C. Since the inner glass surface temperature of 9.4°C higher than the dew point temperature, water vapor will not condense on the glass surface.

Example 8.7 A double-glazed window mounted at 45° has a 20 mm air space between the glass panes. The inner surfaces of the panes have a coatings with an emissivity of 0.67. The outer surfaces of the panes have an emissivity of 0.9. The inside and outside air temperatures are 24°C and 2°C respectively. The wind speed is 6.7$\mathrm{ms^{-1}}$. Calculate (i) the center-of-glass heat transfer coefficient, and (ii) the inside air relative humidity at which condensation will commence.

Solution Consider unit area (A=1$\mathrm{m^2}$) of the window. The effective emissivity for the air-filled cavity between the glass panes is given by

$$\frac{1}{\varepsilon_{eff}} = \frac{1}{\varepsilon_g} + \frac{1}{\varepsilon_g} - 1 = \frac{2}{0.67} - 1 = 1.98$$

Hence the effective emissivity, ε_{eff} of the air gap is 0.5.

As an initial guess we assume the inner and outer glass surface temperatures to be 16°C and 4°C respectively. Therefore the mean air space temperature is 10°C and the temperature difference is 12°C. Table 3 on page 26.13 of the *ASHRAE Handbook - 2013 Fundamentals* [1] gives the thermal resistance of air spaces for different values of the effective emissivity, the inclination of the air space, the mean air space temperature, and the temperature difference. We shall use the values in the above table to estimate the thermal resistance for a 20 mm air gap inclined at 45°. Interpolating the data in Table 3 [1] we obtain the thermal resistance as 0.207 m^2KW^{-1}.

The air film thermal resistance for upward heat flow from still air to an adjacent surface of emissivity 0.9, inclined at 45° is given in Table 10 on page 26.20 of the *ASHRAE Handbook* - 2013 *Fundamentals* [1] as 0.11 m^2KW^{-1}. The outside air film thermal resistance at a wind speed of 6.7ms^{-1} is 0.03 m^2KW^{-1}.

The overall thermal resistance for the center of the window is

$$R_o = 0.11 + 0.207 + 0.03 = 0.347 \text{ m}^2\text{KW}^{-1}$$

Therefore the initial estimate of the overall thermal resistance is 0.347 m^2KW^{-1}. The total heat transfer rate per unit area through the center of the window is given by

$$Q = \frac{(T_i - T_o)}{R_o} = \frac{22}{0.347} = 63.4 \text{ Wm}^{-2}$$

We now obtain an improved estimate of the two glass surface temperatures by applying the network analogy. Hence we have

$$T_{g1} = 24 - 63.4 \times .11 = 17°\text{C}$$

$$T_{g2} = 2 + 63.4 \times 0.03 = 3.9°\text{C}$$

Therefore the new mean air space temperature and temperature difference are 10.4°C and 13.1°C respectively. From Table 3 on page 26.13 of Ref. [1] we obtain the new air space thermal resistance as 0.203 m^2KW^{-1}.

Hence the overall thermal resistance of the center of the window is

$$R_o = 0.11 + 0.203 + 0.03 = 0.343 \text{ m}^2\text{KW}^{-1}$$

The total heat transfer rate per unit area through the center of the window is

$$Q = \frac{(T_i - T_o)}{R_o} = \frac{22}{0.343} = 64.14 \text{ Wm}^{-2}$$

(ii) The inner glass surface temperature is

$$T_{g1} = 24 - 64.14 \times 0.11 = 16.9°\text{C}$$

Water vapor will condense on the inner glass surface if the inside air dew point is below 16.9°C. The relative humidity of the inside air at a dry-bulb temperature of 24°C and a dew-point of 16.9°C is obtained from the psychrometric chart as 64%. Therefore if the inside air relative humidity exceeds 64% water vapor will condense on the glass surface.

Example 8.8 A fixed double-glazed vertical window, has two 3 mm thick glass panes with a 13 mm air space between them. The area of a glazing is 2 m × 2 m. The window frame of width 65 mm is made of aluminum without a thermal break. The inside and outside air temperatures are 21°C and −18°C respectively. The wind speed is 6.7 ms^{-1}. Calculate (i) the overall heat transfer coefficient, and (ii) the rate of heat loss through the window.

Solution The overall heat transfer coefficient includes three different components contributed by the center of the glass panes, the edge of the glass panes, and the frame. The values of these components are given in Table 4 on page 15.8 of the *ASHRAE Handbook* - 2013 *Fundamentals* [1] which are based on winter conditions of −18°C outdoor temperature and 21°C indoor temperature with a wind speed of 6.7 ms^{-1}. The following values are obtained from Table 4 [1].

For the center of the glass panes, $U_{cg} = 2.73\ Wm^{-2}K^{-1}$, for the edge of the glass panes, $U_{eg} = 3.36\ Wm^{-2}K^{-1}$, for the fixed aluminum frame without a thermal break, $U_{fr} = 3.54\ Wm^{-2}K^{-1}$.

The total area of the glazing is $4m^2$. The edge of the glass has a constant width of 63.5mm according to the data in the *ASHRAE Handbook* - 2013 *Fundamentals* [1]. Therefore the area of the center of the glass is

$$A_{cg} = (2 - 0.127) \times (2 - 0.127) = 3.508\ m^2$$

The area of the edge of the glass panes is

$$A_{eg} = 4 - 3.508 = 0.492\ m^2$$

The area of the frame is

$$A_{fr} = 2.13 \times 2.13 - 4 = 0.54 \text{ m}^2$$

Now the overall heat transfer coefficient, U_o is given by Eq. (8.20) as

$$U_0 = \frac{A_{cg}U_{cg}+A_{eg}U_{eg}+A_{fr}U_{fr}}{A_{cg}+A_{eg}+A_{fr}}$$

Substituting numerical values in the above equation we have

$$U_0 = \frac{(3.508\times2.73)+(0.492\times3.36)+(0.54\times3.54)}{3.508+0.492+0.54} = 2.9 \text{ Wm}^{-2}\text{K}^{-1}$$

The total heat transfer rate through the window is given by

$$Q_{tot} = A_{tot}U_o(T_i - T_o) = 4.54 \times 2.9 \times 39 = 513 \text{ W}$$

Example 8.9 A sealed 6 mm wide space between the two glass panes of a 3 m × 3 m, double-glazed, fixed window is filled with argon. The inner surfaces of the glazings have coatings with an emissivity of 0.4. The window frame of width 70mm is made of aluminum with a thermal break. The indoor and outdoor air temperatures are 21°C and −18°C respectively. The wind speed is 6.7 ms^{-1}. Calculate (i) the average heat transfer coefficient of the window, and (ii) the rate of heat loss to the ambient.

Solution The overall heat transfer coefficient includes three different components contributed by the center of the glass panes, the edge of the glass panes, and the frame. The values of these components are given in Table 4 on page 15.8 of the *ASHRAE Handbook* - 2013 *Fundamentals* [1] which are based on winter conditions of −18°C outdoor temperature and 21°C indoor temperature, with a wind speed of 6.7 ms^{-1}. The following values for an argon gas filled window with a gap of 6 mm, and glass surface emissivity 0.4 are obtained from Table 4 [1].

For the center of the glass panes, $U_{cg} = 2.44$ Wm^{-2}K^{-1}, for the edge of the glass panes, $U_{eg} = 3.16$ Wm^{-2}K^{-1}, for the fixed aluminum frame with a thermal break, $U_{fr} = 2.94$ Wm^{-2}K^{-1}.

The total area of the glazing is 9 m^2. The edge of the glass has a constant width of 63.5 mm according to the data in the *ASHRAE Handbook* - 2013 *Fundamentals* [1]. Therefore the area of the center of the glass is

$$A_{cg} = (3 - 0.127) \times (3 - 0.127) = 8.25 \text{ m}^2$$

The area of the edge of the glass panes is

$$A_{eg} = 9 - 8.25 = 0.75 \text{ m}^2$$

The area of the frame is

$$A_{fr} = 3.14 \times 3.14 - 9 = 0.86 \text{ m}^2$$

Now the overall heat transfer coefficient, U_o is given by Eq. (8.20) as

$$U_0 = \frac{A_{cg}U_{cg} + A_{eg}U_{eg} + A_{fr}U_{fr}}{A_{cg} + A_{eg} + A_{fr}}$$

Substituting numerical values in the above equation we have

$$U_0 = \frac{(8.25 \times 2.44) + (0.75 \times 3.16) + (0.86 \times 2.94)}{8.25 + 0.75 + 0.86} = 2.54 \text{ Wm}^{-2}\text{K}^{-1}$$

The total heat transfer rate through the window is given by

$$Q_{tot} = A_{tot}U_o(T_i - T_o) = 9.86 \times 2.54 \times 39 = 977 \text{ W}$$

Example 8.10 The wall of a basement has a total horizontal length of 90 m and extends 1.9 m below grade (see Fig. 8.4a). The thickness of the wall is 200 mm and it is made of concrete ($k = 2.1$ Wm^{-1}K^{-1}). The wall is insulated on the inside with a layer of polystyrene insulation ($k = 0.038$ Wm^{-1}K^{-1}) of thickness 50 mm and finished with plywood paneling ($k = 0.13$ Wm^{-1}K^{-1}) of thickness 9 mm. The soil surrounding the wall has a thermal conductivity of 1.4 Wm^{-1}K^{-1}. The inside air temperature is 18°C and the outside ambient temperature is 3°C. The inside and outside heat transfer coefficients are 8 Wm^{-2}K^{-1} and 28 Wm^{-2}K^{-1} respectively. Calculate (i) the average heat transfer coefficient for the basement wall, and (ii) the total rate of heat loss through the basement wall.

Solution The thermal resistances of the wall elements and the inside and outside air films are as follows:

$$R_{wall} = \frac{200 \times 10^{-3}}{2.1} = 0.095, \quad R_{ins} = \frac{50 \times 10^{-3}}{0.038} = 1.316$$

$$R_{fac} = \frac{9 \times 10^{-3}}{0.13} = 0.07, \quad R_{oa} = \frac{1}{28} = 0.036, \quad R_{ia} = \frac{1}{8} = 0.125$$

Hence the total additional thermal resistance is

$$R_a = 0.095 + 1.316 + 0.07 + 0.036 + 0.125 = 1.64$$

The expression for the average heat transfer coefficient for the basement wall is given by Eq. (8.28) as

$$U_{bw} = \left(\frac{2k_s}{\pi z_f}\right) ln\left(\frac{z_f+2k_sR_a/\pi}{2k_sR_a/\pi}\right)$$

Substituting numerical values in the above equation we have

$$U_{bw} = \left(\frac{2\times1.4}{\pi\times1.9}\right) ln\left(\frac{1.9+2\times1.4\times1.64/\pi}{2\times1.4\times1.64/\pi}\right) = 0.39$$

Hence the overall heat transfer coefficient for heat flow through the wall and the surrounding soil is 0.39 $Wm^{-2}K^{-1}$.

The total rate of heat loss across the wall and soil is given by

$$Q_{to} = U_{bw}(Lz_f)(t_{ia} - t_{oa})$$

Substituting numerical values we have

$$Q_{to} = 0.39 \times 90 \times 1.9 \times (18 - 3) = 1000\text{W}$$

Example 8.11 The wall of a basement extends 1.8 m below grade (see Fig. 8.4b). The length and breadth of the floor of the basement are 15 m and 10m respectively. The floor has a thickness of 150 mm and is made of concrete (k = 2.4 $Wm^{-1}K^{-1}$). The entire floor is carpeted. The carpet has a thickness of 9 mm and a thermal conductivity of 0.08 $Wm^{-1}K^{-1}$. The soil surrounding the basement has a thermal conductivity of 1.4 $Wm^{-1}K^{-1}$. The inside air temperature is 16°C and the outside ambient temperature is 4°C. The inside and outside heat transfer coefficients are 8 $Wm^{-2}K^{-1}$ and 28 $Wm^{-2}K^{-1}$ respectively. Calculate (i) the average heat transfer coefficient for the basement floor, and (ii) the total rate of heat loss through the basement floor.

Solution The thermal resistances of the carpet, the floor , and the inside and outside air films are as follows:

$$R_{carp} = \frac{9\times10^{-3}}{0.08} = 0.1125, \quad R_{flo} = \frac{150\times10^{-3}}{2.4} = 0.0625$$

$$R_{oa} = \frac{1}{28} = 0.036\,, \quad R_{ia} = \frac{1}{8} = 0.125$$

Hence the total additional thermal resistance is

$$R_a = 0.1125 + 0.0625 + 0.036 + 0.125 = 0.336$$

The expression for the average heat transfer coefficient for the basement floor, U_{bf} is given by Eq. (8.35) as

$$U_{bf} = \left(\frac{2k_s}{\pi w_b}\right) ln\left(\frac{\frac{w_b}{2}+\frac{z_f}{2}+\frac{k_S R_a}{\pi}}{\frac{z_f}{2}+\frac{k_S R_a}{\pi}}\right)$$

Substituting numerical values in the above equation we have

$$U_{bf} = \left(\frac{2\times1.4}{\pi\times10}\right) ln\left(\frac{\frac{10}{2}+\frac{1.8}{2}+\frac{1.4\times0.336}{\pi}}{\frac{1.8}{2}+\frac{1.4\times0.336}{\pi}}\right) = 0.156$$

Hence the overall heat transfer coefficient for heat flow through the floor and the surrounding soil is 0.156 $Wm^{-2}K^{-1}$.

Now the total rate of heat loss through the floor is given by

$$Q_{to} = U_{bf}(Lw)(t_{ia} - t_{oa})$$

Substituting numerical values in the above equation we have

$$Q_{to} = 0.156 \times 10 \times 15 \times (16 - 4) = 281\text{W}$$

Example 8.12 The walls of a building are made of 200 mm blocks with a brick facing. The perimeter of the slab-on-grade floor is 325 m. The insulation at the edge of the floor has a thermal resistance of 0.95 m^2KW^{-1}. The indoor and outdoor air temperatures are 20°C and 4°C respectively. Calculate the rate of heat loss from the floor. If the edge of the slab is not insulated what would be the rate of heat loss?

Solution The heat transfer coefficient, F_p for the heat loss through the edge of a slab is given in Table 24 on page 18.31 of the *ASHRAE Handbook - 2013 Fundamentals* [1]. For the insulated (R = 0.95 m^2KW^{-1}) 200mm block wall with brick facing we obtain F_p from Table 24 [1] as 0.86 $Wm^{-1}K^{-1}$. The rate of heat loss is given by Eq. (8.36) as

$$q_{sl} = pF_p\Delta t$$

Substituting numerical values in the above equation we have

$$q_{sl} = 325 \times 0.86 \times (20 - 4) = 4.47 \text{ kW}$$

For a slab without insulation, Table 24 [1] gives F_p as 1.17 $\text{Wm}^{-1}\text{K}^{-1}$.

Hence the rate of heat loss from the edge is

$$q_{sl} = 325 \times 1.17 \times (20 - 4) = 6.08 \text{ kW}$$

Example 8.13 The height of a building is 150 m and it has openings distributed uniformly along its height. The inside and outside air temperatures are 22°C and −12°C respectively. The ambient pressure is 101 kPa, and the mean wind speed is 7 ms^{-1}. (i) Calculate the theoretical pressure difference between the inside and outside, due to the stack effect at heights of 40 m and 120 m. (ii) Assuming the respective wind coefficients for the windward and leeward sides of the building as 0.6 and −0.6, calculate the net pressure differences at heights of 120 m and 40 m.

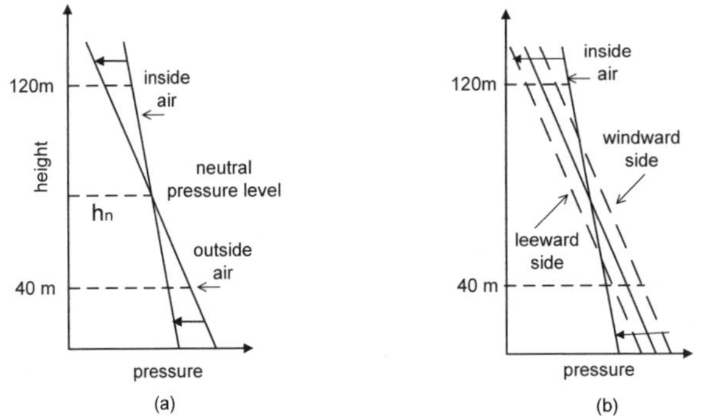

Fig. E8.13.1 (a) Stack pressure, (b) Stack effect and wind effect

Solution (i) The pressure difference due to the stack effect at a height z is given by Eq. (8.50) as

$$\Delta P_{st}(z) = \left(\frac{gP_o}{R_a}\right)(h_n - z)\left(\frac{1}{T_o} - \frac{1}{T_i}\right) \tag{E8.13.1}$$

Since the openings are distributed uniformly along the height of the building, the neutral plane is at mid-height. Therefore h_n = 75 m. The outside and inside air temperatures are, 295K and 261K respectively.

Substituting the relevant numerical data in Eq. (E8.13.1) we have

$$\Delta P_{st}(40m) = \left(\frac{9.81\times101}{0.287}\right)(75-40)\left(\frac{1}{261}-\frac{1}{295}\right) = 53.4 \text{ Pa}$$

$$\Delta P_{st}(120m) = \left(\frac{9.81\times101}{0.287}\right)(75-120)\left(\frac{1}{261}-\frac{1}{295}\right) = -68.6 \text{ Pa}$$

Expressed in cm of water (100Pa = 1.02 cm of water) the above pressures are, 0.54 cm and −0.7 cm respectively. The inside and outside pressures due to the stack effect are depicted in Fig. E8.13.1(a).

(ii) The static pressure due to the wind effect is given by Eq. (8.52) as

$$\Delta P_w = C_p \rho U^2/2 \tag{E8.13.2}$$

We use the ideal gas equation of state to express the density of air, ρ in Eq. (E8.13.2), in terms of the outside pressure and temperature. Hence we have

$$\Delta P_w = C_p P_o U^2/2R_a T_o$$

Substituting numerical values in the above equation we obtain the following wind pressures for the windward side (*ws*) and the leeward side (*ls*) respectively as

$$\Delta P_{w,ws} = 0.6 \times 101 \times \frac{7^2}{2\times0.287\times261} = 19.8 \text{ Pa}$$

$$\Delta P_{w,ls} = -0.6 \times 101 \times \frac{7^2}{2\times0.287\times261} = -19.8 \text{ Pa}$$

The inside and outside pressure distributions due to the stack effect, and the wind pressure on the windward and leeward sides, are depicted in Fig. E8.13.1(b). The net pressure differences at the two heights are given below:

Height	Windward side	Leeward side
40 m	53.4 + 19.8 = 73.2 Pa	53.4-19.8 = 33.6 Pa
120 m	−68.6 + 19.8 = −48.8 Pa	−68.6-19.8 = −88.4 Pa

Example 8.14 A two-story house has a volume of 420 m^3 and an effective air leakage area of 480 cm^2. The outdoor design temperature for

winter conditions is −10°C and the average wind speed is 6.7ms^{-1}. The indoor design temperature is 22°C. The house is sheltered by other houses across the street. Estimate (i) the air infiltration rate, and (ii) the air exchange rate.

Solution We shall use the basic model for residential air infiltration calculations given on page 16.23 of the *ASHRAE Handbook - 2013 Fundamentals* [1] to estimate the air infiltration rate. The expression for the infiltration rate is given by Eq. (8.54) as

$$Q_{in} = \left(\frac{A_L}{1000}\right)\sqrt{C_s|\Delta t| + C_w U^2} \qquad \text{(E8.14.1)}$$

The stack coefficient for a two-story house is obtained from Table 4 on page 16.23 of Ref. [1] as $Cs = 0.00029$.

The description of the location of the house fits the shelter type 3 listed in Table 5 on page 16.23 of Ref. [1]. The wind coefficient for a two-story house of shelter class 3 is obtained from Table 6 on page 16.24 of Ref. [1] as $C_w = 0.000231$. Substituting the above numerical values in Eq. (E8.14.1) we have

$$Q_{in} = \left(\frac{480}{1000}\right)\sqrt{0.00029 \times 34 + 0.000231 \times 6.7^2}$$

(i) Hence the air infiltration rate, $Q_{in} = 0.068\ \text{m}^3\text{s}^{-1}$.

(ii) The air exchange rate is given by

$$I = \frac{3600 Q_{in}}{V} = \frac{3600 \times 0.068}{420} = 0.59 \text{ ach}$$

Example 8.15 The multi-layered wall shown schematically in Fig. E8.15.1 has a thin sheet of vapor retarder material, of negligible thermal resistance, located between the inner layer of plywood ($k = 0.12$ Wm^{-1}K^{-1}) of thickness 6 mm, and the layer of glass fiber insulation ($k = 0.036$ Wm^{-1}K^{-1}) of thickness 140 mm. The thicknesses of the layer of concrete ($k = 2.2$ Wm^{-1}K^{-1}), and the layer of brick ($k = 0.9$ Wm^{-1}K^{-1}) are 200 mm and 100 mm respectively. The temperature and relative humidity of the inside, and outside air, are respectively, 22°C, 30% and

−18°C, 75%. The inside and outside heat transfer coefficients are 8 $Wm^{-2}K^{-1}$ and 28 $Wm^{-2}K^{-1}$ respectively. (i) Calculate the temperatures at the interfaces of the different layers of the wall. (ii) Obtain the saturation vapor pressure at the interfaces. (iii) Calculate the required value of the vapor resistance of the vapor retarder to avoid condensation of water vapor in the wall.

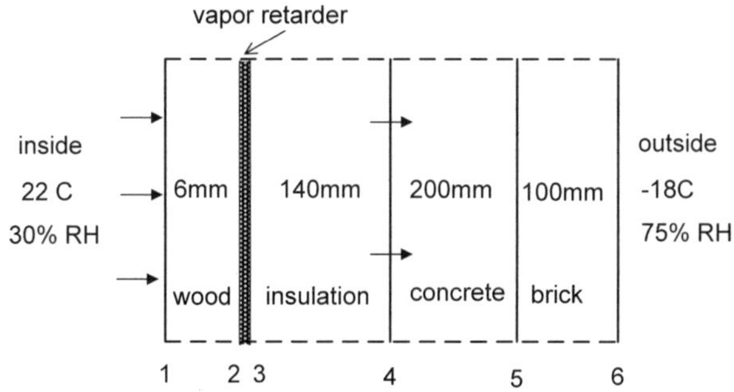

Fig. E8.15.1 Multi-layered wall

Solution (i) Consider unit area (A = 1 m^2) of the wall where the constituent layers are in series. The thermal resistance of a layer is given by

$$R_i = \frac{L_i}{k_i}$$

The thermal resistances of the different layers are given in Table E8.15.1 (a) below.

Table E8.15.1(a) Thermal resistances and vapor resistances

Layer, i	L_i mm	k_i $Wm^{-1}K^{-1}$	R_i m^2KW^{-1}	μ $ngPa^{-1}s^{-1}m^{-1}$	$R_{vap,i}$ $Pa.s.m^2ng^{-1}$
Inside air	-	-	0.125	-	0
Plywood*	6	0.12	0.05	0.65	0.0092
Retarder	small	large	0	-	Z
Insulation	140	0.036	3.9	172	0.00081
Concrete	200	2.2	0.09	4.65	0.043
Brick	100	0.9	0.11	4.54	0.022
Outside air	-	-	0.036	-	0

*k-values extracted from Table 1, Page 26.7; μ-values extracted from Table 6, Page 26.17 - *ASHRAE Handbook* - 2013 *Fundamentals* [1]

The total thermal resistance is

$$R_{tot} = 0.125 + 0.05 + 3.9 + 0.09 + 0.11 + 0.036 = 4.31$$

The total heat flow rate is given by

$$Q_{tot} = \frac{[22-(-18)]}{4.31} = 9.28 \text{ W}$$

Applying the thermal network analogy to the different layers of the wall we obtain the interface temperatures, T_i given in Table E8.15.1(b).

Table E8.15.1(b) Interface temperature and pressure

Interface i	1	2	3	4	5	6
T_i, °C	20.84	20.38	20.38	−15.84	−16.65	−17.67
P_{sat}, Pa	2462	2393	2393	153.6	141.9	128.9
P_{vap}, Pa	793	784.7	153	152.3	113.5	93.6

The saturation pressure of water vapor at the different interface temperatures, obtained from data in Ref. [5], are given in Table 8.15.1(b).

(ii) The unit vapor diffusion resistance of a layer is given by

$$R_{vap,i} = \frac{L_i}{\mu_i}$$

where μ_i is the water vapor permeability of the material and L_i is the thickness.

We have estimated the vapor permeability of the different materials of the wall from the data available in Table 6, page 26.17 of Ref. [1]. The calculated values of the vapor diffusion resistances, $R_{vap,i}$ of the different layers are listed in Table E8.15.1(a).

Let the vapor resistance of the vapor retarder be Z. Neglecting the vapor resistance of the inside and outside air layers, we obtain the total vapor resistance of the wall as

$$R_{vap,tot} = (0.0092 + Z + 0.00081 + 0.043 + 0.022)$$

$$R_{vap,tot} = (0.075 + Z) \text{ Pa.s.m}^2\text{ng}^{-1}$$

Note that 1 ng (nano-gram) = 10^{-9} g.

The inside and outside vapor pressures are given by

$$P_{in} = \phi_{in} P_{sat,in} = 0.3 \times 2644 = 793 \text{ Pa}$$

$$P_{out} = \phi_{out} P_{sat,out} = 0.75 \times 124.9 = 93.7 \text{ Pa}$$

where ϕ is the relative humidity.

Applying Fick's law we obtain the total vapor flow rate as

$$\dot{m}_{vap,tot} = \frac{[P_{in} - P_{out}]}{R_{vap,tot}} = \frac{793 - 93.7}{(0.075 + Z)} = \frac{699.3}{(0.075 + z)} \text{ (ng.s}^{-1}\text{.m}^{-2}\text{)}$$

Applying Fick's law to the brick layer we have

$$\dot{m}_{vap,tot} = \frac{(P_5 - 93.6)}{0.022} = \frac{699.3}{(0.075 + Z)} \qquad \text{(E8.15.1)}$$

Applying Fick's law to the concrete layer we have

$$\dot{m}_{vap,tot} = \frac{(P_4 - 93.6)}{(0.022 + 0.043)} = \frac{699.3}{(0.075 + Z)} \qquad \text{(E8.15.2)}$$

To avoid condensation of water vapor the vapor pressures at interfaces 4 and 5 must be less than the corresponding saturation vapor pressures at these locations. Hence we have: $P_4 < 153.6$ Pa, and $P_5 <$ 141.9 Pa respectively.

Applying these conditions to Eqs. (E8.15.1) and (E8.15.2) we obtain the following conditions for the vapor resistance of the vapor retarder:

$$Z > 0.68 \quad \text{and} \quad Z > 0.24$$

Choosing a value of, $Z = 0.7$ Pa.s.m^{2}ng^{-1} that is larger than the higher allowable value (0.68), we calculate the vapor pressures at the different interfaces by applying Fick's law to the layers. The resulting vapor pressures are given in Table E8.15.1(b). Note that the vapor pressures at all interfaces are below the corresponding saturation vapor pressures. Therefore vapor will not condense at the interfaces.

Problems

P8.1 A wall of a building is 15 m long and 4 m high. It has a 9 mm thick gypsum board ($k = 0.16$ Wm^{-1}K^{-1}) on the inside. Next to it is a 140 mm thick layer of glass fiber insulation ($k = 0.046$ Wm^{-1}K^{-1}) which is followed by a 180 mm thick layer of concrete ($k = 2.1$ Wm^{-1}K^{-1}). The outer face is made brick ($k = 1.35$ Wm^{-1}K^{-1}) with a thickness of 100 mm.

The outside and inside heat transfer coefficients are 28 $Wm^{-2}K^{-1}$ and 8 $Wm^{-2}K^{-1}$ respectively. The inside and outside air temperatures are 23°C and 2°C respectively. Calculate (i) the total thermal resistance, (ii) the overall heat transfer coefficient, and (iii) the total heat transfer rate through the wall.
[Answers: (i) 3.42 m^2KW^{-1}, (ii) 0.293 $Wm^{-2}K^{-1}$, (iii) 369W]

P8.2 A building has an exterior wall with 3.8 cm × 14 cm wood framing (k = 0.15 $Wm^{-1}K^{-1}$) that makes 25 percent of its area. Adjacent to the framing on the inside is a layer of plywood board (k = 0.09 $Wm^{-1}K^{-1}$) of thickness of 8 mm. The space between the framing is filled with fiber glass insulation (k = 0.043 $Wm^{-1}K^{-1}$). Next to it is a 10 mm thick sheathing (k = 0.06 $Wm^{-1}K^{-1}$) which is followed by a 200 mm thick layer of expanded polystyrene (k = 0.035 $Wm^{-1}K^{-1}$), and a 100 mm thick layer of brick (k = 1.1 $Wm^{-1}K^{-1}$) on the outside. The air temperature inside is 22°C and the outside ambient temperature is −8°C. The wind speed is 12 km h^{-1}. The inside heat transfer coefficient is 10 $Wm^{-2}K^{-1}$. Calculate (i) the total thermal resistance, (ii) the average heat transfer coefficient, and (iii) average heat transfer rate through the wall using: (a) the parallel path method, and (b) the isothermal plane method.
[*Answers*: (a) (i) 8.75 m^2KW^{-1}, (ii) 0.114 $Wm^{-2}K^{-1}$, (iii) 3.42Wm^{-2}, (b) (i) 8.2 m^2KW^{-1}, (ii) 0.122 $Wm^{-2}K^{-1}$, (iii) 3.66Wm^{-2}]

P8.3 A *horizontal* section of a vertical wall, made of three-core blocks of concrete, is shown in Fig. P8.3.1. The length, breadth and height of a block are 44 cm, 22 cm and 24 cm respectively. The two faces of a block are 40 mm thick and the thickness of a web is 30 mm. The cores are filled with perlite insulation of thermal conductivity, 0.06 $Wm^{-1}K^{-1}$. The thermal conductivity of concrete is 1.8 $Wm^{-1}K^{-1}$. The outside and inside heat transfer coefficients are 26 $Wm^{-2}K^{-1}$ and 10 $Wm^{-2}K^{-1}$ respectively. Using the isothermal plane method, calculate (i) the average thermal resistance of the wall, and (ii) the average heat transfer coefficient. [*Answers*: (i) 3.038 m^2KW^{-1} (ii) 0.33 m^2KW^{-1}]

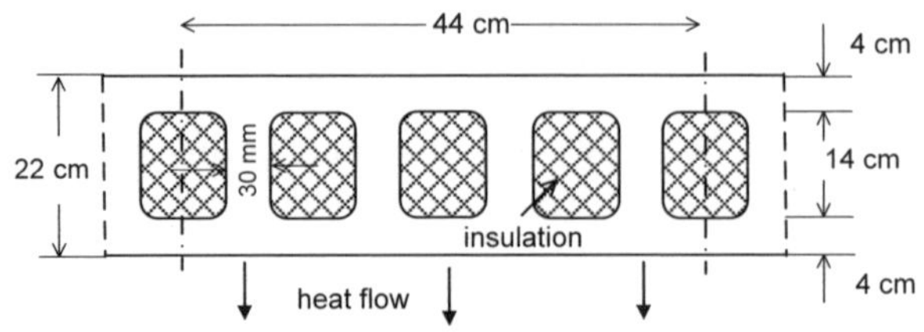

Fig. P8.3.1 Section of a concrete block wall

P8.4 A building has a triple-glazed vertical window with a 13 mm air space between adjacent glasses. The glass thickness is 6 mm and the thermal conductivity is 0.8 $Wm^{-1}K^{-1}$. The emissivity of glass is 0.9. The inside and outside heat transfer coefficients are 9 $Wm^{-2}K^{-1}$, and 29 $Wm^{-2}K^{-1}$ respectively. The indoor and outdoor air temperatures are 22°C and −12°C respectively. (a) Calculate the rate of heat loss through center of the window. (b) What is the minimum indoor relative humidity at which condensation will occur on the inner surface of the window? Will the answer be different if the edge of the glass is considered?
[*Answers*: (a) 67 Wm^{-2}, (b) 62%]

P8.5 A fixed double-glazed vertical window, has two 6 mm thick glass panes. The 13 mm wide space between the panes is filled with argon gas. A coating with an emissivity of 0.6 is applied on the inner surfaces of the glasses, facing the air space. The area of a glazing is 1.8 m × 1.5 m. The window frame of width 55 mm is made of aluminum with a thermal break. The inside and outside temperatures are 21°C and −18°C respectively. The wind speed is 6.7 ms^{-1}. Calculate (i) the overall heat transfer coefficient (see Table 8.4), and (ii) the rate of heat loss through the window.
[*Answers*: (i) 2.49 $Wm^{-2}K^{-1}$, (ii) 299 W]

P8.6 The thickness of a basement wall made of concrete (k = 2.1 $Wm^{-1}K^{-1}$) is 210 mm. It is insulated on the inside with a layer of polystyrene insulation (k = 0.04 $Wm^{-1}K^{-1}$) of thickness 40 mm, and finished with gypsum board paneling (k = 0.16 $Wm^{-1}K^{-1}$) of thickness 8 mm. The total horizontal length of the wall is 80 m and it extends 2 m below grade.

The soil surrounding the wall has a thermal conductivity of 1.3 $Wm^{-1}K^{-1}$. The inside air temperature is 17°C and the outside ambient temperature is −2°C. The inside and outside heat transfer coefficients are 9 $Wm^{-2}K^{-1}$ and 30 $Wm^{-2}K^{-1}$ respectively. Calculate (i) the average heat transfer coefficient for the basement wall, and (ii) the total rate of heat loss through the basement wall.
[*Answers*: (i) 0.436 $Wm^{-2}K^{-1}$, (ii) 1.325kW]

P8.7 The length and breadth of a basement floor are 16 m and 12 m respectively and it is 2 m below grade. The 150 mm thick floor is made of concrete (k = 2.2 $Wm^{-1}K^{-1}$), and it is entirely carpeted. The carpet is 10 mm thick and its thermal conductivity is 0.1 $Wm^{-1}K^{-1}$. The soil surrounding the basement has a thermal conductivity of 1.4 $Wm^{-1}K^{-1}$.

The inside air temperature is 18°C and the outside ambient temperature is −3°C. The inside and outside heat transfer coefficients are 10 $Wm^{-2}K^{-1}$ and 27 $Wm^{-2}K^{-1}$ respectively. Calculate (i) the average heat transfer coefficient for the basement floor, and (ii) the total rate of heat loss through the basement floor.
[*Answers*: (i) 0.136 $Wm^{-2}K^{-1}$, (ii) 548 W]

P8.8 A building of height 160 m has openings distributed uniformly along its height. The ambient pressure is 101 kPa and the mean wind speed is 8 ms^{-1}. The inside and outside air temperatures are 24°C and −14°C respectively. (i) Calculate the theoretical pressure difference between the inside and outside due to the stack effect at heights of 30 m and 140 m. (ii) Assuming the respective wind coefficients for the windward and leeward sides of the building as 0.6 and −0.6, calculate the net pressure differences at heights of 30 m and 140 m.
[*Answers*: (i) 85.3 Pa, −102.3 Pa, (ii) 111.4 Pa, 59.2 Pa, −76.2 Pa, −128.4 Pa]

P8.9 A two-story building of volume of 540 m^3 has an effective air leakage area of 510 cm^2. The outdoor and indoor design temperatures for winter conditions are −10°C and 20°C respectively. The average design wind speed is 7ms^{-1}. The house is sheltered by other houses across the

street. Estimate (i) the air infiltration rate, and (ii) the air exchange rate per hour.
[*Answers*: (i) 0.072 m^3s^{-1}, (ii) 0.48 ach]

P8.10 The exterior wall of a building has 10 mm thick gypsum (k = 0.16 $Wm^{-1}K^{-1}$) board on the inside. Next to it is a vapor retarder, of negligible thermal resistance, that bears on a wood frame (k = 0.15 $Wm^{-1}K^{-1}$), made of 38 mm × 140 mm studs. The space between the studs is filled with fiberglass insulation (k = 0.042 $Wm^{-1}K^{-1}$). This is followed by a 20 mm thick layer of sheathing (k = 0.055 $Wm^{-1}K^{-1}$), 40 mm thick layer of expanded polystyrene (k = 0.035 $Wm^{-1}K^{-1}$) and 100 mm thick layer of brick (k = 0.85 $Wm^{-1}K^{-1}$). The area of the framing is 20% of the area of the wall.

The inside air temperature and relative humidity are respectively 20°C, 30%. The outside air at −15°C is saturated. The inside and outside heat transfer coefficients are 8 $Wm^{-2}K^{-1}$ and 24 $Wm^{-2}K^{-1}$ respectively.

(a) Calculate the average heat transfer coefficient using: (i) the parallel path method and (ii) the isothermal plane method.

(b) Considering the heat flow path through the insulation, calculate the minimum vapor diffusion resistance of the vapor retarder, needed to avoid condensation of water vapor in the wall.
[*Answers*: (a), (i) 0.226$Wm^{-2}K^{-1}$, (ii) 0.247$Wm^{-2}K^{-1}$, (b) 0.64 $Pa.s.m^2ng^{-1}$]

References

1. *ASHRAE Handbook* - 2013 *Fundamentals*, American Society of Heating, Refrigeration and Air Conditioning Engineers, Atlanta, 2013.
2. *ASHRAE Handbook* - 2009 *Fundamentals*, American Society of Heating, Refrigeration and Air Conditioning Engineers, Atlanta, 2009.
3. Kuehn, Thomas H., Ramsey, James W. and Threlkeld, James L., *Thermal Environmental Engineering*, 3rd edition, Prentice-Hall, Inc., New Jersey, 1998.

4. Mitchell, John W. and Braun, James E., *Heating, Ventilation, and Air Conditioning in Buildings,* John Wiley & Sons, Inc., New York, 2013.
5. Stoecker, Wilbert F. and Jones, Jerold W., *Refrigeration and Air Conditioning*, International Edition, McGraw-Hill Book Company, London, 1982.
6. Wijeysundera, N. E., M. N. A. Hawlader and Y. T. Tan, 'Water vapor diffusion and condensation in fibrous insulations', *International Journal of Heat and Mass transfer*, 32(10) (1989):1865-1878.
7. Wijeysundera, N. E., B. F. Zheng, M. Iqbal, and E. G. Hauptmann, 'Effective thermal conductivity of flat-slab and round-pipe insulations in the presence of condensation', *Journal of Thermal Insulation and Building Envelopes*, 17 (July 1993): 55-77.

Chapter 9

Solar Radiation Transfer Through Building Envelopes

9.1 Introduction

Solar radiation incident on the external surfaces of a building contributes significantly to the cooling load of the building. Opaque surfaces like the walls and roofs absorb a fraction of the incident solar radiation, and reflect the rest. A portion of the absorbed radiation is conducted through the wall or the roof, while the rest is lost to the ambient due to convection, and thermal radiation exchange with surrounding surfaces.

In contrast, solar radiation incident on *transparent* surfaces, like glass windows and skylights, usually called *fenestrations*, undergoes reflection, absorption, and transmission. A fraction of the absorbed solar radiation is transferred to the air inside by conduction and convection while the rest is transferred to the surroundings. Solar radiation transmitted *directly* through the fenestration, is absorbed by the floor, the inner walls, and items like the furniture, causing their surface temperatures to increase. Subsequently, the latter surfaces transfer heat to the air in the space by convection. Usually there is a time lag between the solar radiation absorption and heat convection processes, due to the thermal mass of the items receiving the transmitted solar radiation.

In the summer, the above heat flow processes contribute significantly to the cooling load of buildings, whereas in the winter, these same heat flows help to partially balance the heat losses through the building envelope. In order to quantify the aforementioned energy transfer processes we need to know the solar radiation intensity and the direction of incidence, both of which undergo daily and seasonal variations.

In the next section we shall introduce the various equations needed to determine the intensity of solar radiation and the angle of incidence of the radiation beam on an arbitrarily inclined surface. For this purpose we need to first review several fundamental aspects of solar radiation.

9.2 Fundamentals of Solar Radiation

In this section we shall consider several physical aspects of solar radiation including its energy intensity and solar geometry, which are of importance for building cooling load estimation.

9.2.1 *Beam and diffuse solar radiation*

The intensity of direct normal extraterrestrial solar radiation on the Earth's atmosphere is called the *solar constant*. It is the solar radiation striking a unit area normal to the direction of the beam from the sun. Although the solar constant varies from about 1323 Wm^{-2} to 1414 Wm^{-2} due to the slightly elliptical orbit of the earth around the sun, its reference value is treated as constant, and equal to 1367 Wm^{-2}. The *ASHRAE Handbook* - 2013 *Fundamentals* [1] gives the following approximate expression for the variation of the extraterrestrial solar radiation incident on a surface normal to the sun's rays:

$$E_o = E_{sc}\left\{1 + 0.033cos\left[360^0\frac{(n-3)}{365}\right]\right\} \tag{9.1}$$

where E_{sc} = 1367 Wm^{-2}, and n is the day of the year numbered from January 1.

A fraction the solar radiation entering the atmosphere is transmitted to the Earth's surface while the rest is partially absorbed and scattered by the constituents of the atmosphere like air, carbon dioxide, clouds, and chemical molecules. The intensity of solar radiation at the surface of the earth depends on atmospheric conditions, season, time of day, latitude, and orientation.

The solar radiation transmitted directly through the atmosphere without change in direction and striking a surface is called *beam radiation*. Beam radiation is also referred to as *direct radiation*. The solar radiation received from the sun after its direction has been changed due to scattering by the atmosphere is called *diffuse radiation*. It does not have a unique direction. The sum of the beam and diffuse radiation on a surface is the *total radiation.*

On a very clear day the fraction of extraterrestrial radiation absorbed or reflected by the upper atmosphere is about 20%. The beam and diffuse radiation fractions striking the surface of the earth are about 70% and 10% respectively. On cloudy days the solar radiation reaching the earth's surface is almost entirely diffuse.

9.2.2 *Direction of beam radiation*

In this section we shall derive the equations needed to determine the direction of the beam component of solar radiation at any location on the surface of the earth, on any day and time. For this purpose, a brief review of solar geometry is presented below.

The earth revolves about the sun in an orbit that is slightly elliptical. It takes 24 hours to make one rotation about its axis and completes one revolution about the sun in approximately 365.25 days. The axis of rotation of the earth is inclined at an angle of 23.5° to the plane of its orbit around the sun as shown in Fig. 9.1. This tilt of the earth's axis causes the seasons.

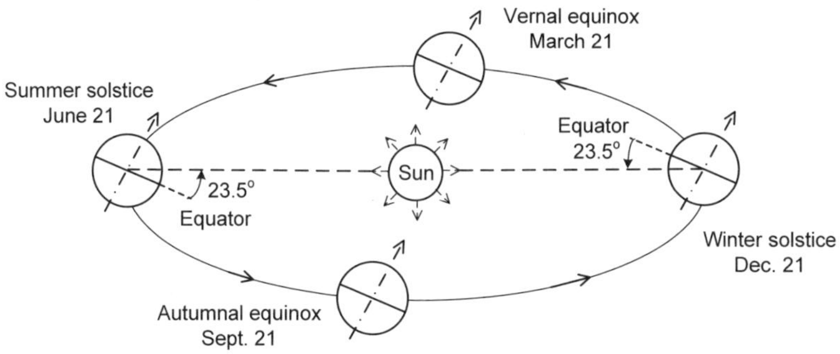

Fig. 9.1 Motion of the earth about the sun

The position of the earth at the start of each of the seasons is depicted in Fig. 9.1. At the vernal equinox and the autumnal equinox the plane of the sun's rays is parallel to the equatorial plane of the earth. At the summer solstice and the winter solstice the plane of the sun's rays is inclined at 23.5° and −23.5° respectively to the equatorial plane.

Since all motion is relative, it is convenient to take the earth as fixed and consider the virtual motion of the sun in relation to any location on earth as depicted in Fig. 9.2. To an observer at P on earth, the plane of the sun's rays appears to swing through an angle of 23.5° about a plane parallel to the equatorial plane of the earth.

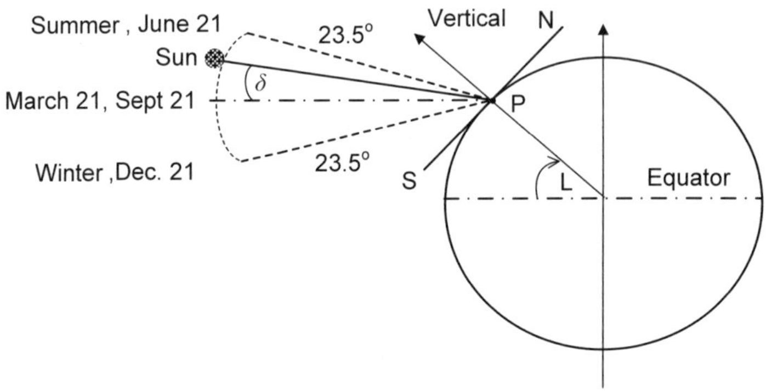

Fig. 9.2 Apparent motion of the sun as observed from a location P on earth

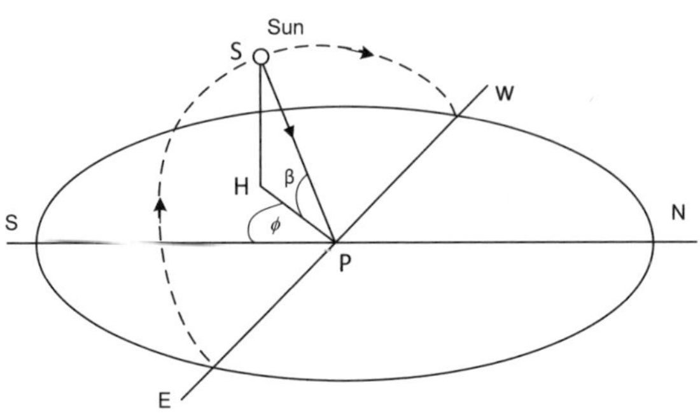

Fig. 9.3 Solar altitude and solar azimuth angles

The location of the sun at any time during the day can be specified by two angles β and ϕ as shown in Fig. 9.3. The angle β, measured from the local horizontal plane upward to the center of the sun is called *solar altitude angle.* It is the angle between the sun's rays, SP and the horizontal plane at P. The angle ϕ between the due south line at P and the projection PH, of PS on the horizontal plane, is called the *solar azimuth angle*. Azimuth angles east of south are positive and west of south are negative by convention.

Although the two angles β and ϕ are convenient in locating the sun's position at any time, they are not fundamental angles. These angles have to be related to the three fundamental angular quantities called: (i) the latitude, (ii) the declination, and (iii) the hour angle.

The latitude, L is the angular distance of a point on the earth measured north or south of the equator as shown in Fig. 9.2.

The declination of the sun, δ is the angle between the sun's rays at any time during the day and the zenith direction (directly overhead) at noon on the earth's equator as indicated in Fig. 9.2. The declination is zero at the autumnal and vernal equinoxes. It is +23.5° and −23.5° respectively at the summer solstice and winter solstice in the northern hemisphere. An approximate expression for the solar declination angle on any day of the year is [1]

$$\delta = 23.45 sin\left[\frac{360(284+N_d)}{365}\right] \tag{9.2}$$

where N_d is the day of the year numbered from January 1.

The *hour angle*, H at a location is the angular displacement of the sun from solar noon, when the sun is directly overhead at the location. The hour angle is zero at solar noon. For each hour away from solar noon the hour angle is 15° because the earth rotates at the rate of 15° of longitude per hour. In the morning the hour angle is negative, and it is positive in the afternoon. Hence we have

$$H = 15(solar\ time - 12) \text{ degrees} \tag{9.3}$$

The *Solar time* depends on the rotation of the earth and therefore varies continuously with longitude. However, *local standard time*, indicated by a clock, is the same for all locations in a *time zone* which covers a finite longitude interval of about 15°. Each time zone has its assigned *standard longitude* which is used to obtain the *standard* time for the zone. When the sun is directly overhead at a location, it is *solar noon*, whereas when the sun is directly overhead at the *standard* longitude of a time zone it is *local noon*. Solar time T_{solar} at a location, with a longitude L_{loc}, and standard time T_{std}, both measured in minutes, are related by the equation [3]

$$T_{solar} = T_{std} + 4(L_{std} - L_{loc}) + E_{time} - DT \tag{9.4}$$

where L_{std} is the longitude used to obtain the standard time for the time zone of the location.

Longitude is measured positive east of Greenwich where the longitude is zero. The term DT is the called the daylight saving time correction, which is the number of hours that the *time is advanced* for daylight saving.

In Eq. (9.4), E_{time} is a correction factor called the '*equation of time*' which accounts for the perturbations in the earth's rate of rotation. It is given by the equation [3]

$$E_{time} = 9.87 sin2B - 7.53 cosB - 1.5 sinB \tag{9.5}$$

where

$$B = \frac{360(N_{day} - 81)}{364} \tag{9.6}$$

and N_{day} (≤ 365) is the day of the year.

The solar altitude β and the solar azimuth ϕ, shown in Fig. 9.3, are related to the latitude L, the solar declination δ, and the hour angle H by the following equations [1]:

$$sin\beta = cosLcosHcos\delta + sinLsin\delta \tag{9.7}$$

$$sin\phi = \frac{cos\delta sinH}{cos\beta} \tag{9.8}$$

$$cos\phi = \frac{(cos\delta sinLcosH - sin\delta cosL)}{cos\beta} \tag{9.9}$$

9.2.3 *Angle of incidence of beam radiation on a surface*

We shall now obtain an expression for the angle of incidence of beam radiation on a plane surface inclined at angle Σ to the horizontal as shown in Fig. 9.4. The direction of the direct beam of solar radiation is PS and its projection on the horizontal plane is PA. The solar altitude and the solar azimuth at the location are β and ϕ respectively. The normal to the surface is PN, and its projection on the horizontal plane PB, makes an angle ψ with the south.

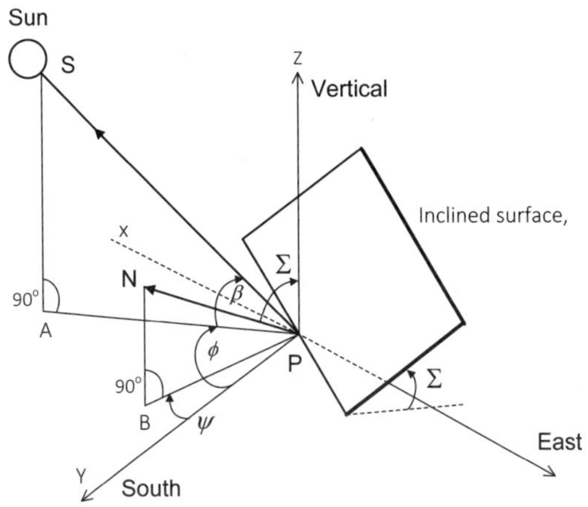

Fig. 9.4 Angle of incidence on an inclined surface

Consider the coordinate system, shown in Fig. 9.4 with the x and y axis directed toward the west and south on the horizontal plane (also see Fig. 9.2) at P. The z-axis is in the vertical direction. The unit vector in the direction PS is given by

$$\bar{I}_s = \begin{pmatrix} cos\beta sin\phi \\ cos\beta cos\phi \\ sin\beta \end{pmatrix} \tag{9.10}$$

The unit vector in the direction of the surface normal PN is

$$\bar{n} = \begin{pmatrix} sin\Sigma sin\psi \\ sin\Sigma\, cos\psi \\ cos\Sigma \end{pmatrix} \tag{9.11}$$

The angle of incidence, θ on the inclined surface is the angle between the direct beam PS and the surface normal PN. This is given by the dot product of the unit vectors $\bar{I}_s$ and $\bar{n}$. Hence we have

$$cos\theta = cos\beta sin\phi sin\Sigma sin\psi + cos\beta cos\phi sin\Sigma\, cos\psi + sin\beta cos\Sigma$$

$$cos\theta = cos\beta sin\Sigma\, cos(\phi - \psi) + sin\beta cos\Sigma \tag{9.12}$$

$$cos\theta = cos\beta sin\Sigma\, cos\gamma + sin\beta cos\Sigma \tag{9.13}$$

The angle, $\gamma = (\phi - \psi)$ is called the *surface–solar azimuth*.

We note from Eq. (9.13) that for a horizontal surface like a roof, $\Sigma = 0$ and angle of incidence is $(90 - \beta)$. For a vertical surface like a wall, $\Sigma = 90°$ and the angle of incidence is given by

$$cos\theta = cos\beta cos\gamma \tag{9.14}$$

The angles β and ϕ in Eq. (9.13) are related to the latitude L, the solar declination δ, and the hour angle H by Eqs. (9.7) and (9.8) respectively.

9.2.4 *Total radiation incident on an inclined surface*

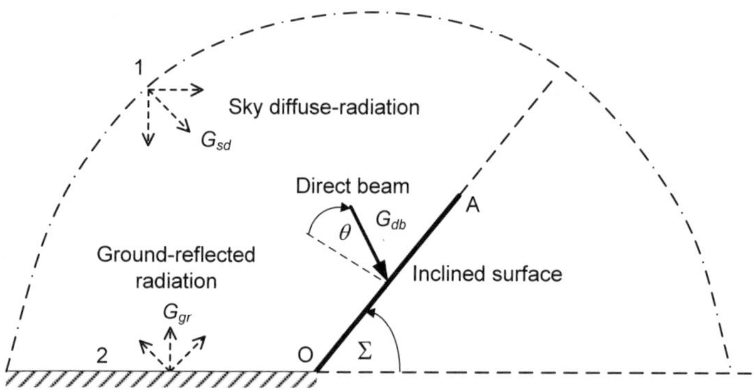

Fig. 9.5 Radiation incident on an inclined surface

The total solar radiation incident on a building surface like a roof, a wall or a window consists of three components. These are: (i) the direct beam solar radiation G_{db}, (ii) the diffuse radiation from the sky G_{sd}, and (iii) the radiation reflected from the ground and the surrounding buildings G_{gr}. The sky-diffuse radiation and the ground-reflected radiation may be treated as isotropic.

Shown schematically in Fig. 9.5 is a three-surface enclosure consisting of a rectangular building surface OA, the sky 1, and the large horizontal ground surface 2. The fraction of diffuse radiation emitted by OA that lands on the ground surface 2, considered an infinite plane, is the view factor $F_{OA\text{-}2}$, which is given by the expression [5]

$$F_{OA-2} = \frac{(1 - cos\Sigma)}{2} \tag{9.15}$$

where Σ is the inclination of the surface OA to the horizontal.

The fraction of diffuse radiation emitted by OA that lands on the sky 1 is given by

$$F_{OA-1} = 1 - F_{OA-2} = \frac{(1+cos\Sigma)}{2} \quad (9.16)$$

Using the reciprocity relation for diffuse view factors [5] we obtain the following expression for the sky radiation striking the surface OA per unit area

$$G_{S-OA} = \frac{(1+cos\Sigma)G_{sd}}{2} \quad (9.17)$$

where G_{ds} is the sky-diffuse radiation incident on a *horizontal* surface per unit area. Note that the view factor from the horizontal surface to the sky is unity (see problem 9.5).

Similarly, the ground-reflected diffuse radiation striking the surface OA per unit area is given by

$$G_{g-OA} = \frac{(1-cos\Sigma)G_{gr}}{2} = \frac{(1-cos\Sigma)\rho_g G_{tg}}{2} \quad (9.18)$$

where G_{tg} is the total solar radiation incident on the ground, which includes the direct and diffuse components. The reflectivity of the ground is ρ_g.

The intensity of total solar radiation incident on the ground G_{tg} is given by

$$G_{tg} = (G_{sd} + G_{db}sin\beta) \quad (9.19)$$

where β is the solar altitude angle, and G_{db} is the intensity of direct-beam radiation.

The direct-beam radiation striking the surface OA per unit area is

$$G_{d-OA} = G_{db}cos\theta \quad (9.20)$$

where θ is the angle of incidence on the surface.

The total radiation incident on the surface OA per unit area G_{t-OA} is the sum of the direct radiation, the sky-diffuse radiation, and the ground-reflected radiation. Hence we have

$$G_{t-OA} = G_{d-OA} + G_{s-OA} + G_{g-OA} \quad (9.21)$$

9.2.5 *Clear-sky model of direct and diffuse solar radiation*

Measured values of hourly averaged direct and diffuse solar radiation intensities are commonly used in building energy simulation software to estimate the hourly cooling loads. However, when such measured data are not readily available, an alternative design approach is to use mathematical correlations that have been developed for 'clear-sky radiation intensities'. The first such solar radiation model for use in cooling load estimation through fenestrations was introduced in the *ASHRAE Handbook* - 1993 *Fundamentals*. The above model has a simple mathematical form and incorporates three coefficients that depend on the month of the year, and are applicable to any location [3,4].

The shortcomings and limitations of the above model are discussed in the study by Gueymard and Thevenard [2]. They developed a new 'clear sky model', which is recommended in the *ASHRAE Handbook* - 2013 *Fundamentals* [1], for computing the intensities of the direct beam component E_b, and diffuse component E_d, of clear-sky solar radiation. The main equations of the new model are [1,2]

$$E_b = E_o exp[-\tau_b m^{ab}] \tag{9.22}$$

$$E_d = E_o exp[-\tau_d m^{ad}] \tag{9.23}$$

where E_o is the extraterrestrial normal radiation intensity, given by Eq. (9.1), and m is the *relative air mass*.

The relative air mass is the ratio of the mass of atmosphere along the actual direct beam to the mass if the sun was directly overhead. The dependence of the air mass on solar altitude angle may be expressed in the form [1]

$$m = [sin\beta + 0.50572(6.07995 + \beta)^{-1.6364}]^{-1} \tag{9.24}$$

The optical depths τ_b and τ_d included in Eqs. (9.22) and (9.23) are *location-specific* and vary during the year. The air mass exponents ab and ad may be expressed in terms of the optical depths as [1]

$$ab = 1.454 - 0.406\tau_b - 0.268\tau_d + 0.021\tau_b\tau_d \tag{9.25}$$

$$ad = 0.507 + 0.205\tau_b - 0.080\tau_d - 0.190\tau_b\tau_d \tag{9.26}$$

The monthly values of τ_b and τ_d for a large number of locations around the world are tabulated in the compact disc accompanying the *ASHRAE Handbook* - 2013 *Fundamentals* [1]. A few representative values are listed in Table 9.1 for purposes of illustration.

Table 9.1 Monthly values of optical depths for selected locations (values extracted from the data CD of the *ASHRAE Handbook* - 2013 *Fundamentals* [1])

Optical depths	τ_b			τ_d		
Location/Month	Jan.	May	Sept.	Jan.	May	Sept.
New York, USA	0.318	0.417	0.402	2.514	2.179	2.326
Dallas, USA	0.332	0.376	0.373	2.588	2.350	2.465
Toronto, Canada	0.294	0.388	0.387	2.414	2.245	2.333
Beijing, China	0.382	0.700	0.511	2.222	1.485	1.883
Bangalore, India	0.370	0.448	0.419	2.526	2.161	2.373
Sydney, Australia	0.414	0.293	0.330	2.561	2.660	2.556

The *direct beam solar radiation* incident on a surface of area A is

$$H_{db} = AE_b cos\theta \tag{9.27}$$

where θ is the angle of incidence.

The *diffuse radiation* falling on a surface is more difficult to determine because of the anisotropy of the radiation from the sky. For a vertical surface the following expression, based on the empirical relations recommended in Ref. [1], may be used to estimate the diffuse radiation incident on the surface:

$$H_{dif} = AE_d Y \tag{9.28}$$

where Y is a function of the angle of incidence θ of the direct beam. It is given by the expression

$$Y = max.[0.45, (0.55 + 0.437cos\theta + 0.313cos^2\theta)] \tag{9.29}$$

To estimate the *diffuse radiation* falling on a surface inclined at an angle Σ to the horizontal the following expression may be used:

$$H_{dif} = AE_d(Ysin\Sigma + cos\Sigma), \text{ for } \Sigma \leq 90^o \tag{9.30}$$

$$H_{dif} = AE_d Ysin\Sigma, \text{ for } \Sigma > 90^o \tag{9.31}$$

The *ground-reflected radiation* falling on a surface is given by

$$H_{gr} = \frac{A(E_d + E_b sin\beta)\rho_{gr}(1 - cos\Sigma)}{2} \tag{9.32}$$

where Σ is the inclination and ρ_{gr} is the ground reflectance, taken as 0.2 for typical ground surfaces.

9.3 Absorption of Solar Radiation by an Opaque Surface

In this section we shall obtain expressions for the rate of solar radiation absorption by external *opaque* surfaces like walls and roofs of buildings. The different energy interactions occurring at an external building surface exposed to solar radiation are depicted in Fig. 9.6.

The total solar radiation incident on the surface G_{ts} consists of beam radiation G_{db}, sky-diffuse radiation G_{sd}, and ground-reflected radiation, G_{gr}. Hence we have

$$G_{ts} = G_{db} + G_{sd} + G_{gr} \tag{9.33}$$

The rate of absorption of solar radiation per unit area may be expressed as

$$q_{sol} = \alpha_s G_{ts} \tag{9.34}$$

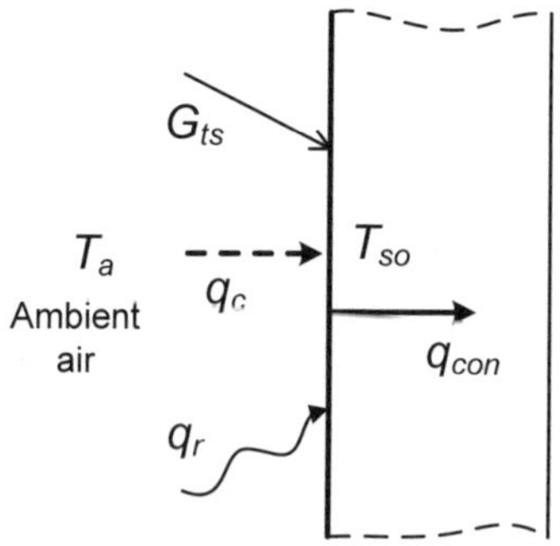

Fig. 9.6 Energy interactions at external surface

Applying the energy balance equation to the surface we have

$$q_{con} = q_{sol} + q_c + q_r \tag{9.35}$$

where q_{con} is rate of heat conduction into the wall, and q_c is the rate of heat convection from the outside air to the surface.

The rate of long-wave radiation exchange q_r between the surface and the surrounding surfaces, including the sky, is a function of the various surface temperatures, and the sky temperature. These temperatures, in general, are different from the ambient air temperature. As an approximation we assume that the surrounding surface temperatures are equal to the ambient air temperature. However, the sky temperature is usually lower than the local ambient temperature.

We now substitute heat transfer rate equations in Eq. (9.35) to obtain

$$q_{con} = q_{sol} + h_c(T_a - T_s) + h_r(T_{sky} - T_s) \tag{9.36}$$

where T_s, T_a and T_{sky} are the surface temperature, the ambient air temperature, and the sky temperature respectively. The convective heat transfer coefficient is h_c. The radiation heat transfer rate can be expressed in terms of a linearized radiation heat transfer coefficient h_r because the temperature differences involved are relatively small. This was demonstrated in section 8.2.4. Hence we express Eq. (9.36) in the form

$$q_{con} = q_{sol} + h_c(T_a - T_s) + h_r(T_a - T_s) - h_r(T_a - T_{sky})$$

$$q_{con} = q_{sol} + (h_c + h_r)(T_a - T_s) - \varepsilon_s \Delta R \tag{9.37}$$

The last term in Eq. (9.37), $\varepsilon_s \Delta R$ may be thought of as a correction factor that accounts for the difference between the ambient air temperature and the sky temperature. We can rearrange the terms in Eq. (9.37) to obtain the following convenient form:

$$q_{con} = (h_c + h_r)(T_{sa} - T_s) = h_o(T_{sa} - T_s) \tag{9.38}$$

where h_o is the sum of the convection and radiation heat transfer coefficients. The fictitious temperature, T_{sa}, called the *sol-air temperature*, is given by

$$T_{sa} = T_a + \frac{q_{sol}}{h_o} - \frac{\varepsilon_s \Delta R}{h_o} \tag{9.39}$$

The sol-air temperature is an effective driving temperature that incorporates the contributions of solar radiation, long wave radiation, and convection to the energy interactions at an external surface.

For roof surfaces that are orientated towards the sky, the correction factor, $(\varepsilon_s \Delta R/h_o)$ in Eq. (9.39) for long-wave radiation is taken to be about 4°C [3].

Usually, ambient air is warmer than the sky, but cooler than the ground. Therefore for vertical wall surfaces, exposed to both the sky and the ground, the contributions to the long-wave radiation correction factor from the sky and the ground tend to cancel out.

9.4 Transmission and Absorption of Solar Radiation

In this section we shall apply the net radiation method [5,6] to obtain expressions for the transmittance of solar radiation through multi-layered fenestrations and the rate of absorption of radiation in each layer.

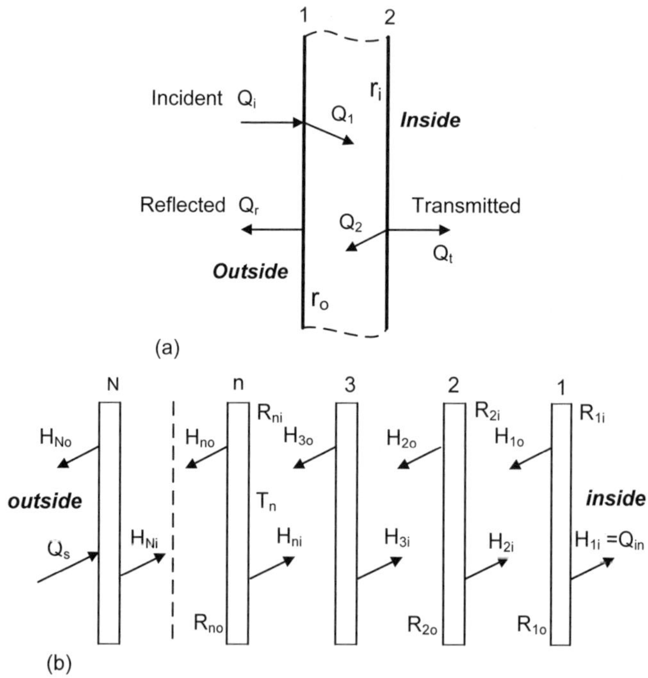

Fig. 9.7 (a) Net radiation fluxes for one layer (b) Multi-layered fenestration

9.4.1 *Effective properties of a single layer*

We first consider the single transparent layer shown in Fig. 9.7(a), which is a representative component of the multi-layered fenestration system depicted in Fig. 9.7(b). The flux of solar radiation incident on surface 1 of the layer is Q_i and the reflected and transmitted net-radiation fluxes are Q_r and Q_l respectively. At surface 2, the respective reflected and transmitted net-radiation fluxes are Q_2 and Q_t. We recall from section 2.8.9 in chapter 2 that the net-radiation flux [5,6] is the algebraic sum of all the different radiation currents that result from multiple reflections at the two faces of the slab.

The reflectivities r_i and r_o of the two surfaces 1 and 2 of the layer are assumed different to allow for the possible presence of surface coatings and thin reflecting films on these surfaces [1]. The transmittance τ of the layer can be computed knowing the angle of incidence of the radiation beam, the refractive index, the extinction coefficient of the material, and the layer thickness [3]. (See worked example 9.10.)

For most fenestration materials, the reflectivity and the transmittance vary with the wavelength of incident radiation. Therefore these spectral properties have to be averaged over the solar radiation spectrum before they are used in the radiation balance equations given below.

We now write the following radiation balance equations that relate the various net radiation fluxes for the layer depicted in Fig. 9.7(a).

$$Q_1 = (1 - r_o)Q_i + \tau r_o Q_2 \tag{9.40}$$

$$Q_2 = \tau r_i Q_1 \tag{9.41}$$

$$Q_t = \tau(1 - r_i)Q_1 \tag{9.42}$$

$$Q_r = r_o Q_i + \tau(1 - r_o)Q_2 \tag{9.43}$$

Eliminating Q_1 and Q_2 in Eqs. (9.40) to (9.43) we obtain the following expressions for the *effective transmittance*, T and the *effective reflectivity* R_o, used to characterize the slab:

$$T = \frac{Q_t}{Q_i} = \frac{\tau(1-r_o)(1-r_i)}{1-\tau^2 r_o r_i} \tag{9.44}$$

$$R_o = \frac{Q_r}{Q_i} = r_o + \frac{\tau r_i(1-r_o)^2}{1-\tau^2 r_o r_i} \tag{9.45}$$

Similarly, for radiation incident on surface 2 of the slab, the effective reflectivity, R_i is

$$R_i = \frac{Q_r}{Q_i} = r_i + \frac{\tau r_o(1-r_i)^2}{1-\tau^2 r_o r_i} \tag{9.46}$$

9.4.2 *Transmittance of a multi-layered fenestration*

We shall now develop a series of general relationships that are applicable to fenestration systems consisting of different types of semi-transparent materials like glass and plastic.

Consider the multi-layered fenestration system, consisting of N partially transparent layers separated by non-absorbing gas spaces, as shown schematically in Fig. 9.7(b). The layer, n (n = 1,N) is characterized by the three independent properties T_n, R_{ni} and R_{no}, given in Eqs. (9.44), (9.45) and (9.46) respectively.

The radiation flux incident on the outer surface of layer N is Q_s and the radiation flux entering the inside through innermost layer 1 is Q_{in}. The net radiation fluxes entering and leaving the different layers are indicated in Fig. 9.7(b). Applying the net radiation balance equation to the different layers we have

$$Q_{in} = H_{1i} = T_1 H_{2i} \qquad (9.47)$$

$$H_{1o} = R_{1o} H_{2i} \qquad (9.48)$$

$$H_{2i} = T_2 H_{3i} + R_{2i} H_{1o} \qquad (9.49)$$

$$H_{2o} = R_{2o} H_{3i} + T_2 H_{1o} \qquad (9.50)$$

$$H_{3i} = T_3 H_{4i} + R_{3i} H_{2o} \qquad (9.51)$$

$$H_{3o} = R_{3o} H_{4i} + T_3 H_{2o} \qquad (9.52)$$

$$H_{ni} = T_n H_{(n+1)i} + R_{ni} H_{(n-1)o} \qquad (9.53)$$

$$H_{no} = R_{no} H_{(n+1)i} + T_n H_{(n-1)o} \qquad (9.54)$$

$$H_{(N+1)i} = Q_s \qquad (9.55)$$

For a fenestration system consisting of N layers, the overall transmittance is defined as the ratio, $Q_{in}/ Q_s = T_{ON}$. An expression for T_{ON} may be obtained by solving simultaneously Eqs. (9.47) to (9.55). However, the regular pattern of the this set of equations enables us to develop the following set of expressions for the overall transmittance.
For 1 layer

$$\frac{1}{T_{o1}} = \frac{1}{T_1} \tag{9.56}$$

For 2 layers

$$\frac{1}{T_{o2}} = \frac{(1-R_{2i}R_{1o})}{T_2}\left(\frac{1}{T_{o1}}\right) \tag{9.57}$$

For 3 layers

$$\frac{1}{T_{o3}} = \frac{(1-R_{3i}R_{2o})}{T_3}\left(\frac{1}{T_{o2}}\right) - \frac{T_2R_{3i}R_{1o}}{T_3}\left(\frac{1}{T_{o1}}\right) \tag{9.58}$$

For 4 layers

$$\frac{1}{T_{o4}} = \frac{(1-R_{4i}R_{3o})}{T_4}\left(\frac{1}{T_{o3}}\right) - \frac{T_3R_{4i}R_{2o}}{T_4}\left(\frac{1}{T_{o2}}\right) - \frac{T_3T_2R_{4i}R_{1o}}{T_4}\left(\frac{1}{T_{o1}}\right) \tag{9.59}$$

For larger number of layers the expression for the overall transmittance could be written down by observing the form of the above expressions. Alternatively, a step-by-step solution procedure using Eqs. (9.47) to (9.55) may be developed to compute the overall transmittance.

9.4.3 *Radiation absorption in multi-layered fenestrations*

The rate of absorption of radiation per unit area, A_n in layer n is obtained by applying the energy balance equation to the layer. Hence we have

$$A_n = H_{(n+1)i} + H_{(n-1)o} - H_{no} - H_{ni} \tag{9.60}$$

Substituting from Eqs. (9.53) and (9.54) in Eq. (9.60) we obtain

$$A_n = (1 - T_n - R_{no})H_{(n+1)i} + (1 - T_n - R_{ni})H_{(n-1)o} \tag{9.61}$$

$$A_n = \alpha_{no}H_{(n+1)i} + \alpha_{ni}H_{(n-1)o} \tag{9.62}$$

where $\alpha_{no} = (1 - T_n - R_{no})$ and $\alpha_{ni} = (1 - T_n - R_{ni})$ (9.63)

are the *effective absorptivities* of layer n for radiation incident on the outer surface and the inner surface respectively.

The fraction of the radiation incident on the outer surface which is absorbed in the different layers, (A_i /Q_s), when a total of N layers are present, may be obtained by applying Eq. (9.62) to each layer of the fenestration system. The expressions for 4 layers are given below:

$$\frac{A_1}{Q_S} = \frac{\alpha_{1o}T_{oN}}{T_{o1}} \tag{9.64}$$

$$\frac{A_2}{Q_S} = \frac{\alpha_{2o}T_{oN}}{T_{o2}} + \frac{\alpha_{2i}R_{1o}T_{oN}}{T_{o1}} \tag{9.65}$$

$$\frac{A_3}{Q_S} = \frac{\alpha_{3o}T_{oN}}{T_{o3}} + \alpha_{3i}T_{oN}\left(\frac{R_{2o}}{T_{o2}} + \frac{R_{1o}T_2}{T_{o1}}\right) \tag{9.66}$$

$$\frac{A_4}{Q_S} = \frac{\alpha_{4o}T_{oN}}{T_{o4}} + \alpha_{4i}T_{oN}\left(\frac{R_{3o}}{T_{o3}} + \frac{R_{2o}T_3}{T_{o2}} + \frac{R_{1o}T_2T_3}{T_{o1}}\right) \tag{9.67}$$

We shall illustrate the application of the various expressions developed in this section to actual fenestration systems in the worked examples to follow in this chapter.

9.5 Overall Energy Transfer through Fenestrations

The energy transfer through a multi-layered fenestration system consists of several coupled processes.

(i) Energy entering at the outer surface due to convection and long-wave radiation is transferred by conduction through the fenestration layers, and by convection and radiation across the separating gas spaces. In section 8.2.5, we analyzed these heat transfer processes under steady conditions.

(ii) Beam solar radiation, sky-diffuse radiation, and ground-reflected diffuse radiation striking the external surface of the fenestration, is partially transmitted through the layers and partially absorbed in the different layers. The expressions for computing the rate of transmission and absorption of solar radiation were developed in section 9.4 above.

In this section we shall combine the above energy transfer processes to develop an expression for the overall energy transfer rate through a fenestration system. Since the fenestration layers are relatively thin, we shall neglect effects due to energy storage in the layers and assume that steady-state conditions prevail.

Solar radiation absorbed in a fenestration layer may be treated as an internal heat source. In section 2.6 we developed a thermal network to represent heat conduction through a slab with the presence of internal heat generation [7].

A three-layer fenestration system subjected to the aforementioned energy transfer processes is shown schematically in Fig. 9.8(a).

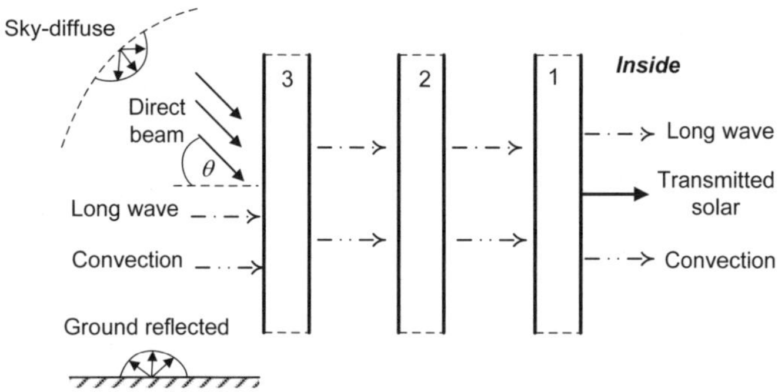

Fig. 9.8 (a) Energy transfer through fenestration system

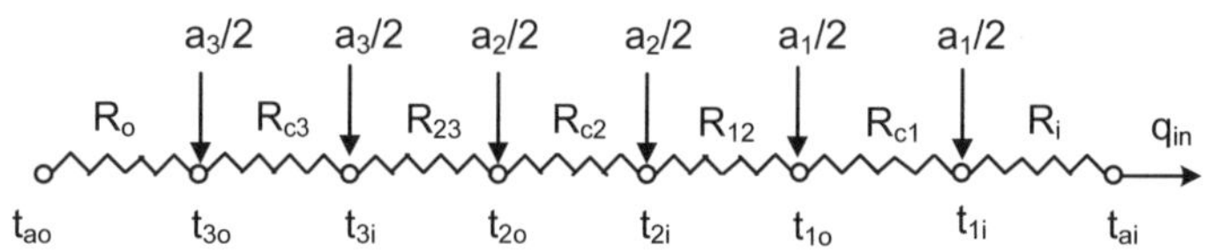

Fig. 9.8 (b) Thermal network for fenestration system

The equivalent thermal network for the fenestration system is depicted in Fig. 9.8(b). The internal heat generation rate in layer i (i =1,2 and 3) due to the absorption of solar radiation is a_i. We showed in section 2.8 that if the internal heat generation within the slab is assumed uniform, then an equivalent network with two concentrated heat sources at the two nodes of the layer may be used to represent the heat transfer through the layer [8]. The *unit* conduction resistance of layer i is given by

$$R_{ci} = \frac{L_i}{k_i}$$

where L_i is the thickness and k_i is the thermal conductivity of the layer.

The *unit* thermal resistances of the two gas spaces between the three layers, R_{12} and R_{23} are given by

$$R_{12} = \frac{1}{h_{c12}+h_{r12}} \quad \text{and} \quad R_{23} = \frac{1}{h_{c23}+h_{r23}} \tag{9.68}$$

The outside and inside air film *unit* thermal resistances are given by

$$R_o = \frac{1}{h_{co}+h_{ro}} \quad \text{and} \quad R_i = \frac{1}{h_{ci}+h_{ri}} \tag{9.69}$$

In Eqs. (9.68) and (9.69), h_c and h_r are the relevant convection and radiation heat transfer coefficients.

For each layer, i the inside and outside surface temperatures are denoted by the subscripts i and o respectively. The respective outside and inside air temperatures are denoted by t_{ao} and t_{ai}. The rate of heat flow into the space inside is q_{in}.

We now apply Ohm's law to the different sections of the thermal network in Fig. 9.8(b) to obtain the following equations:

$$t_{1i} - t_{ai} = q_{in}R_i$$

$$t_{1o} - t_{1i} = (q_{in} - \frac{a_1}{2})R_{c1}$$

$$t_{2i} - t_{1o} = (q_{in} - a_1)R_{12}$$

$$t_{2o} - t_{2i} = (q_{in} - a_1 - \frac{a_2}{2})R_{c2}$$

$$t_{3i} - t_{2o} = (q_{in} - a_1 - a_2)R_{23}$$

$$t_{3o} - t_{3i} = (q_{in} - a_1 - a_2 - \frac{a_3}{2})R_{c3}$$

$$t_{ao} - t_{3o} = (q_{in} - a_1 - a_2 - a_3)R_o$$

The overall energy balance equation is obtained by adding the above set of equations. Hence we have

$$\begin{aligned} t_{ao} - t_{ai} = q_{in}(R_i + R_{c1} + R_{12} + R_{c2} + R_{23} + R_{c3} + R_o) - \\ a_1\left(\frac{R_{c1}}{2} + R_{12} + R_{c2} + R_{23} + R_{c3} + R_o\right) - \\ a_2\left(\frac{R_{c2}}{2} + R_{23} + R_{c3} + R_o\right) - a_3\left(\frac{R_{c3}}{2} + R_o\right) \end{aligned} \tag{9.70}$$

For a fenestration with n layer, Eq. (9.70) may be generalized to the following form:

$$t_{ao} - t_{ai} = q_{in}R_{tot} - \sum_{i=1}^{n} a_i R_{i\to out} \tag{9.71}$$

where R_{tot} is the total thermal resistance from the inside air to the outside air and $R_{i\to out}$ is the thermal resistance from the middle section of the ith fenestration layer to the outside air.

We obtain the heat flow rate into the inside air space per unit area of the fenestration from Eq. (9.71) as

$$q_{in} = \frac{(t_{ao} - t_{ai})}{R_{tot}} + \sum_{i=1}^{n} a_i \left(\frac{R_{i \to out}}{R_{tot}}\right) \tag{9.72}$$

The *inward-flowing fraction* of the solar radiation absorbed in the ith layer is defined as

$$N_i = \left(\frac{R_{i \to out}}{R_{tot}}\right) \tag{9.73}$$

Now the total energy flow to the inside, q_{tot} is due to heat flow, q_{in} given by Eq. (9.72) and the solar radiation transmitted *directly* through fenestration. Hence we have

$$q_{tot} = TE_{dn}cos\theta + \frac{(t_{ao} - t_{ai})}{R_{tot}} + \sum_{i=1}^{n} a_i N_i \tag{9.74}$$

where T is the overall transmittance, E_{dn} is direct normal solar radiation intensity, and θ is the angle of incidence.

Solar radiation absorbed in layer i may be expressed as

$$a_i = f_{si}(\theta)E_{dn}cos\theta \tag{9.75}$$

where f_{si} is the fraction of beam radiation absorbed in the ith layer, which is a function of the angle of incidence.

Substituting from Eq. (9.75) in (9.74) we have

$$q_{tot} = E_{dn}cos\theta[T(\theta) + \sum_{i=1}^{n} f_{si}(\theta)N_i] + \frac{(t_{ao} - t_{ai})}{R_{tot}} \tag{9.76}$$

The quantity within the square bracket in Eq. (9.76) is called the *solar heat gain coefficient*, denoted by SHGC [1]. Therefore

$$SHGC(\theta) = T(\theta) + \sum_{i=1}^{n} f_{si}(\theta)N_i \tag{9.77}$$

Hence we can express the total energy flow rate as

$$q_{tot} = E_{dn}cos\theta SHGC(\theta) + \frac{(t_{ao} - t_{ai})}{R_{tot}} \tag{9.78}$$

In general, the reflectivity, the absorptivity and transmissivity of a fenestration layer depend on the wavelength of the incident radiation. Therefore the solar heat gain coefficient, given by Eq. (9.77), has to be averaged over the *solar radiation wavelength spectrum* to obtain its effective value at any angle of incidence.

For sky-diffuse radiation and ground-reflected radiation incident on the fenestration, we assume that the intensity of radiation is *independent* of the angle of incidence. Hence it is possible to obtain mean values of the solar heat gain coefficient for *diffuse radiation* by integrating Eq. (9.77) over the hemisphere viewed by the outer surface of the fenestration. This procedure is outlined in the *ASHRAE Handbook* - 2013 *Fundamentals* [1].

Solar radiation absorbed on opaque surfaces of fenestration like the frame, any mullion or dividers is partially transferred to the inner air. This heat transfer can be accounted for in the overall solar heat gain coefficient by using the area-weight-SHGC [1] given by

$$SHGC = \frac{(A_g SHGC_g + A_f SHGC_f + A_d SHGC_d)}{A_g + A_f + A_d} \tag{9.79}$$

where A is the area and the subscripts g, f and d denote the glass, the frame and the dividers respectively. The SHGC for the frame is given by

$$SHGC_f = \alpha_f \left(\frac{R_{fo}}{R_f}\right) \tag{9.80}$$

where α_f is the solar absorptivity of the outdoor surface of the frame, R_f is the total thermal resistance of the frame, and R_{fo} is the thermal resistance from the outdoor air film to the surface of the frame. A similar equation is applicable to the dividers.

The SHGC and several other solar-optical properties for a number of fenestration systems are given in Table 10 on page 15.19 of the *ASHRAE Handbook* - 2013 *Fundamentals* [1]. For purposes of illustration we have listed a few representative values in Table 9.2.

Table 9.2 Solar-optical properties of some glass fenestrations* [1]

Fenestration Properties		Angle of incidence, θ 0°	40°	60°	70°	Total window Al- frame fixed	SHGC at θ = 0° Al - frame operable
1-glazing	SHGC	0.81	0.8	0.73	0.62	0.74	0.74
L_g = 6 mm	T	0.77	0.75	0.68	0.58		
	A	0.16	0.17	0.19	0.19		
2 - glazings	SHGC	0.7	0.67	0.58	0.45	0.64	0.64
L_g = 6 mm	T	0.61	0.58	0.48	0.36		
gap=12.7	A_1	0.17	0.18	0.2	0.21		
	A_2	0.11	0.12	0.12	0.1		
3- glazings	SHGC	0.61	0.58	0.48	0.35	0.56	0.56
L_g = 6 mm	T	0.49	0.45	0.35	0.24		
gaps = 12.7	A_1	0.17	0.19	0.21	0.22		
	A_2	0.12	0.13	0.13	0.12		
	A_3	0.08	0.08	0.08	0.06		

*Values extracted from Table 10, Page 15.19, *ASHRAE Handbook* - 2013 *Fundamentals* [1]. T is the overall transmittance, and A_i is the absorbed fraction in layer i.

9.6 Shading of Surfaces from Solar Radiation

The cooling load of a building resulting from solar heat gain through fenestrations could be reduced by installing shading devices like overhangs, awnings, and louvers. Moreover, a window may be partially shaded if it is setback from the external surface of the wall. Shading devices intercept the direct beam from the sun before it reaches the transparent surface of the fenestration. The effectiveness of a shading device, usually defined as the fraction of the fenestration area that is shaded, varies with the position of the sun. In this section we shall develop a general computation procedure to determine the shaded area of a fenestration using a vector approach.

A *vertical* window of height H and width L, with a rectangular overhang projecting out a distance S from the window surface, is shown schematically in Fig. 9.9. The normal to the window surface makes an angle ψ with the south. The projection of the direct solar beam on the horizontal plane makes an angle ϕ with the south. The latter angle is called the solar azimuth angle. The solar altitude angle is β.

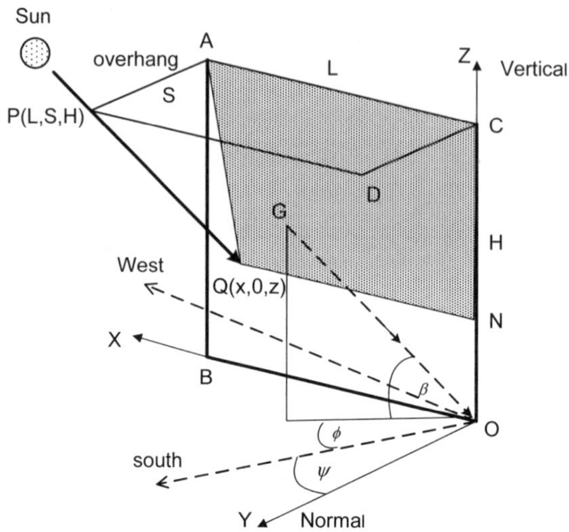

Fig. 9.9 Shading of window by an overhang

In order to represent the various vectors pertinent to the analysis we select a coordinate system x-y-z where the x-direction is along the bottom edge of the window, the y-direction is along the normal to the window and the z-direction is along the vertical edge of the window. The origin is located at the corner O of the window. The unit vector $\bar{I}_d$ in the direction of the direct solar beam, OG in the x-y-z system may be written as

$$\bar{I}_d = \begin{pmatrix} \cos\beta\sin\gamma \\ \cos\beta\cos\gamma \\ \sin\beta \end{pmatrix} \tag{9.81}$$

where $\gamma = (\psi + \phi)$ is the surface–solar azimuth angle.

The edge P of the overhang, with coordinates (L,S,H), casts its shadow at point Q on the window surface with coordinates $(x,0,z)$. Now from the vector triangle OPQ we have

$$\overrightarrow{OQ} + \overrightarrow{QP} = \overrightarrow{OP} \tag{9.82}$$

Expressing Eq. (9.82) in terms of the coordinates of points O, P and Q we obtain

$$\begin{pmatrix} x \\ 0 \\ z \end{pmatrix} + \sigma \bar{I}_d = \begin{pmatrix} L \\ S \\ H \end{pmatrix} \tag{9.83}$$

where σ is the length of the vector QP.

Substituting for $\bar{I}_d$ from Eq. (9.81) in Eq. (9.83) we have

$$\begin{pmatrix} x \\ 0 \\ z \end{pmatrix} + \sigma \begin{pmatrix} cos\beta sin\gamma \\ cos\beta cos\gamma \\ sin\beta \end{pmatrix} = \begin{pmatrix} L \\ S \\ H \end{pmatrix} \tag{9.84}$$

Equating the y-coordinate on both sides of Eq. (9.84) we obtain

$$\sigma = \frac{S}{cos\beta cos\gamma} \tag{9.85}$$

Equating the x and z coordinate on both sides of Eq. (9.84) and substituting for σ from Eq. (9.85) we have

$$x = L - Stan\gamma \tag{9.86}$$

$$z = H - \frac{Stan\beta}{cos\gamma} \tag{9.87}$$

We notice that in Fig. 9.9, the shadow of edge AP of the overhang is formed along the line AQ on the window surface. Similarly, the shadow of the edge PD is formed along the line QN. Therefore the shaded area of the window is the trapezium AQNC and its area may be expressed in the form

$$A_{sh} = \frac{(L+x)(H-z)}{2} \tag{9.88}$$

The vector approach presented above may be applied to find the shaded area of any other shape of overhang by locating the shadows of the points on the edges of the overhang on the window surface using the vector equation (9.84). Applications are considered in worked examples 9.14 and 9.15.

9.7 Worked Examples

Example 9.1 The longitude of a location is 95°W and the standard longitude of the time zone is 90°W. Calculate the solar time and the hour angle at the location when the standard clock time is 3.30 pm on July 14. The clock time has been advanced by one hour for daylight saving.

Solution The solar time in minutes is given by Eq. (9.4) as

$$T_{solar} = T_{std} + 4(L_{std} - L_{loc}) + E_{time} - DT \qquad \text{(E9.1.1)}$$

Now T_{std} = 3.30 pm = 15.5 hrs., DT = 1, L_{std} = 90°W and L_{loc} = 95°W. The equation of time is given by Eq. (9.5) as

$$E_{time} = 9.87sin2B - 7.53cosB - 1.5sinB$$

where
$$B = \frac{360(N_{day}-81)}{364}$$

For July 14, N_{day} = 195. Substituting in the above equations we obtain B as 112.75 and E_{time} = −5.51 min.
Substituting the above data in Eq. (E9.1.1) we have

$$T_{solar} = 15.5 \times 60 + 4(90 - 95) - 5.51 - 60 = 844.5 \text{ min}$$

Therefore the local solar time is 14.08 hr.
The hour angle H is given by Eq. (9.3) as

$$H = 15(solar\ time - 12) = 31.1^0$$

Example 9.2 (a) Calculate the solar altitude angle and the solar azimuth angle at 9 hr. solar time on August 10 for a location with northern latitude of 40°. (b) Calculate the solar time at sunrise and sunset on August 10 at the same location.

Solution (a) The hour angle is given by Eq. (9.3) as

$$H = 15(T_{sol} - 12) = -15 \times 3 = -45^0$$

For August 10, N_{day} = 222. The declination is given by Eq. (9.2) as

$$\delta = 23.45sin\left[\frac{360(284+N_d)}{365}\right] = 15.36^0$$

The solar altitude angle is given by Eq. (9.7) as

$$sin\beta = cosLcosHcos\delta + sinLsin\delta$$

$$sin\beta = cos40\cos(-45)\,cos15.36 + sin40sin15.36 = 0.6926$$

Therefore the solar altitude angle is 43.8 degrees. The solar azimuth angle is given by Eq. (9.8) as

$$sin\phi = \frac{cos\delta sinH}{cos\beta} = \frac{cos15.36 \times \sin(-45)}{cos43.8} = -0.9447$$

Therefore the solar azimuth is 70.85 degrees east of south.

(b) At sunrise the solar altitude angle, β is zero. Hence we have from Eq. (9.7)

$$cosH_{sr} = -tanLtan\delta = -0.2306$$

Therefore $H_{sr} = -103.3$ degrees. The solar time at sunrise T_{sr} is given by Eq. (9.3) as

$$H_{sr} = 15(T_{sr} - 12) = -103.3 \text{ degrees}$$

Hence $T_{sr} = 5.07$ a.m. solar time. Now the hour angles at sunrise and sunset are symmetrical about solar noon. Hence the solar time at sunset is 6.53 p.m.

Example 9.3 Calculate the angle of incidence of the direct solar beam, at 2.00 p.m solar time on August 5, at a location 38° northern latitude on the following surfaces: (i) a window surface facing 30° east of south, and tilted 50° from the horizontal, (ii) a vertical wall facing 30° east of south, and (iii) a flat roof.

Solution The hour angle is given by Eq. (9.3) as

$$H = 15(T_{sol} - 12) = 15 \times 2 = 30 \text{ degrees}$$

For August 5, $N_{day} = 217$. The declination is given by Eq. (9.2) as

$$\delta = 23.45sin\left[\frac{360(217+284)}{365}\right] = 16.83 \text{ degrees}$$

The solar altitude angle is given by Eq. (9.7) as

$$sin\beta = cosLcosHcos\delta + sinLsin\delta$$

$$sin\beta = cos38 \cos 30\, cos16.83 + sin38 sin16.83 = 0.831$$

Therefore the solar altitude angle is 56.25 degrees.
The solar azimuth angle is given by Eq. (9.8) as

$$sin\phi = \frac{cos\delta sinH}{cos\beta} = \frac{cos16.83 \times \sin 30}{cos56.25} = 0.861$$

Therefore solar azimuth angle is 59.48 degrees west of south.
The surface–solar azimuth is

$$\gamma = 30 + 59.48 = 89.48 \text{ degrees.}$$

The tilt angle of the window is, $\Sigma = 50°$. The angle of incidence, θ is given by Eq. (9.13) as

$$cos\theta = \ cos\beta sin\Sigma\, cos\gamma + sin\beta cos\Sigma$$

$$cos\theta = \ cos56.25sin50cos89.48 + sin56.25cos50 = 0.538$$

Hence the angle of incidence on the window surface is 57.4°.

(ii) For a vertical wall, $\Sigma = 90°$. Therefore the angle of incidence is

$$cos\theta = \ cos56.25sin90cos89.48 + sin56.25cos90 = 0.00504$$

Hence the angle of incidence on a vertical wall surface is 89.7°.

(iii) For a horizontal roof, $\Sigma = 0°$. Therefore the angle of incidence is

$$cos\theta = \ cos56.25sin0cos89.48 + sin56.25cos0 = 0.831$$

Hence the angle of incidence on the roof is 33.75°.

Example 9.4 A vertical south facing surface is located at the outer edge of the atmosphere where the latitude is 40°. Calculate the total solar energy falling on the surface per unit area from 8a.m. to 4p.m. solar time, on 28 May.

Solution For May 28, $N_{day} = 148$. The declination is given by Eq. (9.2) as

$$\delta = 23.45sin\left[\frac{360(148+284)}{365}\right] = 21.44 \text{ degrees}$$

The extraterrestrial solar radiation incident on a surface normal to the sun's rays on May 28 is given by Eq. (9.1) as

$$E_o = E_{sc}\left\{1 + 0.033cos\left[360^0\,\frac{(148-3)}{365}\right]\right\}$$

where $E_{sc} = 1367\ Wm^{-2}$ and n is the day of the year.
Hence solar radiation intensity, $E_o = 1330.9\ Wm^{-2}$.

The hour angles at 8 a.m and 4 p.m. solar time are −60° and 60° respectively. Now the rate of incidence of extraterrestrial radiation on the surface per unit area is given by

$$q = E_o cos\theta$$

where θ is the angle of incidence.
The total energy incident on the surface from 8 a.m. to 4 p.m. is given by

$$Q = \int_{8am}^{4pm} qdT = \int_{8am}^{4pm} E_o\ cos\theta dT \qquad \text{(E9.4.1)}$$

where T is solar time in seconds.

For a surface facing south, the surface–solar azimuth, γ is equal to ϕ, the solar azimuth. Substituting for the incidence θ from Eq. (9.13) in Eq. (E9.4.1) we have

$$Q = \int_{8am}^{4pm} E_o\ cos\beta cos\phi dT \qquad \text{(E9.4.2)}$$

Substituting in Eq. (E9.4.2) from Eq. (9.9) we obtain

$$Q = \int_{8am}^{4pm} E_o\ (cos\delta sinLcosH - sin\delta cosL)dT \qquad \text{(E9.4.3)}$$

Substituting the pertinent numerical values in Eq. (E9.4.3) we have

$$Q = \int_{8am}^{4pm} 1330.9\ (cos21.44sin40cosH - sin21.44cos40)dT$$

$$Q = \int_{8am}^{4pm} 1330.9\ (0.598cosH - 0.28)dT \qquad \text{(E9.4.4)}$$

The change in hour angle, dH in radians is related to the change in solar time dT, in seconds by the equation

$$dH = \frac{15\pi dT}{180 \times 3600}$$

Substituting for dT in Eq. (E9.4.4) we obtain the total energy as

$$Q = \left(\frac{180 \times 3600 \times 1330.9}{15 \times \pi}\right) \int_{-\pi/3}^{\pi/3} (0.598cosH - 0.28)\ dH$$

$$Q = \left(\frac{180 \times 3600 \times 1330.9}{15 \times \pi}\right) \times 0.449 = 8.22 \times 10^3\ kJm^{-2}$$

The total solar energy incident per unit area is 8.22 MJm^{-2}.

Example 9.5 Calculate the clear-sky beam and diffuse radiation intensities in Toronto, Canada, for July 31 at 3.00 p.m. solar time. The latitude of Toronto is 43.63 degrees north.

Solution We calculate the solar altitude angle at 3 p.m. on July 31 using the procedure outlined in Example 9.3. Hence we obtain the following quantities:

latitude, $L = 43.63°$; declination, $\delta = 18.17°$, Hour angle, $H = 45°$.

The solar altitude is given by Eq. (9.7) as

$$sin\beta = cos43.63 \cos 45\, cos18.17 + sin43.63sin18.17 = 0.7015$$

Hence the solar altitude angle is 44.55°.
The solar azimuth angle is given by Eq. (9.8) as

$$sin\phi = \frac{cos18.17 \times \sin 45}{cos44.55} = 0.943$$

Hence the azimuth angle is 70.56°.

The air mass is given by Eq. (9.24) as

$$m = [sin\beta + 0.50572(6.07995 + \beta)^{-1.6364}]^{-1}$$

$$m = [sin44.55 + 0.50572(6.07995 + 44.55)^{-1.6364}]^{-1} = 1.4238$$

The extraterrestrial solar radiation incident on a surface normal to the sun's rays is given by Eq. (9.1) as

$$E_o = E_{sc}\left\{1 + 0.033cos\left[360^0 \frac{(n-3)}{365}\right]\right\}$$

where E_{sc} = 1367 Wm^{-2}, and n is the day of the year. Substituting numerical values in the above equation we have

$$E_o = 1367\left\{1 + 0.033cos\left[360^0 \frac{(212-3)}{365}\right]\right\} = 1326.5 \text{ Wm}^{-2}$$

The beam and diffuse radiation optical depths for July 31 have been obtained from the data CD accompanying Ref. [1]. This gives τ_b =0.386 and τ_d =2.282. Substituting these values in Eqs. (9.25) and (9.26) we have

$$ab = 1.454 - 0.406\tau_b - 0.268\tau_d + 0.021\tau_b\tau_d = 0.704$$

$$ad = 0.507 + 0.205\tau_b - 0.080\tau_d - 0.190\tau_b\tau_d = 0.236$$

The intensities of beam and diffuse radiation are given by Eqs. (9.22) and (9.23) respectively. Substituting the relevant numerical values in the above equations we obtain

$$E_b = E_o exp[-\tau_b m^{ab}] = 1326.5 exp[-0.386 \times 1.4238^{0.704}]$$

$$E_d = E_o exp[-\tau_d m^{ad}] = 1326.5 exp[-2.282 \times 1.4238^{0.236}]$$

Hence the intensity of beam radiation is 808.6 Wm^{-2} and the intensity of diffuse radiation is 111.0 Wm^{-2}.

Example 9.6 A vertical window of area 4 m^2 in Toronto, Canada, faces 50° west of south. The latitude of Toronto is 43.63 degrees north. The ground reflectivity at the location is 0.2. For July 31, at 3.00 p.m. solar time, calculate (i) direct beam solar radiation, (ii) sky-diffuse radiation, and (iii) ground-reflected radiation incident on the window surface. Use the clear-sky radiation model.

Solution Now the conditions for this example are the same as those for example 9.5, where we obtained the solar altitude angle, β = 44.55° and the solar azimuth angle ϕ = 70.56°. Hence the surface–solar azimuth angle is

$$\gamma = 70.56 - 50 = 20.56 \text{ degrees.}$$

The angle of incidence of the direct beam on the vertical window surface is given by Eq. (9.13) as

$$cos\theta = cos 44.55 sin 90 cos 20.56 + sin 44.556 cos 90 = 0.667$$

Hence the angle of incidence is 48.15°.

For the conditions in example 9.5, we obtained the beam and diffuse radiation intensities using the clear-sky model as, E_b = 808.6 Wm^{-2} and E_d = 111.0 Wm^{-2} respectively.

(i) The direct radiation incident on the window surface is given by Eq. (9.27) as

$$H_{db} = AE_b cos\theta = 4 \times 808.6 cos 48.15 = 2158.0 \text{ W}$$

(ii) The diffuse radiation incident on the surface is given by Eq. (9.28) as

$$H_{dif} = AE_d Y$$

where Y is a function of the angle of incidence θ of the direct beam. It is given by the expression in Eq. (9.29) as

$$Y = max.\left[0.45, (0.55 + 0.437cos\theta + 0.313cos^2\theta)\right]$$

Substituting the pertinent numerical data in the above expressions we have

$$Y = max[0.45, (0.55 + 0.437cos48.15 + 0.313cos^2 48.15)] = 0.981$$

$$H_{dif} = AE_d Y = 4 \times 111.0 \times 0.981 = 435.6 \text{ W}$$

(iii) The ground-reflected radiation falling on a surface is given by Eq. (9.32) as

$$H_{gr} = \frac{A(E_d + E_b sin\beta)\rho_{gr}(1 - cos\Sigma)}{2}$$

$$H_{gr} = 4(111.0 + 808.6sin44.55) \times \frac{0.2(1 - cos90)}{2}$$

Hence H_{gr} = 271.3 W.

Example 9.7 A south-facing glass window of a building at a location with a northern latitude of 30° is inclined at an angle of 75° to the horizontal. (i) Calculate the angle of incidence of the direct solar beam at 2.00 p.m. solar time on 18 June. (i) Determine the number of hours on 18 June during which the glass window is sunlit.

Solution For June 18, N_{day} = 169. The declination is given by Eq. (9.2) as

$$\delta = 23.45sin\left[\frac{360(169+284)}{365}\right] = 23.41 \text{ degrees}$$

The latitude is 30° and the hour angle at 2 p.m. solar time is 30°. The solar altitude angle is given by Eq. (9.7) as

$$sin\beta = cosLcosHcos\delta + sinLsin\delta$$

$$sin\beta = cos30\cos 30\, cos23.41 + sin30sin23.41 = 0.8887$$

Therefore the solar altitude angle is 62.49 degrees.
The solar azimuth angle is given by Eq. (9.8) as

$$sin\phi = \frac{cos\delta sinH}{cos\beta} = \frac{cos23.41\times\sin 30}{cos62.49} = 0.993$$

Therefore solar azimuth is 83.4 degrees west of south.
The surface-solar azimuth angle is

$$\gamma = 0 + 83.4 = 83.4 \text{ degrees.}$$

The angle of incidence, θ is given by Eq. (9.13) as

$$cos\theta = \ cos\beta sin\Sigma\, cos\gamma + sin\beta cos\Sigma$$

where Σ is the tilt angle of the glass wall.

$$cos\theta = \ cos62.49sin75cos83.4 + sin62.49cos75 = 0.281$$

Hence the angle of incidence is 73.69°.

(ii) Now the surface–solar azimuth angle for a south-facing wall is ϕ. When the direct beam *just strikes* the wall, $\theta = 90°$. Therefore we have

$$cos\theta = \ cos\beta sin\Sigma\, cos\phi + sin\beta cos\Sigma\ = 0$$

Substituting from Eqs. (9.9) and (9.7) in the above equation we have

$$\sin\Sigma\,(cos\delta sinLcosH - sin\delta cosL)$$
$$= -cos\Sigma\,(cosLcosHcos\delta + sinLsin\delta)$$

The above equation may be rearranged to the form

$$cosH = tan\delta tan\big(\Sigma\ - L\big)$$

Substituting the relevant numerical values in the above equation we have

$$cosH = tan23.41tan(75 - 30) = 0.4329$$

The hour angle at which the beam just strikes the wall is H= 64.34°, which corresponds to 4.29 hours before solar noon. Hence the total number of hours during which the wall is sunlit is 8.58 hours. The sunlit period is therefore from 7.43 a.m. to 4.17 p.m. solar time.

Example 9.8 The measured direct beam and diffuse solar radiation intensities on June 10 at 1 p.m. solar time at a location with a northern latitude of 35° are 620 Wm^{-2} and 182 Wm^{-2} respectively. The ambient temperature is 26°C. For a flat roof with an average emissivity of 0.9 at the location, calculate (i) the sol-air temperature, and (ii) the surface temperature. The external heat transfer coefficient is 32 Wm^{-2}K^{-1}. Assume that the roof is well insulated so that the heat flow rate into the building is negligible. Neglect the heat capacity of the roof.

Solution For June 10, N_{day} = 161. The declination is given by Eq. (9.2) as.

$$\delta = 23.45 sin\left[\frac{360(161+284)}{365}\right] = 23.01 \text{ degrees}$$

The latitude is 35° and the hour angle is 15°.
The solar altitude angle is given by Eq. (9.7) as

$$sin\beta = cosLcosHcos\delta + sinLsin\delta$$

$$sin\beta = cos35\cos 15\, cos23.01 + sin35sin23.01 = 0.9525$$

Therefore the solar altitude angle is 72.27 degrees.
The solar radiation falling on unit area of the horizontal roof is given by

$$H_{dif} = (E_d + E_b sin\beta) = 182 + 620sin72.27 = 772.54 \text{ Wm}^{-2}$$

The sol-air temperature is given by Eq.(9.39) as

$$T_{sa} = T_a + \frac{q_{sol}}{h_o} - \frac{\varepsilon_S \Delta R}{h_o} = 26 + \frac{0.9\times 772.54}{32} - 4 = 43.7°\text{C}$$

Note that the correction factor in the above equation for a horizontal surface is taken as 4°C [3].

Since the thermal capacity of the roof is negligible and the roof is well insulated the net heat flow rate into the inside air is zero. Therefore

$$q_{con} = h_o(T_{sa} - T_s) = 32\left(43.72 - T_{roof}\right) = 0$$

Hence temperature of the roof is 43.7°C.

Example 9.9 A thin vertical metal wall of a building at a location with a northern latitude of 40° faces 35° east of south. The measured direct beam and diffuse solar radiation intensities at the location, on August 15 at 10 a.m. solar time are 580 Wm^{-2} and 148 Wm^{-2} respectively. The average emissivity of the wall surface is 0.85, and the reflectivity of the ground surrounding the wall is 0.3. The ambient temperature and the inside air temperature are 28°C and 23°C respectively. The overall external and internal heat transfer coefficients are 35 $Wm^{-2}K^{-1}$ and 8.5 $Wm^{-2}K^{-1}$ respectively. Calculate (i) sol-air temperature, and (ii) the temperature of the wall. Assume that the heat capacity and the thermal resistance of the wall are negligible.

Solution For August 15, N_{day} = 227. The declination is given by Eq. (9.2) as

$$\delta = 23.45 sin\left[\frac{360(227+284)}{365}\right] = 13.78 \text{ degrees}$$

The latitude is 40° and the hour angle is −30°.
The solar altitude angle is given by Eq. (9.7) as

$$sin\beta = cosLcosHcos\delta + sinLsin\delta$$

$$sin\beta = cos40\cos(-30)\, cos13.78 + sin40sin13.78 = 0.7974$$

Therefore the solar altitude angle is 52.88 degrees.
The solar azimuth angle is given by Eq. (9.8) as

$$sin\phi = \frac{cos\delta sinH}{cos\beta} = \frac{cos13.78 \times \sin(-30)}{cos52.88} = -0.8046$$

The solar azimuth angle is −53.58°.
The surface–solar azimuth of the wall is

$$\gamma = 53.58 - 35 = 18.58 \text{ degrees.}$$

The angle of incidence, θ is given by Eq.(9.13) as

$$cos\theta = \ cos\beta sin\Sigma\, cos\gamma + sin\beta cos\Sigma$$

where Σ is the tilt angle of the wall.

$$cos\theta = \ cos52.88sin90cos18.58 + sin52.88cos90 = 0.572$$

Hence the angle of incidence is 55.1°.
The given radiation intensities are: E_b = 580 Wm^{-2} and E_d = 148 Wm^{-2}.

(i) The direct radiation incident on unit area of the wall surface is

$$H_{db} = E_b cos\theta = 580cos55.1 = 331.8 \text{ Wm}^{-2}$$

(ii) The diffuse radiation incident on unit area is given by Eq. (9.28) as

$$H_{dif} = E_d Y$$

where Y is a function of the angle of incidence θ of the direct beam. It is given by the expression in Eq. (9.29) as

$$Y = 0.55 + 0.437cos\theta + 0.313cos^2 \theta$$

Substituting the pertinent numerical data in the above expressions we have

$$Y = 0.55 + 0.437cos55.1 + 0.313cos^2 (55.1) = 0.9025$$

$$H_{dif} = E_d Y = 148 \times 0.9025 = 133.57 \text{ Wm}^{-2}$$

(iii) The ground-reflected radiation falling on unit area of surface is given by Eq. (9.32) as

$$H_{gr} = \frac{(E_d + E_b sin\beta)\rho_{gr}(1 - cos\Sigma)}{2}$$

$$H_{gr} = (133.57 + 580sin52.88) \times \frac{0.3(1 - cos90)}{2}$$

Hence $H_{gr} = 89.4 \text{ Wm}^{-2}$.

Total solar radiation incident on unit area of the wall is

$$H_{tot} = 331.8 + 133.57 + 89.4 = 554.8 \text{ Wm}^{-2}$$

The sol-air temperature is given by Eq. (9.39) as

$$T_{sa} = T_a + \frac{q_{sol}}{h_o} - \frac{\varepsilon_s \Delta R}{h_o} = 28 + \frac{0.85 \times 554.8}{35} = 41.47°\text{C}$$

Note that the correction factor for a vertical surface in the above equation is taken as zero [3].

Since the thermal capacity and the thermal resistance of the wall are negligible, the net heat flow rate may be expressed as

$$q_{con} = h_o(T_{sa} - T_w) = h_i(T_w - T_{in})$$

$$35(41.47 - T_w) = 8.5(T_w - 23)$$

Hence temperature of the wall is 37.86°C.

Example 9.10 The refractive index and extinction coefficient of a double-strength glass sheet of thickness 3.2 mm are 1.526 and 20 m^{-1} respectively. (a) Calculate the effective transmissivity, the reflectivity, and the absorptivity of the glass sheet for solar radiation at normal incidence. (b) Calculate the overall transmissivity, reflectivity, and the fraction of energy absorbed in each layer of a double-glazed window made of two of the above glass sheets.

Solution The surface reflectivity at normal incidence is given in Ref. [3] as

$$r = \left(\frac{n-1}{n+1}\right)^2 = \left(\frac{0.526}{2.526}\right)^2 = 0.0434$$

The absorptivity of the glass layer is a function of the extinction coefficient, K and the thickness L of the layer [3]. The intrinsic transmittance of the glass layer is given by

$$\tau = exp(-KL) = exp(-20 \times 3.2 \times 10^{-3}) = 0.938$$

The reflectivities of the two surfaces of the glass sheet are equal. Therefore $r_o = r_i = r = 0.0434$. The effective transmittance of the glass sheet is given by Eq. (9.44) as

$$T = \frac{\tau(1-r_o)(1-r_i)}{1-\tau^2 r_o r_i} = \frac{0.938\times(1-0.0434)^2}{1-0.938^2\times 0.0434^2} = 0.86$$

The effective reflectivity of the glass sheet is given by Eq. (9.45) as

$$R = r_o + \frac{\tau r_i (1-r_o)^2}{1-\tau^2 r_o r_i} = 0.0434 + \frac{0.0434\times 0.938\times(1-0.0434)^2}{1-0.938^2\times 0.434\times 0.0434} = 0.080$$

The effective absorptivity is given by

$$A = 1 - T - R = 1 - 0.86 - 0.0807 = 0.0593$$

(b) Now consider a double-glazed window consisting of two glass sheets. We use Eqs. (9.56) and (9.57) to determine the overall transmittance. Hence we have

$$\frac{1}{T_{01}} = \frac{1}{T_1} = \frac{1}{0.8598}$$

$$\frac{1}{T_{o2}} = \frac{(1-R_{2i}R_{1o})}{T_2}\left(\frac{1}{T_{o1}}\right) = \frac{(1-0.0807\times0.0807)}{0.8598\times0.8598} = 1.3439$$

Therefore the overall transmittance of two glass sheets is, $T_{o2} = 0.744$.

The fractions of the energy absorbed in the two layers are given by Eqs. (9.64) and (9.65). Hence we have

$$\frac{A_1}{Q_S} = \frac{\alpha_{1o}T_{ON}}{T_{o1}} = \frac{0.0593\times0.744}{0.8598} = 0.0513$$

$$\frac{A_2}{Q_S} = \frac{\alpha_{2o}T_{ON}}{T_{o2}} + \frac{\alpha_{2i}R_{1o}T_{ON}}{T_{o1}} = 0.0593 + \frac{0.0593\times0.0807\times0.744}{0.8598} = 0.0634$$

The overall reflectivity of two glass sheets is given by

$$R_{o2} = 1 - (0.744 + 0.0513 + 0.0634) = 0.1413$$

Example 9.11 Direct beam solar radiation, of intensity 590 $\mathrm{Wm^{-2}}$, is incident on a triple-glazed window of area 3.5 $\mathrm{m^2}$ at an angle of 50°. The spectrally-averaged properties of the three glass panes of the window for solar radiation incident at 50° are listed in Table E9.11.1.

Table E9.11.1 Averaged properties of glass panes at 50°

Pane	Thickness	Transmittance	Inner reflectance	Outer reflectance
Outer	6mm	0.73	0.09	0.09
Middle	3mm	0.80	0.10	0.10
Inner	3mm	0.59	0.08	0.08

Calculate (i) the rate of transfer of beam radiation energy through the window, and (ii) the rate of absorption of beam radiation in each glass pane.

Solution The transmittance of a triple-glazed window system is given by the set of Eqs. (9.56) to (9.58) as

$$\frac{1}{T_{o1}} = \frac{1}{T_1}$$

$$\frac{1}{T_{o2}} = \frac{(1-R_{2i}R_{1o})}{T_2}\left(\frac{1}{T_{o1}}\right)$$

$$\frac{1}{T_{o3}} = \frac{(1-R_{3i}R_{2o})}{T_3}\left(\frac{1}{T_{o2}}\right) - \frac{T_2R_{3i}R_{1o}}{T_3}\left(\frac{1}{T_{o1}}\right)$$

From the given property data in Table E9.11.1 we obtain the following quantities applicable to the set of equations above:

$$T_1 = 0.59, \quad R_{1i} = R_{1o} = 0.08, \quad \alpha_{1i} = \alpha_{1o} = 0.33$$

$$T_2 = 0.80, \quad R_{2i} = R_{2o} = 0.10, \quad \alpha_{2i} = \alpha_{2o} = 0.10$$

$$T_3 = 0.73, \quad R_{3i} = R_{3o} = 0.09, \quad \alpha_{3i} = \alpha_{3o} = 0.18$$

Substituting these values in the above equations we obtain

$$\frac{1}{T_{o1}} = \frac{1}{T_1} = \frac{1}{0.59} = 1.6949$$

$$\frac{1}{T_{o2}} = \frac{(1-0.1\times0.08)}{0.8}\left(\frac{1}{0.59}\right) = 2.1017$$

$$\frac{1}{T_{o3}} = \frac{(1-0.09\times0.1)}{0.73}(2.1017) - \frac{0.8\times0.09\times0.08}{0.73}(1.6949) = 2.8397$$

Hence the overall transmittance of the three layers is $T_{o3} = 0.352$.

The fraction of energy absorbed in each layer is given by Eqs. (9.64) to (9.66) as

$$\frac{A_1}{Q_s} = \frac{\alpha_{1o}T_{o3}}{T_{o1}}$$

$$\frac{A_2}{Q_s} = \frac{\alpha_{2o}T_{o3}}{T_{o2}} + \frac{\alpha_{2i}R_{1o}T_{o3}}{T_{o1}}$$

$$\frac{A_3}{Q_s} = \frac{\alpha_{3o}T_{o3}}{T_{o3}} + \alpha_{3i}T_{o3}\left(\frac{R_{2o}}{T_{o2}} + \frac{R_{1o}T_2}{T_{o1}}\right)$$

Substituting pertinent numerical data in the above equations we have

$$\frac{A_1}{Q_s} = \frac{0.33\times0.352}{0.59} = 0.1969$$

$$\frac{A_2}{Q_s} = \frac{0.1\times0.352}{0.4758} + \frac{0.1\times0.08\times0.352}{0.59} = 0.07875$$

$$\frac{A_3}{Q_s} = \frac{0.18\times0.352}{0.352} + 0.18\times0.352\left(\frac{0.1}{0.4758} + \frac{0.08\times0.8}{0.59}\right) = 0.200$$

The rate of transfer of direct beam radiation is given by

$$Q_{tr} = A_w I_d T_{o3} cos\theta = 3.5\times590\times0.352\times cos50 = 467.2 \text{ W}$$

The radiation absorbed in the three glass layers are given by

$$Q_{ab1} = A_w A_1 I_d cos\theta = 3.5\times590\times0.1969\times cos50 = 261.4 \text{ W}$$

$$Q_{ab2} = A_w A_2 I_d cos\theta = 3.5\times590\times0.07875\times cos50 = 104.5 \text{ W}$$

$$Q_{ab3} = A_w A_3 I_d cos\theta = 3.5 \times 590 \times 0.2 \times cos50 = 265.47 \text{ W}$$

Example 9.12 A double-glazed window consists of two glass sheets of thickness 3.5 mm and thermal conductivity 0.85 $Wm^{-1}K^{-1}$. At 11 a.m. on a certain day, direct beam radiation of intensity 580 Wm^{-2} is incident at an angle of 62° on the window. The fractions of the incident direct beam radiation absorbed in the inner and outer glass sheets are 0.052 and 0.065 respectively. The overall transmittance of the window is 0.76. The overall heat transfer coefficients for the outside air film, the air gap between the glasses, and the inside air film are respectively 30 $Wm^{-2}K^{-1}$, 7.2 $Wm^{-2}K^{-1}$ and 8.2 $Wm^{-2}K^{-1}$. The inside and outside air temperatures are 20°C and 33°C respectively. Calculate (i) solar heat gain coefficient (SHGC) for the direct beam at 11 a.m., and (ii) the total rate of energy transfer into the inside space.

Solution Let subscripts 1 and 2 denote the inner and outer glazing respectively. We first compute the following thermal resistances using the given data:

$$R_i = \frac{1}{h_i} = \frac{1}{8.2} = 0.122, \quad R_{12} = \frac{1}{h_{12}} = \frac{1}{7.2} = 0.139,$$

$$R_{g1} = \frac{L_{g1}}{k_g} = \frac{3.5\times10^{-3}}{0.85} = 0.0041, \quad R_{g2} = \frac{L_{g2}}{k_g} = \frac{3.5\times10^{-3}}{0.85} = 0.0041,$$

$$R_o = \frac{1}{h_o} = \frac{1}{30} = 0.033$$

Total thermal resistance is

$$R_{tot} = 0.122 + 0.0041 + 0.139 + 0.0041 + 0.033 = 0.3022$$

The solar heat gain coefficient is given by Eq. (9.77) as

$$SHGC(\theta) = T(\theta) + \sum_{i=1}^{n} f_{si}(\theta) N_i$$

The overall transmittance, T is 0.76. The fractions of beam radiation absorbed in the inner and outer glass sheets are: f_{s1} = 0.052 and f_{s2} = 0.065. The thermal resistance ratios are given by Eq. (9.73) as

$$N_i = \left(\frac{R_{i\to out}}{R_{tot}}\right)$$

Applying the above equation to the two glazings we have (see Fig. 9.8b)

$$N_1 = \left(\frac{R_{1\to out}}{R_{tot}}\right) = \frac{0.5\times 0.0041+0.139+0.0041+0.033}{0.3022} = 0.5895$$

$$N_2 = \left(\frac{R_{2\to out}}{R_{tot}}\right) = \frac{0.5\times 0.0041+0.033}{0.3022} = 0.116$$

Hence the solar heat gain coefficient (SHGC) is obtained as

$$SHGC(62^o) = 0.76 + 0.052 \times 0.5895 + 0.065 \times 0.116 = 0.798$$

The total energy flow rate into the inner space is given by Eq. (9.78) as

$$q_{tot} = E_{dn} cos\theta SHGC(\theta) + \frac{(t_{ao}-t_{ai})}{R_{tot}}$$

$$q_{tot} = 580cos62 \times 0.798 + \frac{(33-20)}{0.3022} = 260.3 \text{ Wm}^{-2}$$

Example 9.13 Consider the triple-glazed window described in example 9.11. The following overall heat transfer coefficients have been estimated using the procedures developed in chapter 8:

h_i = 8.5 $\text{Wm}^{-2}\text{K}^{-1}$, h_{12} = 5.5 $\text{Wm}^{-2}\text{K}^{-1}$, h_{23} = 6.0 $\text{Wm}^{-2}\text{K}^{-1}$, h_o = 30 $\text{Wm}^{2}\text{K}^{-1}$

where subscript 1 denotes the innermost glazing. The thermal conductivity of glass is 0.8 $\text{Wm}^{-1}\text{K}^{-1}$. The inside and outside air temperatures are 23°C and 32°C respectively. For the conditions stated in worked example 9.11, calculate (i) the solar heat gain coefficient (SHGC) for direct beam radiation, and (ii) the total energy transfer rate per unit area into the inside space.

Solution Let subscripts 1, 2 and 3 denote the inner, the middle, and the outer glazing respectively. We first compute the following thermal resistances using the given data:

$$R_i = \frac{1}{h_i} = \frac{1}{8.5} = 0.118, \quad R_{12} = \frac{1}{h_{12}} = \frac{1}{5.5} = 0.182,$$

$$R_{g1} = \frac{L_{g1}}{k_g} = \frac{3\times 10^{-3}}{0.8} = 0.00375, \quad R_{g2} = \frac{L_{g2}}{k_g} = 0.00375,$$

$$R_{g3} = \frac{L_{g3}}{k_g} = \frac{6\times 10^{-3}}{0.8} = 0.0075, \quad R_{23} = \frac{1}{h_{23}} = \frac{1}{6} = 0.167$$

$$R_o = \frac{1}{h_o} = \frac{1}{30} = 0.033$$

The total thermal resistance is

$$R_{tot} = 0.118 + 0.00375 + 0.182 + 0.00375 + 0.167 + 0.0075 + 0.033 = 0.515$$

The solar heat gain coefficient is given by Eq. (9.77) as

$$SHGC(\theta) = T(\theta) + \sum_{i=1}^{n} f_{si}(\theta) N_i$$

We use the quantities computed in example 9.11 to obtain the following values. The overall transmittance, $T = 0.352$. The fractions of beam radiation absorbed in the inner, middle, and outer glass sheets are: $f_{s1} = 0.197$, $f_{s2} = 0.0788$ and, $f_{s3} = 0.2$. The thermal resistance ratios are given by Eq. (9.73) as

$$N_i = \left(\frac{R_{i\to out}}{R_{tot}}\right)$$

Applying the above equation to the three glazings we have (see Fig. 9.8b)

$$N_1 = \left(\frac{R_{1\to out}}{R_{tot}}\right) = \frac{0.5\times 0.00375 + 0.182 + 0.00375 + 0.167 + 0.0075 + 0.033}{0.515} = 0.767$$

$$N_2 = \left(\frac{R_{2\to out}}{R_{tot}}\right) = \frac{0.5\times 0.00375 + 0.167 + 0.0075 + 0.033}{0.515} = 0.4066$$

$$N_3 = \left(\frac{R_{3\to out}}{R_{tot}}\right) = \frac{0.5\times 0.0075 + 0.033}{0.515} = 0.0713$$

Substituting in Eq. (9.77), the solar heat gain coefficient is

$$SHGC(50^o) = 0.352 + 0.197 \times 0.767 + 0.0788 \times 0.4065 + 0.2 \times 0.0713 = 0.549$$

The total energy flow rate into the inner space is given by Eq. (9.78) as

$$q_{tot} = E_{dn} cos\theta SHGC(\theta) + \frac{(t_{ao} - t_{ai})}{R_{tot}}$$

$$q_{tot} = 590 cos 50 \times 0.549 + \frac{(32-23)}{0.515} = 225.7 \text{ Wm}^{-2}$$

Example 9.14 A south facing wall of a building, at a location with a northern latitude of 35°, has a window of height 2 m and width 1.25, flush with the outer surface of the wall.

(a) Determine the width, the breadth, and height above the edge of the window of an overhang that would: (i) completely shade the window on 28 May at 10 a.m. solar time, and (ii) leave the window completely unshaded on 2 December at solar noon.

(b) Calculate sunlit fraction of the window area on 10 February at 2 p.m. solar time.

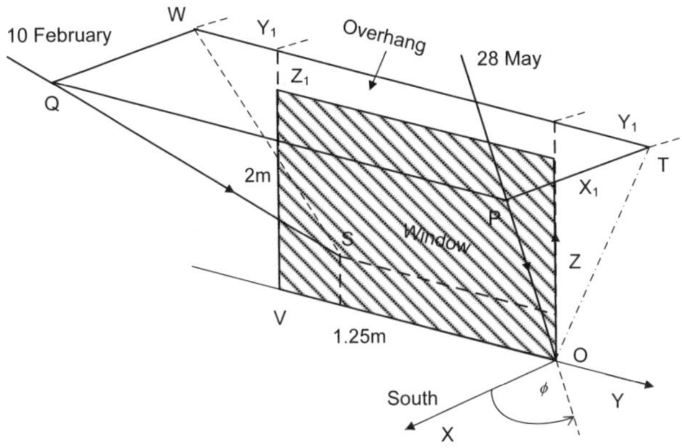

Fig. E9.14.1 Window with overhang

Solution For May 28, N_{day} = 148. The declination is given by Eq. (9.2) as

$$\delta = 23.45 sin\left[\frac{360(148+284)}{365}\right] = 21.43 \text{ degrees}$$

The latitude is 35°. The hour angle at 10 a.m. is −30°.
The solar altitude angle is given by Eq. (9.7) as

$$sin\beta = cosLcosHcos\delta + sinLsin\delta$$

$$sin\beta = cos35\cos 30\, cos21.431 + sin35sin21.43 = 0.8699$$

Therefore the solar altitude angle is 60.45 degrees.
The solar azimuth angle is given by Eq. (9.8) as

$$sin\phi = \frac{cos\delta sinH}{cos\beta} = -\frac{cos21.43\times\sin 30}{cos60.45} = -0.9437$$

Therefore solar azimuth is 70.69 degrees east of south.

The window with an overhang of width x_1, located at a height z_1 above its top edge, is depicted in Fig. E9.14.1. It extends a distance of y_1 from either vertical edge of the window. At 10 a.m. on May 28 the window is fully shaded. For this to happen the shadow of the corner P has to fall on the corner O of the window. Then the shadows of the edges TP and PQ of the overhang will be formed along the lines TO and OV respectively. Hence the window is fully shaded at this time.

Consider the coordinate system x-y-z with its origin at O. The x and y directions point towards the south and the east respectively. The z direction is along the vertical. The unit vector in the direction of the direct solar beam is given by Eq. (9.81) as

$$\bar{I_s} = \begin{pmatrix} cos\beta cos\phi \\ cos\beta sin\phi \\ sin\beta \end{pmatrix}$$

The vector OP may be expressed in the form

$$\begin{pmatrix} x_1 \\ y_1 \\ 2+z_1 \end{pmatrix} = \lambda \begin{pmatrix} cos\beta cos\phi \\ cos\beta sin\phi \\ sin\beta \end{pmatrix}$$

where λ is the length vector OP. Substituting numerical values in the above equation and equating the coordinates we have

$$y_1 = x_1 tan\phi = 2.854x_1 \qquad \text{(E9.14.1)}$$

$$2 + z_1 = \frac{x_1 tan\beta}{cos\phi} = 5.334x_1 \qquad \text{(E9.14.2)}$$

Using the procedure outlined earlier, we obtain the following quantities for 2 December at solar noon when the hour angle is zero.

$$\delta = -22.23°, \quad \beta = 32.77°, \quad \text{and} \quad \phi = 0°$$

At solar noon on 2 December, the window is completely unshaded.

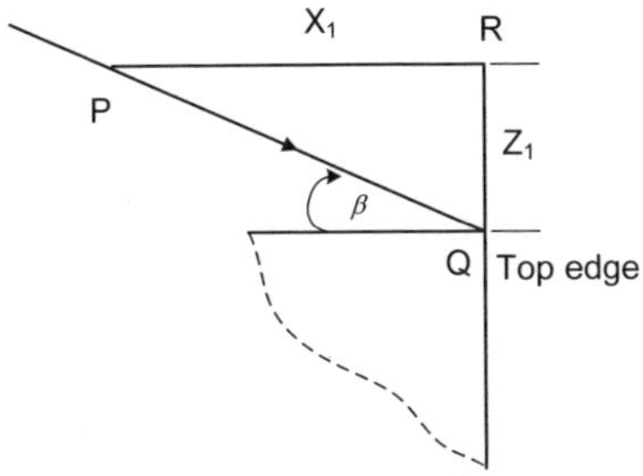

Fig. E9.14.2 Situation on December 2

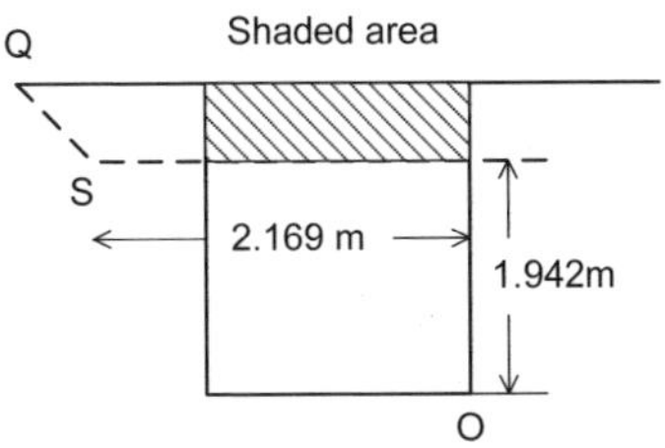

Fig. E9.14.3 Situation on February 10

Therefore the shadow of the top edge of the overhang is formed on the top edge of the window as shown in Fig. E9.14.2. Also, at solar noon the vertical plane containing the direct solar beam is normal to the window surface. From triangle PQR in Fig. E9.14.2 we have

$$z_1 = x_1 tan\beta = 0.6437x_1 \qquad \text{(E9.14.3)}$$

Solving Eq. (E9.14.1) to Eq. (9.14.3) simultaneously we obtain

$$z_1 = 0.274 \text{ m}, \quad x_1 = 0.426 \text{ m} \quad \text{and} \quad y_1 = 1.216 \text{ m}$$

(b) Using the procedure described section 9.2.2, we obtain the following quantities for 2 p.m. on 10 February:

$$\delta = -14.9°, \quad H = 30°, \quad \beta = 32.55° \quad \text{and} \quad \phi = -34.98°$$

The shadow of the corner Q of the overhang is formed at point S on the window as indicated in Fig. E9.14.1. The coordinates of S are $(0, y_2, z_2)$.

The unit vector I_s in the direction of the direct solar beam may be represented as

$$\bar{I_s} = \begin{pmatrix} cos\beta cos\phi \\ cos\beta sin\phi \\ sin\beta \end{pmatrix}$$

Considering the vector triangle OSQ we can write the following vector equation:

$$\overrightarrow{OQ} = \overrightarrow{OS} + \lambda\bar{I_s}$$

where λ is the length QS.

Expressing the vectors in the above equation in terms of their coordinates in the x-y-z system we have

$$\begin{pmatrix} x_1 \\ -(1.25+y_1) \\ 2+z_1 \end{pmatrix} = \begin{pmatrix} 0 \\ y_2 \\ z_2 \end{pmatrix} + \lambda \begin{pmatrix} cos\beta cos\phi \\ cos\beta sin\phi \\ sin\beta \end{pmatrix}$$

Substituting the pertinent numerical values and equating the x, y and z coordinates on the two sides of the above equation we obtain

$$x_1 = 0.426 = \lambda cos\beta cos\phi = 0.691\lambda$$

$$-2.467 = y_2 - \lambda cos\beta sin\phi = y_2 - 0.483\lambda$$

$$2.274 = z_2 + \lambda sin\beta = z_2 + 0.538\lambda$$

Solving the three equations above simultaneously we have

$$z_2 = 1.942 \text{ m} \quad \text{and} \quad y_2 = -2.169\text{m}$$

The shape of the shaded area of the window is depicted in Fig. E9.14.3. Note that the shadow S of the corner Q is formed outside the window area. Therefore the sunlit area of the window is a rectangle of width 1.25 m and height 1.942 m. The sunlit fraction of the area is 97%.

Example 9.15 A vertical window of a building at a location with a northern latitude of 40° faces 20° east of south. The height and width of the window are 1.8 m and 1.25 m respectively. It is set back 0.25 m from the outer surface of the wall of the building. Calculate the following quantities at 1 p.m. solar time on January 20 when the direct beam intensity is 420 Wm^{-2}.

(i) The angle of incidence of the direct solar beam on the window.
(ii) The fraction of the window that is unshaded.
(iii) The total direct beam solar radiation falling on the window.

Solution (a) For January 20, $N_{day} = 20$. The declination is given by Eq. (9.2) as

$$\delta = 23.45sin\left[\frac{360(20+284)}{365}\right] = -20.34 \text{ degrees}$$

The latitude of the location is 40°N. The hour angle at 1.00 p.m., solar time is 15°. The solar altitude angle is given by Eq. (9.7) as

$$\sin\beta = \cos L \cos H \cos\delta + \sin L \sin\delta$$

$$\sin\beta = \cos 40 \cos 15 \cos 20.34 - \sin 40 \sin 20.34 = 0.47$$

Therefore the solar altitude angle is 28.06 degrees.
The solar azimuth angle is given by Eq. (9.8) as

$$\sin\phi = \frac{\cos\delta \sin H}{\cos\beta} = \frac{\cos 20.34 \times \sin 15}{\cos 28.06} = 0.275$$

Therefore solar azimuth is 15.96 degrees west of south.
The surface-solar azimuth of the window is

$$\gamma = 20 + 15.96 = 35.96 \text{ degrees.}$$

The angle of incidence, θ for a *vertical* surface is given by Eq. (9.13) as

$$\cos\theta = \cos\beta \cos\gamma$$

$$\cos\theta = \cos 28.06 \cos 35.96 = 0.7143$$

Hence the angle of incidence is 44.4 degrees.

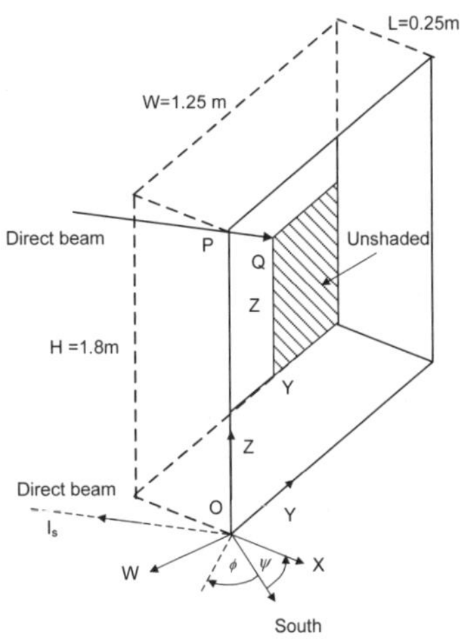

Fig. E9.15.1 Window with setback in wall

(b) The window set back in the wall is shown schematically in Fig. E9.15.1. We choose a coordinate system x-y-z with its origin O at the edge of the window to represent the various vectors involved. The x-axis is along the normal to the window surface, the y-axis is parallel the bottom edge and the z-axis is in the vertical direction as indicated in Fig. E9.15.1.

The direct beam makes an angle β, which is the solar altitude angle, with the horizontal plane x-y. The projection of the direct beam on the horizontal plane makes an angle ϕ with the South, which is the solar azimuth angle. The normal to the window makes an angle ψ with the South. The unit vector I_s in the direction of the direct solar beam may be represented as (see Fig. E9.15.1)

$$\bar{I}_s = \begin{pmatrix} cos\beta cos\gamma \\ -cos\beta sin\gamma \\ sin\beta \end{pmatrix} \quad \text{(E9.15.1)}$$

where the angle, $\gamma = (\phi + \psi)$ is the surface–solar azimuth angle.

Let Q be the point on the window surface where the shadow of the front corner P of the window-cavity falls. We observe that the rectangular area with Q as one of its corners is unshaded. The rest of the window is shaded by the left vertical surface and the top horizontal surface of the window cavity. Considering the vector triangle OPQ we can write the following vector equation:

$$\overrightarrow{OP} = \overrightarrow{OQ} + \lambda \bar{I}_s \quad \text{(E9.15.2)}$$

where λ is the length QP.

Expressing the vectors in Eq. (E9.15.2) in terms of their coordinates in the x-y-z system we have

$$\begin{pmatrix} 0 \\ 0 \\ H \end{pmatrix} = \begin{pmatrix} -L \\ y \\ z \end{pmatrix} + \lambda \begin{pmatrix} cos\beta cos\gamma \\ -cos\beta sin\gamma \\ sin\beta \end{pmatrix}$$

Equating the x, y and z coordinates on the two sides of the above equation we obtain

$$L = \lambda cos\beta cos\gamma$$

$$y = \lambda cos\beta sin\gamma$$

$$z = H - \lambda sin\beta$$

Eliminating λ between the above equations we have

$$y = Ltan\gamma = 0.25tan35.96 = 0.18 \text{ m}$$

$$z = H - \frac{Ltan\beta}{cos\gamma} = 1.8 - \frac{0.25tan28.06}{cos35.96} = 1.635 \text{ m}$$

The unshaded area of the window is (see Fig. E9.15.1)

$$A_{us} = z(W - y) = 1.635 \times (1.25 - 0.18) = 1.75 \text{ m}^2$$

The total area of the window is, $A_w = 2.25$ m^2.
Hence the unshaded fraction of the area is 1.75/2.25 = 77.8%.

(c) Total direct solar radiation striking the window surface is given by

$$Q_{dir} = A_{us}I_{ds}cos\theta = 1.75 \times 420 \times cos44.4 = 525 \text{ W}$$

Problems

P9.1 The longitude of a location is 93°W and the standard longitude of the time zone is 90°W. The northern latitude of the location is 35°. Calculate (i) the solar time, (ii) the hour angle, (iii) the solar altitude angle, and (iv) the solar azimuth angle at the location when the standard clock time is 11.30 a.m. on August 10. The clock time is advanced by one hour for daylight saving.
[*Answers*: (i) 10.12 a.m., (ii) −26.85°, (iii) 58.9°, (iv) −57.48°]

P9.2 (a) Calculate the solar altitude angle and the solar azimuth angle at 2 p.m. solar time on January 15 for a location with northern latitude 35°. (b) Calculate the solar time at sunrise and sunset on January 15 at the same location.
[*Answers*: (a) 26.9°, 31.5°, (b) 7.03 a.m., 4.57 p.m.]

P9.3 A window facing south-east is tilted at 60° from the horizontal. The northern latitude of the location is 42°. Calculate the angle of incidence of the direct solar beam on the window at 1.00 p.m. solar time on September 8.
[*Answer*: 54.3°]

P9.4 On July 15 at 2 p.m. solar time, the direct beam and diffuse solar radiation intensities at a location, with a northern latitude of 40°, are 580

Wm^{-2} and 165 Wm^{-2} respectively. The ambient temperature is 31°C. The thickness and thermal conductivity of a flat roof at the location are 55 mm and 0.12 $Wm^{-1}K^{-1}$ respectively. The emissivity of the roof surface is 0.8. The inside air temperature is 24°C. The overall heat transfer coefficients for the outside and inside air films are respectively 28 $Wm^{-2}K^{-1}$ and 8.5 $Wm^{-2}K^{-1}$. Calculate (i) the sol-air temperature, (ii) the surface temperature of the roof, and (iii) the heat flow rate into the building. Assume that the heat capacity of the roof is negligible.
[*Answers*: (i) 45.85°C, (ii) 44.57°C, (iii) 35.7 Wm^{-2}]

P9.5 Derive the expressions in Eq. (9.17) and Eq. (9.18) by considering diffuse radiation exchange between the inclined flat surface, the sky, and the horizontal ground surface shown in Fig. 9.5.

P9.6 At a location with a northern latitude of 35°, the measured direct beam and diffuse solar radiation intensities, on August 15 at 10 a.m. solar time, are 605 Wm^{-2} and 156 Wm^{-2} respectively. A thin vertical wall at the location faces south-east. The emissivity of the wall surface is 0.85 and its thermal resistance is 0.2 m^2KW^{-1}. The reflectivity of the ground surrounding the wall is 0.2. The ambient temperature and the inside air temperature are 32°C and 24°C respectively. The overall external and internal heat transfer coefficients are 30 $Wm^{-2}K^{-1}$ and 9 $Wm^{-2}K^{-1}$ respectively. Calculate (i) sol-air temperature, (ii) the temperature of the wall surface, and (iii) the heat flow rate into the building. Assume that the heat capacity of the wall is negligible.
[*Answers*:(i) 47.13°C, (ii) 44.9°C, (iii) 67.2 Wm^{-2}]

P9.7 A double-glazed window consists of two glass sheets, each of thickness 6 mm and thermal conductivity 0.8 $Wm^{-1}K^{-1}$. The transmittance and reflectance of each glass sheet for beam radiation incident at 40° are 0.75 and 0.08 respectively. Calculate the overall transmittance of the window and the fraction of radiation absorbed in each glass sheet.
[*Answers*: 0.567, 0.128, 0.1388]

P9.8 A triple-glazed window has three glass sheets, each of thickness 3 mm and thermal conductivity 0.85 $Wm^{-1}K^{-1}$. The transmittance and reflectance of each glass sheet for beam radiation incident at 50° are 0.8 and 0.1 respectively. Calculate the overall transmittance of the window and the fraction of radiation absorbed in each glass sheet.
[*Answers*: 0.526, 0.0657, 0.0879, 0.1134]

P9.9 Consider the double-glazed window described in problem P9.7. The overall heat transfer coefficients for the outside air film, the air gap between the glasses, and the inside air film are respectively 28 $Wm^{-2}K^{-1}$, 6.8 $Wm^{-2}K^{-1}$ and 7.8 $Wm^{-2}K^{-1}$. The inside and outside air temperatures are 22°C and 32°C respectively. Calculate (i) solar heat gain coefficient (SHGC) for the direct beam, and (ii) the total rate of energy transfer per unit area into the inside space if the direct beam radiation intensity is 520 Wm^{-2}.
[*Answers*: (i) 0.66, (ii) 293.6 Wm^{-2}]

P9.10 Consider the triple-glazed window described in problem P9.8. The following pertinent overall heat transfer coefficients have been estimated: $h_i = 8.0$ $Wm^{-2}K^{-1}$, $h_{12} = 6$ $Wm^{-2}K^{-1}$, $h_{23} = 7$ $Wm^{-2}K^{-1}$, $h_o = 28$ Wm^2K^{-1}, where subscript 1 denotes the innermost glazing. The inside and outside air temperatures are 21°C and 33°C respectively. For the conditions stated in problem P9.8, calculate (i) the solar heat gain coefficient (SHGC) for direct beam radiation, and (ii) the total energy transfer rate per unit area into the inside space if the direct beam radiation intensity is 480 Wm^{-2}.
[*Answers*: (i) 0.617, (ii) 215 Wm^{-2}]

P9.11 A south facing window of height 2.2 m and width 1.3 m is flush with the outside surface of a wall. A solid overhang of length 1.8 m is located symmetrically, 0.16 m above the top edge of the window. The overhang extends 0.5 m out of the wall. The northern latitude of the location is 40°. Calculate the unshaded area of the window at 11 a.m. solar time on June 10.
[*Answer*: 0.9025 m^2]

P9.12 A vertical window of height 2 m and width 1.5 m faces south-east. It is set back 0.3 m from the outer surface of a wall. The northern latitude of the location is 35°. Calculate the following quantities at solar noon on February 10 when the direct beam intensity is 480 Wm^{-2}: (i) the angle of incidence of the direct solar beam on the window, (ii) the unshaded fraction of the window, and (iii) the total direct beam solar radiation falling on the window.
[*Answers*: (i) 57.3°, (ii) 0.66, (ii) 510.8W]

References

1. *ASHRAE Handbook* - 2013 *Fundamentals*, American Society of Heating, Refrigeration and Air Conditioning Engineers, Atlanta, 2013.
2. Gueymard, Christian A. and Thevenard Didier, 'Monthly average clear-sky broadband irradiance database for worldwide solar heat gain and building cooling load calculations', *Solar Energy*, 83(2009), 1998-2018.
3. Kuehn, Thomas H., Ramsey, James W. and Threlkeld, James L., *Thermal Environmental Engineering*, 3rd edition, Prentice-Hall, Inc., New Jersey, 1998.
4. Mitchell, John W. and Braun, James E., *Principles of Heating, Ventilation and Air Conditioning in Buildings*, John Wiley & Sons, Inc., New York, 2013.
5. Siegel, Robert and Howell, John R., *Thermal Radiation Heat Transfer*, Hemisphere Publishing Corporation, Washington, 1992.
6. Wijeysundera, N. E., 'A net radiation method for the transmittance and absorptivity of a series of parallel regions'. *Solar Energy*, 17(1975), 75-77.
7. Wijeysundera, N. E., 'Application of the network analogy to one - dimensional systems with internal heat generation', *Applied Energy*, 12 (1982), 229-236.

Chapter 10

Cooling and Heating Load Calculations

10.1 Introduction

The rate of heat input needed to maintain the indoor temperature and humidity of a building within specified limits, during the winter heating season, is called the *space heating load.* The heating load includes: (i) the heat loss to the outside ambient across the building envelope, and (ii) the heat required to raise the temperature of any cold air entering the space through openings in doors, windows, and other structural components. The *unintended* entry of ambient air into a conditioned space is called *infiltration.*

Under typical summer conditions heat flows into a building because the outside ambient temperature is usually higher than the indoor air temperature. In addition, a building may receive heat inputs from sources such as: (i) solar radiation entering through windows, (ii) artificial lights, (iii) occupants, and (iv) equipment like computers and motors within the building.

The indoor air humidity may increase due to infiltration of ambient air, and moisture inputs from sources within the building. For summer air conditioning systems, the total rate of energy input needed to maintain the space at a constant temperature and relative humidity is called the *space cooling load.*

All the aforementioned energy flows into a building, are time-varying due to several factors. The heat flow through the building envelope is time dependent due to the thermal mass of structural components like the walls, the windows and the roof. Furthermore, the transient nature of the external weather conditions, such as the solar radiation intensity, the wind speed and ambient temperature, makes the heat flow unsteady. The internal heat inputs from occupants, lights and equipment are time dependent because of varying occupancy and usage schedules.

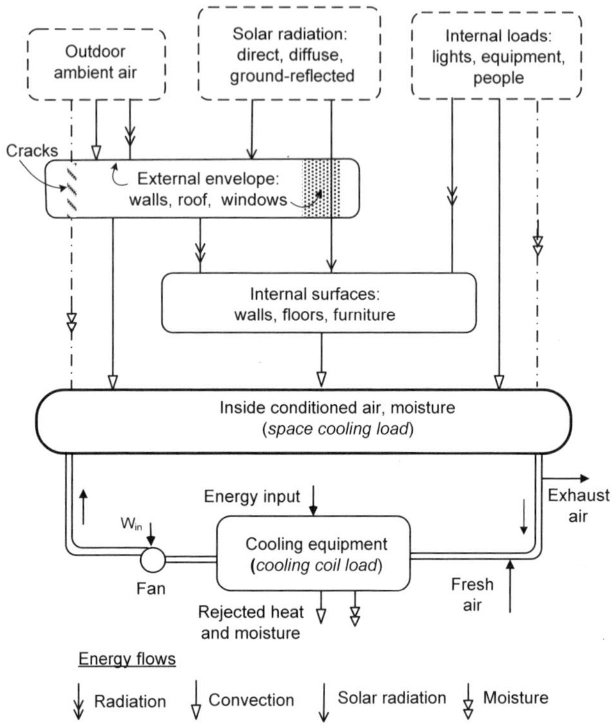

Fig.10. 1 Schematic display of energy and moisture flows

The convective and radiative heat transfer processes, and moisture transfer processes, in a typical building, are depicted schematically in Fig. 10.1, using different arrow symbols.

The external surfaces of the building envelope receive energy by several mechanisms. These include: (i) heat flow from ambient air by convection, (ii) long-wave heat radiation from the sky, (iii) direct beam solar radiation, (iv) diffuse solar radiation from the sky, and (v) reflected solar radiation from the ground.

A fraction of the total energy *absorbed* by the building envelope is stored in the structural materials and the rest enters the inner space by convection and long-wave radiation. The convected heat is absorbed immediately by the indoor air, and becomes a part of the *space cooling load.* The long-wave radiation, on the other hand, is first absorbed by surfaces within the space like the walls, the floor and the furniture. This stored energy is later transferred to the indoor air by convection, thus becoming a portion of the instantaneous cooling load.

Similarly, the direct and diffuse solar radiation transmitted through the fenestration surfaces, like the windows in the building envelope, is first absorbed by the surfaces within the space and later released to the indoor air by convection, thus contributing to the cooling load.

The lights and equipment located within a building, and the people occupying the space, exchange heat by convection and radiation. The convected heat becomes a portion of the cooling load immediately, while the radiation is first absorbed by surfaces within the space and released to the indoor air by convection at a later time.

Moisture enters the inside of a building through openings and cracks in the building envelope. People occupying the building, and appliances like coffee makers and water heaters, are internal moisture sources. The moisture entering the space constitutes the *latent cooling load* of the space.

The cooling system installed in a building maintains the inner space at a constant dry bulb temperature and relative humidity by supplying conditioned air to the space as shown in Fig. 10.1. For *ventilation* purposes, a portion of the return air is exhausted and replaced with an equal amount of fresh ambient air. The rate at which energy is removed by the cooling equipment is called the *cooling coil load.* The space cooling load differs from the cooling coil load because of heat gain from the fan, heat gain through the duct walls, and heat gain due to the introduction of fresh ambient air.

It is clear that a detailed analysis of all the transient energy and moisture transfer processes, indicated in Fig. 10.1, is indeed complex and tedious. Therefore in real design situations, such load estimations are now carried out mostly using computer software programs. However, our aim in this introductory textbook is to develop the physical principles that form the basis of these computerized procedures of load estimation. The *ASHRAE Handbook* - 2013 *Fundamentals* [1], includes several simplified approaches to heating and cooling load estimation, deemed more suitable for introductory courses in heating and air conditioning. We shall adopt these simplified approaches in this book.

For winter heating load estimation, a conservative design approach, recommended in the *ASHRAE Handbook* - 2013 *Fundamentals* [1], is to ignore the heat inputs to a building resulting from solar radiation entering through fenestrations, lighting, occupants, and equipment. These energy flows, in general, help reduce the external heat input required from the heating system. Similarly, transient effects due to the thermal mass of the building envelope are ignored because they usually decrease the heating load.

Therefore, for estimating the *winter heating load* we shall consider only *steady* heat loss through building envelope and the heating load due to *infiltration* of cold air. Such steady heat and moisture flow processes were analyzed in chapter 8.

10.2 Outdoor Design Conditions

The heating and cooling systems of buildings are usually designed to meet the extreme weather conditions expected at a location. The weather records for a large number of locations around the world have been complied and processed to develop a series of weather related outdoor design parameters. These parameters are tabulated in an appendix in Chapter 14 of the *ASHRAE Handbook* - 2013 *Fundamentals* [1]. Representative data for five cities are presented in Tables 10.1(a) and (b) for purposes of illustration.

The data in Table 10.1(a) may be used to design heating systems for winter conditions. The first three columns list the location, its latitude, and its longitude respectively. The next two columns give the dry bulb temperatures during the coldest month of the year with respective annual cumulative frequencies of 99.6% and 99%.

Table 10.1(a) Design weather data for selected locations*

Location	Lat. Deg.	Long. Deg.	DB(°C) 99.6%	DB(°C) 99%	WS 1%	WS 5%
New York, USA	40.7N	73.8W	−13.0	−10.2	12.1 ms^{-1}	9.2 ms^{-1}
Dallas, USA	32.9N	97.04W	−5.0	−2.6	11.6 ms^{-1}	9.2ms^{-1}
Toronto, Canada	43.68N	79.63W	−16.1	−13.3	13.3 ms^{-1}	10.5 ms^{-1}
Sydney, Australia	33.93S	151.18E	6.1	7.1	12.9 ms^{-1}	10.2 ms^{-1}
Bejing, China	39.93N	116.28E	−11.0	−9.1	9.8 ms^{-1}	6.7 ms^{-1}
Bangalore, India	12.97N	77.58E	15.2	15.9	5.5 ms^{-1}	4.1 ms^{-1}

Table 10.1(b) Design weather data for selected locations*

Location	DB/MCWB(°C) 0.4%	DB/MCWB(°C) 1.0%	DB/MCWB(°C) 2.0%
New York, USA	29.8/22.4	27.8/21.6	26.6/21.1
Dallas, USA	38.0/23.6	37/23.7	35.8/23.9
Toronto, Canada	28.5/21.8	26.8/21.4	25.3/20.8
Sydney, Australia	32.9/19.5	30.1/20.1	28.2/20.0
Beijing, China	35/22.0	33.2/22.5	32/22.4
Bangalore, India	34.2/19.8	33.4/19.8	32.6/19.8

*Values extracted from Appendix, Chapter 14, *ASHRAE Handbook* - 2013 *Fundamentals* [1]

The values listed under 99.6% is the ambient temperature exceeded 99.6% of the time during an year. In the case of New York this implies that, on average, the low temperature of −13°C is exceeded during 8725 hours out of a total of 8760 hours. Similarly, the temperature of −10.2°C, listed under 99% is exceeded on average during 8672 hours.

The extreme annual wind speeds are listed in the last two columns of Table 10.1(a). The values listed under 1% and 5% are the wind speeds exceeded, on average, during 1% (88 hours) and 5% (438 hours) of the time during an year.

The average difference of about 2°C to 4°C between the temperatures listed under the 99.6% and 99% levels is small compared to the average difference between the indoor and outdoor temperatures. Therefore heating systems are usually designed using the 99.6% level outdoor temperature.

The ambient temperature parameters that may be used to design summer cooling systems are listed in Table 10.1(b) for the same locations listed in Table 10.1(a). The dry bulb (DB) temperatures under 0.4%, 1%, and 2% levels are the high temperatures exceeded, on average, 0.4% (35 hours), 1% (88 hours) and 2% (175 hours) of the time during a year.

The 'mean coincident wet-bulb temperature' (MCWB) listed in Table 10.1(b) is the average value of the wet-bulb temperature at the corresponding dry-bulb temperature. The difference in dry-bulb temperatures for the different levels is about 3°C to 4°C. For design purposes, the outdoor temperature for the 0.4% level is often selected to ensure that the cooling system has adequate capacity.

10.3 Thermal Comfort and Indoor Design Conditions

10.3.1 *Heat transfer from the human body*

Air conditioning systems are designed and operated with the aim of providing a comfortable and healthy environment to the occupants of a building. The indoor design conditions are the parameters that ensure such an indoor environment.

Chapter 9 of the *ASHRAE Handbook* - 2013 *Fundamentals* [1] provides a detailed discussion on human thermal comfort. Complete chapters on thermal comfort are also included in Refs. [2] and [3]. Here we shall briefly review several aspects of thermal comfort that help us identify the desirable indoor design conditions for heating and air conditioning systems.

The human body converts chemical energy from food into heat and mechanical work by a process called *metabolism*. The metabolic energy generation rate, usually expressed in MET units (1 MET= 58.2 Wm^{-2}), depends on the activity level, the age, and the health of an individual, among others.

Temperature regulation of the body is achieved through the control of blood flow rate to the skin. As the environmental temperature goes up, the blood flow rate to the skin increases to raise the skin temperature, which in turn, increases the heat transfer rate to the environment. This process is effective until the skin temperature reaches the core body temperature of 37°C. At this stage sweating, which transfers metabolic heat to the surroundings by evaporation, is initiated.

In contrast, when the temperature of the environment decreases the blood flow to the skin is reduced to lower the skin temperature, and consequently the rate of heat loss from the skin.

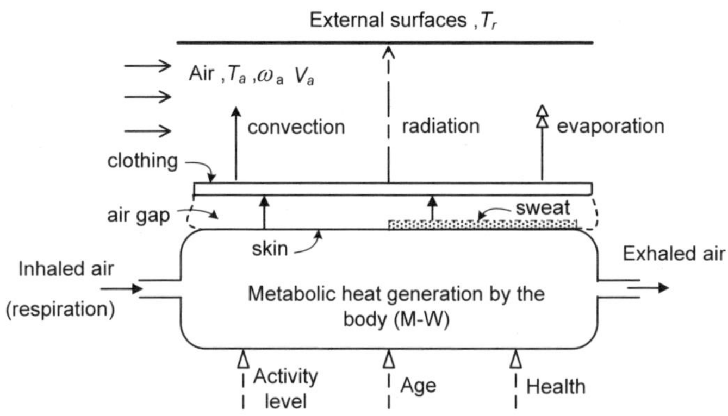

Fig. 10.2 Thermal interaction of the human body with the environment

The various energy interactions between the human body and the external environment are depicted schematically in Fig. 10.2. The metabolic heat (M-W) and sweat reaching the skin flows across the air gap between the skin and the inner surface of the clothing. Heat and moisture then diffuse through the clothing layer to the outer surface. The above heat and moisture transfer processes depend respectively on the effective thermal resistance, and mass transfer resistance of the clothing layer.

Heat transfer from the clothing surface to the ambient occurs by convection and thermal radiation. The rate of convection is governed by the ambient air temperature and the speed of air movement around the clothing surface. The radiation transfer, on the other hand, is governed by the temperature of the surrounding surfaces, which may be different from the air temperature.

The *operative temperature*, which takes into account the difference between the air temperature and the surface temperature, is defined as a *fictitious* air temperature that gives an effective heat loss equal to the sum of the actual convective and radiative heat losses.

The rate of energy transfer to the ambient due to the evaporation of sweat is dependent on the ambient temperature, the relative humidity, and the speed of air movement.

Heat is also transferred from the body to the ambient due to *respiration*. The rate of energy transfer in this case depends on the average air flow rate into the lungs and the temperature and relative humidity of ambient air.

Two important physiological variables affecting thermal comfort are the skin temperature and the evaporation rate due to sweating. These variables, in turn, depend on the activity and clothing level of the person, and the temperature, the humidity ratio, and the velocity of the surrounding air.

For people engaged in *light activity*, Mitchell and Braun [3] gives the relationship in Eq. (10.1), between the optimal air temperature, $T_{a,opt}$ (°C) and the thermal resistance of the clothing, R_c in 'clo', where 1 clo = 0.155 m^2KW^{-1}. It is applicable when the relative humidity of air is 50% and air velocity less than 0.15ms^{-1}.

$$T_{a,opt} = 27.2 - 0.4R_c \tag{10.1}$$

Typical insulation values for clothing ensembles are listed in Table 8, Chapter 9 of the *ASHRAE Handbook* - 2013 *Fundamentals* [1].

For higher activity levels the following relationship is recommended in Ref. [3]:

$$T_{a,opt} = 27.2 - 5.9R_c - 3.0(1 + R_c)(\dot{M} - 1.2) \tag{10.2}$$

The metabolic rate $\dot{M}$ is in mets, where 1 met = 105 W. Typical metabolic heat generation rates for various activities are given in Table 1, Chapter 9 of the *ASHRAE Handbook* - 2013 *Fundamentals* [1].

10.3.2 *Indoor design conditions*

Several comfort indices have been developed to correlate human thermal comfort to the surrounding environmental conditions. One such index, called the '*effective temperature*', is based on the trade-off between air temperature and relative humidity for the same total heat transfer [2].

By analyzing the energy and moisture transfer processes indicated in Fig. 10.2 it is possible to derive an expression for the heat loss rate from the skin, $\dot{Q}_{sk}$ to the surrounding ambient at temperature t_o and humidity ratio ω. The main simplifying assumptions are: (i) the energy loss due to respiration is negligible, (ii) the *operative* temperature is equal to the air temperature, and (iii) the wetted area of the skin is constant [2]. Thus we have

$$Q_{sk} = c_{sen}(t_{sk} - t_o) + c_{lat}(\omega_{s,sk} - \omega) \tag{10.3}$$

where t_{sk} and $\omega_{s,sk}$ are the skin temperature and the saturation humidity ratio at the skin temperature respectively.

The sensible heat parameter c_{sen}, in Eq. (10.3) depends on the thermal resistance of the clothing, and the overall heat transfer coefficient from the clothing to the ambient. The latent heat parameter, c_{lat} depends on the mass transfer resistance of the clothing, the mass transfer coefficient from the clothing to the ambient, and the latent heat of vaporization.

If we select an arbitrary air temperature, t_o and a relative humidity of 50%, the rate of heat transfer given by Eq. (10.3) becomes

$$Q_{sk,o} = c_{sen}(t_{sk} - t_o) + c_{lat}(\omega_{s,sk} - 0.5\omega_{s,t_o}) \tag{10.4}$$

Note that at 50% relative humidity, the humidity ratio is approximately half the saturation humidity ratio, $\omega_{s,to}$.

The rate of heat loss, $Q_{sk,o}$ given by Eq. (10.4), is now a unique function of the temperature t_o, commonly called the '*effective temperature*', and denoted ET*. Other pairs of values of the air temperature t_o and humidity ratio ω that result in the same heat loss rate, $Q_{sk,o}$ may be obtained by substituting for Q_{sk} in Eq. (10.3).

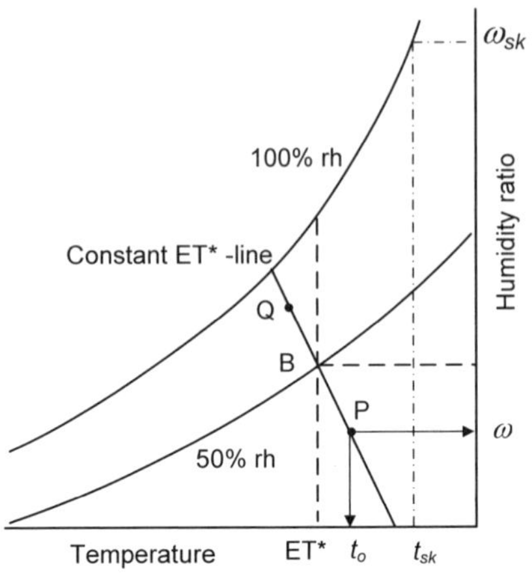

Fig. 10.3 Line of constant heat loss and effective temperature, ET*

If we assume that the dry-bulb temperature and humidity ratio scales of the psychrometric scale are approximately linear, then for a constant heat transfer rate or effective temperature ET^*, the graph of Eq. (10.3) is a straight line through the point of intersection B, of the 50% - relative humidity line and the ET* - constant temperature line, as shown in Fig. 10.3.

If the conditions of the environment change from B to a state with a higher temperature, such as P, then the sensible heat transfer rate decreases. To maintain the same total heat loss the latent heat transfer due to sweating has to be increased by lowering the humidity ratio. Conversely, when the state of the environment changes to point Q with a lower ambient temperature the humidity ratio has to be increased to maintain the same total heat loss.

Depicted in Fig. 10.4 is a sketch of a thermal comfort chart, based on the concept of effective temperature, ET^*, available in Chapter 9 of the *ASHRAE Handbook* - 2013 *Fundamentals* [1].

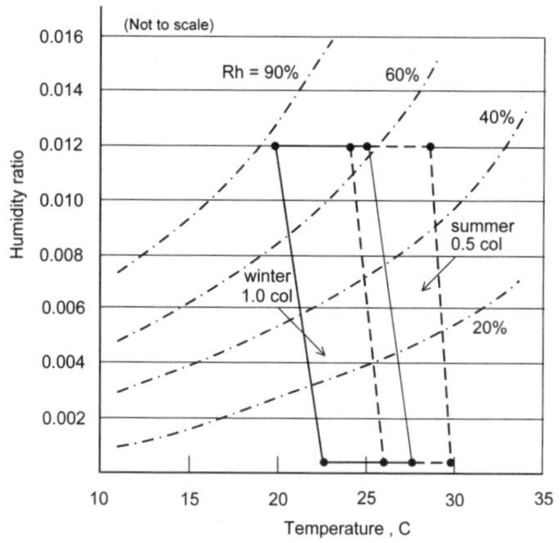

Fig. 10.4 *ASHRAE* - summer and winter comfort zones [1]

The indicated comfort zones for winter and summer are for people performing office type work, and wearing clothing with thermal resistances of 1.0 clo and 0.5 clo respectively, representative of typical clothes worn during these two seasons. The speed of air movement is less than 0.2 ms^{-1}. In the middle of a zone, a person wearing the prescribed clothing would have a neutral thermal sensation. The middle of the winter and summer comfort zones are 22°C and 50% rh and 25°C and 50% rh respectively.

The upper limit of the humidity ratio recommended in Ref. [1] is 0.012 $kgkg^{-1}$. However, there is no recommended lower limit for humidity ratio. The recommended temperature range [3] for winter is between 20°C and 25°C, and for summer it is between 24°C and 28°C. For a more complete discussion on human thermal comfort the reader is referred to Refs. [1-3].

10.3.3 *Indoor air quality*

The quality of indoor air plays a vital role in maintaining a healthy and comfortable indoor environment in buildings. Pollutants enter indoor air from sources within the conditioned space, and in the surrounding ambient. Some pollutants may be removed by the use of filters. Moreover, contaminant levels in the indoor air can be diluted by introducing fresh ambient air, commonly called *ventilation air*. The locations of these fresh air intakes should not face areas where pollution levels are already high, like roads with heavy traffic.

The fresh air introduced for ventilation purposes has to be cooled and dehumidified to the condition of the supply air by the air conditioning system as shown in Fig.10.1. This process contributes to the *cooling coil load* of the building. *ASHRAE* has developed guidelines for the desirable flow rates of ventilation air, taking into consideration the impact on the cooling energy consumption.

The flow rate of fresh air depends on the number of people occupying a zone and the type activities they are engaged in. Typical recommended values on fresh air flow rates for a few representative applications, extracted from the *ASHRAE Ventilation Standard* 62.1-2010 [3], are listed in Table 10.2.

Table 10.2 Minimum *ASHRAE*-recommended ventilation rates*

Application	Function	Design Occupancy/100m^2	Ventilation air flow Rate per person, Ls^{-1}
Office	Offices	7	10 Ls^{-1}
	Conference room	50	10 Ls^{-1}
Restaurant	Lounge	100	15 Ls^{-1}
	Dining room	70	10 Ls^{-1}
	Kitchen	20	7.5 Ls^{-1}
Retail store	Shops, malls	20	1 Ls^{-1}m^{-2}
Sport area	Ballrooms	100	13 Ls^{-1}
	Gymnasiums	30	10 Ls^{-1}

*Representative values extracted from Table 7.3, Mitchell and Braun [3]

In most large commercial buildings, a single air handling unit (AHU) could be supplying conditioned air to several zones. These zones may be used for different functions and may have varying occupancy levels for which the recommended ventilation air flow rates may be different.

It is often possible to identify a zone, called the *critical zone*, that has the highest ventilation air requirement. If the total ventilation flow rate is selected to satisfy the needs of this critical zone, then the other zones may receive significantly more ventilation air than the recommended quantity. Such a design would have an adverse effect on the energy consumption of the cooling system.

The *ANSI/ASHRAE Ventilation Standard* 62.1-2010 [3] recommends a procedure to select the ventilation rates for multi-zone spaces supplied by a single air handling unit. Let the circulation air flow rate for zone *n*, based on the cooling load, be $\dot{V}_{cn}$ and the corresponding *recommended ventilation air flow* rate be, $\dot{V}_{rn}$. The parameter *X* is defined as

$$X = \frac{\sum_{n=1}^{N} \dot{V}_{rn}}{\sum_{n=1}^{N} \dot{V}_{cn}} \tag{10.5}$$

where *N* is the total number of zones.

Let *Z* be the *maximum value* of the ratio, $[\dot{V}_{rn}/\dot{V}_{cn}]$ amongst all the zones. Then the ratio of the *design ventilation air flow rate,* $\dot{V}_{vn}$ to zone *n,* to the circulation flow rate $\dot{V}_{cn}$, is given by

$$Y = \frac{\dot{V}_{vn}}{\dot{V}_{cn}} = \frac{X}{X+1-Z} \tag{10.6}$$

The application of above design procedure will be illustrated in the worked examples to follow in this chapter.

10.4 Internal Heat Sources in Buildings

Cooling systems of air conditioned buildings have to remove the heat generated within the building by sources such as people, lights and equipment. Therefore these sources contribute to the summer cooling load of the building. However, during the winter season these same energy sources help reduce the load on the heating system, and therefore they could be ignored in heating load calculations.

10.4.1 *Heat gain from people*

As was discussed in section 10.2, people occupying buildings produce energy due to metabolic processes. Metabolic heat is released to the air both as sensible heat, and latent heat due to sweating. The sensible heat includes a convective component, and a radiative component due to thermal radiation exchange between the occupants and the surrounding surfaces. Convective heat and latent heat are released to the indoor air immediately and therefore become part of the cooling load. The radiative component, however, is first absorbed by the surrounding surfaces and later released to the air by convection. This time delay has to be accounted for in cooling load calculations.

Typical heat and moisture release rates by people engaged in different activities are given in Table 1, Chapter 18 of the *ASHRAE Handbook - 2013 Fundamentals* [1]. For purposes of illustration we have listed a few representative values in Table 10.3.

Table 10.3 Heat and moisture given off by people performing different activities*

Activity Level	Total Heat, W		Sensible Heat, W	Latent Heat, W	Radiant/Sensible	Heat Percent
	Adult male	Adjusted[a] M/F			Low V[b] 0.5 ms^{-1}	High V 2.0 ms^{-1}
Seated at theater	115	95	65	30	60	27
Moderate office work	140	130	75	55	58	38
Walking, standing	160	145	75	70	58	38
Light bench work, factory	235	220	80	140	49	35
Heavy work, factory	440	425	170	255	54	19

(a) - Adjusted heat gain based on normal percentage of men, women and children for the particular activity listed. (b) V is the air speed

*Representative values extracted from Table 1, Page 18.4, *ASHRAE Handbook* - 2013 *Fundamentals* [1]

10.4.2 *Heat gain from lighting*

The lights installed in a building can make a significant contribution to the cooling load. The radiant energy emitted by lights is absorbed by different surfaces in the room, especially the floor. This energy is later released to the air as sensible heat. Lights also transfer sensible heat directly to the air by convection due to their higher operating temperatures. The rate of heat gain, q_{el} due to electric lighting may be expressed as [1]

$$q_{el} = WF_{ul}F_{sa} \tag{10.7}$$

where W is the total light wattage, F_{ul} is the lighting use factor, and F_{sa} is the lighting special allowance factor.

The total light wattage, W is obtained from the ratings of all the lights installed in the space. The energy released by ballasts are not included in W. The lighting use factor F_{ul} is the fraction of the wattage in use. The special allowance factor F_{sa} is the ratio of the actual energy released by the lighting fixture to the power consumption of the lamps. For fluorescent lights F_{sa} accounts for the power consumed by the ballast.

Furthermore, the heat gain from lights depends on the type of installation. For instance, all the heat released by pendant lights and floor lamps enter the conditioned space. However, for recessed lights located in ceilings, only a portion of the heat released enters the room air while the rest is transferred to the unconditioned space above the ceiling. The fraction of the total heat that enters the room is called the *space fraction.*

Heat transfer to the room due to lighting occurs by convection and radiation. The convective fraction of the heat is transferred to the indoor air immediately. The heat entering by radiation, usually called the *radiative fraction*, is first absorbed by the surrounding surfaces and later released to the air by convection.

Lighting heat gain parameters for typical operating conditions are given in Table 3, Chapter 18 of the *ASHRAE Handbook - 2013 Fundamentals* [1]. For purposes of illustration we have listed a few representative values in Table 10.4.

Table 10.4 Lighting heat gain parameters*

Luminaire type	Space fraction	Radiative Fraction
Recessed fluorescent without lens	0.64 to 0.74	0.48 to 0.68
Recessed fluorescent with lens	0.40 to 0.50	0.61 to 0.73
Downlight compact fluorescent	0.12 to 0.24	0.95 to 1.0
Downlight incandescent	0.70 to 0.80	0.95 to 1.0
Non-in-ceiling fluorescent	1.0	0.50 to 0.57

*Representative values extracted from Table 3, Page 18.6, *ASHRAE Handbook - 2013 Fundamentals* [1]

10.4.3 *Heat gain from equipment*

The heat emanating from equipment such as computers, fax machines, and printers located within a conditioned space, contribute significantly to the cooling load. The estimation of the cooling load due to equipment is made difficult by the wide variety available, and their varying operating schedules. In most instances, the data on the nameplate of the equipment is the only source of information available. However, use of this nameplate data directly could significantly overestimate the cooling load.

Tables 8, 9 and 10, Chapter 18 of the *ASHRAE Handbook* - 2013 *Fundamentals* [1] give the nameplate data and the corresponding recommended design values of heat gain from typical computer equipment, laser printers, copiers, and other miscellaneous office equipment. A few of these values are listed in Table 10.5.

Table 10.5 Recommended heat gain from typical office equipment*

Equipment	Nameplate power Consumption W	Average power consumption, W
Desktop computer	1200-480	97-73
Laptop computer	130-50	36-12
Flat-panel monitor	383-240	90-19
Laser printer, office type	1344-430	130-137
Copy machine	1850-1750	1060-800
Fax machine	936-40	90-20

*Representative values extracted from Tables 8 and 9, Page 18.11, *ASHRAE Handbook* - 2013 *Fundamentals* [1].

10.5 Transient Effects in Building Energy Transfer

The *steady* heat flow processes across walls, roofs, fenestrations, and basement floors of buildings were analyzed in chapter 8. We shall use the expressions thus obtained to calculate the winter heating load of buildings [1].

However, for calculating summer cooling loads of buildings we need to consider the *time dependent* nature of the heat gains by the indoor air. The use of steady-state expressions to estimate heat gains under summer conditions would significantly overestimate the cooling load, and hence lead to the selection of an oversized cooling system.

The energy transfer processes in a building are, in general, unsteady due to: (i) variations in the solar radiation intensity, the ambient temperature, and the wind speed, (ii) the thermal mass of the building envelope elements like the walls and the roof, (iii) the thermal mass of the surfaces within the indoor space that absorb and store incident solar radiation and radiant energy from sources such as lights, (iv) variations in the usage schedule of lights and equipment, and (v) variations of the occupancy schedule of people.

Figure 10.5(a) depicts a simplified schematic of a building where a few of the above energy transfer processes are illustrated. Solar radiation incident on the external wall of the building is partially absorbed. A portion of this energy is transferred to the external ambient by convection and radiation and the rest is conducted through the wall.

Solar radiation transmitted through the window, on the other hand, is absorbed on the table top and later transferred to the room by convection and radiation as shown in Fig. 10.5(c). Similarly, when the light is switched on the radiant energy landing on the table is partially absorbed and later released to the room by convection and radiation.

In this section we shall develop models to analyze the aforementioned transient energy transfer processes that contribute to the cooling load of the building.

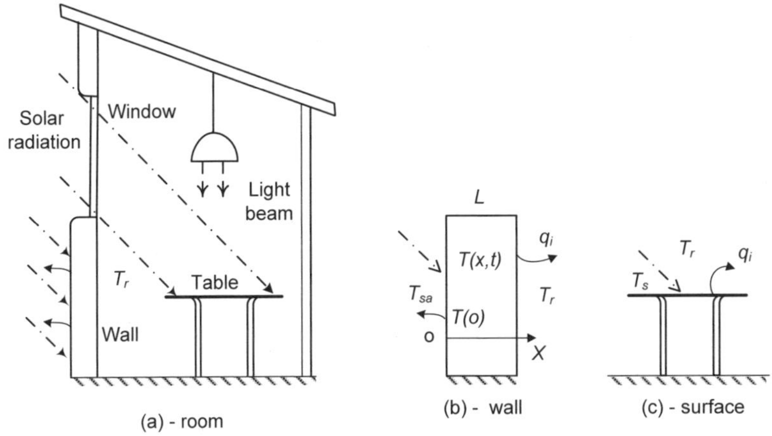

Fig. 10.5 Transient heat transfer processes in building components

10.5.1 *Transient heat conduction through walls*

Heat conduction through a wall (see Fig. 10.5b) is governed by Fourier's law of heat conduction. The transient temperature distribution, is obtained by solving the *one-dimensional* heat diffusion equation:

$$\rho c \frac{\partial T}{\partial t} = k \frac{\partial^2 T}{\partial x^2} \tag{10.8}$$

where $T(x,t)$ is the temperature at distance x from the outer wall surface at time t. The density is ρ, the specific heat capacity is c, and the thermal conductivity is k.

The boundary condition at the outer wall may be written in terms of the sol-air temperature as was demonstrated in section 9.3. Hence we have

$$-k\frac{\partial T}{\partial x}\bigg]_{x=0} = h_o[T_{sa}(t) - T(0)] \tag{10.9}$$

where $T_{sa}(t)$ is the sol-air temperature given by Eq. (9.39). The overall heat transfer coefficient at the outer surface is h_o. The sol-air temperature is, in general, time dependent because the solar radiation intensity and the ambient temperature vary with time.

At the inner surface of the wall the boundary condition is

$$-k\frac{\partial T}{\partial x}\bigg]_{x=L} = h_i[T(L) - T_r] \tag{10.10}$$

where T_r is the indoor air temperature, and h_i is the overall heat transfer coefficient at the inner surface.

Analytical methods for solving Eq. (10.8) are documented in most textbooks on heat transfer, including those listed in chapter 2. A finite *difference method* based on the thermal network representation is described in Ref. [2] and Ref. [3]. An analytical approach, called the *transfer function* method, that yields a convenient expression for the heat flux at the inner wall is presented in Ref. [2]. Chapter 18 of the *ASHRAE Handbook* - 2013 *Fundamentals* [1] also lists the main equations of the transfer function approach. A complete discussion of the above solution procedures, now carried out commonly using computer software programs, is beyond the scope of the present book.

The transfer function method has the advantage that it can be used to develop a convenient set of parameters, called *transfer function coefficients*, to compute the heat flux at the inner surface of the wall, which affects the cooling load directly [2]. The transfer function coefficients relate the heat flux, $Q_{i,t}$ at time t to the heat flux, $Q_{i,\tau}$, the sol-air temperature, $T_{sa,\tau}$, and the indoor temperature, $T_{r,\tau}$ at an earlier time τ. If we express the time in terms of a series of time intervals of *equal duration* Δt, then the time difference is given by

$$t - \tau = n\Delta t$$

where n is the number of intervals.

Hence the heat flux may be written in the form [2]

$$Q_{i,t} = \sum_{n=0}^{\infty}\left[b_n T_{as,(t-n\Delta t)} - c_n T_{r,(t-n\Delta t)}\right] - \sum_{n=1}^{\infty} d_n Q_{i,(t-n\Delta t)} \quad (10.11)$$

The transfer function coefficients, b_n, c_n and d_n have been determined for a large number of typical wall and roof sections using a time interval, Δt of 1 hour. These coefficients are tabulated in Ref. [2]. The derivation of Eq. (10.11) from first principles requires the application of several transformation techniques like the Laplace transform method and the Z-transform method [2].

10.5.2 *Heat gain by a thin surface*

In this section we shall consider the energy absorption by a *thin well-conducting surface*, like a metal table top (see Fig. 10.5c) for which a *lump capacitance* model may be applied. For this simplified situation we could derive the surface heat flux and the *transfer function coefficients* in a straightforward manner.

Since metal table top has a high thermal conductivity we assume that its temperature T_s is *uniform*. The surface absorbs a fraction α of the incident solar radiation, $I_{s,in}$ and transfers energy to the room by convection and long-wave radiation, for which the overall heat transfer coefficient is h_s. We assume that the *room temperature, T_r is constant* and that all other heat transfers rates are negligible. Subject to the above assumptions, the time dependent energy balance equation for the table top may be written as

$$\rho c l A \frac{dT_s}{dt} = \alpha A I_{s,in}(t) - A h_s(T_s - T_r) \quad (10.12)$$

where A is the surface area, l is the thickness, c is the specific heat capacity, and ρ is the density.

The *heat flux* per unit area entering the room from the surface is

$$q_i = h_s(T_s - T_r) \quad (10.13)$$

Substituting from Eq. (10.13) in Eq. (10.12) we have

$$\frac{dq_i}{dt} + \lambda q_i = \beta I_{s,in}(t) \tag{10.14}$$

The constant parameters of Eq. (10.14) are defined as

$$\lambda = \frac{h_s}{\rho c l} \quad \text{and} \quad \beta = \frac{h_s \alpha}{\rho c l} = \alpha\lambda \tag{10.15}$$

The solution of the first order differential Eq. (10.15) is obtained by introducing an integrating factor, exp (λt). Hence we have

$$e^{\lambda t}\frac{dq_i}{dt} + \lambda e^{\lambda t} q_i = \alpha\lambda e^{\lambda t} I_{s,in}(t)$$

$$\frac{d}{dt}\left(e^{\lambda t} q_i\right) = \alpha\lambda e^{\lambda t} I_{s,in}(t) \tag{10.16}$$

Integrating both sides of Eq. (10.16) we obtain the heat flux at time t as

$$q_i(t) - q_{io}e^{-\lambda t} = \alpha\lambda \int_0^t I_{s,in}(t')e^{-\lambda(t-t')}\,dt' \tag{10.17}$$

where q_{i0} is the heat flux at time zero.

We can interpret the expression within the integral on the RHS of Eq. (10.17) as the contribution to the heat flux, q_i at time t, by the heat gain due to solar radiation absorption at an earlier time t'.

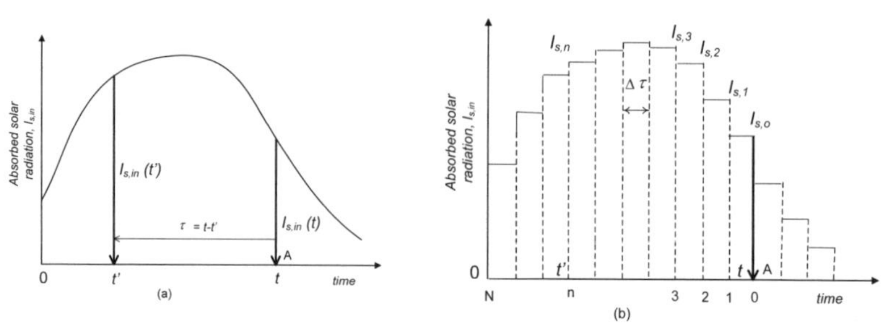

Fig. 10.6 Variation of heat gain due to radiation absorption with time

The graph in Fig. 10.6(a) shows the variation of the heat gain due to solar radiation absorption with time. The physical interpretation of Eq. (10.17) can be further clarified by shifting the origin of the time axis to point A where the time is t.

We transform Eq. (10.17) by introducing the variable τ defined by

$$\tau = t - t' \quad \text{and} \quad d\tau = dt'$$

Equation (10.17) now takes the form

$$q_i(t) - q_{io}e^{-\lambda t} = \alpha\lambda \int_0^t I_{s,in}(\tau)e^{-\lambda\tau}\, d\tau \qquad (10.18)$$

Equation (10.18) shows how the *current* heat flux is affected by the heat gain that occurred at a time interval τ before the present time.

Measured solar radiation data are usually available in the form of hourly averaged values. Therefore the variation of the heat gain due to solar radiation absorption may be represented as a series of step functions of constant width, $\Delta\tau$, usually one hour, as shown in Fig. 10.6(b). We now express the integral in Eq. (10.18) as a summation over these finite time intervals. For the nth time interval *before* the present time, let the solar radiation absorption rate be $I_{s,n}$.

The integral on the RHS of Eq. (10.18) can be evaluated directly for each time interval, [$n\Delta\tau$ to $(n+1)\Delta\tau$] because the rate of heat gain, $I_{s,n}$ is assumed *constant* over the time interval (1 hour). Hence we have

$$q_i(t) - q_{io}e^{-\lambda N\Delta\tau} = \alpha \sum_{n=0}^{(N-1)} \left[e^{-n\lambda\Delta\tau} - e^{-(n+1)\lambda\Delta\tau}\right] I_{s,n} \qquad (10.19)$$

$$q_i(t) = q_{io}e^{-\lambda N\Delta\tau} + \sum_{n=0}^{(N-1)} A_n\, \alpha I_{s,n} \qquad (10.20)$$

where N is the total number of intervals such that, $t = N\Delta\tau$.

In Eq. (10.20) we have defined the terms within the square brackets in Eq. (10.19) as a series of *transfer function coefficients*, A_n that relate the heat flux, q_i at time t to the past heat absorption rates, $\alpha I_{s,n}$. These coefficients are given by

$$A_n = e^{-n\lambda\Delta\tau} - e^{-(n+1)\lambda\Delta\tau}, \ (n = 0, N\text{-}1) \qquad (10.21)$$

The lumped capacity model, which is only applicable to highly conducting thin layers of material, is included here mainly to explain the physical meaning of *transfer function coefficients*.

However, for actual walls and roofs, the transient heat transfer analysis needed to obtain the transfer function coefficients must include both the spatial and the time variations of the temperature distribution. The mathematical methods required to perform such an analysis is beyond the scope of this book.

10.6 Cooling Load Calculation Methods

In the proceeding sections of this chapter we have identified several heat gain processes contributing to the cooling load of a space. These include heat gains from people, lights and equipment which are partly convective and partly radiative. Heat entering from the surfaces of walls, roofs and windows also include convective and radiative components.

Although we have analyzed each of the above heat gain processes separately, in actual fact, these processes are coupled. A closer look at the simple schematic in Fig. 10.5(a) shows that the six surfaces of the indoor space and the surfaces within the space exchange heat by radiation. Moreover, the same surfaces transfer heat to indoor air by convection.

The *heat balance method* (HBM) described in Chapter 18 of the *ASHRAE Handbook* - 2013 *Fundamentals* [1], is a computational procedure to analyze the coupled heat transfer processes mentioned above. This method has been used to predict the cooling load of test rooms and the predictions were found to agree well with experimentally determined cooling loads [1].

10.6.1 *Heat balance method (HBM)*

The main assumptions of the HBM are: (i) the indoor air of a zone is well mixed, (ii) the temperatures of the surfaces of the walls and windows are uniform, (iii) long-wave and short-wave radiation distributions on surfaces are uniform, (iv) all surfaces are diffuse emitters of long-wave radiation, and (v) heat conduction through walls is one-dimensional.

The overall heat balance for a zone consists of three separate heat balances.

(a) The heat balance at each *exterior* surface of the building envelope (see Fig. 10.5a) includes: (i) absorbed direct and diffuse radiation, $q_{a,sol}$ (ii) long-wave radiation exchange with surrounding surfaces, q_{olw}, (iii) convective heat exchange with outdoor ambient air, $q_{o,con}$, and (iv) conductive heat flux into the surface, q_{ko}. Heat balance at the *exterior* surface gives

$$q_{a,sol} + q_{olw} + q_{o,con} - q_{ko} = 0 \qquad (10.22)$$

Recall that in section 9.3 the sol-air temperature was used to express the energy balance at an exterior surface in a compact manner.

(b) The heat balance at each *interior* surface of a zone (see Fig. 10.5a) includes: (i) the net long- wave radiant heat flux, q_{ilw}, (ii) the net short-wave radiant heat flux from lights in the zone, q_{lsw}, (iii) the long-wave radiant heat flux from equipment in the zone, q_{elw}, (iv) the transmitted solar radiation absorbed at inner surfaces, $q_{i,sol}$, (v) the convective heat exchange with indoor air, $q_{i,con}$, and (vi) the conductive heat flux into the surface, q_{ki}. Heat balance at the *interior* surface gives

$$q_{ilw} + q_{lsw} + q_{elw} + q_{i,sol} + q_{ki} + q_{i,con} = 0 \qquad (10.23)$$

Heat conduction through a wall is governed by the transient conduction equation, given as Eq. (10.8). The latter equation may be solved numerically or by using the transfer function method. The boundary conditions are the specified exterior and interior surface temperatures, T_o and T_i respectively. The conductive heat fluxes at the surfaces, q_{ko} and q_{ki} are given by the product of the wall thermal conductivity and the respective temperature gradients at the wall surfaces.

A coupled analysis of the long-wave radiation exchange between the interior surfaces and the simultaneous convective heat transfer from the surfaces to the indoor air is quite complex. A simplified approach is to split the total heat flux from a surface into a radiative component and a convective component using preassigned fractions. In Table 10.4 we have listed such fractions for the energy released from lights. Items within the zone like furniture (see Fig. 10.5a) increase the surface area for radiation exchange and the thermal mass of the zone.

Short-wave radiation from lights is spread over the interior surfaces using a distribution function dependent on the type of light fixture or lamp [1]. Long-wave radiation from lights is included as a preassigned fraction of the total energy output of lights. The same approach is used for long-wave radiation from people and equipment.

A convenient parameter to calculate the net energy gain through windows, is the solar heat gain coefficient (SHGC), which was derived in section 9.5. Although the SHGC has been used as a practical guide to rate and characterize glazing systems, it is not used directly in the heat balance method (HBM), but as two separate terms. Now the net heat gain through a window is given by Eq. (9.76) as

$$q_{tot} = E_{dn}cos\theta[T(\theta) + \sum_{i=1}^{n} f_{si}(\theta)N_i] + \frac{(t_{ao}-t_{ai})}{R_{tot}} \quad (10.24)$$

The SHGC is defined by the expression (see section 9.5)

$$SHGC(\theta) = T(\theta) + \sum_{i=1}^{n} f_{si}(\theta)N_i \quad (10.25)$$

Rearranging Eqs. (10.24) and (10.25) we obtain

$$q_{tot} = TE_{dn}cos\theta + [E_{dn}cos\theta(SHGC - T) + \frac{(t_{ao}-t_{ai})}{R_{tot}}] \quad (10.26)$$

The first term in Eq. (10.26) is the solar radiation transmitted through a window. It is included as, $q_{i,sol}$ in the heat balance equation (10.23) for interior surfaces on which the solar radiation is incident. The expression inside the square bracket in Eq. (10.26), is the total heat flux at the interior surface of the window, consisting of the inward flowing component of the solar radiation absorbed in the glazings, and the heat conducted through the window. This quantity is included as, q_{ki} in the heat balance equation (10.23) for windows.

(c) The heat balance for *indoor air* includes: (i) convective heat flow from interior surfaces, q_{conv}, (ii) convective heat flows, $q_{is,c}$ from internal sources, such as lights, people and equipment, taken as preassigned fractions of the total heat production rates of these sources, (iii) heat required to change the temperature of infiltration air to the zone air temperature, q_{inf}, and (iv) the heat removed by the cooling system, q_{sys}. Neglecting the thermal capacity of the air and assuming quasi-steady conditions, the energy equation for zone air may be written as

$$q_{conv} + q_{is,c} + q_{inf} - q_{sys} = 0 \quad (10.27)$$

The HBM is applied separately to each thermal zone, maintained at a fixed temperature by the cooling system. A typical zone consists of 4 walls, a roof or ceiling and a floor. Each wall and roof could have a fenestration area. In addition, a surface, called '*a thermal mass surface*', is included to account for any surfaces within the zone. This surface exchanges thermal radiation with the interior surfaces of the zone and also adds extra thermal mass to the zone.

The heat balance equations listed above for the different surfaces of a zone are solved simultaneously for a 24-hour steady-periodic condition (see Fig. 10.7) to obtain the inside and outside surface temperatures. Hence the time variation of the cooling load of the zone is calculated. The solution procedure, the input information required, and other useful details on the HBM are available in Chapter 18 of the *ASHRAE Handbook* - 2013 *Fundamentals* [1].

10.6.2 *Radiant time series (RTS) method*

The radiant time series (RTS) is a simplified method of cooling load estimation, derived from the heat balance method (HBM). It is presented in Chapter 18 of the *ASHRAE Handbook* - 2013 *Fundamentals* [1] as a replacement for earlier manual methods, such as, the transfer function method (TFM), the cooling load temperature difference (CLTD) method, the cooling load factor (CLF) method, and the total equivalent temperature difference (TETD) method.

Although simple in concept, the RTS method still requires more computational effort than the manual methods mentioned above. However, the RTS method could be implemented using a computerized spread sheet.

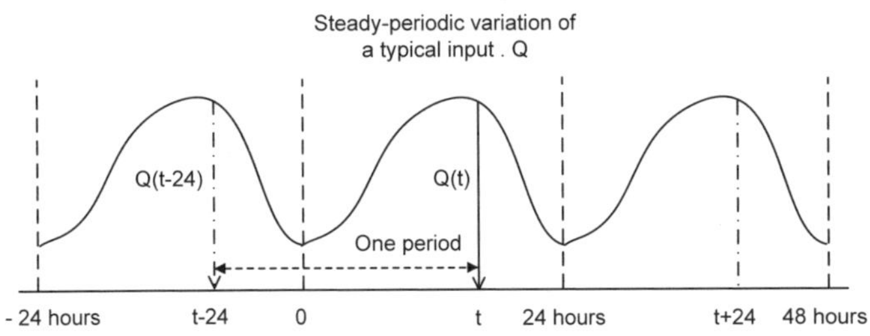

Fig. 10.7 24- hour steady-periodic conditions

In the RTS method the cooling loads are calculated assuming 24-hour steady–periodic conditions for all inputs as shown graphically in Fig. 10.7. This means that the patterns of variation of inputs like the weather conditions, the lighting schedule, and the occupancy schedule are repeated every 24 hours.

The main differences between the earlier manual methods and the RTS method are summarized below.

(i) The time delay caused by the thermal mass of walls and roofs are taken into account by introducing a series of coefficients called the *conduction time factors* (CTF).

(ii) The total heat gain from interior surfaces of walls and roofs, and internal sources like lights and people are split into convective and radiative components using preassigned fractions.

(iii) The radiation from various internal sources, and transmitted solar radiation, does not contribute to the cooling load immediately because of the thermal mass of the surfaces that absorb the radiation. The absorbed radiation is later released to the air by convection. This time delay is taken into account by using a series of coefficients called *radiant time factors* (RTF). The conduction time factors and the radiant time factors are calculated using the HBM for 24-hour steady–periodic conditions.

We shall now extend the lumped capacity model, developed in section 10.5.2, to illustrate the application of a 24-hour steady–periodic input. As seen from Fig. 10.7, for a steady 24-hour period, the time axis extends from its origin, O at (t-24) to the current time, t. Due to the periodic nature of the inputs, the instantaneous heat flux, q_i is also repeated over 24-hour periods. Hence we have

$$q_i(t) = q_i(t-24) = q_{io} \tag{10.28}$$

Applying the above condition to Eq. (10.19) we obtain

$$q_i(t)\left(1 - e^{-\lambda N\Delta\tau}\right) = \alpha \sum_{n=0}^{(N-1)} \left[1 - e^{-\lambda\Delta\tau}\right] e^{-n\lambda\Delta\tau} I_{s,n} \tag{10.29}$$

$$q_i(t) = \sum_{n=0}^{(N-1)} e^{-n\lambda\Delta\tau} \left[\frac{1-e^{-\lambda\Delta\tau}}{1-e^{-\lambda N\Delta\tau}}\right] (\alpha I_{s,n}) \tag{10.30}$$

When Eq. (10.30) is applied to a 24-hour period, N = 24 and $\Delta\tau = 1$ hour. We now rewrite Eq. (10.30) in terms of a series of *radiant time factors*, r_n defined by

$$r_n = \left[\frac{1-e^{-\lambda\Delta\tau}}{1-e^{-\lambda N\Delta\tau}}\right] e^{-n\lambda\Delta\tau} \tag{10.31}$$

Hence the instantaneous heat flux (cooling load) at time t may be written as

$$q_i(t) = \sum_{n=0}^{(N-1)} r_n\,(\alpha I_{s,n}) \tag{10.32}$$

From Eq. (10.31) we obtain the sum of the radiant time factors for the 24-hour period as

$$\sum_{n=0}^{N-1} r_n = \left[\frac{1-e^{-\lambda\Delta\tau}}{1-e^{-\lambda N\Delta\tau}}\right] \sum_{n=0}^{(N-1)} e^{-n\lambda\Delta\tau} = 1 \tag{10.33}$$

Therefore the sum of the hourly radiant time factors for the 24 hours is equal to 1.

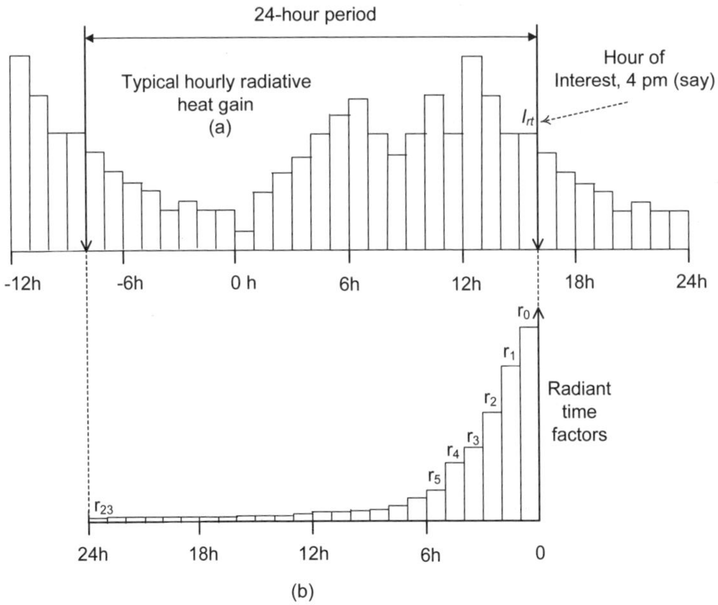

Fig. 10.8 Application of the radiant time factors

A representative distribution of hourly radiative heat gains into a zone is depicted in Fig. 10.8(a). A time series of the hourly radiant time factors, r_i (i = 0,23) for the particular type of heat gain is shown in Fig. 10.8(b). These factors are functions of the thermal mass of the participating surfaces and properties such as the thermal conductivity and the heat transfer coefficient.

Once computed for a particular surface arrangement, these coefficients are fixed and they can be used to calculate the time delay effects of different radiative heat gain distributions from the same source. For example, if the radiative heat gains shown in Fig. 10.8(a), are from people in a zone, then different occupancy schedules will produce different radiant heat gain distributions. However, the radiative time factors shown in Fig. 10.8(b) for people are the same. Similar considerations apply to different lighting schedules and the corresponding radiant time factors.

Applying Eq. (10.32) to the distributions shown in Fig. 10.8 we obtain the expression for the instantaneous heat gains at hour t as

$$q_{i,t} = r_0 I_{r,t} + r_1 I_{r,(t-1)} + r_2 I_{r,(t-2)} + r_3 I_{r,(t-3)} + \cdots + r_{23} I_{r,(t-23)}$$

$$q_{i,t} = \sum_{n=0}^{23} r_n I_{r,(t-n)} \tag{10.34}$$

where $I_{r,(t-n)}$ is the radiant heat gain rate during the nth hour before the current hour t, and r_n is the radiant time factor for hour n.

The time delay due to the thermal mass of walls and roofs are accounted for by using *conduction time factors*. The procedure is identical to that described above for applying the radiant time factors. Hence the hourly heat gain due to conduction, $q_{ic,t}$ at time t is

$$q_{ic,t} = \sum_{n=0}^{23} c_n Q_{e,(t-n)} \tag{10.35}$$

where c_n is the *conduction time factor* for hour n. $Q_{e,(t-n)}$ is the heat input at the exterior surface of the wall during the nth hour before the current hour t, which is given by

$$Q_{e,(t-n)} = UA\left[t_{sa,(t-n)} - t_{rc}\right] \tag{10.36}$$

In Eq. (10.36), U is the overall heat transfer coefficient for the wall and A is the surface area. The sol-air temperature during the nth hour before the current hour t is $t_{sa,(t-n)}$. The indoor air temperature, t_{rc} is assumed to remain constant. A lumped capacity model of a wall or a roof may be used to clarify the meaning of Eqs. (10.35) and (10.36) (see problem P10.11).

10.6.3 *Application of the RTS method and the CTS method*

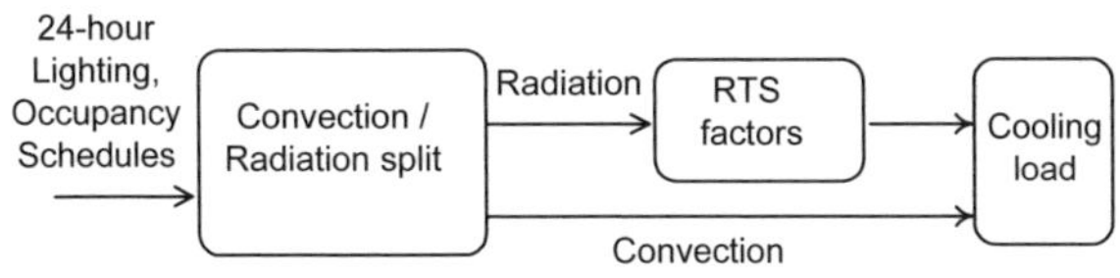

Fig. 10.9(a) Computation of cooling load due to people, lights and equipment

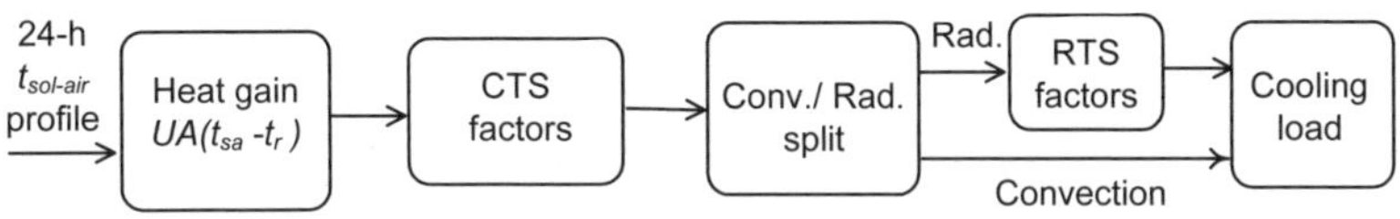

Fig. 10.9(b) Computation of cooling load due conduction heat gains

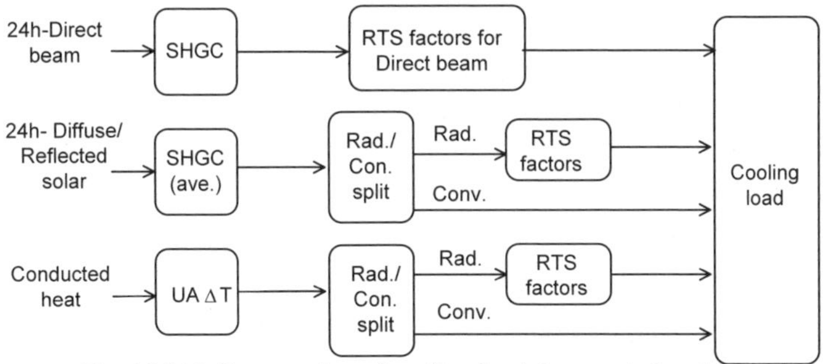

Fig. 10.9(c) Computation of cooling load due to window heat gains

In the preceding section we described the RTS-method and the CTS-method to convert heat gains to cooling loads by accounting for the time delays caused by the thermal mass of walls, roofs and surfaces within a zone. The computational procedures for different types of heat gains are illustrated using block diagrams in Figs. 10.9(a), (b) and (c).

The conduction time factors for a number of different walls and roofs are presented in Tables 16 and 17 in Chapter 18 of the *ASHRAE Handbook* - 2013 *Fundamentals* [1]. Radiant time factors for non-solar and solar heat gains are listed in Tables 19 and 20 respectively [1]. Representative values of conduction and radiant time factors, extracted from Ref. [1], are given in Tables 10.6(a), (b), (c) and (d).

Table 10.6(a) Conduction time series (CTS) for curtain wall*

Mass = 20.9 kgm^{-2}, Thermal capacity = 20.4 kJm^{-2}K^{-1}, U = 0.429Wm^{-2}K^{-1}												
Hour	0	1	2	3	4	5	6	7	8	9	10	11
C_n%	25	57	15	3	0	0	0	0	0	0	0	0
Hour	12	13	14	15	16	17	18	19	20	21	22	23
C_n%	0	0	0	0	0	0	0	0	0	0	0	0

*Values extracted from Table 16, Page 18.24, *ASHRAE Handbook* - 2013 *Fundamentals* [1]

Table 10.6(b) Conduction time series (CTS) for brick wall*

Mass = 304 kgm^{-2}, Thermal capacity = 253.5kJm^{-2}K^{-1}, U = 0.348Wm^{-2}K^{-1}												
Hour	0	1	2	3	4	5	6	7	8	9	10	11
C_n%	2	2	2	3	5	6	7	7	7	7	6	6
Hour	12	13	14	15	16	17	18	19	20	21	22	23
C_n%	5	5	5	4	4	3	3	3	3	2	2	1

*Values extracted from Table 16, Page 18.24, *ASHRAE Handbook* - 2013 *Fundamentals* [1]

Table 10.6(c) Conduction time series (CTS) for metal deck roof*

Mass = 57.6 kgm^{-2}, Thermal capacity = 57.2 kJm^{-2}K^{-1}, U = 0.297Wm^{-2}K^{-1}												
Hour	0	1	2	3	4	5	6	7	8	9	10	11
C_n%	0	10	22	20	14	10	7	5	4	3	2	1
Hour	12	13	14	15	16	17	18	19	20	21	22	23
C_n%	1	1	0	0	0	0	0	0	0	0	0	0

*Values extracted from Table 17, Page 18.26, *ASHRAE Handbook* - 2013 *Fundamentals* [1]

Table 10.6(d) Nonsolar radiation time series (RTS) values*

Light construction, With carpet, 50 percent glass												
Hour	0	1	2	3	4	5	6	7	8	9	10	11
C_n%	50	18	10	6	4	3	2	1	1	1	1	1
Hour	12	13	14	15	16	17	18	19	20	21	22	23
C_n%	1	1	0	0	0	0	0	0	0	0	0	0

*Values extracted from Table 19, Page 18.28, *ASHRAE Handbook* - 2013 *Fundamentals* [1]

Table 10.6(e) Solar radiation time series (RTS) values*

Heavy construction, With carpet, 10 percent glass												
Hour	0	1	2	3	4	5	6	7	8	9	10	11
C_n%	47	11	6	4	3	2	2	2	2	2	2	2
Hour	12	13	14	15	16	17	18	19	20	21	22	23
C_n%	2	2	2	1	1	1	1	1	1	1	1	1

*Values extracted from Table 20, Page 18.29, *ASHRAE Handbook* - 2013 *Fundamentals* [1]

In the computational schemes shown in Fig. 10.9, the total heat gain from lights, equipment and occupants, and the conduction heat gains through walls, roofs and windows are split into convective and radiative fractions. Table 14 in Chapter 18 of the *ASHRAE Handbook* - 2013 *Fundamentals* [1] gives a series of recommended values for the above fractions. We have listed a few representative values in Table 10.7.

Table 10.7 Radiative and convective fractions for various heat gains*

Type of heat gain	Radiative fraction	Convective fraction
Occupants, office conditions	0.6	0.4
Office equipment	0.1 to 0.8	0.9 to 0.2
Lights	See Table 11.4	
Conduction through walls	0.46	0.54
Conduction through roofs	0.6	0.4
Conduction through windows		
SHGC>0.50	0.33	0.67
SHGC<0.5	0.46	0.54

*Values extracted from Table 14, Page 18.22, *ASHRAE Handbook* - 2013 *Fundamentals* [1]

We shall illustrate the application of the RTS method to cooling load calculations in the worked examples to follow in this chapter. Comprehensive design calculations for estimating 24-hour cooling load distributions due to people, wall conduction, windows, and windows with overhangs are carried out using the MATLAB codes listed in Appendix A10.1 to A10.4 respectively.

10.7 Heating Load Calculation Methods

The methods for calculating the winter heat losses are in principle the same as those used for estimating the heat gains under summer conditions. However, the conservative design approach recommended in the *ASHRAE Handbook* - 2013 *Fundamentals* [1] allows us to ignore the several heat gains because their contributions partially balance the heating load on the building. These include: (i) heat inputs from lights, people and equipment, and (ii) solar heat gain through fenestrations. Furthermore, we could neglect the thermal energy storage in walls and roofs, and the resulting time delays, because they effectively decrease the heat load estimated under steady conditions.

The simplified approach for estimating the heating load of a building involves the following steps.

(i) Selection of the outdoor design conditions: temperature, humidity and wind speed. The climatic data needed for this purpose are discussed in section 10.2.

(ii) Selection of the design temperature and relative humidity for the indoor environment. The recommended indoor conditions for thermal comfort are discussed in section 10.3.

(iii) Estimation of the steady heat losses through walls, roofs, windows, floors and below-grade surfaces. The governing heat transfer rate equation may be expressed in the general form

$$Q_{loss} = AU_{eff}(T_r - T_{amb})$$

where Q_{loss} is the rate of heat loss, A is the area, U_{eff} is the overall indoor–outdoor heat transfer coefficient, T_r is the indoor temperature, and T_{amb} is the outdoor ambient temperature.

The methods for estimating the above heat losses are discussed in sections 8.2, 8.3 and 8.4. Moreover, the worked examples presented in section 8.7 illustrate the computational procedures to determine these heat losses.

(iv) Estimation of the heat load due to the infiltration of cold air into the building. The heat input, Q_{inf} required to raise the temperature of the infiltration air from T_{amb} to the indoor temperature T_r is given by

$$Q_{inf} = \rho \dot{V}_{inf} c_p (T_r - T_{amb})$$

where $\rho, \dot{V}_{inf}$ and c_p are the density, the volume flow rate and the specific heat capacity of infiltration air. The techniques for calculating the infiltration heat load are discussed in section 8.5, and numerical examples are presented in section 8.7.

(v) Computation of the total heating load by adding the steady heat losses through the structural elements and the infiltration heat loss.

A simplified procedure for estimating the heating and cooling loads of *residential buildings* is presented in Chapter 17 of the *ASHRAE Handbook* - 2013 *Fundamentals* [1]. It is based on the residential load factor (RLF) method. RLF values have been obtained from detailed simulations of a variety of residential buildings across a range of climates, using the heat balance (HB) method. The empirical relations used to estimate different load components involve cooling and heating load factors (CFs and HFs). These factors, in turn, incorporate a series of tabulated coefficients obtained from detailed simulations and statistical regression techniques.

Within its range of applicability, the loads predicted by the RLF method agreed within about 10% of those given by the HB method. The calculation procedure and the use of tabulated data are illustrated in Ref. [1] using a comprehensive worked example.

10.8 Worked Examples

Example 10.1 Three persons occupy an air conditioned office. One of them, wearing a suit with thermal resistance of 1.0 clo, is engaged in managerial activities for which the metabolic energy production rate is 1.6 mets. The other two persons, wearing lightweight clothing with thermal resistance of 0.6 clo, are working at computer terminals for which the metabolic rate is 1.1 mets. Determine the optimal air temperature for the office.

Solution The optimal air temperature for comfort is given by Eq. (10.2).

$$T_{a,opt} = 27.2 - 5.9R_c - 3.0(1 + R_c)(\dot{M} - 1.2)$$

Substituting the values of R_c and $\dot{M}$ for the manager we have

$$T_{a,opt,1} = 27.2 - 5.9 \times 1.0 - 3.0(1 + 1.0)(1.6 - 1.2) = 19°\text{C}$$

Similarly, for persons at the computer terminals

$$T_{a,opt,2} = 27.2 - 5.9 \times 0.6 - 3.0(1 + 0.6)(1.1 - 1.2) = 24°\text{C}$$

There is a difference of about 5°C between the indoor temperatures that provide optimal comfort for the different individuals. A possible solution is for the manager to modify his clothing level such that, R_c = 0.7 clo. This would bring the comfortable temperature for the manager to 21°C. However, providing a single comfortable temperature for occupants in the same zone wearing different types of clothing is a general problem faced by building designers.

Example 10.2 The effective temperature, ET* for a particular indoor ambient is 25°C. The ratio of sensible heat parameter c_{sen}, to the latent heat parameter c_{lat}, is 0.0038. The average skin temperature is 33.7°C. (i) Calculate the ratio of the sensible heat transfer to the latent heat transfer to the ambient. (ii) Calculate the ambient relative humidity if the temperature changes to 23°C and 27°C for the same ET*.

Solution The sensible and latent heat components of the total heat transfer are obtained from Eq. (10.3),

$$Q_{sk} = c_{sen}(t_{sk} - t_o) + c_{lat}(\omega_{s,sk} - \omega) \qquad \text{(E10.2.1)}$$

The skin temperature, t_{sk} =33.7°C. The saturation humidity ratio at the skin temperature, obtained from tabulated data in Ref. [2], is 0.03406. The effective temperature is 25°C. The humidity ratio, at 25°C and 50% relative humidity is 0.01.
The sensible heat to latent heat transfer ratio, r is given by

$$r = \frac{c_{sen}(t_{sk}-t_o)}{c_{lat}(\omega_{s,sk}-\omega_{s,o})} = \frac{0.0038(33.7-25)}{(0.03406-0.01)} = 1.37$$

From Eq. (E10.2.1) it follows that along a *constant-ET* line*

$$c_{sen}t_o + c_{lat}\omega = constant$$

Applying the above equation for, $t_o = ET^* = 25$°C and $t_o = 23$°C we have

$$c_{sen} \times 25 + c_{lat} \times 0.01 = c_{sen} \times 23 + c_{lat} \times \omega$$

Substituting numerical values in the above equation we obtain, $\omega =$ 0.0176. Hence the relative humidity at 23°C is 96%.
Similarly by substituting $t_o = ET^* = 25$°C and $t_o = 27$°C we have

$$c_{sen} \times 25 + c_{lat} \times 0.01 = c_{sen} \times 27 + c_{lat} \times \omega$$

Therefore $\omega = 0.0024$ and the relative humidity at 27°C is 12%.

Example 10.3 An office building has three zones A, B and C. The occupancy levels, the activities, and the circulation air flow rates based on the cooling loads of the zones are tabulated below. The summer outdoor conditions are 34°C and 65% relative humidity.

Table E10.3.1 Design data

Zone	Occupancy level	Office Activity	Air flow rate, Ls^{-1}
A	25	Light (1.0 met)	2200
B	40	Moderate (1.25 met)	4000
C	30	Moderate(1.25 met)	3500

Determine (i) the indoor design temperatures for the zones, (ii) the design ventilation air flow rates for the zones, and (iii) the total ventilation cooling load.

Solution (i) We compute the optimal air temperatures for the zones using Eq. (10.2),

$$T_{a,opt} = 27.2 - 5.9R_c - 3.0(1 + R_c)(\dot{M} - 1.2)$$

Substituting R_c = 0.4 clo, for summer clothing, and $\dot{M}$ =1.0 mets, for light office work, the air temperature for zone A is

$$T_{a,opt,A} = 27.2 - 5.9 \times 0.4 - 3.0(1 + 0.4)(1.0 - 1.2) = 25.7°\text{C}$$

Substituting R_c = 0.4, for summer clothing and $\dot{M}$ = 1.25 mets, for moderate office work, the air temperature for zones B and C is

$$T_{a,opt,B} = 27.2 - 5.9 \times 0.4 - 3.0(1 + 0.4)(1.25 - 1.2) = 24.6°\text{C}$$

Since the optimum temperatures for the zones are close we shall use a mean value of 25°C. The indoor design relative humidity is 50%.

(ii) From the data in Table 10.2, the recommended ventilation air flow rate for light and moderate office work is 10Ls^{-1} person.

Table E10.3.2 Calculated values

Zone	N_p	$V_{rec,n}$, Ls^{-1}	$V_{cir,n}$, Ls^{-1}	$V_{rec,n}/V_{cir,n}$	$V_{des,n}$, Ls^{-1}
A	25	250	2200	0.1136	219
B	40	400	4000	0.10	398
C	30	300	3500	0.0857	348

Now the parameter X is given by Eq. (10.5) as

$$X = \frac{\sum_{n=1}^{N} \dot{V}_{rn}}{\sum_{n=1}^{N} \dot{V}_{cn}} = \frac{250+400+300}{2200+4000+3500} = 0.098$$

From the ratios of $(V_{rec,n}/V_{cir,n})$ in Table E10.3.2 for the different zones we see that zone A is the *critical zone*. The ratio of the *design ventilation air flow rate,* $\dot{V}_{vn}$ to zone *n* to the circulation flow rate $\dot{V}_{cn}$ is given by Eq. (10.6) as

$$Y = \frac{\dot{V}_{vn}}{\dot{V}_{cn}} = \frac{X}{X+1-Z} = \frac{0.098}{0.098+1-0.1136} = 0.0995$$

Hence the design ventilation flow rates are 0.0995 times the circulation flow rates of the zones. The computed design ventilation flow rates are given in Table E10.3.2 above.

(iii) The enthalpy of outdoor air at 34°C and 65% relative humidity is 91 kJkg^{-1} and the enthalpy of indoor air at 25°C and 50% relative humidity is 51 kJkg^{-1}. The total flow rate of ventilation air to the three zones is 965Ls^{-1}. The specific volume of outdoor air is 0.902 m^{3}kg^{-1}. The ventilation cooling load is given by

$$\dot{Q}_{ven} = \left(\dot{V}_{ven}/v_{sp}\right)(h_{out} - h_{in})$$

Substituting numerical values in the above equation we have

$$\dot{Q}_{ven} = (0.965/0.902)(91 - 51) = 42.8 \text{ kW}$$

Example 10.4 The zone of a building is occupied by people engaged in moderate office work. The occupancy schedule is given in Table E10.4.1. Each person generates 75 W of sensible heat and 55 W of latent heat (see Table 10.3). The radiative fraction of the sensible component is 0.6 (see Table 10.7). The interior of the zone is of light construction for which the appropriate non-solar RTS values from Table 19, Chapter 18 of the *ASHRAE Handbook* - 2013 *Fundamentals* [1] are given in Table E10.4.2.

Table E10.4.1 Occupancy schedule of zone

*Hour from midnight, N_p = number of people

Hour*	1	2	3	4	5	6	7	8	9	10	11	12
N_p	0	0	0	0	0	0	0	0	4	12	12	12
Hour*	13	14	15	16	17	18	19	20	21	22	23	24
N_p	12	12	12	12	12	12	5	0	0	0	0	0

Table E10.4.2 Non-solar RTS values

Light construction, With carpet, 50 percent glass												
Hour*	0	1	2	3	4	5	6	7	8	9	10	11
C_n%	50	18	10	6	4	3	2	1	1	1	1	1
Hour	12	13	14	15	16	17	18	19	20	21	22	23
C_n%	1	1	0	0	0	0	0	0	0	0	0	0

*Hour 0 is the current hour in the *ASHRAE* data table

(a) Calculate the cooling load due to people for (i) hour 11, and (ii) hour 15.

(b) Obtain the daily cooling load profile due to people.

Solution The latent heat gain and the convective component of the sensible heat gain from people contribute to the cooling load instantaneously. Hence for hour 11 we have

$$q_{cl1} = N_p(q_{lat} + 0.4q_{sen}) = 12(55 + 0.4 \times 75) = 1020 \text{ W}$$

The radiant components of the sensible heat gains during the previous 24 hours contribute to the cooling load at 11 a.m. These hourly contributions are obtained by multiplying the hourly radiant heat gains by the radiant time factors for the 24 hours. The total contribution is the sum of the hourly contributions as seen from Eq. (10.32). The radiative heat gain per person is $(0.6 \times 75 = 45W)$. The computed quantities are summarized in Table E10.4.3 below.

Table E10.4.3 Computation of radiant contribution for hour 11

*24-hours before hour 11												
Hour*	1	2	3	4	5	6	7	8	9	10	11	12
C_n%	50	18	10	6	4	3	2	1	1	1	1	1
N_p	12	12	4	0	0	0	0	0	0	0	0	0
q_{rad}	540	540	180	0	0	0	0	0	0	0	0	0
$C_n q_{rad}$	270	97.2	18	0	0	0	0	0	0	0	0	0
Hour*	13	14	15	16	17	18	19	20	21	22	23	24
C_n%	1	1	0	0	0	0	0	0	0	0	0	0
N_p	0	0	0	0	5	12	12	12	12	12	12	12
q_{rad}	0	0	0	0	225	540	540	540	540	540	540	540
$C_n q_{rad}$	0	0	0	0	0	0	0	0	0	0	0	0

Table E10.4.4 Computation of radiant contribution for hour 15

*number of hours before hour 15

Hour*	1	2	3	4	5	6	7	8	9	10	11	12
C_n%	50	18	10	6	4	3	2	1	1	1	1	1
N_p	12	12	12	12	12	12	4	0	0	0	0	0
q_{rad}	540	540	540	540	540	540	180	0	0	0	0	0
C_nq_{rad}	270	97.2	54	32.4	21.6	16.2	3.6	0	0	0	0	0
Hour*	13	14	15	16	17	18	19	20	21	22	23	24
C_n%	1	1	0	0	0	0	0	0	0	0	0	0
N_p	0	0	0	0	0	0	0	0	5	12	12	12
q_{rad}	0	0	0	0	0	0	540	0	225	540	540	540
C_nq_{rad}	0	0	0	0	0	0	0	0	0	0	0	0

The total contribution from radiant heat gains during the 24 hours prior to hour 11 is given by

$$q_{cl2} = 270 + 97.2 + 18 + 0 + 0 \ldots = 385.2 \text{ W}$$

Hence the total cooling load due to people for hour 11 is

$$q_{pt} = q_{cl1} + q_{cl2} = 1020 + 385.2 = 1405.2 \text{ W}$$

The total contribution from radiant heat gains during the 24 hours prior to hour 15 is given by

$$q_{cl2} = 270 + 97.2 + 54 + 32.4 + 21.6 + 16.2 + 3.6 + 0 = 495\text{W}$$

Hence the total cooling load due to people at hour 15 is

$$q_{pt} = q_{cl1} + q_{cl2} = 1020 + 495 = 1515 \text{ W}$$

The maximum hourly cooling load due to people may be found by repeating the above calculation procedure for all the hours and then selecting the largest value. The MATLAB code in Appendix 10.1 was used for this purpose. The results are summarized in Table E10.4.5.

Table E10.4.5 Cooling load due to people

Time	1	2	3	4	5	6	7	8	9	10	11	12
Q_{lat}+Q_{sen}	0	0	0	0	0	0	0	0	340	1020	1020	1020
Q_{rad}	42	35	29	24	19	13	8	2	90	302	385	432
Q_{tot}	42	35	29	24	19	13	8	2	430	1322	1405	1452
Time	13	14	15	16	18	18	19	20	21	22	23	24
Q_{lat}+Q_{sen}	1020	1020	1020	1020	1020	1020	425	0	0	0	0	0
Q_{rad}	461	481	495	504	509	514	363	199	132	96	74	55
Q_{tot}	1480	1501	1515	1524	1529	**1534**	788	199	132	96	74	55

We notice from the values in Table E10.4.5 that the highest cooling load due to people is 1534 W and it occurs at hour 18.

Example 10.5 The zone of a building has recessed fluorescent lights with a total output of 620 W. The space fraction is 0.76 and all the lights are operated from 8 a.m. to 8 p.m. The radiant fraction of the heat released by the lights is 0.65. The interior of the zone is of medium construction for which the appropriate non-solar RTS values from Table 19, Chapter 18 of the *ASHRAE Handbook* - 2013 *Fundamentals* [1] are tabulated below:

Table E10.5.1 Non-solar RTS values

Medium construction, No carpet, 10 percent glass

Hour	0	1	2	3	4	5	6	7	8	9	10	11
C_n%	31	17	11	8	6	4	4	3	3	2	2	2
Hour	12	13	14	15	16	17	18	19	20	21	22	23
C_n%	1	1	1	1	1	1	1	0	0	0	0	0

Table E10.5.2 Lighting schedule based on wattage

*Hours from midnight, W_L = Total lighting wattage

Hour*	1	2	3	4	5	6	7	8	9	10	11	12
W_L	0	0	0	0	0	0	0	0	620	620	620	620
Hour	13	14	15	16	17	18	19	20	21	22	23	24
W_L	620	620	620	620	620	620	620	620	0	0	0	0

Calculate (i) the cooling load due to lights at hour 16, and (ii) the daily cooling load profile due to lights.

Solution The convective component of the heat gain due to lights contributes to the cooling load immediately. The radiant component is delayed due to absorption by surfaces in the zone. We use the radiant time series to account for this time delay. Since the wattage is 620 W, the space fraction is 0.76, and the radiant fraction is 0.65, the convective cooling load is

$$q_{c1} = W f_{sp} f_{con} = 620 \times 0.76 \times 0.35 = 165 \text{ W}$$

The rate of radiant heat gain from lights is

$$q_{rad} = 620 \times 0.76 \times 0.65 = 306 \text{ W}$$

The radiant components of the heat gains from lights during the previous 24 hours contribute partially to the cooling load at hour 16.

These hourly contributions are obtained by multiplying the hourly radiant heat gains by the radiant time factors for the 24 hours. The total contribution is the sum of the hourly contributions as seen from Eq. (10.32). The computational steps are summarized in Table E10.5.3 below.

Table E10.5.3 Computation of radiant contribution for hour 16

*number of hours before hour 16

Hour*	1	2	3	4	5	6	7	8	9	10	11	12
C_n%	31	17	11	8	6	4	4	3	3	2	2	2
q_{rad}	306	306	306	306	306	306	306	306	0	0	0	0
$C_n q_{rad}$	95	52	33.7	24.5	18.4	12.2	12.2	9.2	0	0	0	0
Hour*	13	14	15	16	17	18	19	20	21	22	23	24
C_n%	1	1	1	1	1	1	1	0	0	0	0	0
q_{rad}	0	0	0	0	0	0	0	0	306	306	306	306
$C_n q_{rad}$	0	0	0	0	0	0	0	0	0	0	0	0

The total contribution from the radiant heat gains during the previous 24 hours to the cooling load at hour 16 is given by

$$q_{cl2} = 95 + 52 + 33.7 + 24.5 + 18.4 + 12.2 + 12.2 + 9.2 = 257 \text{ W}$$

Hence the total cooling load due to lights at hour 16 is

$$q_{lt} = q_{cl1} + q_{cl2} = 165 + 257 = 422 \text{ W}$$

The maximum hourly cooling load due to lights may be found by repeating the above calculations for all the hours and then selecting thc largest value. The MATLAB code given in Appendix 10.1 could be easily modified to perform the computation by replacing the occupancy schedule with the lighting schedule. This is left as an exercise to the reader. The results are summarized in Table E10.5.4

Table E10.5.4 Cooling load profile due to lights

Hour	1	2	3	4	5	6	7	8	9	10	11	12
Q_{co}	0	0	0	0	0	0	0	0	165	165	165	165
Q_{ra}	77	67	58	49	40	34	28	21	113	162	193	214
Q_{tot}	77	67	58	49	40	34	28	21	278	327	358	379
Hour	13	14	15	16	17	18	19	20	21	22	23	24
Q_c	165	165	165	165	165	165	165	165	0	0	0	0
Q_{ra}	230	239	248	257	266	273	279	285	193	144	113	92
Q_{tot}	395	404	413	422	431	438	444	**450**	193	144	113	92

We notice from the values in Table E10.5.4 that the highest cooling load due to lights is 450 W and it occurs at hour 20.

Example 10.6 A vertical wall, at a location with a northern latitude of 40°, faces 30° west of south. (i) Calculate the sol-air temperature at 2 p.m. solar time on 21 August. (ii) Obtain the sol-air temperature profile for the 24-hour period on August 21. The ground reflectivity is 0.25. The overall external heat transfer coefficient is $28 Wm^{-2}K^{-1}$. The solar absorptivity of the external wall surface is 0.8. Assume that the optical depth parameters for the location for the month of August as, $\tau_b = 0.697$ and $\tau_d = 1.51$. The design ambient temperature profile for the location is given below in Table E10.6.1.

Table E10.6.1 Ambient temperature profile

Hour	1	2	3	4	5	6	7	8	9	10	11	12
T_{amb}, °C	21	20.2	19.6	19.2	18.8	18.6	18.8	19.6	20.6	22	23.5	25.6
Hour	13	14	15	16	17	18	19	20	21	22	23	24
T_{amb}, °C	27.2	29.1	30	30	29.5	28.8	27.2	25.6	24.5	23.3	22	21.3

Solution The hour angle is

$$H = 15(T_{sol} - 12) = 15 \times 2 = 30 \text{ degrees}$$

For August 21, $N_{day} = 233$. The declination is given by

$$\delta = 23.45 sin\left[\frac{360(233+284)}{365}\right] = 11.75 \text{ degrees}$$

Latitude, $L = 40°$, declination, $\delta = 11.75°$, hour angle, $H = 30°$.
The solar altitude angle is given by

$$sin\beta = cos40 \cos 30\, cos11.75 + sin40 sin11.75 = 0.780$$

Hence the solar altitude angle is 51.3°.
The solar azimuth angle is given by

$$sin\phi = \frac{cos11.75 \times \sin 30}{cos51.3} = 0.783$$

Hence the azimuth angle is 51.54°.
The air mass is given by Eq. (9.24) as

$$m = [sin\beta + 0.50572(6.07995 + \beta)^{-1.6364}]^{-1}$$

$$m = [sin51.3 + 0.50572(6.07995 + 51.3)^{-1.6364}]^{-1} = 1.28$$

The extraterrestrial solar radiation incident on a surface normal to the sun's rays is given by Eq. (9.1) as

$$E_o = E_{sc}\left\{1 + 0.033cos\left[360^0 \frac{(n-3)}{365}\right]\right\}$$

where E_{sc} = 1367 Wm^{-2} and n is the day of the year.

$$E_o = 1367\left\{1 + 0.033cos\left[360^0 \frac{(233-3)}{365}\right]\right\} = 1336.1\ Wm^{-2}$$

The beam and diffuse radiation optical depths for August 21 are given as τ_b = 0.697 and τ_d = 1.51. Substituting these values in Eqs. (9.25) and (9.26) we have

$$ab = 1.454 - 0.406\tau_b - 0.268\tau_d + 0.021\tau_b\tau_d = 0.7884$$

$$ad = 0.507 + 0.205\tau_b - 0.080\tau_d - 0.190\tau_b\tau_d = 0.3291$$

The intensities of beam and diffuse radiation are given by Eqs. (9.22) and (9.23) respectively. Substituting the relevant numerical values in the above equations we obtain

$$E_b = E_o exp[-\tau_b m^{ab}] = 1336.1exp[-0.697 \times 1.28^{0.7884}]$$

$$E_d = E_o exp[-\tau_d m^{ad}] = 1336.1exp[-1.51 \times 1.28^{0.3291}]$$

Hence the radiation intensities are: E_b = 572.9 Wm^{-2} and E_d = 259.7 Wm^{-2}.

The surface-solar azimuth of the wall is

$$\gamma = 51.3 - 30 = 21.3 \text{ degrees.}$$

The angle of incidence, θ is given by Eq. (9.13) as

$$cos\theta = \ cos\beta sin\Sigma\ cos\gamma + sin\beta cos\Sigma$$

where Σ is the tilt angle of the wall.

$$cos\theta = \ cos51.3sin90cos21.3 + sin51.3cos90 = 0.5825$$

Hence the angle of incidence is 54.4°.

The direct radiation incident on unit area of the wall surface is

$$H_{db} = E_b cos\theta = 572.9cos54.4 = 333.5\ Wm^{-2}$$

The diffuse radiation incident on unit area is given by Eq. (9.28) as

$$H_{dif} = E_d Y$$

where Y is a function of the angle of incidence θ of the direct beam. It is given by the expression

$$Y = 0.55 + 0.437cos\theta + 0.313cos^2\theta$$

Substituting the pertinent numerical data in the above expressions we have

$$Y = 0.55 + 0.437cos54.4 + 0.313cos^2(54.4) = 0.9104$$

$$H_{dif} = E_d Y = 259.7 \times 0.9104 = 236.4 \text{ Wm}^{-2}$$

The ground-reflected radiation falling on unit area of surface is given by Eq. (9.32) as

$$H_{gr} = \frac{(E_d + E_b sin\beta)\rho_{gr}(1 - cos\Sigma)}{2}$$

$$H_{gr} = (259.7 + 572.9sin51.3) \times 0.25 \times \frac{(1 - cos90)}{2}$$

Hence $H_{gr} = 88.35 \text{Wm}^{-2}$.
Total solar radiation incident on unit area of wall is

$$H_{tot} = 333.5 + 236.4 + 88.35 = 658.25 \text{ Wm}^{-2}$$

The sol-air temperature is given by Eq. (9.39) as

$$T_{sa} = T_a + \frac{q_{sol}}{h_o} - \frac{\varepsilon_s \Delta R}{h_o} = 29.1 + \frac{0.8 \times 658.25}{28} = 47.9°\text{C}$$

Note that the correction factor for a vertical surface is zero.

(ii) The time at sun rise is obtained from Eq. (9.7) as

$$sin\beta = cos40 \cos H_{sr} \, cos11.75 + sin40sin11.75 = 0$$

Hence, the hour angle at sunrise, H_{sr} = 100° and the number of hours from noon to sunrise is, 100/15 = 6.67 h. The solar times at sunrise and sunset are symmetrical about solar noon and therefore the respective times are 5.20 a.m. and 6.40 p.m. The solar radiation intensities from midnight to 5.20 a.m., and from 6.40 p.m. to midnight are zero. For these periods the hourly sol-air temperatures are equal to the corresponding ambient temperatures.

We will now repeat the calculation steps shown above for hourly intervals from 5 a.m. to 7 p.m. (approximately) when the sun is above the horizon. The first section of the MATLAB code in Appendix A10.2 is used to compute the sol-air temperature profile.

The results are summarized in Table E10.6.2.

Table E10.6.2 Sol-air temperature profile

Hour	1	2	3	4	5	6	7	8	9	10	11	12
T_{amb}, °C	21	20.2	19.6	19.2	18.8	18.6	18.8	19.6	20.6	22	23.5	25.6
Q_{inc}	0	0	0	0	0	42	91.5	142	193	295	447	573
T_{sa}	21	20.2	19.6	19.2	18.8	19.8	21.4	23.7	26	30.4	36.3	41.9
Hour	13	14	15	16	17	18	19	20	21	22	23	24
T_{amb}, °C	27.2	29.1	30	30	29.5	28.8	27.2	25.6	24.5	23.3	22	21.3
Q_{inc}	648	658	594	460	274	110	0	0	0	0	0	0
T_{sa}	45.7	47.9	47.0	43.1	37.3	32	27.2	25.6	24.5	23.3	22	21.3

Direct beam radiation is received by the wall only when the angle of incidence is less than 90°. However, if the sun is above the horizon, diffuse radiation and ground-reflected radiation will be incident on the wall.

Example 10.7 The vertical wall described in worked example 10.6 is made of brick and a layer of insulation board. It has an overall heat transfer coefficient (*U*-value) of 0.58 $Wm^{-2}K^{-1}$. The area of the wall is $14m^2$. Assume that the inside air temperature is maintained constant at 23°C. Calculate (i) the hourly rates of conductive heat input at the exterior surface of the wall on the 21 August, (ii) the rate of conductive heat input at the interior surface of the wall at 5 p.m. solar time on 21 August, and (iii) the hourly heat gain profile at the inner surface of the wall on 21 August.

Table E10.7.1 Wall conduction time series (CTS)

Hour	0	1	2	3	4	5	6	7	8	9	10	11
C_n%	1	1	2	5	8	9	9	9	8	7	7	6
Hour	12	13	14	15	16	17	18	19	20	21	22	23
C_n%	5	4	4	3	3	2	2	2	1	1	1	0

Solution Form Table 16 on page 18.24 of the *ASHRAE Handbook - 2013 Fundamentals* [1] we obtain the conduction time factors appropriate for the wall. These values are listed in Table E10.7.1 above.

The rate of heat gain at the outer surface is given by Eq. (10.36) as

$$Q_e = AU_o(T_{sa} - T_r) \tag{E10.7.1}$$

where $U_o = 0.58$ Wm^{-2}K^{-1} and the inside air temperature $T_r = 23$°C. We obtained the hourly values of the sol-air temperature for 21 August in worked example 10.6 and these values are listed in Table E10.6.2. Applying Eq. (E10.7.1) to each hour we obtain the heat flow rates at the exterior surface listed in Table 10.7.2 below.

Table E10.7.2 Heat input at exterior surface, Q_{in} [W]

Hour	1	2	3	4	5	6	7	8	9	10	11	12
Q_e	−16.2	−22.7	−27.6	−30.9	−30.6	−26	−12.9	5.4	25.2	60.3	107.8	154
Hour	13	14	**15**	16	17	18	19	20	21	22	23	24
Q_e	184.5	202	194.6	163.5	116.3	72.6	41.2	21.1	12.2	2.4	−8.1	−13.8

(ii) We compute the heat flow rate at the interior surface by applying the conduction time factors as shown schematically in Fig. 10.9(b). The heat flow rate is given by Eq. (10.35) as (see Table E10.7.3)

$$q_{ic,t} = \sum_{n=1}^{24} c_n Q_{e,(t-n)} \tag{E10.7.2}$$

where c_n is the *conduction time factor* for hour n and $Q_{e,(t-n)}$ is the heat input at the exterior surface during the nth hour before the current hour t.

Table E10.7.3 Heat flow rate at 3 p.m. solar time

*number of hours before hour 15

Hour*	1	2	3	4	5	6	7	8	9	10	11	12
C_n%	1	1	2	5	8	9	9	9	8	7	7	6
Q_e	195[1]	202	184.5	154	107	60	25	5.4	−13	−26	−30.6	−31
$100C_nQ_e$	195	202	369	770	856	540	225	48.6	−104	−182	−214	−186
Hour*	13	14	15	16	17	18	19	20	21	22	23	24
C_n%	5	4	4	3	3	2	2	2	1	1	1	0
Q_e	−28	−23	−16	−14	−8	2.4	12	21	41	73	116	163
$100C_nQ_e$	−140	−92	−64	−52	−24	4.4	24	42	34	47	113	0

[1]note that the values are rounded-off to the first whole number.

The different terms of the summation in Eq. (E10.7.2) for hour 15 are listed in Table E10.7.3. The summation gives the heat flow rate for hour 15 as 25.0 W.

The heat flow rates at the interior surface for a 24-hour period is obtained by applying the above procedure for each hour beginning at midnight. The second section of the MATLAB code in Appendix A10.2 was used for the computation.

The results are given in Table E10.7.4.

Table E10.7.4 Heat input at the interior surface, Q_{in} [W]

Hour	1	2	3	4	5	6	7	8	9	10	11	12
Q_{in}	80	71	61.5	52	42.7	33.7	25.2	17.5	10.7	5.6	2.7	2.6
Hour	13	14	15	16	17	18	19	20	21	22	23	24
Q_{in}	5.9	13.2	24.7	39.5	55.8	71.8	85.2	94.2	98.5	98.2	94.3	87.9

Example 10.8 Consider the vertical wall described in worked examples 10.6 and 10.7. Calculate (i) cooling load due to conduction heat flow through the wall at 3 pm solar time and (ii) the 24-hour cooling load profile due to conduction heat flow through the wall.

Solution The rate of heat gain at the *interior surface* of the wall consists of a convective component and a radiative component. Form Table 14 on page 18.22 of the *ASHRAE Handbook* - 2013 *Fundamentals* [1] we obtain the convective and radiative fractions as 0.54 and 0.46 respectively. The convective component contributes immediately to the cooling load while the radiative component is first absorbed by the surfaces within the space and later released to the air by convection.

This time delay is accounted for by using the radiant time series (RTS) given in Table 19 on page 18.28 of the *ASHRAE Handbook* - 2013 *Fundamentals* [1]. We select the values for medium construction with 50% glass and no carpet which are listed in Table E10.8.1 below.

Table E10.8.1 Non-solar RTS values for conduction heat flow

	Medium construction, No carpet, 50 percent glass											
Hour	0	1	2	3	4	5	6	7	8	9	10	11
C_n%	33	16	10	7	5	4	3	3	2	2	2	2
Hour	12	13	14	15	16	17	18	19	20	21	22	23
C_n%	1	1	1	1	1	1	1	1	1	1	1	0

(ii) The cooling load due to the radiant component of the heat gain at the interior surface is obtained by applying the non-solar radiant time factors as shown schematically in Fig. 10.9(b). The cooling load is given by Eq. (10.35) as (see Table E10.8.2)

$$q_{ic,t} = f_{rad} \sum_{n=1}^{24} c_n Q_{in,(t-n)} \tag{E10.8.1}$$

where c_n is the *radiant time factor* for hour n and $f_{rad} Q_{in,(t-n)}$ is the radiant component of the heat gain at the interior surface during the nth hour before the current hour t.

The different terms of the summation in Eq. (E10.8.1) for hour 15 are listed in Table E10.8.2. Hence we obtain the sum as 24.19 W.

Table E10.8.2 Heat flow rate at 3 p.m. solar time

*number of hours before hour 15

Hour*	1	2	3	4	5	6	7	8	9	10	11	12
C_n%	33	16	10	7	5	4	3	3	2	2	2	2
Q_{in}	24.7	13.2	5.9	2.6	2.7	5.6	10.7	17.5	25.2	33.7	42.7	52
C_nQ_{in}	8.15	2.1	0.59	0.06	0.18	0.22	0.32	0.53	0.5	0.67	0.85	1.04
Hour*	13	14	15	16	17	18	19	20	21	22	23	24
C_n%	1	1	1	1	1	1	1	1	1	1	1	0
Q_{in}	61.6	71.1	80	87.9	94.3	98.2	98.5	94.2	85.1	71.7	55.8	39.5
C_nQ_{in}	0.62	0.71	0.8	0.88	0.94	0.98	0.98	0.94	0.85	0.72	0.56	0

Applying Eq. (E10.8.1) for hour 15 we obtain the cooling load as (see Table E10.8.2)

$$Q_{c,15} = f_{con} Q_{in,15} + f_{rad} \sum_{1}^{24} c_n Q_n \tag{E10.8.2}$$

$$Q_{c,15} = 0.54 \times 24.7 + 0.46 \times 24.19 = 24.5 \text{ W}$$

We now apply the above procedure to a 24-hour cycle. The results are summarized in Table E10.8.3. The third section of the MATLAB code listed in Appendix A10.2 was used to perform the calculations.

Table E10.8.3 Cooling load profile

Hour	1	2	3	4	5	6	7	8	9	10	11	12
Q_{in}	80	71	61.5	52	42.7	33.7	25.2	17.5	10.7	5.6	2.7	2.6
Q_{conv}	43.2	38.4	33.2	28.1	23.1	18.2	13.6	9.4	5.8	3.0	1.5	1.4
Q_{rad}	34.7	33.0	30.9	28.4	25.7	22.9	20	17.2	14.5	12.2	10.3	9.1
Q_{tot}	77.9	71.4	64.1	56.5	48.8	41.1	33.6	26.6	20.3	15.2	11.8	10.5
Hour	13	14	15	16	17	18	19	20	21	22	23	24
Q_{in}	5.8	13.2	24.7	39.5	55.8	71.8	85	94	98.5	98.2	94.3	87.9
Q_{con}	3.2	7.1	13.4	21.3	30.1	38.8	46.0	50.9	53.2	53.0	50.9	47.5
Q_{rad}	8.7	9.3	11.2	14.2	18.0	22.4	26.7	30.6	33.5	35.3	36	35.7
Q_{tot}	11.9	16.4	**24.6**	35.5	48.1	61.2	72.7	81.5	86.7	88.3	86.9	83.2

Example 10.9 A building at a location with a northern latitude of 34° has a flat roof of area 80 m^2, made of concrete and an insulation layer, similar to roof number 14 in Table 17 on page 18.26 of the *ASHRAE Handbook* - 2013 *Fundamentals* [1]. The conduction time factors for the roof obtained from Table 17 are listed in Table E10.9.2. The *U*-factor of the roof is 0.304 Wm^{-2}K^{-1}. The absorptivity of the roof surface is 0.75.

The radiant time series for the interior which is of medium construction with 50% glass and no carpet, obtained from Table 19 the *ASHRAE Handbook* - 2013 *Fundamentals* [1], is listed in Table E10.9.2.

On June 15, calculate the following: (i) the hourly heat gains at the exterior surface of the roof, (ii) the hourly heat gains at the interior surface of the roof, and (iii) the hourly cooling load due to conduction through the roof. The external heat transfer coefficient of the roof is 24 Wm^{-2}K^{-1}. The hourly ambient temperature profile for the location is given in Table E10.9.1. Assume that the optical depth parameters for the location for the month of June as $\tau_b = 0.337$ and $\tau_d = 2.36$.

Table E10.9.1 Ambient temperature profile

Hour	1	2	3	4	5	6	7	8	9	10	11	12
T_{amb}, °C	21.4	20.8	19.8	19.5	18.8	19.2	19.4	19.6	21.6	22.4	23.5	25.6
Hour	13	14	15	16	17	18	19	20	21	22	23	24
T_{amb}, °C	27.2	29.1	30.5	30.5	29.5	28.8	28.2	26.2	24.5	23.3	22.2	21.6

Table E10.9.2 Roof conduction time series (CTS) [1]

Hour	0	1	2	3	4	5	6	7	8	9	10	11
C_n%	1	2	8	11	11	10	9	7	6	5	5	4
Hour	12	13	14	15	16	17	18	19	20	21	22	23
C_n%	3	3	3	2	2	2	1	1	1	1	1	1

Table E10.9.3 Radiant time series (RTS) for interior [1]

Hour	0	1	2	3	4	5	6	7	8	9	10	11
C_n%	33	16	10	7	5	4	3	3	2	2	2	2
Hour	12	13	14	15	16	17	18	19	20	21	22	23
C_n%	1	1	1	1	1	1	1	1	1	1	1	0

Solution Using the given value of the northern latitude of the location and the date we obtain the following quantities. The day number = 166, the declination = 23.31°. The solar times at sunrise and sunset, approximated to the nearest hour, are 5 a.m. and 7 p.m. respectively. The beam and diffuse components of solar radiation on a horizontal roof surface are given by

$$E_b = E_o exp[-\tau_b m^{ab}] = 1336.1exp[-0.337 \times m^{ab}]$$

$$E_d = E_o exp[-\tau_d m^{ad}] = 1336.1exp[-2.36 \times m^{ad}]$$

where the exponents are:

$$ab = 1.454 - 0.406\tau_b - 0.268\tau_d + 0.021\tau_b\tau_d = 0.7014$$

$$ad = 0.507 + 0.205\tau_b - 0.080\tau_d - 0.190\tau_b\tau_d = 0.2361$$

The air mass m is given by

$$m = [sin\beta + 0.50572(6.07995 + \beta)^{-1.6364}]^{-1}$$

where β is the solar altitude angle.
Now for the horizontal roof surface, the angle of the incidence of the direct solar beam is

$$\theta = \frac{\pi}{2} - \beta$$

The ground reflected radiation is zero. Therefore the total solar radiation intensity is

$$q_{sol} = E_b cos\theta + E_d = E_b sin\beta + E_d$$

The sol-air temperature at hour t is given by Eq. (9.39) as

$$T_{sa,t} = T_{a,t} + \frac{q_{sol,t}}{h_o} - \frac{\varepsilon_s \Delta R}{h_o}$$

The correction factor, $\varepsilon_s \Delta R = 63$ Wm^{-2} according to the values recommended in Ref. [1]. The rate of heat gain at the outer surface at hour t is given by

$$q_{o,t} = AU_o(T_{sa,t} - T_r)$$

where T_r is the inside air temperature and U_o is the overall U-factor of the roof.

The heat gain at the interior surface is given by Eq. (10.35) as

$$q_{ic,t} = \sum_{n=0}^{23} c_{cn} q_{o,(t-n)}$$

where c_{cn} is the *conduction time factor* for hour n and $q_{o,(t-n)}$ is the heat input at the exterior surface during the nth hour before the current hour t.

The cooling load due to the radiant component of the heat gain is given by

$$q_{icr,t} = \sum_{n=0}^{23} c_{rn} f_{rad} q_{ic,(t-n)}$$

where c_{rn} is the *radiant time factor* for hour n and $f_{rad} q_{ic,(t-n)}$ is the radiant component of the heat gain at the interior surface during the nth hour before the current hour t.

Hence the total cooling load at hour t is

$$Q_{c,t} = f_{con} q_{ic,t} + q_{icr,t}$$

The MATLAB computer code, given in Appendix 10.2 was suitably modified to compute the hourly conduction cooling load for a horizontal roof. The important quantities obtained as output are summarized in Table E10.9.3 below.

Table E10.9.3 Summary of results: Roof cooling load (W) computation

Hour	1	2	3	4	5	6	7	8	9	10	11	12
Q_{ext}	−63.2	−77.8	−102	−109	−182	−60	106.6	274.3	467	599	697	772
Q_{int}	218.9	182	148	117	88.5	62.5	36.6	20.9	24.6	50	95.7	157
Q_{conv}	118	98.5	80	63.3	47.8	33.7	19.8	11.3	13.3	27	51.7	85
Q_{rad}	126	114	102	90.6	80	69	58.7	50	45	45	51	62
Q_{tot}	244	212.5	182	153.9	127.8	102.7	78.5	61.3	58.3	72	102.7	147
Hour	13	14	15	16	17	18	19	20	21	22	23	24
Q_{ext}	787	762	683.6	539	352	173	46.5	53.5	12	−17	−43.7	−58.4
Q_{int}	226	296	360	413	452	470	463	435	391	345	301	259
Q_{con}	122	160	194	223	244	253.7	250	234.7	211	186	162	139.6
Q_{rad}	77.8	96	115	133.8	150	163	170	171	167	159	149	138
Q_{tot}	199.8	256	309	256.8	394	416.7	420	405.7	378	345	311	277.6

Example 10.10 A double-glazed vertical window of area 6.5 m^2, at a location with a northern latitude of 30°, faces 15° east of south. The ground reflectivity is 0.25. At 11 a.m. solar time on April 12, calculate the following quantities: (i) the angle of incidence of the direct beam on the window, (ii) the intensity of direct and diffuse solar radiation incident on the window, and (iii) the direct beam, the diffuse radiation and the conductive heat gains through the window.

The solar heat gain factors (SHGC) for the window, obtained from Table 10, page 15.19, of the *ASHRAE Handbook* - 2013 *Fundamentals* [1] are listed in Table E10.10.1. The overall heat transfer coefficient for the window is 3.5$Wm^{-2}K^{-1}$. Assume that the optical depth parameters for the location for the month of June as τ_b = 0.337 and τ_d = 2.36. The outdoor and indoor air temperatures are 32°C and 23°C respectively.

Table E10.10.1 Solar heat gain coefficients for window [1]

Inc. angle	0	40	50	60	70	80	diffuse
SHGC	0.7	0.67	0.64	0.58	0.45	0.23	0.6

Solution The solar time is 11 am and therefore the hour angle is

$$H = 15(T_{sol} - 12) = -15 \times 1 = -15 \text{ degrees}$$

For 12 April, N_{day} = 102. The declination is given by

$$\delta = 23.45 sin\left[\frac{360(102+284)}{365}\right] = 8.294 \text{ degrees}$$

Latitude, L = 30°; declination, δ = 8.29°, Hour angle, H = −15°.
The solar altitude angle is given by

$$sin\beta = cos30\cos(-15)\,cos8.29 + sin30 sin8.29 = 0.9$$

Hence the solar altitude angle is 64.14°.
The solar azimuth angle is given by

$$sin\phi = \frac{cos8.29 \times \sin(-15)}{cos64.14} = -0.587$$

Hence the azimuth angle is −35.95°.

The air mass is given by Eq. (9.24) as

$$m = [sin\beta + 0.50572(6.07995 + \beta)^{-1.6364}]^{-1}$$

$$m = [sin64.14 + 0.50572(6.07995 + 64.14)^{-1.6364}]^{-1} = 1.11$$

The extraterrestrial solar radiation incident on a surface normal to the sun's rays is given by Eq. (9.1) as

$$E_o = E_{sc}\left\{1 + 0.033cos\left[360^0\frac{(n-3)}{365}\right]\right\}$$

where E_{sc} = 1367 Wm^{-2} and n is the day of the year.

$$E_o = 1367\left\{1 + 0.033cos\left[360^0\frac{(102-3)}{365}\right]\right\} = 1361 \text{ Wm}^{-2}$$

The beam and diffuse radiation optical depths for 10 April at the location are given as τ_b =0.337 and τ_d =2.36. Substituting these values in Eqs. (9.25) and (9.26) we have

$$ab = 1.454 - 0.406\tau_b - 0.268\tau_d + 0.021\tau_b\tau_d = 0.7014$$
$$ad = 0.507 + 0.205\tau_b - 0.080\tau_d - 0.190\tau_b\tau_d = 0.2361$$

The intensities of beam and diffuse radiation are given by Eqs. (9.22) and (9.23). Substituting the relevant numerical values in the above equations we obtain

$$E_b = E_o exp[-\tau_b m^{ab}] = 1361 exp[-0.337 \times 1.11^{0.7014}]$$
$$E_d = E_o exp[-\tau_d m^{ad}] = 1361 exp[-2.36 \times 1.11^{0.2361}]$$

Hence the radiation intensities are: $E_b = 947.0$ Wm^{-2} and $E_d = 121$ Wm^{-2}
The surface-solar azimuth of the wall is

$$\gamma = -35.95 + 15 = -20.95 \text{ degrees.}$$

The angle of incidence, θ is given by

$$cos\theta = cos\beta sin\Sigma \, cos\gamma + sin\beta cos\Sigma$$

where Σ is the tilt angle of the window.

$$cos\theta = cos64.14 sin90 cos20.95 + sin64.14 cos90 = 0.4073$$

Hence the angle of incidence is 65.96°.
The direct radiation incident on unit area of the window is

$$H_{db} = E_b cos\theta = 947 cos65.96 = 385.78 \text{ Wm}^{-2}$$

The diffuse radiation incident on unit area is

$$H_{dif} = E_d Y$$

where Y is a function of the angle of incidence θ of the direct beam. It is given by the expression

$$Y = 0.55 + 0.437 cos\theta + 0.313 cos^2\theta$$

Substituting the pertinent numerical data in the above expressions we have

$$Y = 0.55 + 0.437 cos65.96 + 0.313 cos^2(65.96) = 0.78$$
$$H_{dif} = E_d Y = 121 \times 0.78 = 94.4 \text{ Wm}^{-2}$$

The ground-reflected radiation falling on unit area of the window is given by

$$H_{gr} = \frac{(E_d + E_b sin\beta)\rho_{gr}(1 - cos\Sigma)}{2}$$

$$H_{gr} = (121 + 947sin64.14) \times \frac{0.25(1-cos90)}{2}$$

Hence H_{gr} = 121.6 Wm^{-2}.

The total diffuse radiation intensity is

$$H_{dif,to} = H_{dif} + H_{gr} = 94.4 + 121.6 = 216 \text{ Wm}^{-2}$$

The window heat gain includes the beam radiation component, q_b, total diffuse radiation component, q_d, and the overall conduction component, q_c. The beam component is given by

$$q_b = AH_{db}SHGC(\theta) = 6.5 \times 385.78 \times SHGC(65.97)$$

$$q_b = 6.5 \times 385.78 \times 0.5137 = 1288.1 \text{ W}$$

Note that the SHGC, for the incidence angle of 65.97°, obtained by interpolation (see Table E10.10.1) is 0.5137.

The diffuse component is given by

$$q_d = AH_{dif,to}\langle SHGC\rangle_{dif} = 6.5 \times 216 \times 0.6 = 842.4 \text{ W}$$

Note that the SHGC for diffuse radiation from Table E10.10.1 is 0.6.

The conductive heat gain through the window is

$$q_c = AU_o(T_{amb} - T_{in}) = 6.5 \times 3.5 \times (32 - 23) = 204.7 \text{ W}$$

Example 10.11 For the window described in worked example 10.10, obtain the 24-hour cooling load profile due to beam radiation, diffuse radiation and conduction heat entering through the window. All conditions listed in worked example 10.10 are applicable. The design ambient temperature profile is given in Table E10.11.1.

The solar RTS values, and the non-solar RTS values for medium construction with carpet and 50% glass, are obtained from Tables 20 and 19 respectively, in chapter 18 of the *ASHRAE Handbook - 2013 Fundamentals* [1]. These values are given in Tables E10.11.2 and E10.11.3 respectively.

Table E10.11.1 Ambient temperature profile

Hour	1	2	3	4	5	6	7	8	9	10	11	12
T_{amb}, °C	22.3	21.7	20.8	20.5	19.8	20.2	20.4	20.6	22.7	23.6	24.6	26.7
Hour	13	14	15	16	17	18	19	20	21	22	23	24
T_{amb}, °C	28.3	30.2	31.6	31.6	30.7	29.9	29.4	27.3	25.6	24.4	23.3	22.6

Table E10.11.2 Solar radiant time series (RTS) [1]

Hour	0	1	2	3	4	5	6	7	8	9	10	11
C_n%	54	16	8	4	3	2	1	1	1	1	1	1
Hour	12	13	14	15	16	17	18	19	20	21	22	23
C_n%	1	1	1	1	1	1	1	0	0	0	0	0

Table E10.11.3 Non-solar radiant time series (RTS) [1]

Hour	0	1	2	3	4	5	6	7	8	9	10	11
C_n%	49	17	9	5	3	2	2	1	1	1	1	1
Hour	12	13	14	15	16	17	18	19	20	21	22	23
C_n%	1	1	1	1	1	1	1	1	0	0	0	0

Solution The rate of heat gain at the interior surface of the window by conduction and diffuse radiation transmission consists of a convective component and a radiative component. Form Table 14 on page 18.22 of the *ASHRAE Handbook* - 2013 *Fundamentals* [1] (see Table 10.7) we obtain the convective fraction, f_{con} and the radiative fraction, f_{rad} for conduction heat gain. For diffuse solar radiation transmitted through the window, $f_{con} = 0.67$ and $f_{rad} = 0.33$ if SHGC > 0.5 and $f_{con} = 0.54$ and $f_{rad} = 0.46$ if SHGC < 0.5.

The convective component contributes immediately to the cooling load while the radiative component is first absorbed by the surfaces within the space and later released to the air by convection. This time delay is accounted for by using the *non-solar* radiant time series (RTS) values given in Table 19 on page 18.28 of the *ASHRAE Handbook* - 2013 *Fundamentals* [1]. We select the values for medium construction with 50% glass and carpet which are listed in Table E10.11.2.

The *direct beam radiation* transmitted through the window is absorbed by the surfaces within the space and later released by convection to the air inside. This time delay is accounted for by using the *solar* RTS values listed in Table 20 on page 18.29 of the *ASHRAE Handbook* - 2013 *Fundamentals* [1]. We select the values for medium construction with 50% glass and carpet, which are listed in Table E10.11.3.

The different heat gains through the window at each hour are computed by applying the steps outlined in worked example 10.10. The calculation procedure to convert the different window heat gains to cooling loads is illustrated schematically in Fig. 10.9(c).

The cooling load due to the *radiative component of the conduction and diffuse radiation heat gains* through the window is given by

$$q_{icr,t} = \sum_{n=0}^{23} (nsc)_{rn} f_{rad} q_{ic,(t-n)}$$

where $(nsc)_{rn}$ is the *non-solar radiant time factor* for hour n and $f_{rad} q_{ic,(t-n)}$ is the radiant component of the heat gain during the nth hour before the current hour t.

Hence the total *cooling load* at hour t due to conduction and diffuse radiation transmission is

$$q_{c,t} = f_{con} q_{ic,t} + q_{icr,t}$$

The first term in the above equation is the convective fraction of the heat gain through the window.

The *cooling load* due to the *transmitted direct beam radiation* is given by

$$q_{tb,t} = \sum_{n=0}^{23} (sc)_{rn} (qs)_{(t-n)}$$

where $(sc)_{rn}$ is the *solar radiant time factor* for hour n and $(qs)_{(t-n)}$ is the direct beam radiation transmitted during the nth hour before the current hour t.

A MATLAB computer code based on the above equations was used to compute the hourly cooling loads due to conduction, diffuse radiation transmission, and direct beam radiation transmission through the window. The listing of the code is given in Appendix A10.3. The important quantities computed are summarized in Table E10.11.4 below.

Table E10.11.4 Summary of results: window cooling load computation

Hour	1	2	3	4	5	6	7	8	9	10	11	12
q^*_{sol}	67.6	64.2	56.4	45	31.7	146.5	229	497	780	983	1046	949
$qc_{dif+con}$	−8.6	−16	−27	−30.7	−39.3	5.8	115	228	430	528	589	505
$qr_{dif+con}$	32	27	19	13	4.8	18.6	66.6	128	158	195	224	305
q_{tot}	91	75	48	26.8	−2.8	171	411	852	1368	1706	1859	1759
Hour	13	14	15	16	17	18	19	20	21	22	23	24
q_{sol}	714	427	221	139	100	81	72	70	70	70	70	70
$qc_{dif+con}$	498	468	411	322	215	114	79	53	32	17	3.7	−4.9
$qr_{dif+con}$	334	338	320	280	222	160	120	93	72	57	45	37
q_{tot}	1546	1233	952	741	537	355	271	216	174	144	119	102

*Cooling loads(W): q_{sol} - due to *direct* solar beam; $qc_{dif+con}$ - due to convective component of *diffuse* radiation and conduction; $qr_{dif+con}$ - due to radiative component of *diffuse* radiation and conduction; q_{tot} - total window cooling load.

Example 10.12 The double-glazed window described in worked examples 10.10 and 10.11 is 3.25 m wide and 2 m high. An overhang of width 0.75 m and length 3.25 m is to be installed 0.3 m above the top edge of the window. For the conditions stated in worked example 10.10 calculate (i) the unshaded area of the window at 11 a.m. solar time, and (ii) the 24-hour the cooling load profile due to the direct beam.

Solution We assume that the overhang only affects the heat gain due to direct beam transmission by changing the sunlit area of the window. The analysis of the shading of a window due to the presence of an overhang was developed in section 9.6. In this example we shall use some of the equations obtained in section 9.6, directly.

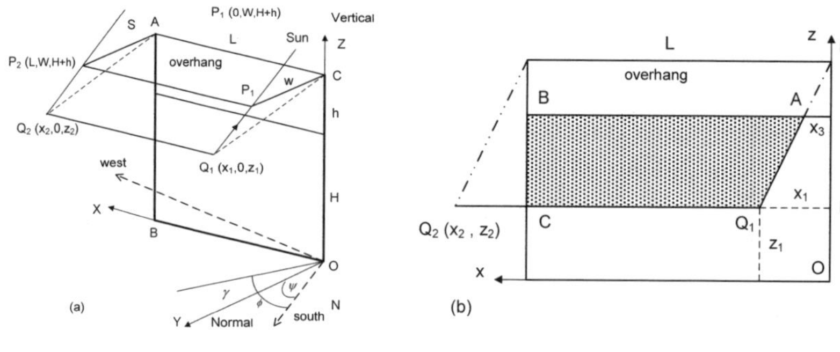

Fig. E10.12.1 (a) Geometrical parameters of overhang, (b) shaded area of window

Shown in Fig. E10.12.1(a) is a schematic diagram of a window with an overhang located a distance *h* above the top edge of the window. The surface-solar azimuth of the window, $\gamma = (\phi - \psi)$. The shadow cast on the window surface by the two corners P_1 and P_2 of the overhang are indicated in Fig. E10.12.1(b) as Q_1 and Q_2 respectively. The shaded area of the window is a trapezium $ABCQ_1$. As the sun moves across the sky, the points Q_1 and Q_2 will move on the window surface thus changing the sunlit area. The shaded area of the window is given by

$$A_{sh} = \frac{[(L-x_1)+(L-x_3)](H-z_1)}{2}$$

Following the analysis developed in section 9.6, we obtain the following expressions for the coordinates x_1, z_1, x_2 and z_2:

$$x_1 = -w\tan\gamma$$

$$z_1 = (H+h) - \frac{w\tan\beta}{\cos\gamma}$$

$$x_2 = L - w\tan\gamma$$

$$z_2 = (H+h) - \frac{w\tan\beta}{\cos\gamma}$$

From the geometry of triangle EQ_1D it follows that:

$$x_3 = \frac{hx_1}{(H+h-z_1)}$$

Now the coordinates (x_1,z_1), and (x_2,z_2) depend on the solar altitude angle β and the solar azimuth angle ϕ, which in turn, are dependent on the hour angle and the latitude of the location.

We shall illustrate the use of the above equations by applying the data obtained in worked example 10.10 for 11 a.m. solar time. The pertinent parameters are: length, $L = 3.25$ m, height, $H = 2$ m, height of overhang above top of window, $h = 0.3$ m, width of overhang, $w = 0.75$ m, window surface azimuth, $\psi = -15°$.

In worked example 10.10 we obtained the following quantities at 11 a.m. solar time: solar azimuth, $\phi = -35.95°$, surface-solar azimuth, $\gamma = (\phi - \psi) = -20.95°$, solar altitude angle, $\beta = 64.14°$, angle of incidence of direct beam, $\theta = 65.96°$.

Substituting numerical data in the equations listed above we obtain the following quantities at 11 a.m. solar time:

$$x_1 = -w tan\gamma = 0.75 \times tan20.95 = 0.287 \text{ m}$$

$$z_1 = (H + h) - \frac{w tan\beta}{cos\gamma} = 2.3 - 0.75 \times \frac{2.063}{0.9338} = 0.643 \text{ m}$$

$$x_3 = \frac{h x_1}{(H+h-z_1)} = 0.3 \times \frac{0.287}{(2.3-0.643)} = 0.052 \text{ m}$$

The shaded area of the window is given by

$$A_{sh} = \frac{[(L-x_1)+(L-x_3)](H-z_1)}{2} = \frac{6.161 \times 1.657}{2} = 4.18 \text{ m}^2$$

Hence the sunlit area (unshaded area) of the window is (6.5-4.18) = 2.32 m^2.

We apply the above procedure at each of the sunlit hours to obtain the hourly variation of the unshaded area of the window. The solar time at sunrise and sunset, rounded to an hour, are 6 a.m. and 6 p.m. respectively. The MATLAB code, listed in Appendix 10.4, was used to obtain the cooling load profile. The computed results are summarized in Table E10.12.1, where the cooling load due to the direct beam radiation is q_{sol}.

Table E10.12.1 Summary of results: unshaded area with overhang

Hour	1	2	3	4	5	6	7	8	9	10	11	12
x_1, m	-	-	-	-	-	1.83	2.76	1.76	1.16	0.71	0.29	−0.2
z_1, m	-	-	-	-	-	2.15	1.42	1.2	1.02	0.85	0.64	0.35
x_2, m	-	-	-	-	-	5.09	6.01	5.0	4.41	3.96	3.54	3.05
A_{ush}*	-	-	-	-	-	6.5	5.69	4.78	4.03	3.25	2.32	1.33
q_{sol}	30.7	27.8	22.0	14.9	8.3	131	199	377	508	538	448	295
Hour	13	14	15	16	17	18	19	20	21	22	23	24
x_1, m	−0.92	−2.53	−15.2	6.2	2.83	5.47	-	-	-	-	-	-
z_1, m	−0.16	−1.4	−11.7	5.9	3.2	2.7	-	-	-	-	-	-
x_2, m	2.33	0.72	−12	9.47	6.08	8.7	-	-	-	-	-	-
A_{ush}	0.98	1.78	2.8	6.5	6.5	6.5	-	-	-	-	-	-
q_{sol}	186	127	73	49.4	39.1	35.3	33.8	33.4	33.1	33.1	33.1	33.1

A_{ush} (m^2) = unshaded area of window, q_{sol} (W) = hourly cooling load due to transmitted *direct* beam solar radiation.

Care has to be exercised when the expression derived above is used to compute the unshaded area because the shadow line Q_1Q_2 in Fig. E10.12.1(b) could, under some conditions be outside the window area (z_1 < 0 and z_1> 2 m). Such situations occur during hours 13 to 18 as seen from the data in Table E10.12.1. In the computer code in Appendix 10.4

the expression for the shaded area has been suitably modified for these cases.

Example 10.13 The wall of a building is 8 m long and 3 m high. It has two equal sized windows of length 2.5 m and height 2 m located in it. The design of the wall is similar to that described in worked example 8.2 and the design of the double-glazed windows are similar to that described in worked example 8.6. The indoor and outdoor design temperatures are −10°C and 20°C respectively. Calculate the total heat load due to heat loss through the wall.

Solution In worked example 8.2, we obtained the overall heat transfer coefficient for the wall using the parallel flow method (0.234 $\mathrm{Wm^{-2}K^{-1}}$) and the isothermal plane method (0.258 $\mathrm{Wm^{-2}K^{-1}}$). We shall use the average value (0.246 $\mathrm{Wm^{-2}K^{-1}}$) to estimate heat load through the wall. The area of the wall without the two windows is

$$A_{wall} = (3 \times 8) - 2(2.5 \times 2) = 14\ \mathrm{m^2}$$

The heat loss due to wall conduction is given by

$$\dot{Q}_{wall} = A_{wall}\overline{U}_{wall}(T_{wall} - T_{amb})$$

$$\dot{Q}_{wall} = 14 \times 0.246 \times [20 - (-10)] = 103.3\ \mathrm{W}$$

In worked example 8.6 we obtained the overall heat transfer coefficient for the window as 3.02 $\mathrm{Wm^{-2}K^{-1}}$. Hence the heat loss through the window is

$$\dot{Q}_{window} = A_{window}\overline{U}_{window}(T_{window} - T_{amb})$$

$$\dot{Q}_{window} = 10 \times 3.02 \times [20 - (-10)] = 906\ \mathrm{W}$$

Therefore the total heating load due to wall heat loss is given by

$$\dot{Q}_{load} = 103.3 + 906 = 1009.6\ \mathrm{W}$$

Example 10.14 The length and breadth of the floor of a basement are 12 m and 9 m respectively. The wall of the basement extends 1.5 m below grade. The carpeted, concrete (k = 2.2 $\mathrm{Wm^{-1}K^{-1}}$) floor of the basement has a thickness of 135 mm. The carpet has a thickness 9 mm and a thermal conductivity of 0.085 $Wm^{-1}K^{-1}$. The soil surrounding the basement has a thermal conductivity of 1.4 $Wm^{-1}K^{-1}$. The design temperatures for inside air and outdoor air are 20°C and 2°C respectively. The inside and outside heat transfer coefficients are 8 $Wm^{-2}K^{-1}$ and 26 $Wm^{-2}K^{-1}$ respectively. Calculate the heat load due to conduction heat loss through the basement floor.

Solution The thermal resistances of the carpet, the floor, and the inside and outside air films are as follows:

$$R_{carp} = \frac{9\times10^{-3}}{0.085} = 0.1059, \quad R_{flo} = \frac{135\times10^{-3}}{2.2} = 0.0614$$

$$R_{oa} = \frac{1}{26} = 0.0385, \quad R_{ia} = \frac{1}{8} = 0.125$$

Hence the total additional thermal resistance is

$$R_a = 0.1059 + 0.0614 + 0.0385 + 0.125 = 0.3308$$

The expression for the average heat transfer coefficient of the basement floor, U_{bf} is given by Eq. (8.35) as

$$U_{bf} = \left(\frac{2k_s}{\pi w_b}\right) ln \left(\frac{\frac{w_b}{2}+\frac{z_f}{2}+\frac{k_s R_a}{\pi}}{\frac{z_f}{2}+\frac{k_s R_a}{\pi}}\right)$$

Substituting numerical values in the above equation we have

$$U_{bf} = \left(\frac{2\times1.4}{\pi\times9}\right) ln \left(\frac{\frac{9}{2}+\frac{1.8}{2}+\frac{1.4\times0.3308}{\pi}}{\frac{1.8}{2}+\frac{1.4\times0.3308}{\pi}}\right) = 0.165$$

Hence the overall heat transfer coefficient for heat flow through the soil across the floor is 0.165 $Wm^{-2}K^{-1}$.

Now the total rate of heat loss from the floor to the ambient is given by

$$Q_{to} = U_{bf}(Lw_b)(t_{ia} - t_{oa})$$

Substituting numerical values in the above equation we obtain the heat load due to heat loss through the basement floor as

$$Q_{to} = 0.165 \times 12 \times 9(20 - 2) = 320.8 \text{ W}$$

Example 10.15 For the two-story building described in worked example 8.14, calculate the sensible and latent heat loads due to infiltration of ambient air. The indoor design conditions are 22°C and 40% relative humidity. The outdoor ambient air is saturated at −10°C. Assume that the pressure is 101 kPa.

Solution In worked example 8.14 we obtained the infiltration flow rate as 0.068 m^3s^{-1}. The sensible heat load due to infiltration is given by Eq. (8.41) as

$$Q_s(kW) = 1.23\dot{V}_{in}(t_i - t_o)$$

$$Q_s = 1.23 \times 0.068 \times 10^3[22 - (-10)] = 2676 \text{ W}$$

The latent heat load due to infiltration is given by Eq. (8.42) as

$$Q_l(kW) = 3010\dot{V}_{in}(\omega_i - \omega_o)$$

From the psychrometric chart, $\omega_i = 0.0066$. From the data in Ref. [2], $\omega_o = 0.001606$. Substituting numerical values in the above equation we obtain the latent heat load due to infiltration as

$$Q_l = 3010 \times 0.068 \times 10^3(0.0066 - 0.001606) = 1022 \text{ W}$$

Problems

P10.1 An air conditioned office is occupied by 4 persons. Two of them, wearing suits with a thermal resistance of 1.1 clo, are engaged in managerial activities generating metabolic energy at the rate of 1.6 mets. The other two persons, wearing lightweight clothing with a thermal resistance of 0.65 clo, are working at computer terminals for which the metabolic rate is 1.1 mets. Determine the optimal air temperature for the office.
[*Answers*: 18.2°C, 23.9°C]

P10.2 The occupancy levels and activities of three zones A, B and C of an office building are given in Table P10.2.1. The design air flow rates required to meet the cooling loads of the zones are also tabulated. The summer outdoor conditions are 31°C and 70% relative humidity. The indoor relative humidity is 50%.

Table P10.2.1 Design data for zones A, B and C

Zone	Occupancy level	Office Activity	Air flow rate, Ls^{-1}
A	30	Light (1.0 met)	4000
B	45	Moderate (1.25 met)	5000
C	28	Moderate(1.25 met)	3500

(i) Determine the indoor design temperatures for the three zones.
(ii) Determine the design ventilation air flow rates (Ls^{-1}) for the zones.
(iii) Determine the total ventilation cooling load.
[*Answers*: (i) 25.7°C, 19.6°C, 19.6°C; (ii) 332, 415, 291 (iii) 48.2kW]

P10.3 The occupancy schedule of a building with people engaged in moderate activities is given in Table P10.3.1. Each person generates 75 W of sensible heat and 55 W of latent heat (Table 10.3). The radiative fraction of the sensible heat is 0.6 (Table 10.7).

Table P10.3.2 gives the non-solar RTS values from Table 19, Chapter 18 of the *ASHRAE Handbook* - 2013 *Fundamentals* [1] that are appropriate for the building. Calculate (i) the cooling load due to people at 11 a.m., and (ii) the maximum cooling load and the time when it occurs.
[*Answers*: (i) 1680 W, (ii) 1874.8 W at 6 p.m.]

Table P10.3.1 Occupancy schedule of zone

*Hours from midnight, N_p = number of people

Hour*	1	2	3	4	5	6	7	8	9	10	11	12
N_p	0	0	0	0	0	0	0	6	8	15	15	15
Hour	13	14	15	16	17	18	19	20	21	22	23	24
N_p	15	15	15	15	15	15	8	6	0	0	0	0

Table P10.3.2 Non-solar RTS values

Medium construction, No carpet, 10 percent glass

Hour	0	1	2	3	4	5	6	7	8	9	10	11
C_n%	31	17	11	8	6	4	4	3	3	2	2	2
Hour	12	13	14	15	16	17	18	19	20	21	22	23
C_n%	1	1	1	1	1	1	1	0	0	0	0	0

P10.4 The total energy output of recessed fluorescent lights in a zone of a building is 780 W. The space fraction is 0.75. The lighting schedule based on wattage is given in Table P10.4.1. The radiant fraction of the heat released by the lights is 0.65.

The non-solar RTS values appropriate to the zone, obtained from Table 19, Chapter 18 of the *ASHRAE Handbook* - 2013 *Fundamentals* [1] are given in Table P10.4.2. Calculate (i) the cooling load due to lighting 8 p.m., and (ii) the maximum cooling load due to lighting and the time when it occurs.
[*Answers*: (i) 361W, (ii) 580 W at 4 p.m.]

Table P10.4.1 Lighting schedule based on wattage

*Hours from midnight, W_L = Total lighting wattage (W)

Hour*	1	2	3	4	5	6	7	8	9	10	11	12
W_L	0	0	0	0	0	0	0	420	520	780	780	780
Time	13	14	15	16	17	18	19	20	21	22	23	24
W_L	780	780	780	780	580	580	420	380	0	0	0	0

Table P10.4.2 Non-solar RTS values

Light construction, With carpet, 50 percent glass

Hour	0	1	2	3	4	5	6	7	8	9	10	11
C_n%	50	18	10	6	4	3	2	1	1	1	1	1
Hour	12	13	14	15	16	17	18	19	20	21	22	23
C_n%	1	1	0	0	0	0	0	0	0	0	0	0

P10.5 A vertical wall of area 24 m^2, at a location with a northern latitude of 35°, faces 20° west of south. The solar absorptivity of the external wall surface is 0.75. The external heat transfer coefficient and the overall U-value of the wall are 28 Wm^{-2}K^{-1} and 0.628 Wm^{-2}K^{-1} respectively. The ground reflectivity is 0.2. The design ambient temperature profile for the location is given in Table P10.5.1. The design indoor temperature is 24°C.

The conduction RTS values for the wall, obtained from Table 16, Chapter 18 of the *ASHRAE Handbook* - 2013 *Fundamentals* [1] are given in Table P10.5.2. The non-solar RTS values appropriate to the zone, obtained from Table 19, Chapter 18 [1] are given in Table P10.5.3. For 12 June, obtain the cooling load due to heat conduction through the wall at 4 p.m., and (ii) the maximum cooling load and the time when it

occurs. Assume that the optical depth parameters for the location for the month of June as, $\tau_b = 0.337$ and $\tau_d = 2.36$.
[*Answer*: (i) 21.8W, (ii) 95.2 W at 1 a.m.]

Table P10.5.1 Ambient temperature profile

*Hours from midnight

Hour*	1	2	3	4	5	6	7	8	9	10	11	12
T_{amb}, °C	21	20.2	19.6	19.2	18.8	18.6	18.8	19.6	20.6	22	23.5	25.6
Hour	13	14	15	16	17	18	19	20	21	22	23	24
T_{amb}, °C	27.2	29.1	30	30	29.5	28.8	27.2	25.6	24.5	23.3	22	21.3

Table P10.5.2 Wall conduction RTS values for brick wall

Hour	0	1	2	3	4	5	6	7	8	9	10	11
C_n%	2	2	2	4	5	6	6	7	7	6	6	6
Hour	12	13	14	15	16	17	18	19	20	21	22	23
C_n%	5	5	5	4	4	4	3	3	3	2	2	1

Table P10.5.3 Non-solar RTS values

Light construction, With carpet, 50 percent glass

Hour	0	1	2	3	4	5	6	7	8	9	10	11
C_n%	50	18	10	6	4	3	2	1	1	1	1	1
Hour	12	13	14	15	16	17	18	19	20	21	22	23
C_n%	1	1	0	0	0	0	0	0	0	0	0	0

P10.6 A building at a location with a northern latitude of 35° has a flat roof of area 75 m^2, similar to roof number 17 in Table 17 on page 18.26 of the *ASHRAE Handbook* - 2013 *Fundamentals* [1]. The conduction time factors for the roof obtained from Table 17 are listed in Table P10.6.1. The U-factor of the roof is 0.315 Wm^{-2}K^{-1}. The surface absorptivity is 0.8.

The non-solar radiant time series (RTS) for medium construction with 90% glass and no carpet, obtained from Table 19 the *ASHRAE Handbook* - 2013 *Fundamentals* [1], is listed in Table P10.6.2. The external heat transfer coefficient of the roof is 24 Wm^{-2}K^{-1}. The hourly ambient temperature profile for the location is given in Table P10.6.3. The design indoor temperature is 24°C. For June 12, obtain (i) the cooling load due to conduction through the roof at 11 a.m., and (ii) the maximum cooling load and the time when it occurs. Assume that the optical depth parameters for the location for the month of June as $\tau_b = 0.337$ and $\tau_d = 2.36$.

[*Answers*: (i) 165 W, (ii) 330 W at 8 p.m.]

Table P10.6.1 Roof conduction time series (CTS) [1]

Hour	0	1	2	3	4	5	6	7	8	9	10	11
C_n%	2	2	5	6	7	7	6	6	6	5	5	5
Hour	12	13	14	15	16	17	18	19	20	21	22	23
C_n%	4	4	4	4	3	3	3	3	3	3	2	2

Table P10.6.2 Non-solar radiant time series (RTS) [1]

Hour	0	1	2	3	4	5	6	7	8	9	10	11
C_n%	35	15	10	7	5	4	3	3	2	2	2	2
Hour	12	13	14	15	16	17	18	19	20	21	22	23
C_n%	1	1	1	1	1	1	1	1	1	1	0	0

Table P10.6.3 Ambient temperature profile

*Hours from midnight

Hour*	1	2	3	4	5	6	7	8	9	10	11	12
T_{amb}, °C	21.4	20.8	19.8	19.5	18.8	19.2	19.4	19.6	21.6	22.4	23.5	25.6
Hour	13	14	15	16	17	18	19	20	21	22	23	24
T_{amb}, °C	27.2	29.1	30.5	30.5	29.5	28.8	28.2	26.2	24.5	23.3	22.2	21.6

P10.7 A single-glazed vertical window of area 7.5 m^2, at a location with a northern latitude of 35°, faces 20° west of south. The ground reflectivity is 0.2. The solar heat gain coefficients (SHGC) for the window, obtained from Table 10, chapter 15 of the *ASHRAE Handbook - 2013 Fundamentals* [1], are listed in Table P10.7.1. The overall heat transfer coefficient for the window is 5.1 Wm^{-2}K^{-1}. The indoor air temperature is 23°C. The ambient temperature profile is given in Table P10.7.4.

The solar RTS values, and the non-solar RTS values for medium construction with no carpet and 90% glass, are obtained from Tables 20 and 19 respectively, in chapter 18 of the *ASHRAE Handbook - 2013 Fundamentals* [1]. These values are given in Tables P10.7.2 and P10.7.3 respectively. For June 27, obtain (i) the window cooling load at 11 a.m., and (ii) the maximum cooling load and the time when it occurs. Assume

that the optical depth parameters for the location for the month of June as $\tau_b = 0.337$ and $\tau_d = 2.36$.
[Answers: (i) 944 W, (ii) 1833 W at 5 p.m.]

Table P10.7.1 Solar heat gain coefficients for window [1]

Inc. angle	0	40	50	60	70	80	Diffuse
SHGC	0.81	0.80	0.78	0.73	0.62	0.39	0.73

Table P10.7.2 Solar radiant time series [1]

Hour	0	1	2	3	4	5	6	7	8	9	10	11
C_n%	29	15	10	7	6	5	4	3	3	3	2	2
Hour	12	13	14	15	16	17	18	19	20	21	22	23
C_n%	2	2	1	1	1	1	1	1	1	0	0	0

Table P 10.7.3 Non-solar radiant time series [1]

Hour	0	1	2	3	4	5	6	7	8	9	10	11
C_n%	35	15	10	7	5	4	3	3	2	2	2	2
Hour	12	13	14	15	16	17	18	19	20	21	22	23
C_n%	1	1	1	1	1	1	1	1	1	1	0	0

Table P10.7.4 Ambient temperature profile

*Hours from midnight

Hour*	1	2	3	4	5	6	7	8	9	10	11	12
T_{amb}, °C	22.3	21.7	20.8	20.5	19.8	20.2	20.4	20.6	22.7	23.6	24.6	26.7
Hour	13	14	15	16	17	18	19	20	21	22	23	24
T_{amb}, °C	28.3	30.2	31.6	31.6	30.7	29.9	29.4	27.3	25.6	24.4	23.3	22.6

P10.8 The single-glazed window described in problem 10.7 is 3.75 m wide and 2 m high. An overhang of width 0.6 m and length 3.75 m is to be installed 0.35 m above the top edge of the window. For the conditions stated in problem 10.7, calculate (i) the hourly values of the unshaded area (m^2) of the window from 9am to 4 pm, and (ii) the cooling load due to transmission of the direct beam at 11 a.m. and 4 p.m.
[Answers: (i) 7.5, 7.5, 0.81, 0.19, 0.79, 2.52, 3.5, 5.34, (ii) 58 W, 364 W]

P10.9 The wall of a building of length 21 m and height 3.5 m has an 8 mm thick gypsum board ($k = 0.16$ $Wm^{-1}K^{-1}$) on the inside. Next to it is a 130 mm thick layer of glass fiber insulation ($k = 0.046$ $Wm^{-1}K^{-1}$) which is followed by a 150 mm thick layer of concrete ($k = 0.42$ $Wm^{-1}K^{-1}$).

The outer face of the wall is made of brick (k = 0.5 $Wm^{-1}K^{-1}$) of thickness 100 mm. The wall has two 2 m by 1.2 m double-glazed windows. The 13 mm air gap between the two glass (k = 0.8 $Wm^{-1}K^{-1}$) panes of the window has a U-value of 6.1 $Wm^{-2}K^{-1}$. The thickness of the glass is 6mm. The inside and outside air temperatures are 23°C and −2°C respectively. The outside and inside heat transfer coefficients are 28 $Wm^{-2}K^{-1}$ and 8 $Wm^{-2}K^{-1}$ respectively. Calculate total heat load due to heat conduction through the wall and the windows.
[*Answers*: 831 W]

P10.10 The volume of a two-story house is 480 m^3. The number of air changes per hour (ACH) required to balance infiltration of ambient air has been estimated as 0.62. The outdoor air is saturated at −6°C. The indoor temperature and relative humidity are 23°C and 50% respectively. Determine the sensible and latent heat loads due to infiltration.
[*Answers*: 2.95 kW, 1.622 kW]

P10.11 Derive Eqs.(10.35) and (10.36) using a lumped capacity model of a wall.

References

1. *ASHRAE Handbook* - 2013 *Fundamentals*, American Society of Heating, Refrigeration and Air Conditioning Engineers, Atlanta, 2013.
2. Kuehn, Thomas H., Ramsey, James W. and Threlkeld, James L., *Thermal Environmental Engineering*, 3rd edition, Prentice-Hall, Inc., New Jersey, 1998.
3. Mitchell, John W. and Braun, James E., Principles of Heating, Ventilation and Air Conditioning in Buildings. John Wiley and Sons, Inc., New York, 2013.
4. Stoecker, Wilbert F. and Jones, Jerold W., *Refrigeration and Air Conditioning*, International Edition, McGraw-Hill Book Company, London, 1982.

Appendix A10.1 - MATLAB Code for Cooling Load due to People

```
% MATLAB code for computing cooling load due to people
% computes 24 hour cooling load profile
% numerical data from worked example 10.4
frr=0.6     % radiative fraction; frr
qsen=75   % sensible heat per person, (W)
qlat =55    % latent heat per person, (W)
% input the 24-hour occupancy schedule
ocp=[0,0,0,0,0,0,0,0,4,12,12,12,12,12,12,12,12,12,5,0,0,0,0,0]
% input  24-hour radient time series for people (ASHRAE table)
 rf=[50,18,10,6,4,3,2,1,1,1,1,1,1,1,0,0,0,0,0,0,0,0,0,0]
% compute sensible and latent cooling loads
for k=1:1:24;
    qccp(k)=((1-frr)*qsen+qlat)*ocp(k);
 end
% compute the 24-hour radiative fraction of cooling load
   sum1=0;
for i=1:1:24;
    sum1=0;
for j=1:1:24;
    if j<=i;
        x1=i-j+1;
    sum1=sum1+ rf(j)*ocp(i-j+1)* qsen*frr /100;
else
    x1=25-j+i;
    sum1=sum1+rf(j)*ocp(25-j+i)*qsen*frr /100;
end
  qcrd(i)=sum1                   % radiant load, (W)
  qctot(i)=sum1+qccp(i)     % total cooling load, (W)
end
end
```

Appendix A10.2 - MATLAB Code for Cooling Load due to Wall Conduction

```
% MATLAB code for 24-hour wall conduction cooling load profile
% code could be modified for roof conduction cooling load
% numerical data from worked examples 10.6-10.8
% input ambient temperature profile
tambp=[21,20.2,19.6,19.2,18.8,18.6,18.8,19.6,20.6,22,23.5,25.6,27.2,...
   29.1,30,30,29.5,28.8,27.2,25.6,24.5,23.3,22,21.3]
% input conduction time series for wall construction
 rfc=[1,1,2,5,8,9,9,9,8,7,7,6,5,4,4,3,3,2,2,2,1,1,1,0]
% input non-solar radiant time series for interior space
 rf=[33,16,10,7,5,4,3,3, 2,2,2,2,1,1,1,1,1,1,1,1,1,1,1,0]
% input design data
frr=0.46      % radiative fraction from ASHRAE Handbook
fcc=0.54      % convective fraction from ASHRAE Handbook
uof=0.58               % overall U-factor of wall
temrm=23               % room temperature
awall=14               % area of wall sq.m
corr=0                 % sol-air correction factor for vertical wall
hout=28                % outside surface heat transfer coefficient for wall
sabs=0.8               % absorptivity of wall surface
rfgr=0.25              % ground reflectivity
psi=30                 % direction of normal to wall surface
taub=0.697             % factor in ASHRAE clear-sky model
taud=1.51              % factor in ASHRAE clear sky model
pi=3.1415926
nd=233                 % day number
lat=40                 % northern latitude
% section 1 - computes sol-air temperature profile
latr=lat*pi/180;
ang=360*(nd+284)/365 ;      % calculate solar declination
angr=ang*pi/180;            % angle in radians
dec=23.45*sin(angr)         %  declination, deg.
decr=dec*pi/180;            % declination, rad.
% compute sunrise and sunset solar times
```

```
xan1=-tan(latr)*tan(decr);
hsrr=acos(xan1);
hsr=hsrr*180/pi;
tsrh=12-hsr/15
tssh=12+hsr/15
srh=round(tsrh)          % rounded solar time at sunrise
ssh=round(tssh)          % rounded solar time at sunset
% begin the hour angle loop from sunrise to sunset
for i=srh:1:ssh
 tsol=i; % solar time
hang = 15*(tsol-12);
hangr=hang*pi/180;
% compute solar altitude angle
sbta=cos(latr)*cos(hangr)*cos(decr)+sin(latr)*sin(decr);
btar=asin(sbta);
bta=btar*180/pi ;
% compute solar azimuth angle
sphi=cos(decr)*sin(hangr)/cos(btar);
phir=asin(sphi);
phi=phir*180/pi;
% compute air mass
amas=(sin(btar)+0.50572*(6.07995+bta)^(-1.6364))^(-1);
% compute extraterrestrial radiation intensity
esc=1367;
extr=2*(nd-3)*pi/365;
eo=esc*(1+0.033*cos(extr));
% compute beam and diffuse parameters , ab  and ad
ab=1.454-0.406*taub-0.268*taud+0.021*taub*taud ;
ad=0.507+0.205*taub-0.080*taud-0.190*taub*taud;
% compute solar beam and diffuse radiation intensities
xxb=taub*amas^ab;
ebeam=eo*exp(-xxb);
xxd=taud*amas^ad;
ediff=eo*exp(-xxd);
% compute surface-solar azimuth angle for vertical wall
gam= psi-phi;
```

```
gamr=gam*pi/180;
% compute angle of incidence on vertical wall
costh=cos(btar)*cos(gamr);
thetr=acos(costh);
theta=thetr*180/pi
% compute direct beam intensity on wall
if theta <= 90
hdb=ebeam*costh;
else
hdb=0;
end
% compute the diffuse radiation intensity on vertical wall
Y=0.55+0.437*costh+0.313*costh^2;
hdif=ediff*Y;
% computed ground-reflected radiation on wall
hgr=(ediff+ebeam*sin(btar))*rfgr/2;
%  compute total incident radiation on wall
htot=hdb+hdif+hgr;
tsa(i)= tambp(i)+ sabs*htot/hout; % sol-air temperature profile
% section 2 - computes heat gain at inner wall surface
% compute conduction heat gain from sunrise to sunset
qrfg(i)=awall*uof*(tsa(i)-temrm);
end
% period from midnight to sunrise
for j=1:1:(srh-1)
qrfg(j)=awall*uof*(tambp(j)-temrm)
end
% period from sunset to midnight
for k= (ssh+1):1:24
qrfg(k)=awall*uof*(tambp(k)-temrm);
end
for l=1:1:24
qinsc(l)=qrfg(l)    % conduction heat gain for 24-hour period
end
 % account for conduction time delay due to wall thermal mass
  sum1=0;
```

```
for i=1:1:24
   sum1=0;
for j=1:1:24;
   if j<=i;
   sum1=sum1+rfc(j)*qinsc(i-j+1)/100;
   y=sum1;
else
   sum1=sum1+rfc(j)*qinsc(25-j+i)/100;
end
end
 qcin(i)=sum1   % heat flow rate at the inner wall surface
end
% section 3 - cooling load due to wall conduction
% compute convective component of the wall conduction cooling load
for k=1:1:24;
qcll(k)=fcc*qcin(k)
 end
 % compute radiative component of the wall conduction cooling load
 sum1=0;
for i=1:1:24;
   sum1=0;
for j=1:1:24;
   if j<=i;
   sum1=sum1+ rf(j)*qcin(i-j+1)* frr /100;
else
   sum1=sum1+rf(j)*qcin(25-j+i)*frr /100;
end
end
qrcon(i)=sum1
end
% compute total cooling load due to wall conduction
for l=1:1:24
   qcotot(l)=qcll(l)+qrcon(l)
end
```

Appendix A10.3 - MATLAB Code for Cooling Load due to Windows

```
% MATLAB code for 24 hour window cooling load profile
% numerical data from worked example 10.10
% input ambient temperature profile
tambp=[22.3,21.7,20.8,20.5,19.8,20.2,20.4,20.6,22.7,23.6,24.6,26.7,28.3,
... 30.2,31.6,31.6,30.7,29.9,29.4,27.3,25.6,24.4,23.3,22.6]
% input 24-hour solar radiant time series (RTS)
 srfc=[54,16,8,4,3,2,1,1,1,1,1,1,1,1,1,1,1,1,1,0,0,0,0,0];
% input 24-hour non-solar radiant time series (RTS)
 rfns=[49,17,9,5,3,2,2,1,1,1,1,1,1,1,1,1,1,1,1,1,0,0,0,0]
% input SHGC, ysg ,versus angle of incidence, xan
ysg=[0.7,0.67,0.64,0.58,0.45,0.23,0]
xan=[0,40,50,60,70,80,90]
shgcd=0.6       % SHGC for diffuse solar radiation
temrm=23        % room temperature
awin=6.5        % area of window, sq.m
uof=3.5         % overall U-factor of window
rfgr=0.25       % ground reflectivity
psi=-15         % direction of window surface
taub=0.337      % factor in clear-sky model
taud=2.36       % factor in clear sky model
pi=3.141592
nd=102          % day number
lat=30          % northern latitude
latr=lat*pi/180;
% calculate declination
ang=360*(nd+284)/365;
% angle in radians
angr=ang*pi/180;
dec=23.45*sin(angr)
decr=dec*pi/180;
% compute sunrise and sunset solar times
xan1=-tan(latr)*tan(decr);
hsrr=acos(xan1);
hsr=hsrr*180/pi;
```

```
tsrh=12-hsr/15;
tssh=12+hsr/15;
srh=round(tsrh)          % srh: rounded sunrise hour
ssh=round(tssh)          % ssh: rounded sunset hour
% begin the hour angle loop from sunrise to sunset
for i=srh:1:ssh
tsol=i % solar time
hang = 15*(tsol-12);
hangr=hang*pi/180;
% compute solar altitude angle
sbta=cos(latr)*cos(hangr)*cos(decr)+sin(latr)*sin(decr);
btar=asin(sbta);
bta=btar*180/pi ;
% compute solar azimuth angle
sphi=cos(decr)*sin(hangr)/cos(btar);
phir=asin(sphi);
phi=phir*180/pi
% compute air mass
amas=(sin(btar)+0.50572*(6.07995+bta)^(-1.6364))^(-1);
% compute extraterrestrial radiation
esc=1367;
extr=2*(nd-3)*pi/365;
eo=esc*(1+0.033*cos(extr));
% compute beam and diffuse parameters, ab and ad
ab=1.454-0.406*taub-0.268*taud+0.021*taub*taud;
ad=0.507+0.205*taub-0.080*taud-0.190*taub*taud;
% compute beam and diffuse radiation intensities
xxb=taub*amas^ab;
ebeam=eo*exp(-xxb);
xxd=taud*amas^ad;
ediff=eo*exp(-xxd);
% compute surface-solar azimuth angle for vertical window
gam= phi-psi;
gamr=gam*pi/180;
% compute angle of incidence on vertical window
costh=cos(btar)*cos(gamr);
```

```
thetr=acos(costh);
theta=thetr*180/pi;
% compute direct beam intensity on window
if theta <= 90
hdb=ebeam*costh;
% interpolate SHGC for direct beam
zz=theta;
% choose linear or spline interpolation
% linear interpolation
% shgcb(i)=interp1(xan,ysg,xanv)
% cubic spline interpolation
shgcb(i)=spline(xan,ysg,xanv);
else
   hdb=0;
   shgcb(i)=0;
end
% compute the diffuse radiation intensity on window surface
Y=0.55+0.437*costh+0.313*costh^2;
hdif=ediff*Y;
% compute ground-reflected radiation on window surface
hgr=(ediff+ebeam*sin(btar))*rfgr/2;
% compute total diffuse radiation intensity on window surface
hdiffu=hdif+hgr;
% compute direct beam solar heat gain
 qgsol(i) =awin*hdb*shgcb(i);
% compute diffuse radiation heat gain
sdgain=awin*(hdif+hgr)*shgcd;
% compute conduction heat gain through window
qwcon=awin*uof*(tambp(i)-temrm);
%compute diffuse plus conduction heat gain
qrfg(i)=qwcon+sdgain;
end
% period from midnight to sunrise
for j=1:1:(srh-1)
   qwcon=awin*uof*(tambp(j)-temrm);
   qrfg(j)=qwcon;
```

```
    qgsol(j)=0;
    shgcb(j)=0;
end
% period from sunset to midnight
for k= (ssh+1):1:24
qwcon=awin*uof*(tambp(k)-temrm);
qrfg(k)=qwcon;
qgsol(k)=0;
shgcb(k)=0;
end
% select radiation fraction for diffuse plus conduction heat gain
for l=1:1:24
   if shgcb(l)<=0.5;
        frr(l)=0.46;
    else
        frr(l)=0.33;
end
% compute radiant fraction of the diffuse plus conduction heat gain
     qrfgr(l)=qrfg(l)*frr(l);
end
 % account for solar beam time delay using solar-RTS
   sum1=0;
for i=1:1:24;
    sum1=0;
for j=1:1:24;
    if j<=i;
    sum1=sum1+srfc(j)*qgsol(i-j+1)/100
else
    sum1=sum1+srfc(j)*qgsol(25-j+i)/100;
end
end
     qcin(i)=sum1
end
% compute the convective component of window conduction load
for k=1:1:24;
    qcll(k)=(1-frr(k))*qrfg(k);
```

```
 end
 % compute radiative component of window heat gain
  sum1=0;
for i=1:1:24;
   sum1=0;
for j=1:1:24;
   if j<=i;
   sum1=sum1+ rfns(j)*qrfgr(i-j+1) /100;
else
   sum1=sum1+rfns(j)*qrfgr(25-j+i) /100;
end
end
qrcon(i)=sum1;
end
% compute total cooling load due to window heat gain
for l=1:1:24
qcllr(l)=qcll(l)  % cooling load-conv. comp. of cond. plus diff. rad.
qrllr(l)=qcin(l)       % solar radiation cooling load
qrconr(l)=qrcon(l)   % long wave radiation cooling load
qcotot(l)=qcin(l)+qcll(l)+qrcon(l)  % total window cooling load
end
```

Appendix A10.4 - MATLAB Code for Shading of Windows

```
% MATLAB code for window with an overhang
% numerical data from worked example 10.12
% computes unshaded area and direct beam cooling load
wo= 0.75          % width of overhang, m
lw=3.25           % length of window, m
how=0.3           % height of overhang above top of window, m
hw=2              % height of window, m
% input 24-hour solar radiant time series (RTS)
 srfc=[54,16,8,4,3,2,1,1,1,1,1,1,1,1,1,1,1,1,1,0,0,0,0,0];
 % variation of beam SHGF values, ysg, with angle of incidence, xan
 ysg=[0.7,0.67,0.64,0.58,0.45,0.23,0]
 xan=[0,40,50,60,70,80,90]
```

```
shgcd=0.6          % SHGC for diffuse solar radiation
awin=6.5            % area of window, sq.m
psi=-15            % direction of window surface
taub=0.337         % factor in ASHRAE clear-sky model
taud=2.36          % factor in ASHRAE clear sky model
pi=3.141592
nd=102             % day number
lat=30             % northern latitude
latr=lat*pi/180;
% calculate declination
ang=360*(nd+284)/365;
angr=ang*pi/180;
dec=23.45*sin(angr);
decr=dec*pi/180;
% compute sunrise and sunset solar times
xan1=-tan(latr)*tan(decr);
hsrr=acos(xan1);
hsr=hsrr*180/pi;
tsrh=12-hsr/15;
tssh=12+hsr/15;
srh=round(tsrh)          % srh: rounded sunrise hour
ssh=round(tssh)          % ssh: rounded sunset hour
for i=srh:1:ssh
   tsol=i; % solar time
hang = 15*(tsol-12);
hangr=hang*pi/180;
% compute solar altitude angle
sbta=cos(latr)*cos(hangr)*cos(decr)+sin(latr)*sin(decr);
btar=asin(sbta);
bta=btar*180/pi ;
% compute solar azimuth angle
sphi=cos(decr)*sin(hangr)/cos(btar);
phir=asin(sphi);
phi=phir*180/pi;
% compute air mass
amas=(sin(btar)+0.50572*(6.07995+bta)^(-1.6364))^(-1);
```

```
% compute extraterrestrial radiation
esc=1367;
extr=2*(nd-3)*pi/365;
eo=esc*(1+0.033*cos(extr));
% compute beam and diffuse parameters, ab, ad
ab=1.454-0.406*taub-0.268*taud+0.021*taub*taud;
ad=0.507+0.205*taub-0.080*taud-0.190*taub*taud;
% compute beam and diffuse radiation intensities
xxb=taub*amas^ab;
ebeam=eo*exp(-xxb);
xxd=taud*amas^ad;
ediff=eo*exp(-xxd);
% compute surface-solar azimuth angle for vertical window
gam= -psi+phi;
gamr=gam*pi/180;
% compute coordinates of shadows of window-overhang corners
x1=-wo*tan(gamr);
x2=lw-wo*tan(gamr);
z1=hw+how-wo*tan(btar)/cos(gamr);
z2=hw+how-wo*tan(btar)/cos(gamr);
% compute unshaded area of window, aunshd
if x1>=0
  if z1>0
x3=x1*how/(hw+how-z1);
ashd=(hw-z1)*(2*lw-x1-x3)/2;
aunshd=hw*lw-ashd;
end
end
if x1>=0
  if z1<0
x3=x1*(how+hw)/(hw+how-z1)
x4=x1*how/(hw+how-z1)
aunshd=hw*(x3+x4)/2
end
end
if x1>=lw;
```

```
  if z1>0
  x4=(hw+how-z1)*(x1-lw)/x1;
  ashd=0.5*x1*((hw-z1-x4)^2)/(hw+how-z1);
  aunshd=hw*lw-ashd;
end
end
if z1>=hw;
  aunshd=awin;
end
if x2<lw;
  if z1>0
x33=(lw-x2)*how/(hw+how-z2);
ashd=(hw-z2)*(lw-x33+x2)/2;
aunshd=hw*lw-ashd;
end
end
if x2<0;
  x44=(hw+how-z2)*((lw-x1)-lw)/(lw-x1);
  ashd=0.5*x1*((hw-z1-x44)^2)/(hw+how-z2);
  aunshd=hw*lw-ashd;
end
if z2>=hw;
  aunshd=awin;
end
if z2<0
  if x2<lw
 x6=how*(lw-x2)/(hw+how-z2);
 x7=(hw+how)*(lw-x2)/(hw+how-z2);
 ashd=hw*(2*lw-x6-x7)/2;
 aunshd=hw*lw-ashd;
end
end
if z2<0
  if x2>lw
 x6=how*x1/(hw+how-z2);
 x7=(hw+how)*x1/(hw+how-z2);
```

```
 aunshd=hw*(x6+x7)/2;
end
end
awind=aunshd;
% compute angle of incidence on vertical window
costh=cos(btar)*cos(gamr);
thetr=acos(costh);
theta=thetr*180/pi;
% compute direct beam intensity on window
if theta <= 90
hdb=ebeam*costh;
% interpolate SHGC for direct beam
zz=theta;
xanv=[zz];
% choose linear or spline interpolation
% linear interpolation
% shgcb(i)==interp1(xan,ysg,xanv)
% cubic spline interpolation
shgcb(i)=spline(xan,ysg,xanv);
else
   hdb=0;
   shgcb(i)=0;
end
% compute direct beam solar heat gain
qgsol(i)=awind*hdb*shgcb(i);
end
% period from midnight to sunrise
srh1=srh-1
for j=1:1:(srh-1)
   qgsol(j)=0;
   shgcb(j)=0;
end
% period from sunset to midnight
for k= (ssh+1):1:24
qgsol(k)=0;
shgcb(k)=0;
```

```
end
 % account for solar beam time delay using solar RTS values
   sum1=0;
for i=1:1:24;
    sum1=0;
for j=1:1:24;
    if j<=i;
    sum1=sum1+srfc(j)*qgsol(i-j+1)/100;
else
    sum1=sum1+srfc(j)*qgsol(25-j+i)/100;
end
end
     qcsol(i)=sum1   % hourly cooling load due to direct beam
end
```

Chapter 11

Air Distribution Systems

11.1 Introduction

In chapter 5 we discussed a number of different designs of heating and air conditioning plants, including multi-zone systems, dual-duct systems and others. In actual practice, the heating and cooling units of these systems are installed in a central plant room, usually located far from the various spaces being air conditioned. In a typical central air conditioning system, shown schematically in Fig. 3.6, chilled water produced in a refrigeration plant is pumped to an air handling unit (AHU) that cools and dehumidifies the air from the zones. The conditioned air is supplied to the zones through a network of ducts. The economic operation of such systems requires careful design of the duct network distributing the air to the zones and the pumping system conveying the chilled water to the AHU.

The important design aspects of air distribution systems, to be discussed in this chapter, are: (i) computation of pressure losses across ducts and fittings, (ii) arrangement of the duct network, (iii) characteristics of fans, (iv) interaction of fan and duct network, and (v) distribution of air within the conditioned space.

Moreover, the design of duct–fan systems involve a number of interrelated cost considerations, including energy cost, duct system cost, and building space cost. In view of the somewhat tedious and iterative computations involved, duct system design is often carried out using computer programs. In the following sections we shall present the fundamental physical principles needed for designing duct systems.

11.2 Total Pressure Distribution

The duct network of a typical air conditioning system supplying conditioned air to a number of zones is shown schematically in Fig. 11.1(a). At any section of the duct system, the *total mechanical energy per unit mass* of fluid, E_m [Jkg^{-1}], is given by [1,2]

$$E_m = \frac{p}{\rho} + \frac{V^2}{2} + gz \tag{11.1}$$

where

$\frac{p}{\rho}$ = potential energy due to static pressure, p [Nm^{-2}]

$\frac{V^2}{2}$ = kinetic energy due to fluid velocity, V [ms^{-1}]

gz = gravitational potential energy due to vertical height, z [m]

The fluid density and the acceleration due to gravity are ρ [kgm^{-3}] and g [ms^{-2}] respectively.
We define the *total head*, H [m] of the fluid at any section as

$$H = \left(\frac{E_m}{g}\right) = \frac{p}{\rho g} + \frac{V^2}{2g} + z \tag{11.2}$$

The three terms on the RHS of Eq. (11.2) are usually called the static pressure head, the velocity head, and the elevation head respectively.

As the fluid flows through the duct network a fraction of its *mechanical energy* is converted to *internal thermal energy* due to fluid friction. This process results in a small rise of the fluid temperature. We usually identify the *mechanical energy change* between any two sections as an *energy loss*, E_L. It is also convenient to express this mechanical energy loss as a head loss ΔH. Hence we have the following equation for the *head loss* between two sections 1 and 2 of the duct:

$$\Delta H_{12} = \left(\frac{E_{m1}}{g} - \frac{E_{m2}}{g}\right) = \left(\frac{p_1}{\rho g} + \frac{V_1{}^2}{2g} + z_1\right) - \left(\frac{p_2}{\rho g} + \frac{V_2{}^2}{2g} + z_2\right) \tag{11.3}$$

The mechanical energy loss between two sections may also be expressed as *total pressure loss* ΔP [Pa]. This is given by

$$\Delta P_{12} = \rho g \Delta H_{12} = \left(p_1 + \frac{\rho V_1{}^2}{2} + \rho g z_1\right) - \left(p_2 + \frac{\rho V_2{}^2}{2} + \rho g z_2\right) \tag{11.4}$$

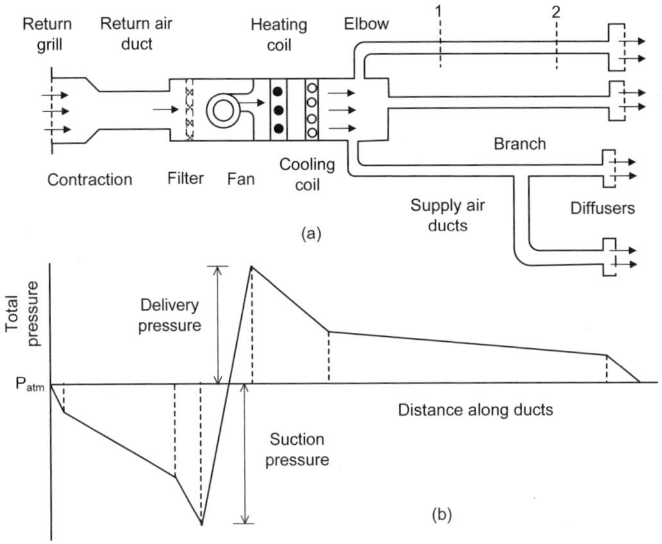

Fig. 11.1 (a) A typical duct network of an air conditioning system, (b) The total pressure (total head) distribution of air.

The variation of the total pressure (total head) of the air as it passes through the duct network is depicted in Fig. 11.1(b). The total pressure of the air entering the return air duct decreases progressively as it passes through the grill, the return duct and the filter. The slope of the graph gives an indication of the total pressure loss in each element. Across the fan the total pressure of air increases sharply due to the work input by the fan.

The flow resistance offered by the heating and cooling coils results in a significant loss of total pressure. The total air flow then divides into three streams that are delivered to the three zones. Each air stream looses total pressure due to friction in the duct, losses in the fittings, and the resistance offered by the diffuser.

In order to design the duct system we need to estimate the loss of total pressure in the ducts and the various fittings like bends, contraction and expansion sections, and branches.

11.3 Pressure Loss in Duct Networks

11.3.1 *Pressure loss in straight ducts*

The frictional pressure loss in a straight duct of circular cross section is given by the Darcy–Weisbach equation [1],

$$\Delta P_f = f\left(\frac{L}{D}\right)\left(\frac{\rho V^2}{2}\right) \tag{11.5}$$

where ΔP_f = frictional pressure loss, Pa
L = length, m
D = diameter, m
V = velocity, ms^{-1}
ρ = density, kgm^{-3}
f = Darcy–Weisbach friction factor

For *laminar flow*, the friction factor is obtained from the relation,

$$f = \frac{64}{Re} \tag{11.6}$$

where the Reynolds number, *Re* is given by

$$Re = \frac{\rho VD}{\mu} \tag{11.7}$$

where μ [Pa.s] is the viscosity of the fluid.

For *turbulent flow*, the friction factor is a function of the Reynolds number and the relative roughness of the duct surface, ε/D, where ε is the absolute surface roughness in meters. The friction factor, f is obtained from Colebrook's equation [2],

$$\frac{1}{\sqrt{f}} = 1.14 + 2log\left(\frac{D}{\varepsilon}\right) - 2log\left[1 + 9.3\left(\frac{D}{\varepsilon}\right)\left(\frac{1}{Re\sqrt{f}}\right)\right] \tag{11.8}$$

Because Eq. (11.8) is a transcendental equation, it has to be solved using an iterative technique. However, f can be obtained directly from the Moody chart, which is available in many sources including Refs. [2,4].

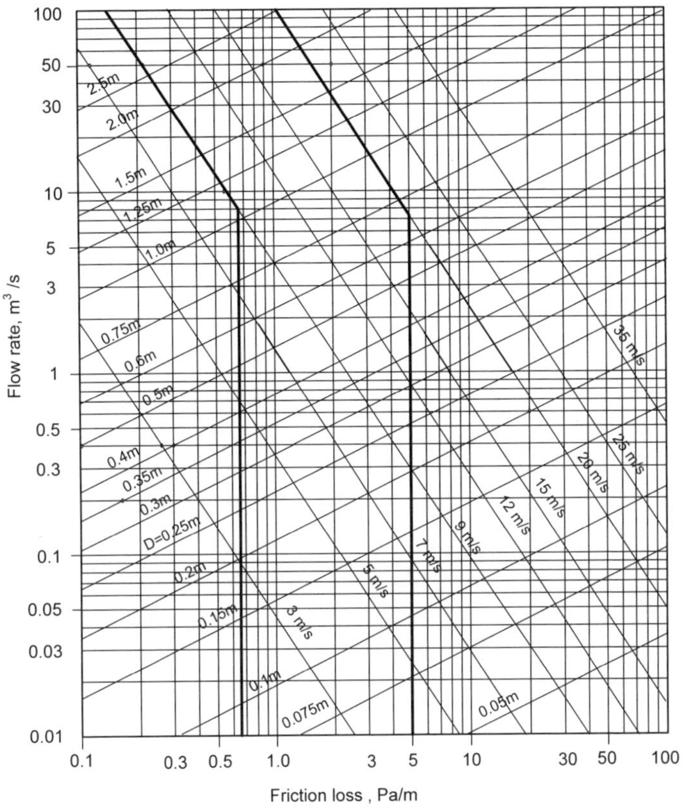

Fig. 11.2 Friction chart for round ducts: curves for (i) constant diameter and (ii) constant velocity. Conditions: Temperature = 20°C, Density = 1.2041 kgm^{-3}, Roughness = 0.09 mm

We now develop a convenient graphical method to obtain the friction pressure loss using Eq. (11.5). The volume flow rate of a fluid through a circular duct may expressed as

$$Q = \frac{\pi D^2 V}{4} \tag{11.9}$$

From Eqs. (11.5) and (11.9) it follows that

$$\frac{\Delta P_f}{L} = \left(\frac{8\rho f}{\pi^2}\right)\frac{Q^2}{D^5} \tag{11.10}$$

Therefore $$log\left(\frac{\Delta P_f}{L}\right) = 2logQ - 5logD + log\left(\frac{8\rho f}{\pi^2}\right) \tag{11.11}$$

The friction chart given in Fig. 11.2 is developed using Eq. (11.11), where the friction factor, *f* is obtained from Eq. (11.8). The first family of curves gives the variation of the friction pressure gradient with the volume flow rate for different values of the duct diameter. The second family of curves is the same variation for different fluid velocities. A friction chart similar to that in Fig. 11.2 is also available in the *ASHRAE Handbook* - 2013 *Fundamentals* [1].

The data in Fig. 11.2 cover a wide range of duct diameters and air flow velocities. However, actual air duct systems are subject to design limits that depend on practical considerations, such as duct cost, fan power and noise level. The design values for diameter and flow velocity, recommended in Ref. [1], fall within the area bounded by the bold lines in Fig. 11.2. Larger duct sizes, usually associated with lower fan power and noise levels, have higher initial cost. Smaller ducts, on the other hand, have higher noise levels and require more fan power.

The data in Fig. 11.2 may be used to determine the friction pressure loss in a rectangular duct whose sides are *a* and *b* by using the equivalent diameter given by [1]

$$D_{eq} = 1.30\frac{(ab)^{0.625}}{(a+b)^{0.25}} \tag{11.12}$$

For a rectangular duct the friction chart in Fig. 11.2 is *only applicable* when the *actual* volume flow rate, Q and the *equivalent* diameter D_{eq}, given by Eq. (11.12), are used.

The friction chart in Fig. 11.2 may also be used to obtain the pressure loss in straight ducts of oval cross section by computing the equivalent diameter of an oval duct using the correlation given in Ref. [1].

11.3.2 *Pressure loss in fittings*

Air distribution systems consist of straight ducts and fittings, such as, contractions, enlargements, elbows, branches, dampers, filters and diffusers. Some of these fittings are shown schematically in Fig. 11.1(a) and Fig. 11.3.

In fittings, mechanical energy losses, or dynamic losses, as these are commonly called, occur as a result of: (i) direction changes, as in elbows,

(ii) flow area changes, as in expansions and contractions, and (iii) mixing, as in branches.

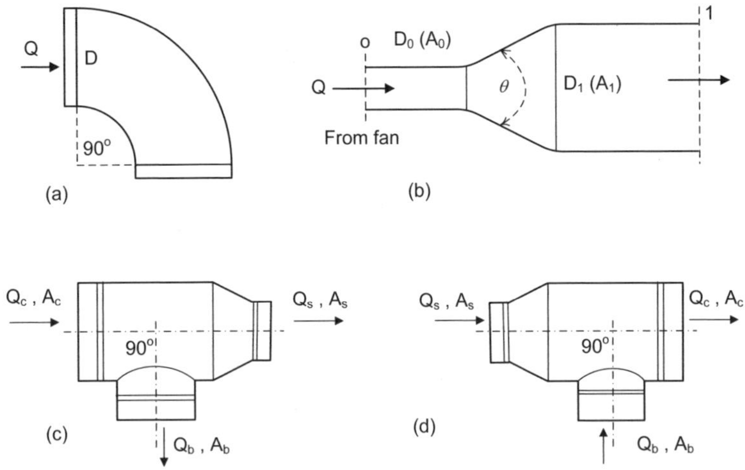

Fig. 11.3 Pressure loss through fittings: (a) 90°-elbow, (b) transition, (c) diverging-tee, (d) converging-tee

The total pressure loss in a fitting is obtained from the following equation:

$$\Delta p_t = c_o p_{vo} = c_o \left(\frac{1}{2}\rho V_o{}^2\right) \tag{11.13}$$

where

p_{vo} = velocity pressure at the selected referenced section, o, [Pa]
c_o = local loss coefficient referenced to section o, dimensionless
V_o = velocity at the selected referenced section, o, [ms^{-1}]

An extensive database of loss coefficients for air duct fittings is available in the *ASHRAE Duct Fitting Database* (2012). This database, available on CD, includes pictorial outlines and loss coefficient tables for more than 200 round, rectangular, and flat oval duct fittings. A more limited list of loss coefficients is tabulated in the *ASHRAE Handbook - 2009 Fundamentals* [2].

We have included representative values for a 90°-elbow (Fig 11.3a), a transition (Fig.11.3b), a diverging-tee (Fig. 11.3c), and a converging-tee (Fig. 11.3d) in Tables 11.1(a), 11.1(b), 11.1(c) and 11.1(d) respectively.

Table 11.1(a) Loss coefficients for 90°-elbow* (Fig. 11.3a)

D, mm	75	100	125	150	180	200	230	250
C_o	0.3	0.21	0.16	0.14	0.12	0.11	0.11	0.11

Table 11.1(b) Loss coefficients for transition, round to round ducts* (Fig. 11.3b)

A_o/A_1	$\theta = 3°$	5°	15°	30°	60°	120°	180°
0.1	0.12	0.09	0.05	0.05	0.08	0.29	0.43
0.25	0.1	0.07	0.04	0.04	0.07	0.27	0.41
0.5	0.07	0.06	0.05	0.05	0.07	0.18	0.24
2.0	0.59	0.51	0.52	1.26	1.30	1.23	1.19
6.0	6.55	6.08	9.14	17.35	27.58	26.32	25.99
10.0	19.50	18.25	27.30	58.50	80.00	84.00	82.70

Table 11.1(c) Loss coefficients for diverging Tee* (Fig. 11.3c)

		C_b values			
A_b/A_c	$Q_b/Q_c = 0.1$	0.2	0.4	0.6	0.8
0.1	1.2	0.62	1.28	2.29	5.44
0.2	4.1	0.72	0.62	0.8	1.28
0.4	15.89	4.10	1.20	0.72	0.62
0.8	63.63	15.89	4.10	1.94	1.20
		C_s values			
A_s/A_c	$Q_s/Q_c = 0.1$	0.2	0.4	0.6	0.8
0.1	0.13	0.24	0.74	0.70	0.60
0.2	0.2	0.13	0.16	0.57	0.74
0.4	2.88	0.20	0.13	0.15	0.16
0.8	26.88	2.88	0.20	0.14	0.13

Table 11.1(d) Loss coefficients for converging-tee*, D_c >250 mm (Fig. 11.3c)

		C_b			
A_s/A_c	A_b/A_c	$Q_b/Q_c = 0.1$	0.2	0.4	0.8
0.5	0.2	−7.26	0.62	0.75	0.93
0.5	0.4	−30.49	−4.67	0.38	0.90
0.5	0.6	−69.03	−11.17	−0.03	0.98
0.5	0.8	−123.3	−20.22	−0.50	1.20
0.8	0.2	−2.9	0.15	0.83	0.95
0.8	0.4	−12.31	−1.34	0.74	0.95
0.8	0.6	−27.50	−3.39	1.09	1.03
0.8	0.8	−48.87	−6.11	1.15	1.25
		C_s			
A_s/A_c	A_b/A_c	$Q_s/Q_c = 0.1$	0.2	0.4	0.8
0.5	0.2	126.36	16.99	2.42	0.38
0.5	0.4	38.84	7.27	1.51	0.34
0.5	0.6	17.98	4.95	1.29	0.33
0.5	0.8	13.78	4.48	1.25	0.33
0.8	0.2	92.27	17.35	3.21	0.27
0.8	0.4	6.75	7.77	2.31	0.23
0.8	0.6	12.05	8.36	2.37	0.23
0.8	0.8	40.23	11.49	2.66	0.24

*Representative values extracted from the *ASHRAE Duct Fittings Database*, Chapter 21, *ASHRAE Handbook* - 2009 *Fundamentals* [2].

The total pressure loss for all fittings, except junctions, is given by Eq. (11.13). For some fittings like transitions, with unequal inlet and outlet areas (Fig.11.3b), the loss coefficient may be changed from section o to section 1 by applying the mass balance equation. Hence we have

$$\Delta p_t = c_1 p_{v1} = c_1 \left(\frac{1}{2}\rho {V_1}^2\right) \tag{11.14}$$

where $$c_1 = c_o \left(\frac{A_1}{A_o}\right)^2$$

For the diverging and converging flow junctions depicted in Figs. 11.3(c) and 11.3(d) respectively, the total pressure loss (mechanical energy loss) through the straight (main) section is given by

$$\Delta p_{t,s} = c_s p_{vs} = c_s \left(\frac{1}{2}\rho {V_s}^2\right) \tag{11.15}$$

where p_{vs} = velocity pressure at the selected referenced section, *s*
c_s = straight section loss coefficient
V_s = velocity at the selected referenced section, *s*

The total pressure loss through the branch section is given by

$$\Delta p_{t,b} = c_b p_{vb} = c_b \left(\frac{1}{2}\rho {V_b}^2\right) \tag{11.16}$$

where p_{vb} = velocity pressure at the selected referenced section, *b*
c_b = branch section local loss coefficient
V_b = velocity at the selected referenced section, *b*

In diverging sections (Fig. 11.3c) the mechanical energy of the fluid decreases both across the straight section and the branch section. Therefore the loss coefficient, which is proportional to the change of mechanical energy per unit mass of fluid, is positive for both sections. However, in converging sections (Fig. 11.3d), mechanical energy is exchanged between two fluid streams moving at different velocities due to the turbulent mixing. Consequently, under some flow conditions the mechanical energy per unit mass of the slower moving stream can increase due to mixing. This results in a negative loss coefficient for the lower velocity stream as seen in Table 11.1(d).

11.3.3 *Total pressure loss in duct sections*

The total pressure loss in a duct section consisting of straight sections and fittings is obtained by combining Eqs. (11.5) and (11.13). Hence we have

$$\Delta p_{tot} = f\left(\frac{L}{D}\right)\left(\frac{\rho V^2}{2}\right) + \sum_n c_n \left(\frac{1}{2}\rho V_n{}^2\right) \tag{11.17}$$

The loss coefficient, c_n for each fitting of the duct section has to be referenced to the appropriate velocity pressure of the section.

A MATLAB program to solve Colebrook's equation iteratively and hence compute the friction pressure loss in circular ducts is given in Appendix A11.1. The code also computes losses through fittings in the duct section, such as elbows and tee joints, when the loss coefficients are supplied as inputs.

11.4 Air Distribution Fans

The distribution of air through a ductwork of an air conditioning system is carried out using a fan. The mechanical work input to the fan generates the pressure necessary to overcome the pressure losses in the ducts and the fittings. Some fans, used mainly for exhaust, are placed at the discharge end of a duct. Fans located near the center of the ductwork, as shown in Fig.11.1, provide air pressure to overcome losses in the return and supply duct systems.

The commonly used fans in air conditioning systems are either *axial flow* fans or *centrifugal* fans.

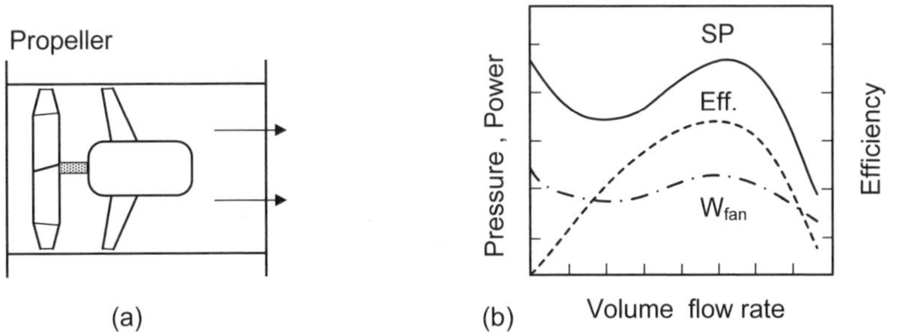

Fig. 11.4 (a) Schematic diagram of axial flow fan, (b) Static pressure rise, ideal fan power, and efficiency for an axial flow fan.

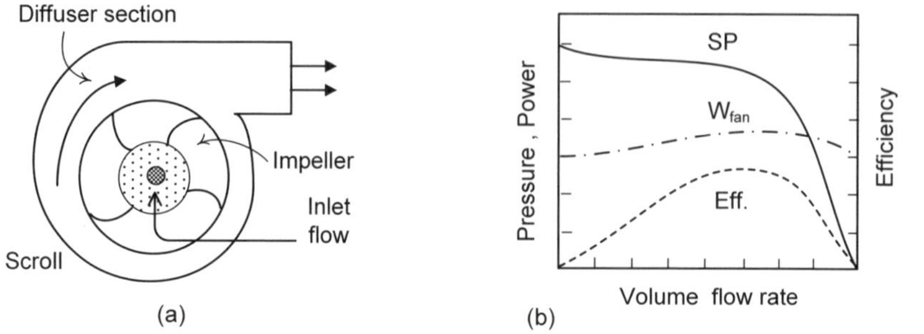

Fig. 11.5 (a) Schematic diagram of centrifugal fan with forward-curved blades, (b) Static pressure rise, ideal fan power, and efficiency for a centrifugal fan.

11.4.1 *Axial flow and centrifugal fans*

An axial flow fan is shown schematically in Fig. 11.4(a). Air is moved axially by the propeller blades, mounted on a concentric hub, rotated by an electric motor. The kinetic energy imparted to the air is converted to a pressure rise by the stationary blades located downstream of the propeller. The stationary blades also help improve the efficiency of the fan by reducing the swirl of the air. Since these fans are located axially in the flow duct, they do not require a change in flow direction of the air. Axial flow fans are used mainly in high volume flow applications.

Shown schematically in Fig. 11.5(a) is a centrifugal fan. It consists of an impeller with a series of blades attached to a hub, driven by an electric motor. Usually, power is transferred from the motor to the fan through a belt drive with pulleys attached to the motor and the impeller. Air enters at the centre of the impeller in a direction normal to the plane of the figure. The centrifugal force created by the rotating blades accelerates the air in the radial direction. The kinetic energy gained by the air is converted to a high static pressure in the diffusing section of the fan.

Four types of blades are used in the impellers of centrifugal fans. These are: radial, forward-curved, backward-curved, and airfoil. Forward-curved blades are commonly used in low pressure systems while the other three types of blades are used mainly in high pressure or high volume flow systems. The forward-curved blades of the impeller shown in Fig. 11.5(a) are curved in the *same* direction as the direction of rotation of the impeller.

11.4.2 *Fan characteristics*

Several parameters are used to rate the overall performance of fans. These are available in tabular form or graphical form in manufacturer's catalogues. The performance data at a given fan speed and volume flow rate usually include the following design parameters.

(i) *Total pressure*: This is the rise in total pressure across the fan, which may be expressed in the form

$$\Delta p_t = p_{t,o} - p_{t,i} \tag{11.18}$$

where i and o denote conditions at the inlet and outlet respectively.

(ii) *Static pressure*: This is the rise in static pressure across the fan, which is given by

$$\Delta p_s = p_{s,o} - p_{s,i} \tag{11.19}$$

(iii) *Ideal power input*: This is the ideal power input required to operate the fan under the given conditions. Assuming the temperature rise of air to be negligible and the density to be constant we have

$$\dot{W}_{id} = \frac{\dot{m}(p_{t,o}-p_{t,i})}{\rho} \tag{11.20}$$

where $\dot{m}$ is the mass flow rate of air.

(iv) *Fan efficiency*: This is the ratio of the *ideal* power required to the *actual* shaft power input of the fan, which may be expressed as

$$\eta_f = \frac{\dot{W}_{id}}{\dot{W}_{sh}} \tag{11.21}$$

Usually fans are driven by electric motors through belt drives. If the efficiency of the electric motor is η_m, then the required electric power input may be expressed as

$$\dot{W}_{el} = \frac{\dot{W}_{id}}{\eta_f \eta_m} \tag{11.22}$$

The typical performance characteristics of an axial fan and a centrifugal fan are shown in Figs. 11.4(b) and 11.5(b) respectively. We have displayed only the general trends of the various performance parameters, extracted from data available in manufacturer's catalogues.

We observe that at very low air flow rates, the fan is able to produce a large increase in the static pressure. As the air flow rate increases, the static pressure rise decreases slowly, initially. This is followed by a rapid decrease in static pressure rise at higher flow rates. The kinetic energy or the velocity pressure of the air increases continuously with the increase of flow rate. The fan power, which is proportional to the product of the static pressure rise and the volume flow rate, increases up to a point, and then decreases at higher flow rates. The ideal power, given by Eq. (11.20), approaches zero at very low and very high flow rates. This results in the efficiency variation depicted in the figures.

11.4.3 *Fan laws*

The *fan laws* are a group of approximate relationships that may be used to predict the effect of certain design and operating variables like, the fan speed, the size of fan, and the conditions of the air, on fan performance. These relationships may be derived using a method commonly known as *dimensional analysis*. Here the volume flow rate, $\dot{Q}$, the total pressure, p_t and the work input, $\dot{W}$ are grouped individually with the density, ρ, the characteristic dimension, D and the rotational speed ω to obtain the following equations:

$$\frac{\dot{Q}}{D^3\omega} = c_1 \tag{11.23}$$

$$\frac{p_t}{\rho D^2\omega^2} = c_2 \tag{11.24}$$

$$\frac{\dot{W}}{\rho D^5\omega^3} = c_3 \tag{11.25}$$

The *dimensionless* constants c_1, c_2 and c_3 in the three relationships in Eqs. (11.23), (11.24) and (11.25) are the same for two *aerodynamically similar* fans. This condition enables us to determine the performance of a particular fan under different operating conditions, or to predict the performance of a fan, aerodynamically similar to one with known performance data. We shall illustrate the application of fan laws in the worked examples to follow in this chapter.

11.5 Fan–Duct Network Interaction

Thus far we have considered the pressure losses in ducts and fittings, and the performance characteristics of fans separately. However, in actual installations any change made either to the duct system or to the fan will affect the performance of the other. We shall demonstrate this interaction between the duct system and the fan by referring to the simple duct network depicted in Fig. 11.6.

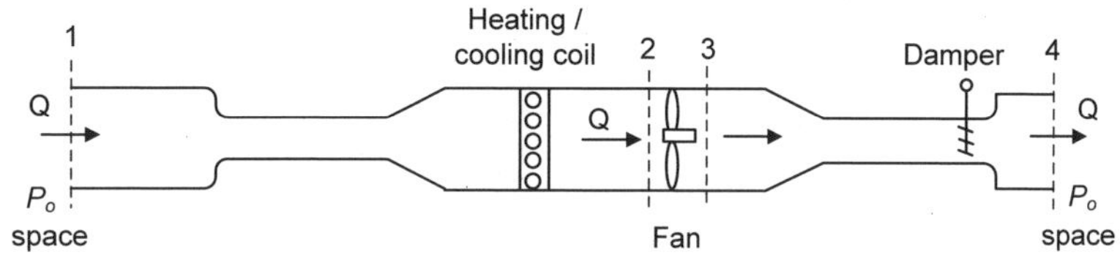

Fig. 11.6 Simple duct network with fan

The duct network consists of duct sections and fittings before and after the fan. Applying Eq. (11.17) to the duct sections 1-2 and 3-4 we obtain the following equations:

$$\Delta p_{1-2} = p_o - p_2 = \left[\sum f\left(\frac{L}{D}\right)\left(\frac{\rho V^2}{2}\right) + \sum c_f\left(\frac{1}{2}\rho V_f{}^2\right)\right]_{12} \tag{11.26a}$$

$$\Delta p_{3-4} = p_3 - p_o = \left[\sum f\left(\frac{L}{D}\right)\left(\frac{\rho V^2}{2}\right) + \sum c_f\left(\frac{1}{2}\rho V_f{}^2\right)\right]_{34} \tag{11.26b}$$

where p_o is the uniform pressure of the space to which the inlet and outlet of the duct system are connected at locations 1 and 4 respectively.

The various fluid velocities involved in Eq. (11.26a) and (11.26b) may be expressed as

$$V = \frac{Q}{A} \tag{11.27}$$

where Q is the flow rate and A is the duct area at the appropriate reference section of the duct. Substituting for the velocities in terms of the flow rates in Eqs. (11.26a) and (11.26b) we obtain

$$\Delta p_{1-2} = P_o - P_2 = c_{12}Q^2 \tag{11.28}$$

$$\Delta p_{3-4} = P_3 - P_o = c_{34}Q^2 \tag{11.29}$$

where c_{12} and c_{34} are terms involving the various friction factors, the duct dimensions and the loss coefficients of the fittings.

Since the friction coefficient, f for the ducts and the loss coefficients, c_f for the fittings and the coil are almost independent of the fluid flow rate, we assume that c_{12} and c_{34} are constant.

From Eqs. (11.28) and (11.29) it follows that

$$P_3 - P_2 = (c_{12} + c_{34})Q^2 \tag{11.30}$$

In Eq. (11.30), the LHS is the pressure rise across the fan, which is a function of the flow rate as shown in Figs. 11.4(b) and 11.5(b). The RHS of Eq. (11.30) is the total pressure loss in the duct system, Δp_{ds} which may be written as

$$\Delta p_{ds} = (c_{12} + c_{34})Q^2 \tag{11.31}$$

The variation of Δp_{ds} with volume flow rate Q, usually called the *system curve*, is shown in Fig. 11.7. Ideally, the system curve is a parabola as seen from Eq. (11.31). Any changes made to the duct system will change the shape of the system curve. For example, if the damper located at the exit of the duct network in Fig. 11.6 is partially closed to reduce the flow rate, the coefficient c_{34} in Eq. (11.31) will increase due to the additional flow resistance. The system curve will then become steeper as seen in Fig. 11.7. The converse occurs if the flow rate is increased by opening the damper.

The operation of the duct–fan system also depends on the *fan characteristic*, shown in Fig. 11.7, which is the variation of the pressure generated by the fan with flow rate, at a *fixed value* of the fan speed. It follows from Eq. (11.30) that the point of intersection of the fan characteristic and the system curve gives the fan pressure and flow rate under *steady* operating conditions.

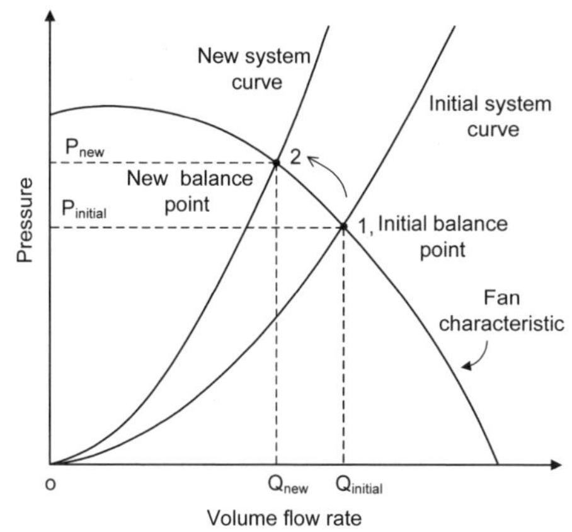

Fig. 11.7 Flow rate control by means of a damper

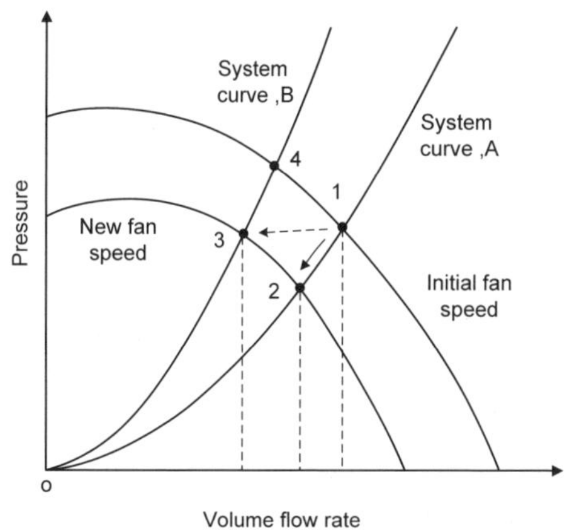

Fig. 11.8 Flow rate control by means of fan speed control

The system curves and the fan characteristics offer a convenient graphical approach to visualize different operational and control conditions of fan–duct systems. We shall discuss a few practical situations here.

A simple method to control the air flow rate to a space under part-load conditions is to use a damper as shown in Fig. 11.6. If the fan speed is held fixed, then the fan pressure for the new balance point 2 at the lower flow rate is higher as seen in Fig. 11.7. Therefore, the new fan power, which is proportional to the product of fan pressure and the flow rate, may not decrease by much from the initial fan power at 1.

A more energy efficient method to decrease the air flow rate for part-load operation is to reduce the fan speed using a variable speed drive. For this situation the system curve, indicated as A in Fig. 11.8 is fixed. However, the fan characteristic at the lower fan speed is below the original fan curve and therefore the fan pressure at the new balance point 2 is lower as seen in Fig. 11.8. Since the flow rate and the pressure are both lower at the new balance point 2, the fan power at the new flow rate is significantly reduced.

Air handling systems of most buildings operate under part-load conditions much of the year except during peak-load conditions. Therefore the yearly operational cost can be significantly reduced by using variable speed fans rather than damper control systems.

A flow control strategy that combines damper control and fan speed control is adopted in variable air volume (VAV) systems of air conditioning plants. At part-load, the dampers in the air supply boxes or VAV boxes close in response to the reduced thermal load, thus moving the system curve in Fig. 11.8 from A to B.

The fan is designed to maintain a constant pressure in the discharge duct regardless of the position of the VAV box. The increase in pressure (4) in the supply duct is used as a signal to reduce the fan speed, which in turn shifts the fan characteristic to the lower curve. Hence the balance point shifts to 3 from 1, as indicated in Fig. 11.8. Significant fan power reductions under part-load conditions are possible with VAV systems.

11.6 Design Methods for Duct Systems

The thermal loads and the ventilation requirements of different zones of a building determine the supply air flow rates to the zones. Duct design principles are used to obtain the appropriate sizes and the layout of the duct network to circulate air between the zones and the processing equipment, like the heating and cooling coils. The main requirements of a duct system are: (i) that it supplies the specified rates of air flow to the different zones, (ii) that it is economical in initial cost, operating cost and the cost of the building space, and (iii) that the level of noise generated is acceptable.

The air pressure in the duct system has to be controlled to avoid leakage. Also, the ductwork should be well insulated to minimize heat exchange with the surroundings. These factors will affect the quantity and the temperature of the air supplied to the zones.

The size of the duct system affects the initial cost and the operating cost in opposing ways. Duct networks of a smaller size have a low initial cost but the fan power required to convey the air is higher due to the larger pressure drop. Moreover, the larger velocities in these systems generate higher noise levels. The fan power needed for larger duct networks is less and they generate less noise. However, larger duct systems have a higher initial cost.

Several methods for the design of duct networks are described in the published literature [1-4]. These include (i) the velocity method, (ii) the equal friction method, (iii) the static regain method, and (iv) the T-method. We shall describe the equal friction method and the static regain method that are included in the *ASHRAE Handbook* - 2013 *Fundamentals* [1].

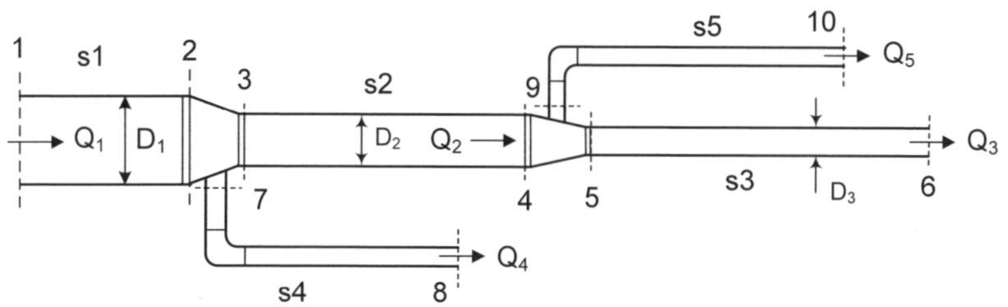

Fig. 11.9 Design of duct networks

11.6.1 *Equal friction method*

In the equal friction method, the *frictional* pressure loss per unit length (unit pressure loss) is assumed to be the same for all sections of the duct network. Pressure losses through any fittings in the duct section are not included in the assumed unit pressure loss. The *ASHRAE Handbook* - 2013 *Fundamentals* [1] recommends that the selected pressure drop be within the region of the duct friction chart in Fig. 11.2, enclosed by the bold lines. When the energy cost to operate the fan is high compared to the initial cost of the duct, it is economical to assume a lower pressure loss per unit length. The converse is true when the duct cost is high compared to the operating cost.

We shall illustrate the *equal friction method* by referring to the simple circular duct system shown in Fig. 11.9. The fan supplies conditioned air at 1 to be delivered to the three zones at 6, 8 and 10.

(i) The flow rates to the zones Q_3, Q_4 and Q_5 are estimated from the thermal loads of the zones. If the density of air is assumed constant, then the flow rate from the fan is

$$Q_1 = Q_3 + Q_4 + Q_5$$

(ii) The layout of the duct network is based on the location of the conditioned spaces and air handling unit (AHU). For each duct section, the length, and the number and type of fittings are specified.

(iii) The pressure for each zone is also specified. Usually, all the zones are at the same pressure.

(iv) As mentioned above, an initial value for the unit frictional pressure loss, p_o is chosen based on the guidelines in Fig. 11.2. This value is applied to all duct sections of the duct network.

(v) Knowing the flow rate, Q and the unit pressure p_o, the diameter and air velocity for each section is obtained directly from Fig. 11.2. Alternatively, these values may be determined by solving Eqs. (11.5)-(11.8), iteratively.

(vi) Since the value of p_o is usually chosen based on past design experience, the duct sizes obtained in step (v) may not always deliver the specified flow rates to the three zones. This has to be checked by calculating the total pressure loss from the entrance 1 to the exits at 6, 8 and 10. For the flow path from 1 to 6, the total pressure loss, $\Delta p_{t,1-6}$ is given by

$$\Delta p_{t,1-6} = \Delta p_{fr,12} + \Delta p_{fit,12} + \Delta p_{tee-s,23} + \Delta p_{fr,34} + \Delta p_{fit,34} + \Delta p_{tee-s,45} + \Delta p_{fr,56} + \Delta p_{fit,56} \quad (11.32)$$

where Δp_{fr} is the friction pressure loss in the duct section, Δp_{tee-s} is the pressure loss across the straight section of the tee-junction and Δp_{fit} is the pressure loss due to any fittings in the duct section. The methods to compute the above pressure losses were discussed in section 11.3.

Similar equations may be written for the total pressure losses, $\Delta p_{t,1-8}$ and $\Delta p_{t,1-10}$ from 1 to 8 and 1 to 10 respectively. To deliver the *specified* flow rates to the 3 zones the total pressure losses must be equal. Therefore

$$\Delta p_{t,1-6} = \Delta p_{t,1-8} = \Delta p_{t,1-10} \tag{11.33}$$

(vii) If Eq. (11.33) is not satisfied, then dampers may have to be installed at the exit of some ducts to increase the pressure drop artificially to achieve the desired flow rates to the zones. An alternative approach is to the change diameter of some branches of the duct network to obtain the desired pressure losses to satisfy Eq. (11.33).

A MATLAB program to carry out the calculations involved in the equal friction method is given in Appendix A11.2.

11.6.2 *Static regain method*

The static regain method is only applicable to supply air duct systems like that shown in Fig. 11.9. The aim of this method is to maintain the same static pressure at all the entry points to the fittings where the air flowing through the network diverges. Maintaining a low static pressure throughout the entire duct system helps minimize duct leakage and also reduce the stresses in the duct wall.

We shall illustrate the design procedure by referring to the section 1-6 of the duct network in Fig. 11.9. The total pressures at locations 2, 4 and 6 may be expressed as

$$p_{t2} = p_{s2} + \frac{\rho {V_1}^2}{2} \tag{11.34}$$

$$p_{t4} = p_{s4} + \frac{\rho {V_2}^2}{2} \tag{11.35}$$

$$p_{t6} = p_{s6} + \frac{\rho {V_3}^2}{2} \tag{11.36}$$

where p_s is the static pressure.

In this design method we assume that the static pressure is the same at locations 2, 4 and 6. Hence we have

$$p_{s2} = p_{s4} = p_{s6} \tag{11.37}$$

The difference between the *total* pressures at any two locations is equal to the sum of the pressure losses due to friction and the losses through the fittings. Therefore

$$p_{t2} - p_{t4} = \Delta p_{tee-s,23} + \Delta p_{fr,34} + \Delta p_{fit,34} \tag{11.38}$$

$$p_{t4} - p_{t6} = \Delta p_{tee-s,45} + \Delta p_{fr,56} + \Delta p_{fit,56} \tag{11.39}$$

Substitute from Eqs. (11.34)–(11.36) in Eqs. (11.38)–(11.39) and apply the conditions given by Eq. (11.37) to obtain the following equations

$$\frac{\rho {V_1}^2}{2} - \frac{\rho {V_2}^2}{2} = \Delta p_{tee-s,23} + \Delta p_{fr,34} + \Delta p_{fit,34} \tag{11.40}$$

$$\frac{\rho {V_2}^2}{2} - \frac{\rho {V_3}^2}{2} = \Delta p_{tee-s,45} + \Delta p_{fr,56} + \Delta p_{fit,56} \tag{11.41}$$

The diameter and velocity in duct sections 2 and 5 are related by the equations

$$Q_2 = \frac{\pi {D_2}^2 V_2}{4} \tag{11.42}$$

$$Q_3 = \frac{\pi {D_4}^2 V_3}{4} \tag{11.43}$$

The main steps of the static regain method are:

(i) The flow rates to the zones Q_3, Q_4 and Q_5 are estimated from the thermal loads of the zones.

(ii) The layout of the duct network is based on the location of the conditioned spaces and air handling unit (AHU). For each duct section, the length, and the number and type of fittings are specified.

(iii) The pressure for each zone is specified. Usually, all the zones are at the same pressure.

(iv) A velocity V_1 is selected for the main duct section 1-2. The pressure loss terms on the RHS of Eq. (11.40) are functions of D_2 and V_2, which are related by Eq. (11.42). Solve Eqs. (11.40) and (11.42) iteratively to obtain V_2 and the diameter D_2 of the section s2.

(v) Similarly, solve Eqs. (11.41) and (11.43) iteratively to obtain V_3 and the diameter D_3 of the section s3.

A MATLAB program to carry out the calculations involved in the static regain method is given in Appendix A11.3. We shall illustrate the application of the above design methods in the worked examples to follow in this chapter.

11.7 Optimization of Duct Systems

Optimization methods based on the minimization of life-cycle cost may be used to design duct systems. The cost of a duct network consists of several components that can be broadly divided into initial costs and operating costs. The initial cost involves the costs of hardware items such as the ducts, the fittings, the thermal insulation, the fan, and the controls. The installation cost, the cost of sound attenuation, and the cost of the space are also included in the initial cost. These costs could vary significantly depending on the equipment suppliers and the installers. Operating costs involve the cost of electricity to operate the fan, and the maintenance cost of the system.

Obtaining complete information on the various items of the total cost could involve considerable time and effort. Moreover, there could be significant uncertainties associated with the estimated costs. Although cost optimization methods are based on well established economic principles, it may be difficult to implement them accurately.

We shall illustrate the optimization procedure by adopting a simplified cost model for a circular duct system which yields an optimum diameter for the duct [4,5]. The total initial cost may be expressed as

$$C_{ini} = (\pi D L \delta_t)\rho_m c_m + C_{fo} \tag{11.44}$$

where D and L are the diameter and the length of the duct. The thickness and density of the duct wall are δ_t and ρ_m respectively. The cost of duct material per unit volume is c_m in dollars per kg. The component of the initial cost which is independent of the duct diameter is C_{fo}.

The power to operate the fan is given by

$$W_{fan} = \frac{Q\Delta P_f}{\eta_f} \tag{11.45}$$

where η_f is the combined efficiency of the electric motor and the drive. The frictional pressure loss in the system is given by Eq. (11.5) as

$$\Delta P_f = f\left(\frac{L}{D}\right)\left(\frac{\rho V^2}{2}\right) \tag{11.46}$$

The volume flow rate is expressed as

$$Q = \frac{\pi D^2 V}{4} \qquad (11.47)$$

The total annual energy cost for operating the fan is obtained by manipulating Eqs. (11.45) to (11.47). Hence we have

$$C_{fan} = \frac{8fLQ^3 N_y c_{ele}}{\pi^2 D^5} \qquad (11.48)$$

where N_y is the number of hours of operation per year and c_{ele} is the cost of electricity in dollars per kWh.

The life-cycle cost of the duct system may be expressed as

$$LCC = P_1 C_{ini} + P_2 C_{fan} \qquad (11.49)$$

where P_1 is the present worth factor for the initial cost of the system. The present worth factor P_2 for the operating cost depends the energy cost escalation rate and other economic factors such as the tax rates. We shall consider a simplified application of Eq. (11.49) in worked example 11.14.

11.8 Air Distribution in Zones

The duct network of an air conditioning system is designed to convey conditioned air to the different zones at the desired temperature, humidity and flow rate. The main function of the air distribution system in a zone is to ensure that the occupants in the zone who receive the conditioned air are comfortable. The outlets through which air enters a zone are usually called *diffusers* and air is removed from the zone through *return air grills*.

11.8.1 *Air flow from diffusers*

The selection of suitable diffusers is important because it determines the attainment of the desired comfort conditions in the space. Shown schematically in Fig. 11.10 is the shape of a *cold air* jet discharging from a diffuser located at the top end of the side wall of a room. The behavior of the jet varies in the horizontal and vertical directions depending on the type of jet, the supply air flow rate, and the temperature difference between supply and room air. Three physical processes affect the interaction between the jet and air in the room. These are: (i) entrainment, (ii) the Coanda effect, and (iii) buoyancy. The jet expands due to air entrained from the room which, in turn, reduces the maximum velocity of the jet.

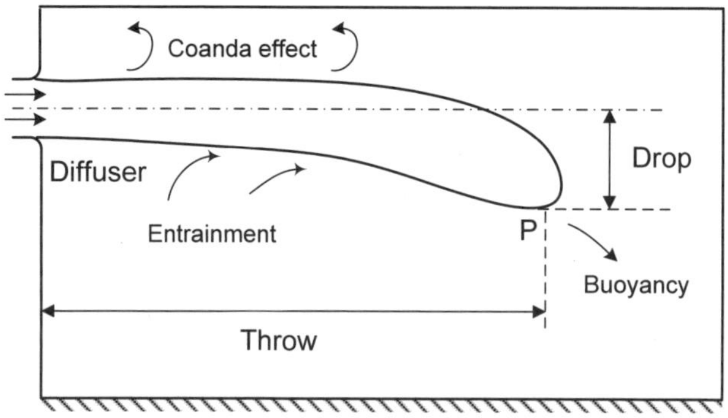

Fig.11.10 Air jet behavior: Entrainment, Coanda effect and buoyancy

When a jet discharges close to a solid surface it has a tendency to adhere to the surface due to the low pressure region between the jet and the surface. This phenomenon is called the *Coanda* effect. When the air discharged from the diffuser is cooler than the air in the room, the jet tends to sink down due to the larger density of the air jet. This is called the *buoyancy* effect. If the air issuing from the jet is warmer than the room air, then the jet will move upwards due to the buoyancy effect.

The overall shape of the expanding jet is determined by the interplay of the aforementioned effects. The position of the jet is usually characterized by two parameters called the throw and the drop, as indicated in Fig. 11.10. (i) The *throw* is the horizontal distance from the diffuser to the point P in the jet where the core velocity has reached the terminal velocity, V_T, which varies from 0.5 to 0.25 ms^{-1}, depending on the type of diffuser. (ii) The *drop* is the vertical distance between point P and the center line of the jet at the diffuser.

External factors affecting the behavior of a jet include the following: interaction with jets from other diffusers in the vicinity, interaction with other air flows like convection currents, and obstacles in the zone. A detailed discussion of the characteristics of jets issuing from different types of diffusers is available in Chapter 20, the *ASHRAE Handbook - 2009 Fundamentals* [2].

11.8.2 *Air diffusion performance index*

The *effective draft temperature* (EDT) at a location in a conditioned space is a *fictitious temperature* that incorporates the effects of the air velocity and dry-bulb temperature on the thermal comfort of occupants in the zone. The EDT (°C) is given by the equation [2]

$$EDT = (t_x - t_c) - 8(V_x - 0.15) \tag{11.50}$$

where t_x (°C) and V_x (ms^{-1}) are the dry-bulb temperature and air velocity at any location x. The set-point dry-bulb temperature of the zone is t_c (°C).

The parameter used as a measure to evaluate the performance of an air distribution system is called the *air diffusion performance index* (ADPI). Its value depends on: (i) the type of outlet diffuser, (ii) dimensions of the zone, (iii) the layout of air inlets and outlets in the zone, (iv) the zone loads, and (v) the throw distances of the air inlets.

The ADPI is defined as the percentage of the measurements taken within an occupied zone (see Fig. 11.10) where the EDT is between −1.5°C and 1°C, with the air velocity less than 0.35 ms^{-1}. Therefore

$$ADPI = \frac{100N_\theta}{N} \tag{11.51}$$

where N_θ is the number of measured EDT values that fall within −1.5°C and 1°C and N is the total number of measured values. ADPI is applicable only for cooling mode operation of systems.

If the ADPI is 100 %, then all the measured points are within the specified range of the EDT, and the thermal condition of the zone is at an acceptable level. Most practical systems are designed to have an ADPI greater than 80%.

11.8.3 *Design aspects of air distribution systems*

We shall briefly outline the design procedure recommended in Ref. [1] for selecting diffusers for air distribution in rooms. The main goal is to ensure that the ADPI for the zone is in the acceptable range.

The volume flow rate of conditioned air through the diffuser is obtained from the zone load. Hence we have

$$\dot{V} = \frac{Q_{tot}}{\rho \Delta h} \tag{11.52}$$

where $\dot{V}$ is the volume flow rate, Q_{tot} is the total design load, ρ is the density, and Δh is the total enthalpy difference between the supply and return air.

The diffuser is usually selected from manufacturer's catalogues that typically include technical data on air flow rate, pressure drop, throw distance, sound power level and others. The number and location of the diffusers are dependent on the dimensions of the room. For example, when circular diffusers are located in the ceiling, they should be centered in a roughly square area of the ceiling.

The *ASHRAE Handbook* - 2009 *Fundamentals* [2] provides guidelines to relate the throw distance, X_{VT}, given in manufacturer's catalogues for different types of diffusers, to the optimum ADPI. The terminal velocity V_T is taken as 0.25 ms^{-1} for all diffusers, except for ceiling slot diffusers for which a value of 0.5 ms^{-1} is used. The data is normalized by adopting a characteristic, L for each type of diffuser. Representative values of L for 3 types of diffusers, are listed in Table 11.2. For the same types of diffusers, the values of (X_{TV}/L) that give the maximum ADPI are listed in Table 11.3.

Table 11.2 Characteristic Length for Diffusers*

Diffuser type	Characteristic length L
High sidewall grills	Distance to wall perpendicular to jet
Circular ceiling pattern diffuser	Distance to closest wall
Ceiling slot diffuser	Distance to wall or mid-plane

*Representative values extracted from Table 2, Page 20.14, *ASHRAE Handbook* - 2009 *Fundamentals* [2]

Table 11.3 Air Diffusion Performance Index (APDI)*

Diffuser Type	Room load Wm^{-2}	$X_{0.25}/L$ for $(ADPI)_{max}$	Maximum (ADPI)	(ADPI) Greater than	Range of $X_{0.25}/L$
High side wall	250	1.8	68	-	-
Grills	125	1.6	78	70	1.2 to 2.3
	65	1.5	85	80	1.0 to 1.9
Circular ceiling	250	0.8	76	70	0.7 to 1.3
Diffuser	125	0.8	88	80	0.5 to 1.5
	65	0.8	93	80	0.4 to 1.7
Ceiling slot	250	0.3	85	80	0.3 to 0.7
Diffusers	125	0.3	91	80	0.3 to 1.1
$(X_{0.5}/L)$	65	0.3	92	80	0.3 to 1.5

*Representative values extracted from Table 3, Page 20.14, *ASHRAE Handbook* - 2009 *Fundamentals* [2]

Once the type, the number and the locations of the diffusers have been decided, the data in Tables 11.2 and 11.3 may be used to determine whether the resulting ADPI for the room is in the acceptable range. We shall illustrate the application of the above diffuser selection procedure in the worked examples to follow in this chapter.

The method outlined above for the design of air distribution systems relies on experimental data and empirical design guidelines. However, the test rooms used to obtain this data may differ significantly from those encountered in actual practice.

Computational Fluid Dynamics (CFD) software offers an alternative approach to simulate the air distribution in actual rooms with most of the details of the actual room included. Chapter 13 of the *ASHRAE Handbook* - 2013 *Fundamentals* [1] presents two common indoor environmental modeling methods to calculate the air flows and contaminant concentration in buildings. These are called CFD and multi-zone network airflow modeling. The modeling software used is widely available but there successful application requires considerable expertise and time.

11.9 Worked Examples

Example 11.1 The supply duct of an air conditioning system, shown in Fig. E11.1.1, delivers 0.8 m^3s^{-1} of air at 20°C dry-bulb temperature. The diameter of the duct is 0.4 m and the total length of the straight duct sections from 1 to 10 is 80 m. The fan generates a total pressure of at section is 125 kPa at section 1. The loss coefficients for each right angle bend is 0.1. The density of the air is 1.2 kgm^{-3}.

Calculate (i) the static pressure and velocity pressure at section 1, (ii) the total frictional pressure loss from 1–10, (iii) the pressure loss in the fittings from 1–10, (iv) the total, static, and velocity pressures at section 10, (v) the static and velocity pressure heads at 1, (vi) the total mechanical energy loss *per unit mass* of air from 1 to 10, and (vii) the *rate* of mechanical energy loss from 1 to 10.

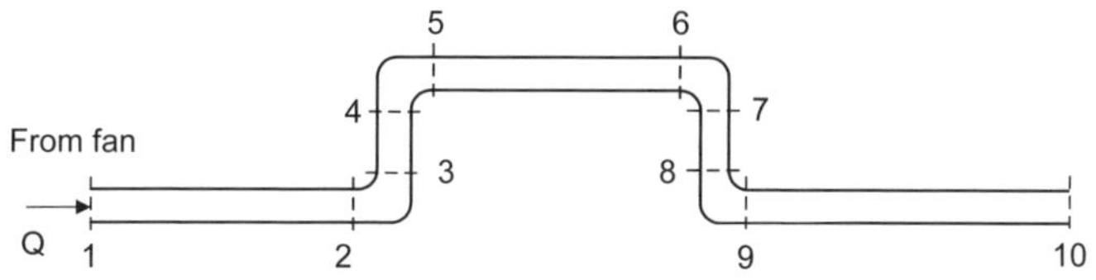

Fig. E11.1.1 Supply air duct section

Solution The total straight length from 1 to 10 is 80 m and the diameter of the duct is 0.4 m. The air velocity through the duct is given by

$$V = \frac{Q}{A} = \frac{4Q}{\pi D^2} = \frac{4\times 0.8}{\pi\times 0.4^2} = 6.37\ \text{ms}^{-1}$$

The velocity pressure is

$$p_{v1} = \frac{\rho {V_1}^2}{2} = \frac{1.2\times 6.37^2}{2} = 24.34\ \text{Pa}$$

The total pressure is

$$p_{t1} = p_{s1} + p_{v1} = 125\ \text{Pa}$$

Hence the static pressure, p_{s1} at 1 is, (125-24.34) = 100.9 Pa

The total frictional and fittings pressure loss is given by Eq. (11.17) as

$$\Delta p_{tot} = f\left(\frac{L}{D}\right)\left(\frac{\rho V^2}{2}\right) + \sum_n c_n\left(\frac{1}{2}\rho {V_n}^2\right) \qquad \text{(E11.1.1)}$$

The first term on the RHS is the frictional pressure loss in the straight sections of the duct, which may be obtained from the friction chart in Fig. 11.2, by knowing the volume flow rate and the duct diameter. Alternatively, we could apply Eq. (11.5) with the friction factor obtained by solving Colebrook's equation (Eq. 11.8) iteratively. A MATLAB code for this purpose is given in Appendix A11.1. The friction factor *f* is obtained from the code as 0.0186.

Substituting numerical values in Eq. (E11.1.1) we have

$$\Delta p_{tot,1-10} = 0.0186 \times \left(\frac{80}{0.4}\right)\left(\frac{1.2 \times 6.37^2}{2}\right) + (4 \times 0.1)\left(\frac{1.2 \times 6.37^2}{2}\right)$$

$$\Delta p_{tot,1-10} = 90.56 + 9.73 = 100.3 \text{ Pa}$$

The duct friction pressure loss obtained directly using the friction chart in Fig. 11.2 agrees closely with the computed value.

The total pressure at section 10 is

$$p_{t,10} = p_{t1} - \Delta p_{tot,1-10} = 125 - 100.3 = 24.7 \text{ Pa}$$

Now the air velocity in the duct is constant. Therefore the velocity pressure at 10 is also 24.34 Pa. Hence the static pressure at 10 is, (24.7-24.34) = 0.36 Pa.

The static pressure head at 1 is given by

$$H_{s1} = \frac{p_{s1}}{\rho g} = \frac{100.9}{1.2 \times 9.81} = 8.57 \text{ m}$$

The velocity head at 1 is

$$H_{v1} = \frac{p_{v1}}{\rho g} = \frac{24.34}{1.2 \times 9.81} = 2.06 \text{ m}$$

The *mechanical energy loss per kg* of air from 1 to 10 is given by

$$\Delta E_{m,1-10} = \frac{\Delta p_{t,1-10}}{\rho} = \frac{100.3}{1.2} = 83.58 \text{ Jkg}^{-1}$$

The *rate of mechanical energy loss* from 1 to 10 is given by

$$\Delta W_{loss} = Q\Delta p_{t,1-10} = 0.8 \times 100.3 = 80.24 \text{ W}$$

The mechanical energy loss or conversion to heat results in a rise in temperature of air of about 0.08°C.

Example 11.2 The dimensions of a duct system, supplying air at 20°C to two zones 5 and 6 are shown in Fig. E11.2.1. The flow rates to the two zones are 1.4 m^3s^{-1} and 1.0 m^3s^{-1} respectively. The fan generates a total pressure of 285 kPa at the entrance section 1. Calculate (a) (i) the velocity pressures in the duct sections, (ii) the total pressures at all sections from 1 to 6. (b) If a damper is installed just upstream of section 5 to make the total pressures at 5 and 6 equal, calculate the required pressure loss through the damper.

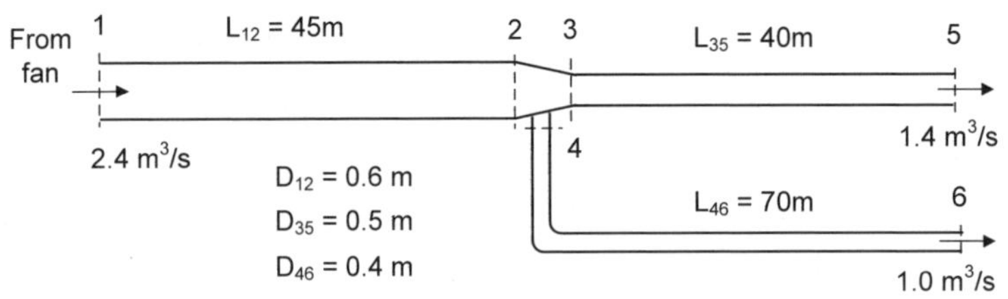

Fig. E11.2.1 Supply air duct system

Solution The air velocities in the duct sections 1-2, 3-5 and 4-6 are given by

$$V_{12} = \frac{Q}{A} = \frac{4Q}{\pi D^2} = \frac{4\times 2.4}{\pi\times 0.6^2} = 8.49 \text{ ms}^{-1}$$

$$V_{35} = \frac{Q}{A} = \frac{4Q}{\pi D^2} = \frac{4\times 1.4}{\pi\times 0.5^2} = 7.13 \text{ ms}^{-1}$$

$$V_{46} = \frac{Q}{A} = \frac{4Q}{\pi D^2} = \frac{4\times 1.0}{\pi\times 0.4^2} = 7.95 \text{ ms}^{-1}$$

The velocity pressures in the duct sections are given by

$$p_{v12} = \frac{\rho V_{12}{}^2}{2} = \frac{1.2\times 8.49^2}{2} = 43.2 \text{ Pa}$$

$$p_{v35} = \frac{\rho V_{35}{}^2}{2} = \frac{1.2\times 7.13^2}{2} = 30.5 \text{ Pa}$$

$$p_{v46} = \frac{\rho V_{46}{}^2}{2} = \frac{1.2\times 7.95^2}{2} = 37.9 \text{ Pa}$$

The frictional pressure loss in the straight duct sections may be obtained directly from the friction chart in Fig. 11.2, by knowing the volume flow rate and the duct diameter. Alternatively, we could apply Eq. (11.5) with the friction factor obtained by solving Colebrook's equation (Eq. 11.8) iteratively. A MATLAB code for this purpose is given in Appendix A11.1. Hence we obtain the following friction pressure losses.

$$\Delta p_{f12} = 53.6 \text{ Pa}, \quad \Delta p_{f35} = 42.8 \text{ Pa}, \quad \Delta p_{f46} = 120.8 \text{ Pa}$$

The loss coefficient for the 90° bend in section 4-6 is obtained from the fittings database in chapter 21, of the *ASHRAE Handbook* - 2009 *Fundamentals* [2] as $C_o = 0.11$. Hence the pressure loss in the bend is

$$\Delta p_{b-46} = \frac{C_o \rho V_{46}^2}{2} = \frac{0.11 \times 1.2 \times 7.95^2}{2} = 4.17 \text{ Pa}$$

The loss coefficients for the diverging tee-junction 2-3-4 is obtained from chapter 21, *ASHRAE Handbook* - 2009 *Fundamentals* [2] (also see Table 11.1c) as, $C_{straight} = 0.14$ and $C_{bend} = 1.2$. The pressure losses through the straight section and the bend are given by

$$\Delta p_{s-23} = \frac{C_s \rho V_{12}^2}{2} = \frac{0.14 \times 1.2 \times 8.49^2}{2} = 6.05 \text{ Pa}$$

$$\Delta p_{b-24} = \frac{C_b \rho V_{12}^2}{2} = \frac{1.2 \times 1.2 \times 8.49^2}{2} = 51.89 \text{ Pa}$$

The total pressures at the different sections are obtained as

$$p_{t1} = 285 \text{ Pa}$$

$$p_{t2} = p_{t1} - \Delta p_{f12} = 285 - 53.6 = 231.4 \text{ Pa}$$

$$p_{t3} = p_{t2} - \Delta p_{s-23} = 231.4 - 6.05 = 225.35 \text{ Pa}$$

$$p_{t4} = p_{t2} - \Delta p_{b-24} = 231.4 - 51.89 = 179.5 \text{ Pa}$$

$$p_{t5} = p_{t3} - \Delta p_{f35} = 225.35 - 42.8 = 182.55 \text{ Pa}$$

$$p_{t6} = p_{t4} - \Delta p_{f46} = 179.5 - 120.8 = 58.7 \text{ Pa}$$

Hence the required pressure loss through the damper at section 5 is

$$\Delta p_{damp} = p_{t5} - p_{t6} = 182.55 - 58.7 = 123.85 \text{ Pa}$$

Example 11.3 The diameters and lengths of the various sections of a circular duct network are indicated in Fig. E11.3.1. The total pressure of the air, supplied by a fan at 1, is 140 Pa above the discharge pressures at 5 and 6, which are equal. (i) Calculate air flow rates, Q_1, Q_2 and Q_3. (ii) Obtain the system curve for the duct network.

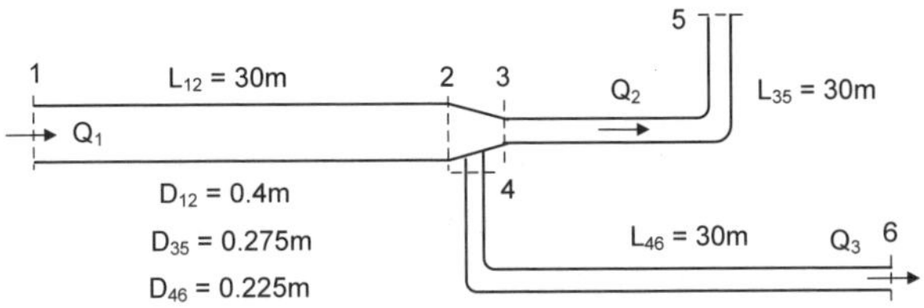

Fig. E11.3.1 Duct system

Solution The total pressure loss from 1-5 consists of: (i) the friction losses in the duct sections 1-2 and 3-5, (ii) the loss in the straight section of the tee-junction 2-3-4, and (iii) the loss in the 90° bend in section 3-5. Hence we have

$$p_{t1} - p_{t5} = \frac{8f_{12}L_{12}\rho Q_1{}^2}{\pi^2 D_{12}{}^5} + \frac{8c_{s23}\rho Q_1{}^2}{\pi^2 D_{12}{}^4} + \frac{8f_{35}L_{35}\rho Q_2{}^2}{\pi^2 D_{35}{}^5} + \frac{8c_{ra}\rho Q_2{}^2}{\pi^2 D_{35}{}^4} = \Delta p \tag{E11.3.1}$$

where the symbols have their usual meaning.

The total pressure loss from 1-6 consists of: (i) the friction losses in the duct sections 1-2 and 4-6, (ii) the loss in the turning section of the tee-junction 2-3-4, and (iii) the loss in the 90° bend in section 4-6. Hence we have

$$p_{t1} - p_{t6} = \frac{8f_{12}L_{12}\rho Q_1{}^2}{\pi^2 D_{12}{}^5} + \frac{8c_{t23}\rho Q_1{}^2}{\pi^2 D_{12}{}^4} + \frac{8f_{46}L_{46}\rho Q_3{}^2}{\pi^2 D_{46}{}^5} + \frac{8c_{ra}\rho Q_3{}^2}{\pi^2 D_{46}{}^4} = \Delta p \tag{E11.3.2}$$

Since the total pressure drops across the duct sections are equal

$$p_{t1} - p_{t5} = p_{t1} - p_{t6} = \Delta p = 140 \text{ Pa} \tag{E11.3.3}$$

Assuming the density of air to be constant, mass balance gives

$$Q_1 = Q_2 + Q_3 \tag{E11.3.4}$$

Equation (E11.3.1) may be expressed in the compact form:

$$Q_2 = \left(\frac{\Delta p - a_1 Q_1{}^2}{b_1}\right)^{1/2} \tag{E11.3.5}$$

Equation (E11.3.2) may be expressed in the compact form:

$$Q_3 = \left(\frac{\Delta p - a_2 Q_1^2}{b_2}\right)^{1/2} \tag{E11.3.6}$$

The constants a_1, b_1, a_2, b_2 in Eqs. (E11.3.5) and (E11.3.6) involve the constant terms in Eqs. (E11.3.1) and (E11.3.2). Substituting from Eqs. (E11.3.5) and (E11.3.6) in Eq. (E11.3.4) we have

$$Q_1 = \left(\frac{\Delta p - a_1 Q_1^2}{b_1}\right)^{1/2} + \left(\frac{\Delta p - a_2 Q_1^2}{b_2}\right)^{1/2} \tag{E11.3.7}$$

The total flow rate Q_1 is obtained by solving Eq. (E11.3.7). A trial and error procedure is used because of the nonlinear form of the equation. The LHS and RHS of Eq. (E11.3.7), as functions of Q_1 are shown graphically in Fig. E11.3.1(b). The intersection of the two curves gives the value of Q_1 that satisfies Eq. (E11.3.7).

We assume the following initial values for the loss coefficients:

$$f_{12} = 0.0195, f_{35} = 0.0195, f_{46} = 0.0195,\ c_{s23} = 0.14,\ c_{b23} = 1.2,\ c_{ra} = 0.11$$

The trial and error solution of Eq. (11.3.7) gives the following air flow rates:

$$Q_1 = 0.792\ \text{m}^3\text{s}^{-1}, \quad Q_2 = 0.518\ \text{m}^3\text{s}^{-1}, \quad Q_3 = 0.274\ \text{m}^3\text{s}^{-1}.$$

The computed flow rates are then used to obtain more accurate estimates of the friction coefficients using the MATLAB program in Appendix A11.1. Hence we obtain $f_{12} = 0.0186$, $f_{35} = 0.0195$, $f_{46} = 0.0209$.

On substitution, the values of the constants are obtained as:

$$a_1 = 58.15, \quad b_1 = 379.38, \quad a_2 = 98.31, \quad b_2 = 1096$$

These values are then used to obtain better estimates of the air flow rates. Hence we have

$$Q_1 = 0.794\ \text{m}^3\text{s}^{-1}, \quad Q_2 = 0.525\text{m}^3\text{s}^{-1}, \quad Q_3 = 0.268\ \text{m}^3\text{s}^{-1}$$

By repeating the above calculation at different pressures Δp we have developed the system curve for the duct network, shown in Fig. E11.3.3.

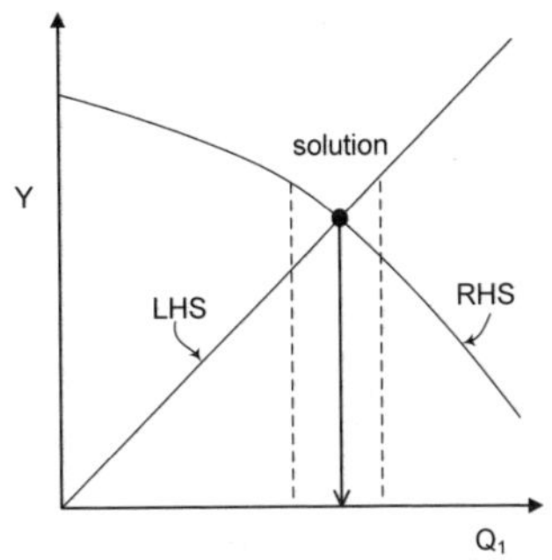

Fig. E11.3.2 Solution procedure

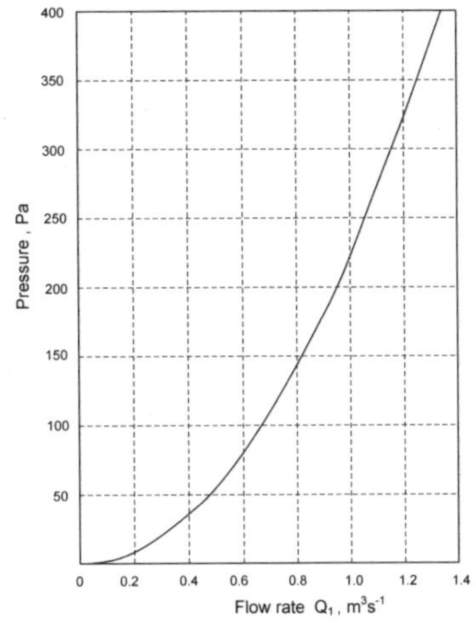

Fig. E11.3.3 System curve

Example 11.4 The breadth and height of the outlet diffuser of a centrifugal fan are 0.62 m and 0.32 m respectively. The fan delivers 1.6 m^3s^{-1} of air when the rotational speed is 205 rpm. The static pressure at the outlet is 540 Pa. (i) Calculate the ideal power input required to operate the fan. (ii) If the efficiency of the motor and drive is 72%, calculate the power input to the motor.

Solution The outlet velocity of the air is given by

$$V_o = \frac{Q}{A_o} = \frac{1.6}{0.62 \times 0.32} = 8.06 \text{ ms}^{-1}$$

The velocity pressure at the outlet of the fan is

$$p_{vo} = \frac{\rho V_o^2}{2} = \frac{1.2 \times 8.06^2}{2} = 39.0 \text{ Pa}$$

The total pressure at the outlet is

$$p_{to} = 39 + 540 = 579 \text{ Pa}$$

The *ideal* power input is given by

$$W_{id} = Q p_{to} = 1.6 \times 579 = 926.4 \text{ W}$$

The efficiency of the motor and drive is 0.72. Therefore the actual power input is 926.4/0.72 = 1287 W.

Example 11.5 Ambient air at 20°C and density 1.2 kgm^{-3} enters a fan rotating at 180 rpm. The total pressure of the air delivered by the fan is 11.5 kPa and the volume flow rate 4.5 m^{3}s^{-1}. (a) Calculate the ideal power input to the fan. (b) Calculate the volume flow rate, the pressure rise and the power input, (i) if the fan speed is increased to 200 rpm with the same density, and (ii) if fan speed remains constant but the ambient air temperature increases to 40°C.

Solution (a) The ideal power input is given by

$$\dot{W}_1 = Q_1 p_{t1} = 4.5 \times 11.5 = 51.75 \text{ kW}$$

(b) (i) From Eq. (11.23), it follows that for the same fan:

$$\frac{\dot{Q}_1}{\dot{Q}_2} = \frac{\omega_1}{\omega_2} = \frac{180}{200} = 0.9$$

Therefore the new flow rate is, (4.5/0.9) = 5.0 m^{3}s^{-1}.

From Eq. (11.24), it follows that for the same fan and constant air density:

$$\frac{p_{t1}}{p_{t2}} = \frac{{\omega_1}^2}{{\omega_2}^2} = \frac{180^2}{200^2} = 0.81$$

Therefore the new pressure is, (11.5/0.81) = 14.2 kPa.

From Eq. (11.25), it follows that for the same fan and constant air density:

$$\frac{\dot{W}_1}{\dot{W}_2} = \frac{{\omega_1}^3}{{\omega_2}^3} = \frac{180^3}{200^3} = 0.729$$

Therefore the new power input is, (51.75/0.729) = 71 kW.

(b) (ii) Assume that air is an ideal gas with the equation of state:

$$p = \rho RT$$

Applying the above equation for air at 20°C and 40°C we have

$$\frac{\rho_2}{\rho_1} = \frac{\rho_2}{1.2} = \frac{(273+20)}{(273+40)} = 0.936$$

The density at 40°C is 1.123 kgm^{-3}. From Eq. (11.23) it follows that the flow rate is independent of the density and therefore remains constant.

From Eq. (11.24) it follows that for the same fan speed

$$\frac{p_{t2}}{p_{t1}} = \frac{p_{t2}}{11.5} = \frac{\rho_2}{\rho_1} = 0.936$$

Therefore the pressure rise is 10.76 kPa.

From Eq. (11.25) it follows that for same fan speed

$$\frac{\dot{W}_2}{\dot{W}_1} = \frac{\dot{W}_2}{51.75} = \frac{\rho_2}{\rho_1} = 0.936$$

Therefore the power input is 48.4 kW.

Example 11.6 A fan delivers ambient air at 18°C at the rate of 4.8 kgs^{-1} through a duct system. The fan speed and the power input to the motor are 180 rpm and 5.2 kW respectively. If the air temperature increases to 48°C, calculate the required fan speed and power input to maintain the same mass flow rate of air through the duct system.

Solution The mass flow rate of air is given by

$$\dot{m} = \rho_1 \dot{Q}_1 = \rho_2 \dot{Q}_2 \qquad \text{(E11.6.1)}$$

Assume that air is an ideal gas with the equation of state

$$p = \rho RT \qquad \text{(E11.6.2)}$$

From Eqs. (E11.6.1) and (E11.6.2) we have

$$\frac{\dot{Q}_2}{\dot{Q}_1} = \frac{\rho_1}{\rho_2} = \frac{(273+48)}{(273+18)} = 1.103$$

From Eq. (11.23) it follows that for the same fan:

$$\frac{\omega_2}{\omega_1} = \frac{\omega_2}{180} = \frac{\dot{Q}_2}{\dot{Q}_1} = 1.103$$

Therefore the required fan speed is 198.5 rpm.

From Eq. (11.25), it follows that for the same fan:

$$\frac{\dot{W}_2}{\dot{W}_1} = \frac{\dot{W}_2}{5.2} = \frac{{\omega_2}^3 \rho_2}{{\omega_1}^3 \rho_1} = \frac{(1.103)^3}{1.103} = 1.2166$$

Therefore the required fan power is 6.32 kW.

Example 11.7 A fan rotating at 200 rpm supplies air to a conditioned space through a duct network. The system curve for the duct network has been obtained by combining the duct friction losses and the pressure losses through the fittings. The resulting relationship has the form

$$P = 207\dot{Q}^2$$

where P is the total pressure loss in Pa and $\dot{Q}$ is the air flow rate in m^3s^{-1}.

The performance curve for the fan at 200 rpm is given in Fig. E11.7.1. (a) Obtain the fan curves at 250 rpm and 150 rpm. (b) Calculate the volume flow rate, the pressure and the ideal fan power at 200 rpm, 250 rpm and 150 rpm.

Solution (a) We use the fan laws to generate the performance curves at 250 rpm and 150 rpm. Consider a point, such as A_o, on the given curve at speed N_o (200 rpm) shown in Fig. E11.7.1. Let the coordinates at A_o be $\dot{Q}_o$ and P_o respectively. From Eq. (11.23) we obtain the volume flow rate at a new fan speed N_1 as

$$\dot{Q}_1 = \dot{Q}_o\left(\frac{N_1}{N_o}\right) = 1.25\dot{Q}_o \qquad \text{(E11.7.1)}$$

From Eq.(11.24) we obtain the fan pressure at a new fan speed N_1 as

$$P_1 = P_o\left(\frac{N_1}{N_o}\right)^2 = 1.25^2 P_o \qquad \text{(E11.7.2)}$$

By substituting values in Eqs. (E11.7.1) and (E11.7.2) we calculate the coordinates of point A_1, which correspond to point A_o. Hence we generate the fan curve for a speed of 250 rpm.

A similar procedure is used to obtain the points, such as A_2, to generate the fan curve for 150 rpm. These curves are plotted in Fig. E11.7.1.

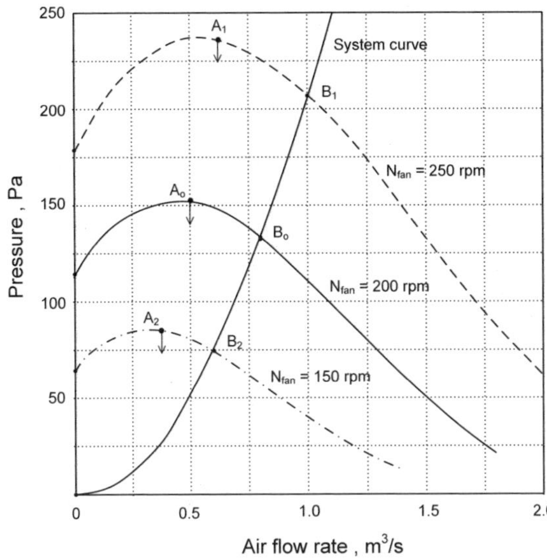

Fig. E11.7.1 Fan curves at three speeds

(b) We now plot the duct system curve given by

$$P = 207\dot{Q}^2 \qquad \text{(E11.7.3)}$$

on Fig. E11.7.1.

The points of intersection of the system curve and the fan curves at 200 rpm, 250 rpm and 150 rpm give the steady operating points, B_o, B_1 and B_2 of the fan–duct system at the three fan speeds. The flow rates and pressures at these points are: B_1 = [1.0, 208], B_o = [0.8, 132], B_2 = [0.6, 75]. The corresponding *ideal* fan power inputs are: 208W, 105.6W, and 45 W.

Example 11.8 The fan–duct system described in worked example 11.7 is equipped with a damper to control the air flow rate to the space. When the fan is operating at 250 rpm, the damper is partially closed in response to a reduction of the space cooling load. When the damper is partially closed, the pressure loss through it varies with air flow rate according to the relation

$$\Delta P_{damp} = 83\dot{Q}^2$$

(a) Determine the air flow rate, the pressure and the ideal fan power input at the new operating point with the same fan speed and the damper partially closed.

(b) If the fan speed is lowered to 200 rpm keeping the damper at the original open position, determine air flow rate, the pressure, and input power at the new fan speed. (c) Determine the fan speed required to obtain the same flow rate as in (a), if the damper is in the open position. Calculate the corresponding pressure and power input.

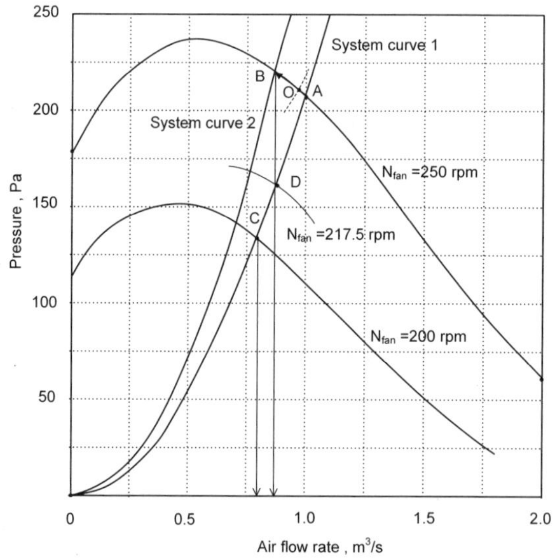

Fig.E11.8.1 Effect of system curves and fan speeds

Solution The relationship for the new system curve after the damper is partially closed is given by [see Eq. (11.31)]

$$P = (207 + 83)\dot{Q}^2 = 290\dot{Q}^2 \qquad \text{(E11.8.1)}$$

(a) The new system curve together with the fan curves at 250 rpm and 200 rpm are plotted in Fig. E11.8.1. The operating point with the damper partially closed is given by the point of intersection B of the new system curve and the fan curve at 250 rpm. The air flow rate and the pressure at B are 0.87 m^3s^{-1} and 220 Pa respectively. Hence the fan power input at B is

$$\dot{W}_B = \dot{Q}_B p_{tB} = 0.87 \times 220 = 191.4 \text{ W}$$

The ideal power at the original operating point A is 208 W (see example 11.7). Hence the decrease in fan power from the original state A is about 8 %.

(b) The fan speed is now lowered to 200 rpm, keeping the damper at the initial open position. The new operating point is given by the point of intersection C of the original system curve and the fan curve at 200 rpm.

The air flow rate and pressure at B are 0.8 m^3s^{-1} and 133.3 Pa respectively. Therefore the fan power input is

$$\dot{W}_C = \dot{Q}_C p_{tC} = 0.8 \times 133.3 = 107 \text{ W}$$

The decrease in fan power from the original state A is about 49%.

(c) Let D be the point of intersection of the original system curve and the fan curve at the required fan speed. An estimate of the required fan speed at D may be obtained by simple graphical interpolation between the curves at 200 rpm and 250 rpm. Alternatively, a more accurate value can be calculated by applying the fan laws given by Eqs. (11.23) and (11.24).

Let the operating point D be similar to point O on the fan curve for 250 rpm. Therefore

$$N_D = N_o\left(\frac{\dot{Q}_D}{\dot{Q}_O}\right) = \frac{250\times0.87}{\dot{Q}_O} \quad \text{(E11.8.2)}$$

$$N_D = N_o\left(\frac{P_D}{P_o}\right)^{1/2} = 250\left(\frac{157}{P_o}\right)^{1/2} \quad \text{(E11.8.3)}$$

Therefore $$\frac{250\times0.87}{\dot{Q}_O} = 250\left(\frac{157}{P_o}\right)^{1/2}$$

$$\dot{Q}_o^{\ 2} = \left(\frac{P_O}{157}\right) \times 0.87^2 \quad \text{(E11.8.4)}$$

The point of intersection of the parabolic curve represented by Eq. (E11.8.4) and the fan curve at 250 rpm gives point O. We then obtain the fan speed at D by substituting the flow rate at O in Eq. (E11.8.2). This gives the fan speed at D as 217 rpm.

The pressure and ideal power input at D are 157 Pa and 137 W respectively. The power input has decreased by about 34% compared to the original state A. This shows that to reduce the air flow rate for part-load operation of air conditioning systems, the use of a variable speed fan is a more energy efficient approach compared to the use of a damper.

Example 11.9 The duct network shown schematically in Fig. E11.9.1 supplies conditioned air to two spaces 4 and 8 at the rate of 3 m^3s^{-1} and 1.2 m^3s^{-1} respectively. (i) Design the sections 1-2 and 3-4 using the equal friction method. (ii) Design the section 5-8 so that the available pressure is used without including a damper.

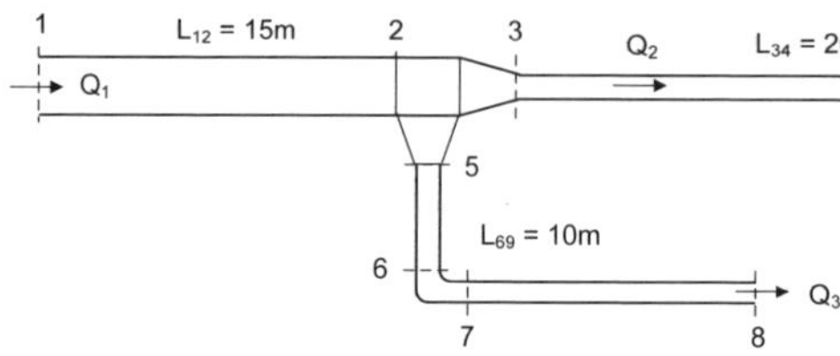

Fig. E11.9.1 Supply air duct network

Solution We shall assume a constant friction pressure loss rate of 4 $Pa.m^{-1}$ for duct sections 1-2 and 3-4. The flow rate in section 1-2 is 4.2 m^3s^{-1}. With this data we obtain the diameter of section 1-2 from the friction chart in Fig. 11.2. Alternatively, we could use the MATLAB code in Appendix A11.2, which solves Colebrook's equation iteratively, to obtain the friction factor and hence determine the duct diameter that gives the assumed unit frictional pressure loss. This gives the diameter, $D_2 = 0.584$ m and the air velocity, $V_2 = 15.7$ ms^{-1}.

For the duct section 3-4 we assume the same pressure loss rate of 4 $Pa.m^{-1}$. The air flow rate is 3 m^3s^{-1}. The duct diameter is obtained as, $D_3 = 0.513$ m and the air velocity, $V_1 = 14.5$ ms^{-1}.

The total pressure loss from 2-4 is given by

$$P_2 - P_4 = \Delta P_{s,23} + \Delta P_{f,34} \tag{E11.9.1}$$

$$P_2 - P_4 = k_s \left(\frac{\rho V_2^{\,2}}{2}\right) + \frac{f_{34} L_{34} \rho V_3^{\,2}}{2D_3} \tag{E11.9.2}$$

From tabulated data on page 21.50 in Ref. [2] we obtain the loss coefficient for the straight section 2-3 of the tee-junction, for the parameters, $(A_3/A_2) = 0.77$ and $(Q_3/Q_2) = 0.714$, as $k_s = 0.14$. Hence the tee-junction pressure loss given by the first term of Eq. (E11.9.2) is 20.7 Pa. The second term of Eq. (E11.9.2), which is the friction pressure loss of section 3-4 is 80 Pa. Therefore the total pressure loss from 2 to 4 is 100.7 Pa. This is the available pressure for the section 2-8.

We need to select the diameter of section 5-8 such that the total pressure loss from 2 to 8 is 100.7 Pa. Now the total pressure loss is given by

$$P_2 - P_8 = k_b\left(\frac{\rho V_2^{\,2}}{2}\right) + \frac{f_{58}L_{58}\rho V_5^{\,2}}{2D_5} + \frac{k_{bend}\rho V_5^{\,2}}{2} = 100.7 \text{ Pa} \quad \text{(E11.9.3)}$$

Since the diameter, D_5 of section 5-8 is not known yet, we need to solve Eq. (E11.9.3) iteratively. The MATLAB code in Appendix A11.1 may be used for this purpose. We begin the computation by assuming an initial guessed value for D_5. Based on this value we obtain the loss coefficient, k_b for the branch of the tee-junction from tabulated data on page 21.50 in Ref. [2]. Similarly, the loss coefficient for the 90° bend, k_{bend} is obtained from tabulated data on page 21.26 in Ref. [2]. The diameter, D_5 is changed in small steps until Eq. (E11.9.3) is satisfied. Hence we have $D_5 = 0.37$ m. The corresponding loss coefficients are: $k_b = 0.38$ and $k_{bend} = 0.11$.

Example 11.10 (a) Use the equal friction method to size the circular supply air duct system shown schematically in Fig. E11.10.1. The flow rates and duct lengths are indicted in the figure. The loss coefficient for the duct exits is 0.6. (b) Determine where dampers should be located to achieve the desired air flow rates to the spaces. (c) Can the pressure losses in the duct runs be balanced by changing the duct diameters?

Solution The flow rates through the different duct sections are as follows:

$$\dot{Q}_{12} = 4.5 \text{ m}^3\text{s}^{-1}, \quad \dot{Q}_{34} = 3.5 \text{ m}^3\text{s}^{-1}, \quad \dot{Q}_{56} = 1.5 \text{ m}^3\text{s}^{-1},$$
$$\dot{Q}_{78} = 1.5 \text{ m}^3\text{s}^{-1}, \quad \dot{Q}_{9,10} = 2.0 \text{ m}^3\text{s}^{-1}, \quad \dot{Q}_{11,12} = 1.0 \text{ m}^3\text{s}^{-1}$$

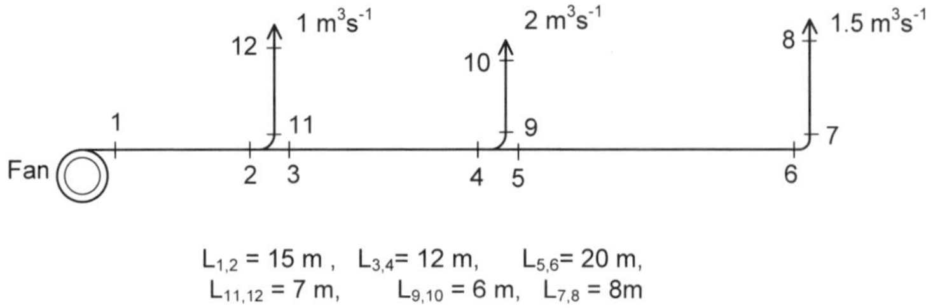

Fig. 11.10.1 Duct network

To apply the equal friction method we shall assume that the air velocity in the section 1-2 is 9 ms^{-1}. From the friction chart in Fig. 11.2 we obtain the duct diameter as 0.8 m and the unit pressure drop as 0.93 $Pa.m^{-1}$. Note that these values are close to the lower bound of the recommended pressure loss [1] indicated by bold lines in Fig. 11.2.

The unit pressure losses for all the other duct sections are taken as 0.93 $Pa.m^{-1}$. The given air flow rates in the different duct sections and the constant unit pressure loss of 0.93 $Pa.m^{-1}$ allow us to obtain the duct diameters of these sections directly from the friction chart.

Alternatively, we could use the MATLAB program in Appendix A11.2 to compute the duct diameter. This program solves the Colebrook equation iteratively to obtain the friction factor and hence determine the duct diameter that gives the assumed unit pressure loss. Thus we compute the velocity pressure, $P_{vel.}$ and the friction loss, P_{loss} in all duct sections separated by fittings. The results are summarized in Table E11.10.1.

Table E11.10.1 Computed quantities

Section	L, m	Q, m^3s^{-1}	D, m	V, ms^{-1}	$P_{vel.}$, Pa	P_{loss}, Pa
1-2	15	4.5	0.8	9.0	48.3	13.95
11-12	7	1.0	0.45	6.2	23.3	6.5
3-4	12	3.5	0.73	8.4	42.7	11.2
9-10	6	2.0	0.59	7.36	32.6	5.58
5-6	20	1.5	0.53	6.87	28.4	18.6
7-8	8	1.5	0.53	6.87	28.4	7.4

Since the flow rates and duct diameters have been computed we can now obtain the loss coefficients for the two diverging tee-junctions (see page 21.49 in Ref. [2]) and the 90° bend from the *ASHRAE duct fittings database* [2]. Representative values are given in Table 11.1(a) and (c). Hence we have

$$c_{2,3} = 0.13, c_{2,11} = 2.4, c_{4,5} = 0.14, c_{4,9} = 1.5, c_{6,7} = 0.11$$

The total pressure losses in the different duct runs are as follows:

$$P_1 - P_8 = P_{1,2} + P_{2,3} + P_{3,4} + P_{4,5} + P_{5,6} + P_{6,7} + P_{7,8} + P_{exit}$$

$$P_1 - P_8 = 13.95 + (0.13 \times 48.3) + 11.2 + (0.14 \times 42.7) + 18.6 + (0.11 \times 28.42) + 7.4 + (0.6 \times 28.42) = 83.6 \text{ Pa}$$

$$P_1 - P_{12} = P_{1,2} + P_{2,11} + P_{11,12} + P_{exit}$$

$$P_1 - P_{12} = 13.95 + (2.4 \times 48.3) + 6.5 + (0.6 \times 23.3) = 150 \text{ Pa}$$

$$P_1 - P_{10} = P_{1,2} + P_{2,3} + P_{3,4} + P_{4,9} + P_{9,10} + P_{exit}$$

$$P_1 - P_{10} = 13.95 + (0.13 \times 48.3) + 11.2 + (1.5 \times 42.7) + 5.58 + (0.6 \times 32.64) = 120.6 \text{ Pa}$$

The largest pressure loss of 150 Pa is between sections 1 and 12. Therefore the fan has to provide a pressure of at least 150 Pa. However, since the pressure losses from 1 to 8 and 1 to 10 are less than 150 Pa, the supply air flow rates to spaces 8 and 10 will exceed the desired flow rates. To balance the pressure losses in the three duct runs, dampers have to be installed along section 7-8 and 9-10 to artificially introduce additional pressure losses. Their magnitudes are: $\Delta P_{7\text{-}8}$ = (150-83.6) = 66.4 Pa and $\Delta P_{9\text{-}10}$ = (150-120.6) = 29.4 Pa.

Note that we could adjust the diameters of ducts 7-8 and 9-10 to increase the pressure losses in the duct runs 1-8 and 1-10 and thereby bring the pressure losses in the three duct runs closer. Exploratory calculations using the MATLAB code in Appendix A11.1 show that if the diameter of section 7-8 is reduced to 0.37 m the pressure loss in the duct run 1-8 becomes 120 Pa and the velocity in section 7-8 becomes 13.4 ms^{-1}.

Similarly, if the diameter of section 9-10 is reduced to 0.42 m, the pressure loss in the duct run 1-10 becomes 145 Pa and the velocity in section 9-10 becomes 14 ms^{-1}. However, these changes bring the unit pressure losses in the sections 7-8 and 9-10 to the boundary of the recommended region in the friction chart in Fig. 11.2.

Example 11.11 A return air duct network, shown schematically in Fig. E11.11.1, is made of round sheet metal ducts. The flow rates and lengths are indicted in the figure. (i) Use the equal friction method to size this circular duct system. (ii) Determine where dampers should be located to achieve the desired air flow rates from the spaces.

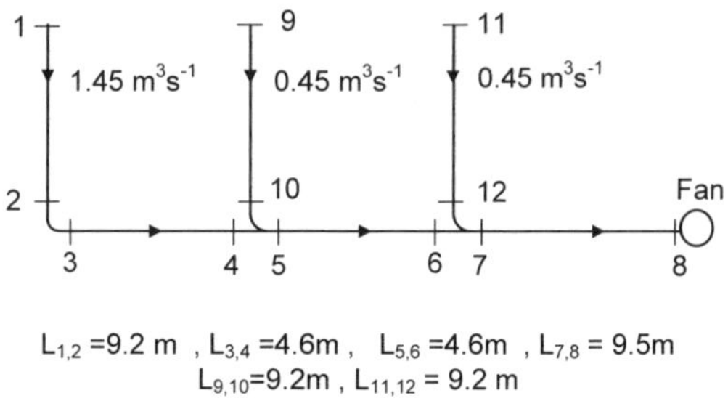

Fig. E11.11.1 Return air duct network

Solution The flow rates through the different duct sections are as follows:

$$\dot{Q}_{12} = 1.45 \text{ m}^3\text{s}^{-1}, \quad \dot{Q}_{34} = 1.45 \text{ m}^3\text{s}^{-1}, \quad \dot{Q}_{56} = 1.9 \text{ m}^3\text{s}^{-1},$$

$$\dot{Q}_{78} = 2.35 \text{ m}^3\text{s}^{-1}, \quad \dot{Q}_{9,10} = 0.45 \text{ m}^3\text{s}^{-1}, \quad \dot{Q}_{11,12} = 0.45 \text{ m}^3\text{s}^{-1}$$

To apply the equal friction method we shall assume that the constant unit pressure drop as 0.9 Pa.m^{-1}. Note that this value is close to the lower bound of the recommended pressure loss indicated by bold lines in Fig. 11.2.

The given air flow rates in the different duct sections and the constant unit pressure loss of 0.9 Pa.m^{-1} allow us to obtain the duct diameters of these sections directly from the friction chart. Alternatively, we could use the MATLAB program in Appendix A11.2 to compute the duct diameter. This program solves the Colebrook equation iteratively to obtain the friction factor and hence determine the duct diameter that gives the assumed unit pressure loss. Thus we compute the velocity pressure, $P_{vel.}$ and the friction loss, P_{loss} in all duct sections separated by fittings. The results are summarized in Table E11.11.1.

Table E11.11.1 Computed quantities

Section	L, m	Q, m^{3}s^{-1}	D, m	V, ms^{-1}	$P_{vel.}$, Pa	P_{loss}, Pa
1-2	9.2	1.45	0.524	6.7	27.2	8.28
3-4	4.6	1.45	0.524	6.7	27.2	4.14
5-6	4.6	1.9	0.581	7.2	31.0	4.14
7-8	9.5	2.35	0.629	7.55	34.4	8.55
9-10	9.2	0.45	0.337	5.04	15.3	8.28
11-12	9.2	0.45	0.337	5.04	15.5	8.28

Since the flow rates and duct diameters have been computed we can now obtain the loss coefficients for the two converging tee-junctions (see page 21.41 in Ref. [2]) and the 90° bend from the *ASHRAE duct fittings database*. Representative values are also given in Tables 11.1(d) and (b) respectively. Hence we have

$$c_{2,3} = 0.11, c_{4,5} = 0.36, c_{10,5} = -0.28, c_{6,7} = 0.25, c_{12,7} = 0.21$$

The total pressure losses in the different duct runs are as follows:

$$P_1 - P_8 = P_{1,2} + P_{2,3} + P_{3,4} + P_{4,5} + P_{5,6} + P_{6,7} + P_{7,8}$$

$$P_1 - P_8 = 8.28 + (0.11 \times 27.2) + 4.14 + (0.36 \times 31) + 4.14 + (0.25 \times 34.4) + 8.55 = 48 \text{ Pa}$$

$$P_9 - P_8 = P_{9,10} + P_{10,5} + P_{5,6} + P_{6,7} + P_{7,8}$$

$$P_9 - P_8 = 8.28 + (-0.28 \times 31) + 4.14 + (0.25 \times 34.4) + 8.55 = 21 \text{ Pa}$$

$$P_{11} - P_8 = P_{11,12} + P_{12,7} + P_{7,8}$$

$$P_{11} - P_8 = 8.28 + (0.21 \times 34.4) + 8.55 = 24 \text{ Pa}$$

The largest pressure loss of 48 Pa is between sections 1 and 8, the fan inlet. Therefore the fan has to provide a suction pressure of at least 48 Pa. However, since the pressure losses from 9 to 8 and 11 to 8 are less than 48 Pa, the return air flow rates from spaces 9 and 11 will be higher than the desired values. To balance the pressure losses in the three duct runs, dampers have to be installed along section 9-10 and 11-12 to artificially introduce additional pressure losses. Their magnitudes are: $P_{9\text{-}10}$ = (48-21) = 27 Pa and $P_{11\text{-}12}$ = (48-24) = 24 Pa.

Note that we could adjust the diameters of ducts 9-10 and 11-12 to increase the pressure losses in the duct runs 9-8 and 11-8 and thereby bring the pressure losses in the three duct runs closer (see worked example 11.10).

Example 11.12 A circular supply air duct system is shown schematically in Fig. E11.12.1. The flow rate through section 1-2 is 40 m^3s^{-1} and the air velocity is 18 ms^{-1}. The section 3-4 with a 90° bend is 12 m long. The flow rate through section 3-4 is 24 m^3s^{-1}. Determine the diameter of section 3-4 using the static regain method.

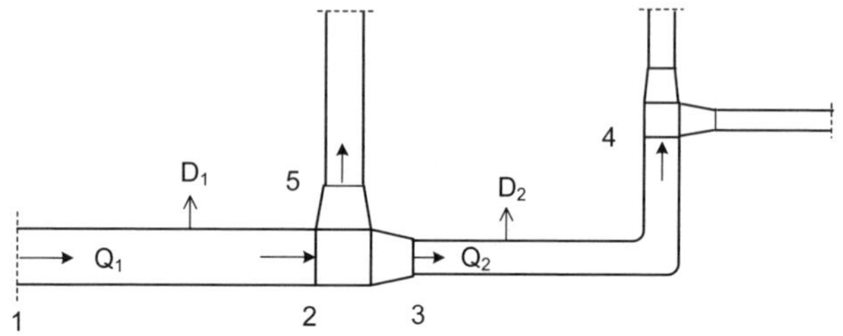

Fig. E11.12.1 Supply air duct network

Solution We shall use the static regain method to determine the diameter D_2 of the round duct section 3-4. The diameter of section 1-2 is given by

$$D_1 = \left(\frac{4\dot{Q}_1}{\pi V_2}\right)^{1/2} = \left(\frac{4\times 40}{\pi\times 18}\right)^{1/2} = 1.68 \text{ m}$$

Applying the energy equation between 2 and 4 we have

$$p_{s2} + \frac{\rho {V_2}^2}{2} = \Delta P_{2-3} + \frac{f_{34}L_{34}\rho {V_3}^2}{2D_3} + \frac{K_b\rho {V_3}^2}{2} + p_{s4} + \frac{\rho {V_3}^2}{2} \qquad \text{(E11.12.1)}$$

where V_2 and V_3 are the air velocities at sections 2 and 3 respectively. The total pressure loss, ΔP_{2-3} across the straight section of the tee-junction 2-3-5, may be written as

$$\Delta P_{2-3} = \frac{k_s\rho {V_2}^2}{2} \qquad \text{(E11.12.2)}$$

The second term on the RHS of Eq. (E11.12.1) is the friction loss in the duct section 3-4 and third term is the pressure loss across the 90° bend.

In the static regain method we assume that the *static pressure remains constant* at the entrance to each fitting across which the flow rates change. Therefore

$$p_{s2} = p_{s4} \tag{E11.12.3}$$

The air velocities in the ducts may be expressed in terms of the respective air flow rates and diameters. Hence we have

$$\dot{Q}_1 = \frac{\pi {D_1}^2 V_2}{4} \tag{E11.12.4}$$

$$\dot{Q}_2 = \frac{\pi {D_2}^2 V_3}{4} \tag{E11.12.5}$$

Substituting from Eqs. (E11.12.2) to (E11.12.5) in Eq. (E11.12.1) we obtain the following equation

$$\frac{(1-k_s)}{{D_1}^4}\left(\frac{\dot{Q}_1}{\dot{Q}_2}\right)^2 = \frac{(1+k_b)}{{D_2}^4} + \frac{f_{34}L_{34}}{{D_2}^5} \tag{E11.12.6}$$

Substituting the given numerical values in Eq. (E11.12.6) we have

$$\frac{(1-k_s)}{1.68^4}\left(\frac{40}{24}\right)^2 = \frac{(1+k_b)}{{D_2}^4} + \frac{12 f_{34}}{{D_2}^5} \tag{E11.12.7}$$

To find the diameter D_2 we need to solve the above non-linear equation (E11.12.7). The loss coefficient, k_s for the tee-junction, which is tabulated on page 21.49 in Ref. [2], (also see Table 11.1c) is a function of two design parameters. These are:

$$\frac{\dot{Q}_2}{\dot{Q}_1} \quad \text{and} \quad \frac{A_2}{A_1} = \left(\frac{D_2}{D_1}\right)^2 \tag{E11.12.8}$$

The loss coefficient k_b for the 90° bend, which is tabulated on page 21.26 in Ref. [2], (also see Table 11.1b) is a function of the diameter D_2.

In view of these functional dependencies, we have to use a trial-and-error approach to solve Eq. (E11.12.6). The MATLAB code used for this purpose is listed in Appendix A11.3.

The main steps of the solution procedure are as follows: (i) Guess an initial value for D_2. (ii) Obtain the loss coefficients k_s and k_b from the appropriate tables in Ref. [2]. (iii) Solve Colebrook's equation to obtain the friction factor, f_{34}. (iv) Substitute numerical values in Eq. (E11.12.6) and compare the RHS with the LHS. (v) Change D_2 in small steps and repeat steps (i) to (iv) until the LHS and the RHS of Eq. (E11.12.6) become equal. The final converged solution gives the diameter, D_2 = 142.6 cm.

The MATLAB code in Appendix A11.3 may be easily modified to apply the static regain method to design situations involving other types of fittings.

Example 11.13 A round supply duct system is shown schematically in Fig. E11.13.1. The loss coefficient for the entrance fitting at 1 is 0.15 and for the diffusers it is 0.1. (i) Size the main branch 1-8 using the static regain method. (ii) Calculate the pressure loss in the main duct run 1-8. (iii) Select the diameters of the branch sections 9-10, 11-12 and 13-14 so that the pressures available to these branches are fully used with no damping.

Solution We determine the duct diameters of sections 1-2, 3-4, 5-6 and 7-8 using the static regain method. The loss coefficients for the tee-junctions are obtained from page 21.50 in Ref. [2]. We assume the air velocity in section 1-2 as 16 ms^{-1}. The diameter of section 1-2 is given by

$$D_{1-2} = \left(\frac{4\dot{Q}_{1-2}}{\pi V_{1-2}}\right)^{1/2} = \left(\frac{4\times 40}{\pi\times 16}\right)^{1/2} = 1.78 \text{ m}$$

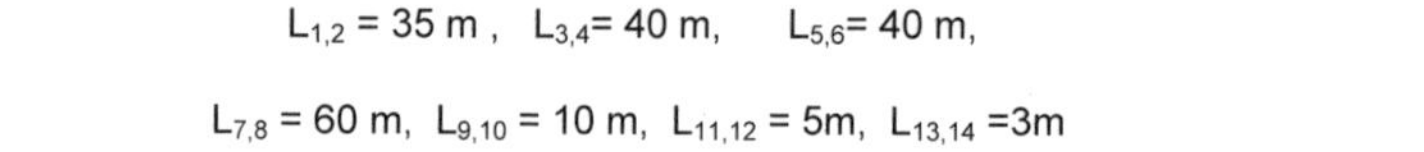

Fig. E11.13.1 Supply air duct network

The diameter of section 3-4 is obtained by applying Eq. (E11.12.6) to section 2-4 of the duct network.

$$\frac{(1-k_s)}{D_2{}^4}\left(\frac{\dot{Q}_2}{\dot{Q}_3}\right)^2 = \frac{1}{D_3{}^4} + \frac{f_{34}L_{34}}{D_3{}^5} \tag{E11.13.1}$$

Substituting numerical values in Eq. (E11.13.1) we have

$$\frac{(1-k_s)}{1.78^4}\left(\frac{40}{30}\right)^2=\frac{1}{D_3{}^4}+\frac{40f_{34}}{D_3{}^5} \qquad \text{(E11.13.2)}$$

where k_s = 0.14. Using the MATLAB code in Appendix A11.3 we obtain, D_3 =1.71 m.

Applying Eq. (E11.13.1) to section 4-6 we have

$$\frac{(1-k_s)}{1.713^4}\left(\frac{30}{20}\right)^2=\frac{1}{D_5{}^4}+\frac{40f_{45}}{D_5{}^5} \qquad \text{(E11.13.2a)}$$

where k_s = 0.137. Using the MATLAB code in Appendix A11.3 we obtain, D_5 =1.56 m.

Applying Eq. (E11.13.1) to section 6-8 we have

$$\frac{(1-k_s)}{1.563^4}\left(\frac{20}{10}\right)^2=\frac{1}{D_7{}^4}+\frac{60f_{45}}{D_7{}^5} \qquad \text{(E11.13.3)}$$

where k_s = 0.14. Using the MATLAB code in Appendix A11.3 we obtain, D_7 =1.30 m.

The following pressure losses in the main duct run 1-8 have been calculated using the MATLAB code in Appendix A11.3:

$$\Delta P_{ent} = 23.13 \text{ Pa}, \quad \Delta P_{fr,12} = 38.22 \text{ Pa},$$

$$\Delta P_{s,23} = 21.42 \text{ Pa}, \quad \Delta P_{fr,34} = 30.9 \text{ Pa},$$

$$\Delta P_{s,45} = 13.86 \text{ Pa}, \quad \Delta P_{fr,56} = 22.6 \text{ Pa},$$

$$\Delta P_{s,67} = 9.25 \text{ Pa}, \quad \Delta P_{fr,78} = 22.4 \text{ Pa},$$

$$\Delta P_{diff} = 3.36 \text{ Pa}$$

The total pressure loss from 1-8 is given by

$$P_{t1}-P_{t8}=23.13+38.22+21.42+30.9+13.86+22.6+9.25+22.4+3.36=185 \text{ Pa}$$

The pressures available for the branch duct runs are given by:

$$P_2-P_{10}=185-(23.13+38.2)=124 \text{ Pa}$$

$$P_4-P_{12}=124-(21.42+30.9)=72 \text{ Pa}$$

$$P_6-P_{14}=72-(13.86+22.6)=36 \text{ Pa}$$

The diameter of duct 9-10 is chosen to have a total pressure loss of 124 Pa between 2 and 10. The computation is carried out using the MATLAB code in Appendix A11.1. An initial value is assumed for the diameter, and based on it the loss coefficient k_b for the branch 2-9 is obtained from the tabulated data on page 21.50 of Ref. [2]. The diameter is adjusted iteratively until the total pressure loss from 2 to 9 is 124 Pa. This gives the diameter of section 9-10 as 0.81 m. A similar procedure gives the diameters of sections 11-12 and 13-14 as 0.83 m and 0.9 m respectively.

Example 11.14 The air flow rate though a straight section of a round duct is 0.9 m^3s^{-1}. The efficiencies of the fan and motor supplying air to the duct are 60% and 85% respectively. Assume the friction factor for the duct as 0.025. The cost of electricity is \$0.15 per kWh. The cost of the sheet metal used to fabricate the duct is \$120 per m^2. The number of hours of operation of the system during the amortization period is 10,000. Calculate the optimum diameter of the duct.

Solution The total pressure loss through the duct is given by

$$\Delta P_t = \frac{fL\rho V^2}{2D} = \frac{8fL\rho Q^2}{\pi^2 D^5}$$

Assuming that the fan pressure is equal to the total pressure loss, the ideal power input to the fan is

$$\dot{W}_f = \Delta P_t Q = \frac{8fL\rho Q^3}{\pi^2 D^5}$$

The actual electrical power input to the motor driving the fan is

$$\dot{W}_{act} = \frac{8fL\rho Q^3}{\pi^2 D^5 \eta_m \eta_f}$$

where η_m and η_f are the efficiencies of the motor and fan respectively.

If the system is operated for H hours, then the total electrical energy consumed, expressed in kWh is

$$E_{tot} = \frac{8fL\rho Q^3 H}{\pi^2 D^5 \eta_m \eta_f 10^3} \quad \text{(E11.14.1)}$$

The total operating cost, which is the total cost of the electrical energy is

$$C_{op} = \frac{8fL\rho Q^3 Hc_e}{\pi^2 D^5 \eta_m \eta_f 10^3} \tag{E11.14.2}$$

where c_e is the unit cost electricity in \$ per kWh.

The initial cost of the duct is given by

$$C_{ini} = \pi DLc_m \tag{E11.14.3}$$

where c_m is the unit cost of the duct, \$ per m^2.

Therefore the total cost of owning and operating the duct system is

$$C_{tot} = C_{ini} + C_{op} = \pi DLc_m + \frac{8fL\rho Q^3 Hc_e}{\pi^2 D^5 \eta_m \eta_f 10^3} \tag{E11.14.4}$$

From Eq. (E11.14.4) we observe that the initial cost increases with diameter while the operating cost decreases with diameter. The diameter that minimizes the total cost is obtained by differentiating Eq. (E11.14.4). Hence we have

$$\frac{dC_{tot}}{dD} = \pi Lc_m - \frac{40fL\rho Q^3 Hc_e}{\pi^2 D^6 \eta_m \eta_f 10^3} = 0$$

Therefore the optimum duct diameter is given by

$$D_{opt} = \left(\frac{40f\rho Q^3 Hc_e}{\pi^2 \eta_m \eta_f c_m 10^3}\right)^{1/6}$$

Substituting the given numerical values in the above expression we have

$$D_{opt} = \left(\frac{40\times 0.025\times 1.2\times 0.9^3\times 10{,}000\times 0.15}{\pi^2\times 0.6\times 0.85\times 120\times 10^3}\right)^{1/6} = 0.36 \text{ m}$$

Therefore the optimum diameter is 36 cm. It should be noted that the foregoing analysis is a highly simplified form of the optimization procedure which is presented mainly to introduce the various costs involved.

Example 11.15 An office room, maintained at 24°C and 50% relative humidity, has a design cooling load of 2.2 kW. The length and breadth of the room are 8m and 4 m respectively. The conditioned air supplied to the room is at 13°C and 90% relative humidity. Determine the required parameters of the diffuser to be used.

Solution We obtain the enthalpies of the room air and the supply air, and the specific volume of the supply air directly from the psychrometric chart. Hence we have

$$h_r = 48\ \text{kJkg}^1, \quad h_i = 34.3\ \text{kJkg}^{-1}, \quad v_i = 0.822\ \text{m}^3\text{kg}^{-1}$$

The volume flow rate of air is given by

$$\dot{Q}_{in} = \frac{v_i q_{cool}}{(h_r - h_i)} = \frac{2.2 \times 0.822}{(48 - 34.3)} = 0.132\ \text{m}^3\text{s}^{-1}$$

We propose to locate two ceiling slot diffusers symmetrically, one at the center of each half of the room. The air flow through each diffuser is 0.066 m^3s^{-1}. Based on the cooling load per unit area of 68.75 Wm^{-2} we select ratio, $(X_{0.5}/L) = 0.3$ from Table 11.3. From Table 11.2, the characteristic length, L is the distance to the wall or the mid-plane. Therefore, $L = 2$ m. Hence the required throw of the diffuser is, $X_{0.5} = 0.6$ m. We now consult a manufacturer's catalogue to select a ceiling slot diffuser with the desired throw at the design air flow rate.

Problems

P11.1 Air flows through a round duct section with a diverging, round to round, transition (see Fig. 11.3b) at the rate of 0.5 m^3s^{-1}. The angle of divergence, $\theta = 30^0$. The lengths and diameters of the duct sections upstream and downstream of the transition are 5 m, 0.25 m and 15 m, 0.35 m respectively. The upstream section includes two 90° bends (see Fig. 11.3a). Assume the friction factor for the straight duct sections as 0.02. Calculate (i) the change in velocity pressure and the total pressure loss across the transition, (ii) the total pressure loss from the entrance to the exit of the duct.

[*Answers*: (i) 46 Pa, 20.3 Pa, (ii) 72.7Pa]

P11.2 A round duct network with a diverging tee-junction 2-3-5 (see Fig. 11.3c) is shown schematically in Fig. P11.2.1, where the lengths and diameters of the sections are indicated. The air flow rates at sections 2 and 3 are 4m^3s^{-1} and 2.5m^3s^{-1} respectively. Assume the friction factor as 0.025. The density of air is 1.2 kgm^{-3}. Calculate (i) the total pressure loss

from 1 to 4, (ii) the total pressure loss from 1 to 6, and (iii) the changes in static pressure from 1 to 4 and 1 to 6.
[*Answers*: (i) 90.45Pa, (ii) 67.8Pa, (iii) 149.8 Pa, 115.4 Pa]

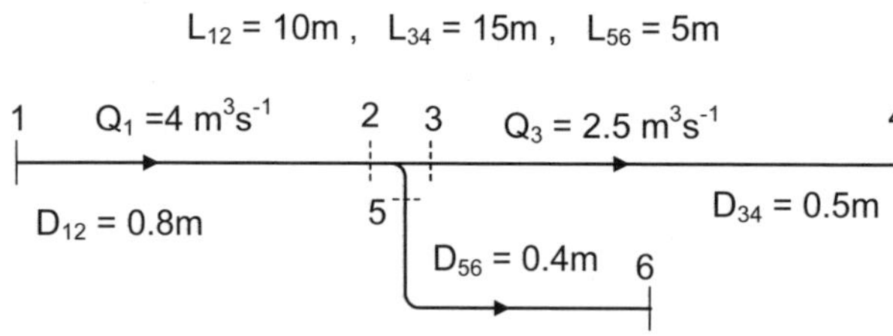

Fig. P11.2.1 Supply air duct network

P11.3 Air flows at the rate of $1.25 m^3s^{-1}$ through a rectangular duct of length 20 m. The cross sectional dimensions of the duct are 0.25 m by 1 m. Assume the friction factor as 0.0176. Calculate (i) velocity pressure of the air, and (ii) the frictional pressure loss.
[*Answers*: (i) 15 Pa, (ii) 14.6 Pa]

P11.4 The design conditions for a fan–duct system are as follows: air temperature = 15°C, fan speed = 200 rpm, volume flow rate =5 m^3s^{-1}, pressure rise = 800 Pa. (i) Calculate the flow rate, the pressure rise, and the power if the speed is decreased to 150 rpm at the same air temperature. (ii) Calculate the flow rate, the pressure rise, and the power if the air temperature is increased to 45°C with the speed at 200 rpm. (iii) Calculate the flow rate, the speed, and the power if the air temperature is increased to 45°C with the pressure rise remaining at 800 Pa.
[*Answers*: (i) 3.75 m^3s^{-1}, 450 Pa, 1.69 kW, (ii) 5 m^3s^{-1} ,724.6 Pa, 3.62kW, (iii) 5.25 m^3s^{-1}, 210 rpm, 4.2 kW]

P11.5 A fan–duct system consists of a fan supplying air through a round duct with a diverging, round to round, transition (see Fig. 11.3b). The divergence angle of the transition is 30°. The lengths and diameters of the duct sections upstream and downstream of the transition are 7 m, 0.25 m and 16 m, 0.35 m respectively. The upstream section includes two 90° bends (see Fig. 11.3a). Assume the friction factor for the straight duct sections as 0.025. (i) Obtain the equation of the system curve for the

duct network. (ii) Use the fan curves given in Fig. E11.8.1 to determine the air flow rates when the fan speeds are 250 rpm and 200 rpm.
[*Answers*: (i) $\Delta P(\text{Pa}) = 383.45\ Q^2$, (ii) 0.77 m^3s^{-1}, 0.62 m^3s^{-1}]

P11.6 A round duct network is shown schematically in Fig. P11.6.1. The lengths of the duct sections and the air flow rates are indicated. (i) Size the ducts using the equal friction method. Assume a constant pressure loss rate of 2.5 $Pa.m^{-1}$. The loss coefficient for the diffusers is 0.2. Estimate the loss coefficients for the fittings from the data given in Table 11.1. (ii) Calculate the total pressure loss from 1 to 7, 1 to 10 and 1 to 13.
[*Answers*: $\Delta P_{1\text{-}7} = 338$ Pa, $\Delta P_{1\text{-}10} = 503$ Pa, $\Delta P_{1\text{-}13} = 555.8$ Pa]

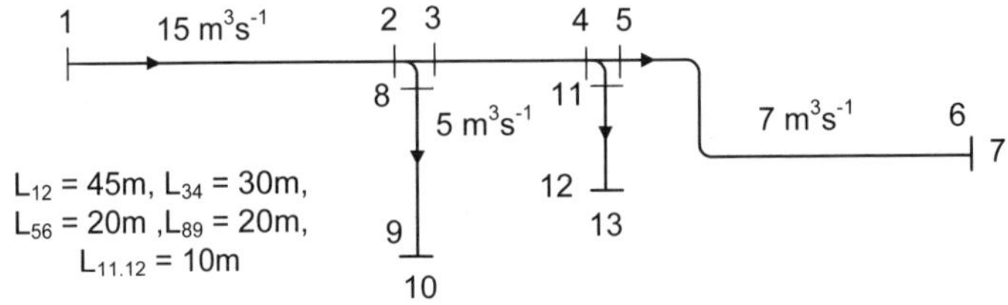

Fig. P11.6.1 Supply air duct network

P11.7 A round duct network is shown schematically in Fig. P11.7.1. The lengths of the duct sections and the air flow rates are indicated. (i) Size the ducts using the equal friction method. Assume a constant pressure loss rate of 3 $Pa.m^{-1}$. The loss coefficient for the diffusers is 0.25. Estimate the loss coefficients for the fittings from the data given in Table 11.1. (ii) Calculate the total pressure loss from 1 to 4a and 1 to 6a.
[*Answers*: $\Delta P_{1\text{-}4a} = 335.8$ Pa, $\Delta P_{1\text{-}6a} = 170.6$ Pa]

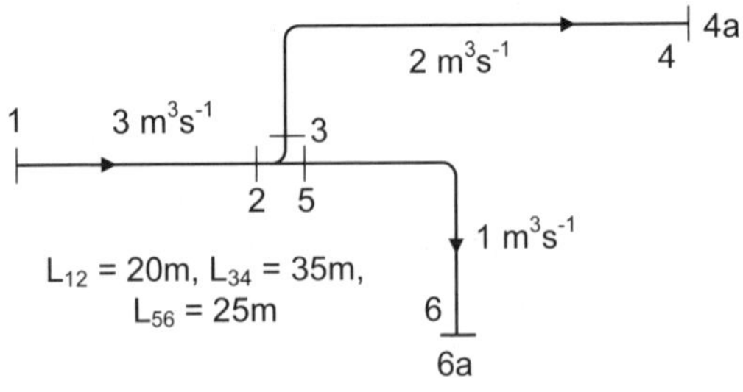

Fig. P11.7.1 Supply air duct network

P11.8 (i) Size the duct sections 1-2, 3-4 and 5-6 in Fig. P11.6.1 using the static regain method. Estimate the loss coefficients from the data in Table 11.1. The loss coefficient for the diffuser is 0.2. Calculate the total pressure loss from 1 to 7.
[*Answers*: $\Delta P_{1\text{-}7} = 245.6$ Pa]

P11.9 (i) Size the duct section 1-6a, in Fig. P11.7.1 using the static regain method. Estimate the loss coefficients from the data in Table 11.1. The loss coefficient for the diffuser is 0.25. Calculate the total pressure loss from 1 to 6a.
[*Answers*: $\Delta P_{1\text{-}6a} = 138$ Pa]

P11.10 The cooling load of an office room, maintained at 22°C and 50% relative humidity, is 10.2 kW. The length and breadth of the room are 16 m and 8 m respectively. The conditioned air supplied to the room is at 11°C and 95% relative humidity. (i) Calculate the supply air flow rate. (ii) Determine the required parameters of the ceiling slot diffusers to be used.
[*Answers*: (i) 0.7 m^3s^{-1}, (ii) $X_{0.5} = 1.2$ m]

References

1. *ASHRAE Handbook* - 2013 *Fundamentals*, American Society of Heating, Refrigeration and Air Conditioning Engineers, Atlanta, 2013.
2. *ASHRAE Handbook* - 2009 *Fundamentals*, American Society of Heating, Refrigeration and Air Conditioning Engineers, Atlanta, 2009.
3. Kuehn, Thomas H., Ramsey, James W. and Threlkeld, James L., *Thermal Environmental Engineering*, 3rd edition, Prentice-Hall, Inc., New Jersey, 1998.
4. Mitchell, John W. and Braun, James E., *Heating, Ventilation, and Air Conditioning in Buildings,* John Wiley and Sons, Inc., New York, 2013.

5. Stoecker, Wilbert F. and Jones, Jerold W., *Refrigeration and Air Conditioning*, International Edition, McGraw-Hill Book Company, London, 1982.

Appendix A11.1 - MATLAB Code for Pressure Loss in Circular Ducts

```
% friction pressure loss in a circular duct with fittings and coil
pi=3.141592;
pconst=0                    % constant pressure drop in coils etc.
kfit = 0                    % constant fittings K -value
dcl=1                       % duct length, m
dh=0.517                    % diameter of duct, m
qa = 1.25                   % air flow rate, cu.m per s
den= 1.2041                 % density kg per cu.m at 20°C
vis=18.178*10^(-6)                 % viscosity, Pa.s at 20°C
roug=0.00015                       % roughness
ren=4*den*qa/(pi*vis*dh);          % Reynolds number
f=64/ren;                          % initial value for iterative solution
x=dh/roug;
for i=1:1:1000       % iterative solution of Colebrook's equation
y=1/f^0.5;
lhs=0.5*(y-1.14);
fu1=1+(9.3*x*y/ren);
rhs=log10(x/fu1);
com=[f,lhs,rhs];   %  for checking the progress of solution
if rhs >= lhs
   fr=f;
   break
end
f=f+0.0001;
end
pred=8*fr*den*dcl*(qa^2)/((pi^2)*(dh^5)); % frictional loss, Pa
prefit=8*kfit*den*(qa^2)/((pi^2)*(dh^4));  %  fittings loss, Pa
ploss=pred+prefit +pconst;      % total loss in duct section, Pa
```

```
disp ('length, diameter, flow rate, fric. factor, p-loss friction, p-loss
fittings, p-loss coil,p-loss total')
codat=[dcl,dh,qa,fr, pred, prefit,pconst, ploss]
end
```

Appendix A11.2 - MATLAB Code for Equal Friction Design Method

```
% equal friction method for duct design
% pressure loss per meter of duct is specified, upl
% numerical data from worked example 11.11
pi=3.141592;
nds=6  % number of duct sections
flo=[1.45,1.45,1.9,2.35,0.45,0.45] % air flow rates in ducts, cu.m/s
lng=[9.2,4.6,4.6,9.5,9.2,9.2]        % lengths of duct sections, m
upl=0.9;              % assumed pressure loss per unit length, Pa/m
den= 1.2041;          % density kg per cu.m
vis=18.178*10^(-6);  %Pa.s
roug=0.00015;          % roughness
disp=('length, flow rate, diameter, velocity, p-velocity, p-loss friction,
assumed p-loss, unit p-loss, rhs, lhs')
for k=1:1:nds         % start loop for different duct sections
pfric=upl*lng(k);    % total friction pressure loss, Pa
leng=lng(k);          % length, m
qa=flo(k);            % air flow rate in duct section
dh=0.4;               % initial guessed value of diameter of duct, m
ren=4*den*qa/(pi*vis*dh);  % initial Raynolds number
f=64/ren;                   % initial value for iterative solution
x=dh/roug;
for i=1:1:4000        % iterative solution of Colebrook's equation
dh1=(8*f*den*(qa^2))/((pi^2)*upl);
dh=dh1^0.2 ;   % improved value of duct diameter, m
ren=4*den*qa/(pi*vis*dh);   % Raynolds number
x=dh/roug;
y=1/f^0.5;
lhs=0.5*(y-1.14);
fu1=1+(9.3*x*y/ren);
```

```
rhs=log10(x/fu1);
ai=i;
com=[ai,f,lhs,rhs];
if rhs >= lhs
   fr=f;
   break
end
f=f+0.00005;
end
vel= 4*qa/(pi*dh^2);        % air velocity in duct section, m/s
vpre=0.5*den*vel^2 ;        % velocity pressure in duct section, Pa
plos= fr*vpre*leng/dh;      % friction pressure loss
fric=f    ;                 % final value of the friction factor
codat=[leng, qa,dh,vel,vpre,plos,pfric,upl,rhs,lhs]
end
```

Appendix A11.3 - MATLAB Code for Static Regain Design Method

```
% static regain method
% numerical data from worked example 11.12
disp('static regain method of duct design')
pi=3.141592;
% loss coefficient for 90-degree bend
y-[0.075,0.1,0.125,0.15,0.18,0.2,0.23,0.25] % duct diameter, m
sd=[0.3,0.21,0.16,0.14,0.12,0.11,0.11,0.11] % loss coefficient
% loss coefficient for straight-section of Tee-junction
x=[0.1,0.2,0.3,0.4,0.5,0.6,0.7,0.8,0.9,1.0]   % ratio of areas
% loss coefficient , ks; 4-9 denote flow rate ratio times ten
sc4=[0.74,0.16,0.14,0.13,0.14,0.14,0.18,0.2,0.3,0.38]
sc5=[0.74,0.28,0.15,0.14,0.13,0.14,0.16,0.15,0.19,0.2]
sc6=[0.7,0.57,0.16,0.15,0.14,0.13,0.14,0.14,0.16,0.17]
sc7=[0.65,0.69,0.2,0.15,0.14,0.14,0.13,0.13,0.15,0.12]
sc8=[0.6,0.74,0.42,0.16,0.15,0.14,0.15,0.13,0.14,0.13]
sc9=[0.56,0.75,0.57,0.34,0.15,0.15,0.14,0.14,0.13,0.14]
% basic design data for duct sections
le2=12    % length of duct 2-3
```

```
q1=40     % flow rate at 2
q2=24      % flow rate at 3
qr=q2/q1;  % flow rate ratio
ve1=18     % flow velocity at 2 , if specified
d1=(4*q1/(pi*ve1))^0.5  % diameter of duct 1-2 , if velocity is specified
%d1=1.682     % diameter of duct 1-2, if obtained from friction chart
d2=1.2           % initial guessed diameter of section 3-4
%kben=interp1(y,sd,d2)  % loss coefficient for bend if d2 is inside the
%range
kben=0.11            % loss coefficient for bend if d2 is outside the range
 ar=(d2^2)/(d1^2)   % area ratio with the initial guess of d2
%ar=0.86             % area ratio at the end of the solution
%ks=interp1(x,sc6,ar)  % interpolated loss coefficient, ks; for the given
%flow-rate ratio
ks=0.14             % loss coefficient obtained manually from tables
den= 1.2041         % density kg per cu.m
vis=18.178*10^(-6)  %Pa.s
roug=0.00015         % roughness
qa=q2
dh=d2
for j=1:1:2000
   aj=j;
  d2=d2+0.0125/64;    % increase duct diameter in small steps
   dh=d2;
ren=4*den*qa/(pi*vis*dh);  % Raynolds number
f=64/ren;               % initial value for iterative solution
x=dh/roug;
% iterative solution of Colebrook's equation
for i=1:1:1000
y=1/f^0.5;
lhs=0.5*(y-1.14);
fu1=1+(9.3*x*y/ren);
rhs=log10(x/fu1);
com=[f,lhs,rhs] ;
if rhs >= lhs
   fr=f;
```

```
   break
end
f=f+0.0001;
end
lhss= (1-ks)/((d1^4)*(qr^2));    % LHS of Eq. (E11.12.6)
rhss=fr*le2/(d2^5)+(1+kben)/(d2^4); % RHS of Eq. (E11.12.6)
if rhss<=lhss
   break
end
eqs=[ks,d1,d2, lhss,rhss]      % check output during computation
if d2>=d1
   break
end
end
ar=(d2^2)/(d1^2);                % area ratio of ducts
disp('number of iterations, flow rates, diameters, area ratio, k-loss, lhs,
rhs')
eqsf=[aj,q1,q2,d1,d2,ar, ks, lhss,rhss]  % output at the end of iterations
% compute pressure parameters
ve1=4*q1/(pi*d1^2) ;             % velocity at section 1
ve2=4*q2/(pi*d2^2) ;             % velocity section 2
vp1=0.5*den*ve1^2  ;             % velocity pressure at section 1
vp2=0.5*den*ve2^2  ;             % velocity pressure at section 2
dp12=ks*vp1   ;                  % total pressure loss across junction
dpfr23=(fr*le2/d2)*vp2   ;   % friction pressure loss in section 2-3
dpft23=kben*vp2   ;     % pressure loss due to any fittings in section 2-3
dsp23=vp2+dp12+dpft23+dpfr23-vp1 ; % static pressure rise from 2 to 3
disp('V1,V2,Pv1,Pv2,P-loss [1-2],P-fric. [2-3],P-fit. [2-3],P-static [1-3]')
predat=[ve1,ve2,vp1,vp2,dp12,dpfr23,dpft23,dsp23
```

Chapter 12

Water Distribution Systems

12.1 Introduction

Water is used widely as a medium to transfer heat between different equipment in heating and air conditioning systems. For example, in the central air conditioning system depicted in Fig. 3.6, a pump circulates chilled water between the evaporator of the refrigeration plant and the air handling units (AHU). A second cold water circuit transfers heat from the condenser of the refrigeration plant to a cooling tower where heat is rejected to the atmosphere. Often, commercial buildings are heated during winter by pumping hot water from a central boiler to terminal units located in various zones.

Water is an excellent fluid for energy transport because it is nontoxic, has good heat transfer properties, is readily available, and inexpensive. The temperature range of operation of water at ambient pressure is from 0°C to 100°C. However, this range can be extended by adding components such as ethylene or propylene glycol.

Moreover, water distribution systems have several practical advantages over the air distribution systems described in chapter 11. These are: (i) the size of the heat source is smaller, (ii) water pipes require less space than air ducts, and (iii) for heating applications, higher temperatures are more practical with water than with air.

The relative merits of air and water as conveying media has resulted in the following choices. Air distribution is commonly used in residential and small commercial applications where the distances from the heat source to the conditioned spaces are usually short. On the other hand, for large central air conditioning systems, it is more common to use hot water and chilled water as distribution media.

In this chapter we shall discuss the design methods and some applications of water distribution systems, which are also called *hydronic* systems [1]. The physical principles introduced in chapter 11 for the design of air distribution systems can be applied to water distribution systems with some minor changes in the details.

12.2 Energy Equation for Hydronic Systems

In section 11.2 we discussed several alternative representations of the basic energy conservation equation for fluids flowing through ducts. For water distribution systems, it is more common to use the form given in Eq. (11.2), based on static pressure head, velocity pressure head, and elevation head. Total pressure head is defined as

$$H = \left(\frac{E_m}{g}\right) = \frac{p}{\rho g} + \frac{V^2}{2g} + z \tag{12.1}$$

To illustrate the application of Eq. (12.1), we shall consider two types of water distribution systems, called *open-loop* systems and *closed-loop* systems, commonly used in heating and air conditioning applications. These are shown schematically in Figs. 12.1(a) and (b) respectively.

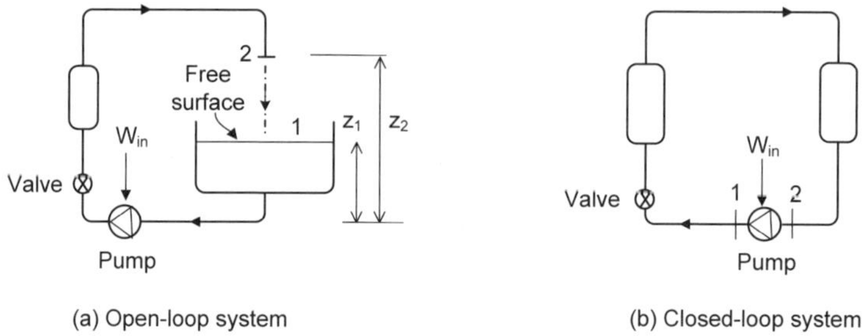

Fig. 12.1 Water distribution systems: (a) open-loop system, (b) closed-loop system

Applying Eq. (11.3) between the two points 1 and 2 of the open-loop system in Fig. 12.1(a) we have

$$\frac{W_{in}}{g} + \left(\frac{p_1}{\rho g} + \frac{V_1^2}{2g} + z_1\right) = \left(\frac{p_2}{\rho g} + \frac{V_2^2}{2g} + z_2\right) + \frac{W_{loss}}{g} \tag{12.2}$$

where W_{in} (Jkg^{-1}) is the work input by the pump *per unit mass* of fluid. The mechanical energy loss *per unit mass*, W_{loss}, includes frictional losses in the pipe and dynamic losses in the fittings.

We note that in the open-loop system in Fig. 12.1(a) one part of the loop is in direct contact with air, as in the cooling tower water loop of the air conditioning system depicted in Fig. 3.8. In contrast, in the closed-loop system shown in Fig. 12.1(b), the water returns to the suction side 2 of the pump without leaving the piping system. The chilled water loop in Fig. 3.8 is an example of a closed-loop system.

Applying Eq. (11.3) between the two points 1 and 2 of the closed-loop system in Fig. 12.1(b) we have

$$\frac{W_{in}}{g} = \left(\frac{p_1}{\rho g} + \frac{V_1^2}{2g}\right) - \left(\frac{p_2}{\rho g} + \frac{V_2^2}{2g}\right) = \frac{W_{loss}}{g} \tag{12.3}$$

Note that in closed-loop systems, the change in elevation, (z_2-z_1), between the inlet and discharge of the pump is usually negligible.

12.3 Head Losses in Hydronic Systems

The total head loss in a hydronic system includes: (i) friction losses in straight pipes, and (ii) dynamic losses through bends, branches like tee-joints, valves, entry fittings, and items like heat exchangers. In this section we shall outline the methods to determine the above head losses.

12.3.1 *Friction head loss in pipes*

The frictional pressure loss in straight pipes of hydronic systems is given by the Darcy–Weisbach equation [1], discussed earlier in section 11.3.1. Expressing the friction pressure loss in terms of a *frictional head loss* we obtain the equation

$$\Delta h_f = \frac{\Delta P_f}{\rho g} = f\left(\frac{L}{D}\right)\left(\frac{V^2}{2g}\right) \tag{12.4}$$

The various terms of the RHS of Eq. (12.4) were defined in section 11.3.1. The velocity V, and flow rate Q, are related by the equation

$$V = \frac{4Q}{\pi D^2} \tag{12.5}$$

The friction head loss chart for round pipes, shown in Fig. 12.2, was developed using Eqs. (12.4) and (12.5), and the analytical procedure is similar to that described in section 11.3.1 for air ducts. The friction head loss chart offers a quick and direct method to determine the unit pressure loss through pipes.

Alternatively, the MATLAB code in Appendix A12.1, which solves Colebrook's equation (11.2) iteratively, may be used to compute the friction head loss in pipes. The code is developed for a length of pipe having fittings of the same diameter, such as elbows and valves. It includes a branch fitting, such as a tee-joint, at the end of the pipe length. The code can be applied to a complex network by dividing it into individual sections of the form described above.

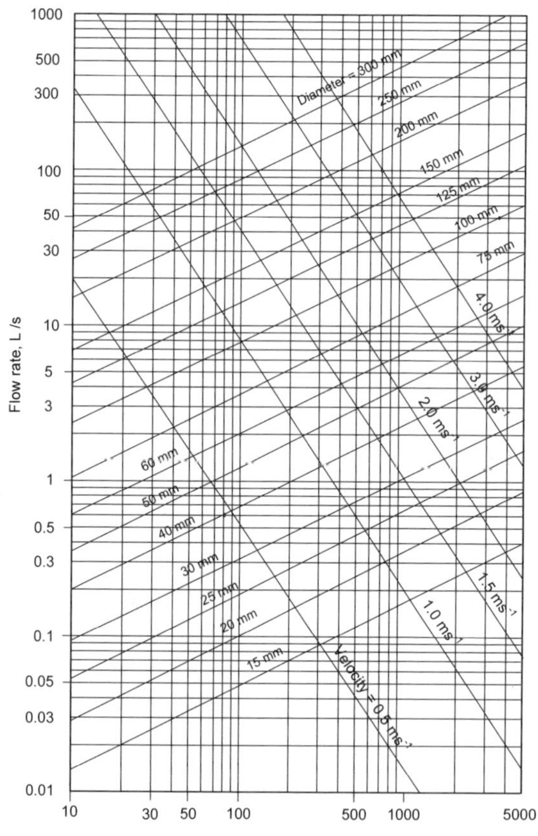

Fig. 12.2 Friction pressure loss chart for water flowing in Schedule 40 steel pipes. Water temperature = 20°C, Absolute roughness = 0.000046.

12.3.2 *Dynamic head loss in fittings*

Two interrelated parameters are used to determine the head loss through fittings. These are: (i) the loss coefficient, and (ii) the equivalent length.

The head loss is expressed in terms of the *loss coefficient*, *k*, using the equation

$$\Delta h_{1-2} = k\left(\frac{V^2}{2g}\right) \tag{12.6}$$

Representative values of *k* for some of the more common pipe fittings, obtained from the *ASHRAE Handbook* - 2013 *Fundamentals* [1], are listed Table 12.1.

The total head loss in a pipe section due to friction in the pipe and the dynamics losses in fittings may be expressed as

$$\Delta h_{loss} = f\left(\frac{L}{D}\right)\left(\frac{V^2}{2g}\right) + \Delta h_{1-2} = \left[f\left(\frac{L}{D}\right) + k\right]\left(\frac{V^2}{2g}\right) \tag{12.7}$$

Table 12.1 Loss coefficients for threaded pipe fittings*

D_{pipe}, mm	90°-Elbow	45°-Elbow	Tee-Straight	Tee-Branch	Globe Valve	Gate Valve	Bell-mouth Inlet
10	2.5	0.38	0.9	2.7	20	0.4	0.05
15	2.1	0.37	0.9	2.4	14	0.33	0.05
20	1.7	0.37	0.9	2.1	10	0.28	0.05
25	1.5	0.34	0.9	1.8	9	0.24	0.05
32	1.3	0.33	0.9	1.7	8.5	0.22	0.05
40	1.2	0.32	0.9	1.6	8	0.19	0.05
50	1.0	0.31	0.9	1.4	7	0.17	0.05
65	0.85	0.30	0.9	1.3	6.5	0.16	0.05
80	0.80	0.29	0.9	1.2	6	0.14	0.05
100	0.70	0.28	0.9	1.1	5.7	0.12	0.05

*Representative values extracted from, Page 22.2, Chapter 22, *ASHRAE Handbook* - 2013 *Fundamentals* [1].

The *equivalent length*, L_{eq} is defined as the length of a straight pipe, of the same nominal diameter as the fitting, and having the same frictional head loss. From Eqs. (12.4) and (12.6) it follows that

$$L_{eq} = \frac{kD}{f} \tag{12.8}$$

The equivalent length is commonly expressed in terms of the number of pipe diameters. The total head loss through a pipe section with fittings can now be found by adding the actual length of the pipe to the equivalent lengths of the fittings attached to it. The frictional head loss per unit length may be obtained directly from the friction chart in Fig. 12.2 or computed using Eq. (12.4).

12.4 Pump Characteristics

The distribution of water through pipe networks is carried out using pumps. Therefore the selection of the appropriate pump to achieve the desired flow rate and overcome the head losses in the pipe network is an important aspect of the design of water distribution systems. The most commonly used pumps are centrifugal pumps, similar in operation to the centrifugal fans discussed in section 11.4.

A schematic diagram of a centrifugal pump is shown in Fig. 12.3. The impeller with blades is driven in the clockwise direction by electric motor.

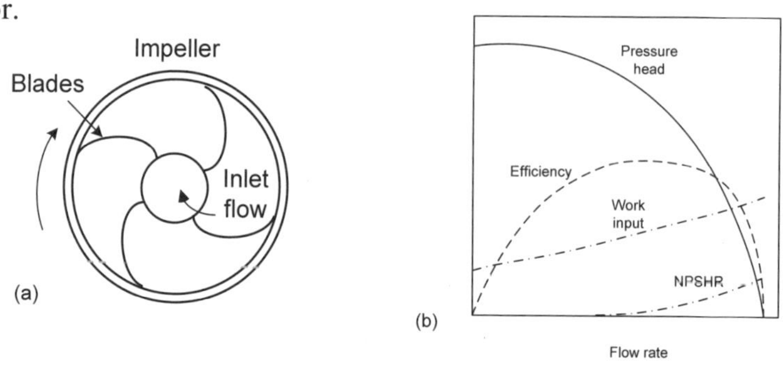

Fig. 12.3 (a) Schematic of centrifugal pump, (b) Performance curves for a typical centrifugal pump

The rotating blades impart a high radial velocity to the water entering the impeller at the center. The diffusing section at the periphery of the impeller converts the high velocity head of water to a static pressure rise.

Shown in Fig. 12.3(b) are typical variations of total pressure head, work input, and efficiency with flow rate, for a centrifugal pump running at a fixed speed. The head across the pump is a maximum when the flow rate is zero. As the flow rate increases the head decreases progressively until at the maximum flow rate the head becomes zero. The power input at zero flow rate is used to stir the fluid in the pump and is all converted into thermal internal energy of the fluid. The power input increases nearly linearly with flow rate.

The *ideal work input* to the fluid *per unit mass* of fluid by the pump follows from Eq. (12.3) as

$$W_{ideal} = \frac{\Delta P_t}{\rho} = g\Delta h_t \tag{12.9}$$

where ρ is the density of the fluid, and Δh_t is the total head.

From Eq. (12.9) we obtain the *ideal power input* to the fluid by the pump as

$$\dot{W}_{ideal} = \frac{\dot{m}\Delta P_t}{\rho} = \dot{Q}\Delta h_t \rho g \tag{12.10}$$

The actual work input *per unit mass*, W_{act} is larger than the ideal work because of frictional losses in the pump and energy losses in the motor. We define the efficiency of the pump as

$$\eta_t = \frac{W_{ideal}}{W_{act}} = \frac{g\Delta h_t}{W_{act}} = \frac{\dot{Q}\Delta h_t \rho g}{\dot{W}_{act}} \tag{12.11}$$

The efficiency of a pump is directly proportional to the *ideal power input*, which is the product of the fluid flow rate and the pressure rise (see Eq. 12.10). Therefore the efficiency is zero when the flow rate is zero. At the maximum flow rate, the pressure rise is zero as seen in Fig. 12.3(b) and hence the efficiency is again zero. The efficiency attains a maximum value between these two extremes of flow rate.

An alternative method to represent the characteristics of a centrifugal pump is to plot contours of constant efficiency as shown in Fig. 12.4. The ideal *power input* (W) at any point on the plot is given by the product of the pressure, $(\Delta h \rho g)$ (Pa) and the flow rate, Q (m^3s^{-1}). The actual power input can be determined by knowing the ideal power and the efficiency at any point on the chart. The head versus flow rate curves in Fig. 12.4 are plotted for two impeller sizes using the data extracted from similar curves in Ref. [2].

With liquid pumps, difficulties arise if the pressure inside the pump becomes very low. At low pressures liquids vaporize and pockets of vapor may be formed if the pressure falls below the saturation vapor pressure corresponding to the prevailing temperature. Vapor bubbles are carried along with the liquid until a region of high pressure is reached, where they suddenly collapse. If vapor bubbles are near to a solid surface when they collapse, the forces exerted by the liquid rushing into the cavities create very high local pressures causing serious erosion of the solid surfaces. This phenomenon is known as *cavitation*.

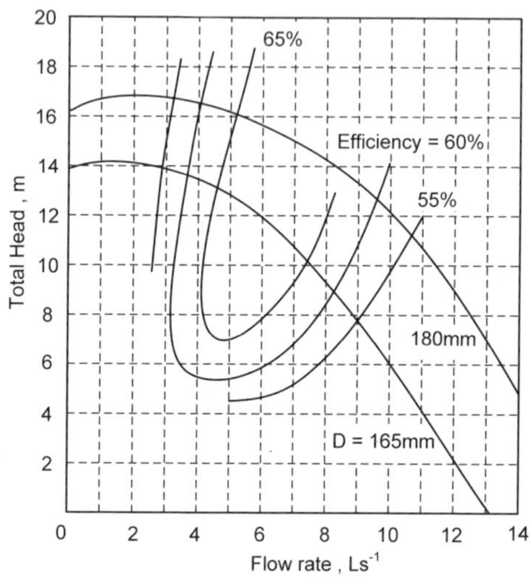

Fig. 12.4 Typical characteristics of a centrifugal pump

In any application the available suction head has to exceed a parameter called the *Net Positive Suction Head (NPSH)*, to avoid cavitation. This parameter is usually provided by pump manufacturers as shown in Fig. 12.3(b). The available NPSH depends on the design of the pumping system, in particular, on the location of the impeller entrance with respect to the free liquid surface, where the absolute pressure is known. This is given by

$$NPSH_{ava.} = h_s + \frac{{v_s}^2}{2g} - h_v \qquad (12.12)$$

where h_s and v_s are the head and the velocity at the impeller entrance and h_v is the vapor pressure of the liquid at the prevailing temperature.

Apply the energy equation per unit mass (Bernoulli's equation) between the pump inlet i, and the free liquid surface o, where the pressure is usually atmospheric, neglecting the velocity head at the free surface. Hence we have

$$h_o + z_o = h_s + \frac{v_s^2}{2g} + z_i + h_f \quad (12.13)$$

where z_o and z_i are the elevations at the free surface and the pump inlet. The total head loss between the free surface and the pump inlet is h_f.

From Eqs. (12.12) and (12.13) it follows that

$$NPSH_{ava.} = (h_o + z_o) - (z_i + h_f + h_v) \quad (12.14)$$

12.5 System–Pump Interaction and Flow Control

In section 11.5 we introduced a graphical procedure to determine the operating point of a fan–duct system using known characteristics of the fan and the duct network. The same method can be used to obtain the operating point or the balance point of a pump and a pipe network.

From Eqs. (12.5) and (12.7) it follows that total head loss due to friction in a pipe network and dynamic loss in the fittings may be expressed in the form

$$\Delta H_t = \left(\frac{fL}{D} + k\right)\left(\frac{V^2}{2g}\right) = \left(\frac{fL}{D} + k\right)\left(\frac{8Q^2}{g\pi^2 D^4}\right) = CQ^2 \quad (12.15)$$

where C is a function of the friction coefficient f of the pipe, loss coefficients k of the fittings, the diameter D and the length L of the pipe.

The operating point or balance point of a pump–piping system can be determined by plotting the relationship given by Eq. (12.15), and pump characteristic from Fig. 12.4, on the same graph. Such a plot is shown in Fig. 12.6 for three different speeds of the pump.

Ideally, the piping system curves are parabolas according to Eq. (12.15) and they apply to closed-loop systems of the type shown in Fig. 12.1(b). From Eq. (12.3) it follows that under steady conditions, the pressure head produced by the pump is equal to the total head loss in the pipe network. Hence the pressure and flow rate under steady conditions are given by the point of intersection, A of the pump curve corresponding to the speed and the piping system curve, as indicated in Fig. 12.5.

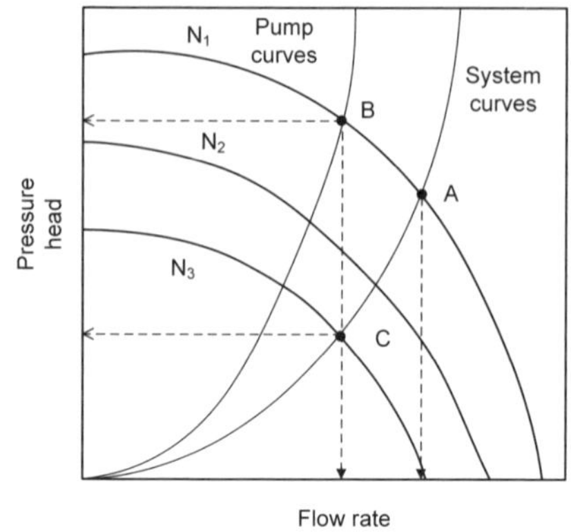

Fig. 12.5 Pump and piping network characteristics

The simplest method of flow control in a piping system is to use a valve as shown in Fig. 12.1(b). When the valve is partially closed to decrease the flow rate, the new system curve is steeper because the coefficient C in Eq. (12.15) is now larger due to the additional loss coefficient of the partially closed valve. Since the pump speed is maintained constant at N_1, the balance point moves to B, where the new system curve intersects the pump curve. Although the flow rate at B is less than at A, the pressure head at B is higher. Therefore the work input, given by the product of the pressure head and the flow rate, would only change marginally.

A more energy efficient method of flow control is to use a variable speed pump to achieve the desired decrease in the flow rate. In this case the system curve is unchanged. The new operating conditions are given by the point of intersection, C of the pump curve at speed N_3 and the original system curve. At C, both the flow rate and the pressure head are lower and therefore the work input to the pump is significantly reduced.

A third method of flow control, unique to water pumping systems, is to run a series of identical pumps in parallel as shown schematically in Fig. 12.6(a). Under design conditions all the pumps are operated to achieve the desired flow rate through the pipe network. When there is a need to decrease the flow rate through the system, pumps are switched off in stages to achieve the desired reduction in flow rate.

The pump and system curves for *two identical* pumps operating in parallel are depicted in Fig. 12.6(b).

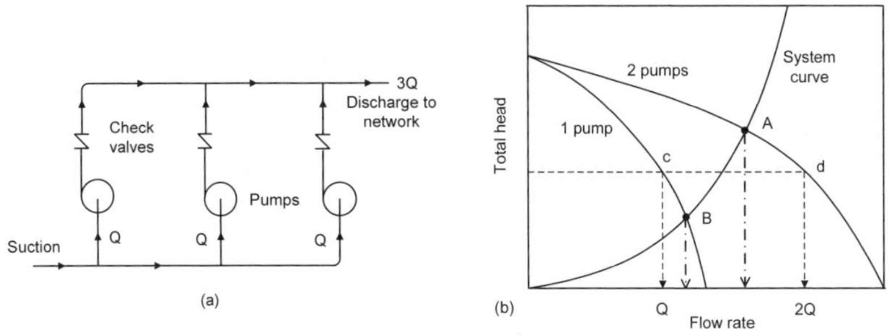

Fig. 12.6 (a) Flow control using pumps in parallel, (b) Pump–system curves

The equivalent pump curve for the two-pump system may be generated from the known curve for one of the pumps as demonstrated in Fig. 12.6(b).

Consider a point c on the given pump curve where the flow rate is Q. The second *identical* pump produces the same head at a flow rate of Q. Therefore the flow rate at the discharge to the pipe network is 2Q at the same head. Thus we locate point d on the two-pump curve, where the flow rate is 2Q and the head is the same. This procedure is repeated at other points to generate the entire two-pump curve.

Imagine a situation when both pumps are running and the operating conditions are given by A where the two-pump curve intersects the system curve for the pipe network. If one of the pumps is switched off to reduce the flow rate, the new operating conditions are given by point B, where the same system curve intersects the curve for a single pump. It is important to note that the flow rate at B is not half the flow rate at A because of the nonlinear form of the curves.

12.6 Design of Water Distribution Systems

A typical water distribution system consists of straight pipes, fittings like bends and tee-joints, valves, heat exchangers, an expansion tank, and pumps. The design of the system includes the arrangement of the above items and the determination of their sizes.

12.6.1 *Direct-return and reverse-return systems*

In section 12.2 we discussed two basic types of water distribution systems called open-loop systems and closed-systems, shown schematically in Fig. 12.1. Two arrangements of closed-loop systems, commonly known as (i) two-pipe direct-return systems, and (ii) two-pipe reverse-return systems are depicted in Figs. 12.7(a) and 12.7(b) respectively. These are called two-pipe systems because there are separate supply and return pipe loops.

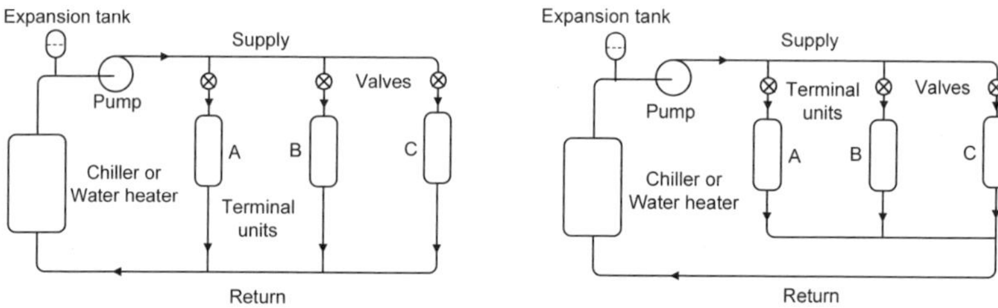

Fig. 12.7 Pipe networks: (a) direct-return system, (b) reverse -return system

In the direct-return pipe system shown in Fig. 12.7(a), the terminal units in the zones A, B and C are supplied with hot water from a boiler or chilled water from a chiller plant, depending on the application. The drawback of this system is that the available pressure to the three terminal units are different due to the different lengths of the supply and return pipes to and from the units. Therefore more water would tend to flow through A than through B or C. If the available pressures differ significantly, then it would be difficult to ensure that the desired flow rates are supplied to the three units A, B and C, even with the help of control valves.

In the reverse-return system, shown in Fig. 12.7(b), the total lengths of the piping from the boiler or the chiller to the three terminal units, including the supply and return pipe lengths, are approximately equal. Therefore the available pressures to the three units will not differ by much, and the flow rates to the units would be more balanced. The disadvantage of the reversed-return system is the additional pipe required compared to the direct-return system.

Open-loop systems of the type depicted in Fig. 12.1(a) have at least one section where the water is in contact with atmospheric air. Therefore the pressure at any other point in the pipe network can be determined by knowing pressure difference between the point and the air–water contact surface.

Closed-loop systems shown in Fig. 12.7 are normally operated with the entire system above atmospheric pressure to prevent air leakage into the system. Moreover, the total volume of closed-loop systems change with temperature due to the different coefficients of thermal expansion of water and the pipe material. An *expansion tank* is installed in closed-loop systems to accommodate the above volume changes of the system and to provide a reference pressure to prevent cavitation in the pump (see Fig. 12.7). The design procedure for expansion tanks is given Ref. [2] and Ref. [4].

12.6.2 *Design of pipe networks*

A number of guidelines for the design of water, steam and refrigerant pipe lines are available in the *ASHRAE Handbook* - 2013 *Fundamentals* [1]. The factors that determine the maximum and minimum velocities in water pipes are: (i) noise level, (ii) erosion level, (iii) installation cost, and (iv) operating cost.

If the pipes are too small, the resulting high velocities lead to unfavorable noise levels, erosion levels and pumping costs. If the pipes are too large, the installation costs are excessive. For pipes of nominal diameter 50 mm or less, the recommended velocity limit is 1.2 ms^{-1}. For pipe diameters greater than 50 mm, a unit pressure loss limit of 400 $Pa.m^{-1}$ is recommended [1]. Other guidelines, based on the type of application or service, or the annual hours of operation are included in Ref. [1].

The main steps in the design of pipe networks are listed below:

1. Determine the layout of the pipe network with all the necessary fittings.
2. Size the pipes in all sections of the system based on the anticipated maximum flow rate.

3. Determine the head loss through the different loops of the pipe network.
4. Obtain the head requirement of the pump based on the largest head loss from step 3 above.
5. Adjust the pipe diameters of the branches so that all the loops in step (3) have the same head loss as that used to select the pump in step (4).

12.7 Worked Examples

Example 12.1 Water at 70°C flows at the rate of 3.2 Ls^{-1} through a horizontal steel pipe of nominal diameter of 50 mm and length 45 m. The absolute roughness of commercial steel is 0.000046. The viscosity and density of water at 70°C are 0.406 mPa.s and 977.7 kgm^{-3}. Calculate (i) the total head loss through the pipe, (ii) the velocity head, and (iii) the change in static head.

Solution From the data in Ref. [4], the actual inner diameter of the pipe is 52.5 mm. We could use the friction loss chart in Fig. 12.2 to obtain the unit pressure loss directly by knowing the diameter (52.5 mm) and the flow rate (3.2Ls^{-1}).

However, this value would be approximate because the chart is developed for a water temperature of 20°C. Therefore we shall solve Eq. (12.3) using the given property data. The MATLAB code in Appendix A12.1 is used to solve Colebrook's equation to obtain the friction factor. Hence we have, $f = 0.0206$.

The water velocity is given by Eq. (12.5) as

$$V = \frac{4Q}{\pi D^2} = \frac{4\times 3.2\times 10^{-3}}{\pi\times(52.5\times 10^{-3})^2} = 1.478 \text{ ms}^{-1}$$

The friction head loss is given by Eq. (12.4) as

$$\Delta H_{fr} = \left(\frac{fL}{D}\right)\left(\frac{V^2}{2g}\right) = \frac{0.0206\times 45\times 1.478^2}{52.5\times 10^{-3}\times 2\times 9.81} = 1.97 \text{ m}$$

The *constant* velocity head in the pipe is

$$H_v = \left(\frac{V^2}{2g}\right) = \frac{1.478^2}{2\times 9.81} = 0.111 \text{ m}$$

From Eq. (12.3) the change in *total head* is equal to the *friction head loss*. However, the *velocity head* is constant.
Therefore the change in *static head* is

$$\Delta H_{st} = \Delta H_{fr} = 1.97 \text{ m}$$

Example 12.2 Water at 20°C is distributed through the pipe network shown in Fig. E12.2.1. The lengths, diameters, and flow rates of the different pipe sections are indicated in the figure. (i) Calculate the total pressure losses from 1-6, 1-8 and 1-10. (ii) If the pressures at the pump suction 0 and the exit sections 8, 6 and 10 are atmospheric, estimate the required delivery pressure of the pump at 1. (iii) If the efficiency of the pump is 75%, calculate the required power input to the pump.

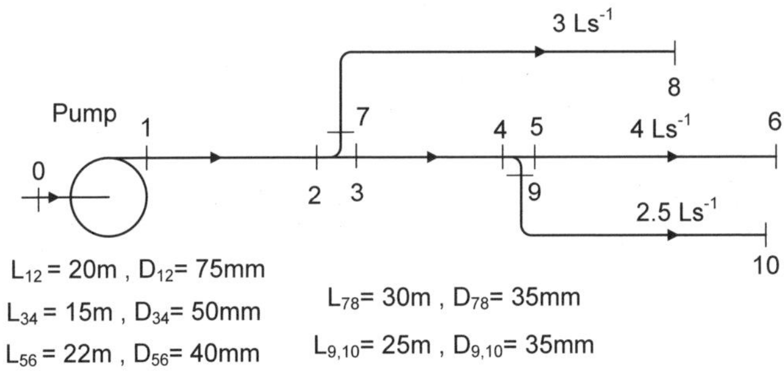

Fig. E12.2.1 Pipe network

Solution The velocity of water in a pipe section is given by Eq. (12.5) as

$$V = \frac{4Q}{\pi D^2} \text{ ms}^{-1} \tag{E12.2.1}$$

The friction head loss in the section is given by Eq. (12.4) as

$$\Delta H_{fr} = \left(\frac{fL}{D}\right)\left(\frac{V^2}{2g}\right) \text{ m} \tag{E12.2.2}$$

We use the MATLAB code in Appendix A12.1 to solve Colebrook's equation to obtain the friction factor, *f*. The water velocity is obtained from Eq. (E12.2.1) by knowing the flow rate and the diameter of the section. The friction loss is calculated by substituting in Eq. (E12.2.2). The numerical results obtained are summarized in Table E12.2.1.

Table E12.2.1 Friction head losses in pipe sections

Section	L, m	D, mm	Q, Ls^{-1}	V, ms^{-1}	H_{vel}, m	$f_{friction}$	$H_{frict.}$, m
1-2	20	75	9.5	2.15	0.236	0.0198	1.244
3-4	15	50	6.5	3.3	0.558	0.021	3.517
5-6	22	40	4	3.18	0.515	0.0223	6.336
7-8	30	35	3	3.12	0.495	0.0231	9.809
9-10	25	35	2.5	2.6	0.345	0.0234	5.755

The loss coefficients for the fittings are obtained from the data given in Table 12.1. Hence for the two tee-junctions we have

$$k_{2,3} = 0.9, \quad k_{2,7} = 1.23, \quad k_{4,5} = 0.9, \quad k_{4,9} = 1.4$$

For the 90° elbows:
In section 7-8, $D = 35$ mm; $k = 1.26$.
In section 9-10, $D = 35$ mm; $k = 1.26$.
The dynamic head loss in a fitting is given by Eq. (12.6) as

$$\Delta H_{fit} = k\left(\frac{V^2}{2g}\right) \tag{E12.2.3}$$

The velocity head in Eq. (E12.2.3) is that in the inlet pipe to the tee-junction.

The total head loss for section 1-6, 1-8 and 1-10 are obtained by adding the friction losses in the pipe sections and the dynamic losses in the fittings. Hence we have

$$H_{t1} - H_{t6} = \Delta H_{12} + \Delta H_{23} + \Delta H_{34} + \Delta H_{45} + \Delta H_{56}$$

$$H_{t1} - H_{t6} = 1.244 + 0.9 \times 0.236 + 3.517 + 0.9 \times 0.558 + 6.336$$

$$H_{t1} - H_{t6} = 11.8 \text{ m}$$

$$H_{t1} - H_{t8} = \Delta H_{12} + \Delta H_{27} + \Delta H_{elbow} + \Delta H_{78}$$

$$H_{t1} - H_{t8} = 1.244 + 1.23 \times 0.236 + 1.26 \times 0.495 + 9.809$$

$$H_{t1} - H_{t8} = 11.97 \text{ m}$$

$$H_{t1} - H_{t,10} = \Delta H_{12} + \Delta H_{23} + \Delta H_{34} + \Delta H_{49} + \Delta H_{elbow} + \Delta H_{9,10}$$

$$H_{t1} - H_{t,10} = 1.244 + 0.9 \times 0.236 + 3.517 + 1.4 \times 0.558 + 1.26 \times 0.345 + 5.775$$

$$H_{t1} - H_{t,10} = 11.96 \text{ m}$$

(ii) We notice that the total head loss through the three sections are nearly equal. Therefore we could select the pump to deliver 9.5 Ls^{-1} of water at a total head of about 12 m. The ideal work input under these operating conditions is given by

$$\dot{W}_{ideal} = (\Delta H \rho g)\dot{Q} = 12 \times 998 \times 9.81 \times 9.5 \times 10^{-3} = 1116 \text{ W}$$

Since the efficiency of the pump is 75%, the required power input is (1116/0.75) = 1488 W

Example 12.3 A 50 m long section of a pipe network, conveying water at the rate 35 Ls^{-1}, has four 90° elbows. The diameter of the pipe is 102 mm. The equivalent length of an elbow is given as 12 pipe diameters. Assuming that the friction factor is 0.019, calculate the total pressure loss through the pipe section.

Solution From the given data we obtain the length of the straight pipe with the same head loss as a 90° elbow as

$$L_{el} = 12D_p = 12 \times 102 = 1.224\text{m}$$

The total equivalent length of the pipe section is

$$L_{eq} = 4L_{el} + L_p = 4 \times 1.224 + 50 = 54.9 \text{ m}$$

The flow velocity is given by Eq. (12.5) as

$$V = \frac{4Q}{\pi D^2} = \frac{4 \times 35 \times 10^{-3}}{\pi \times 0.1224^2} = 2.97 \text{ ms}^{-1}$$

The head loss for the pipe section is given by Eq. (12.4) as

$$\Delta h_f = f\left(\frac{L}{D}\right)\left(\frac{V^2}{2g}\right) = 0.019 \times \left(\frac{54.9}{0.102}\right)\left(\frac{2.97^2}{2 \times 9.81}\right) = 4.6 \text{ m}$$

Example 12.4 (i) Obtain the system curve for the pipe section 1-2, of length 74 m and diameter 65 mm, shown in Fig. E12.4.1. The loss coefficient of the initially-open gate valve is 0.16. (ii) When the valve at the exit, 2 is partially closed its loss coefficient is 8.8. Obtain the new system curve.

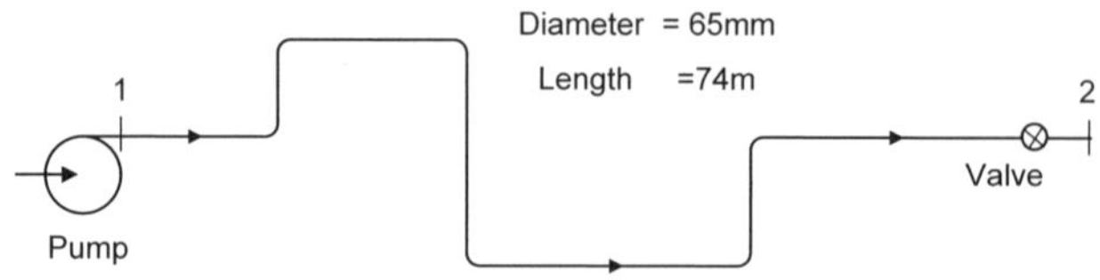

Fig. E12.4.1 Pipe network

Solution The total head loss from 1-2 consists of: (i) the friction loss in the duct sections 1-2, (ii) the head loss in the six 90°-elbows in section 1-2, and (iii) the head loss in the valve. Hence we have

$$H_{t1} - H_{t2} = \Delta H = \frac{8f_{12}L_{12}{Q_o}^2}{g\pi^2 {D_{12}}^5} + \frac{6\times 8K_{el}{Q_o}^2}{g\pi^2 {D_{12}}^4} + \frac{8k_v{Q_o}^2}{g\pi^2 {D_{12}}^4} \qquad \text{(E12.4.1)}$$

where the symbols have their usual meaning. We assume an average friction coefficient of 0.021. The loss coefficients for a 90° elbow and the open gate valve are obtained from Table 12.1 as, $k_{elb} = 0.85$ and $k_v = 0.16$.

Substituting the given numerical values in Eq. (E12.4.1) we obtain the system equation as:

$$\Delta H = \frac{8\times 0.021\times 74{Q_o}^2}{9.81\times\pi^2\times 0.065^5} + \frac{6\times 8\times 0.85{Q_o}^2}{9.81\times\pi^2\times 0.065^4} + \frac{8\times 0.16{Q_o}^2}{9.81\times\pi^2\times 0.065^4} \qquad \text{(E12.4.2)}$$

$$\Delta H = (23.9 + 5.1 + 0.16)\times 4628.8{Q_o}^2$$

$$\Delta H = 135\times 10^3{Q_o}^2 \qquad \text{(E12.4.3)}$$

where ΔH is in m and Q_o is in m^3s^{-1}.

When the valve is partially closed the loss coefficient is, $k_v = 8.8$. Substituting in Eq. (E12.4.2) we have

$$\Delta H = (23.9 + 5.1 + 8.8)\times 4628.8{Q_o}^2$$

$$\Delta H = 175\times 10^3{Q_o}^2 \qquad \text{(E12.4.4)}$$

The system curves given by Eqs. (E12.4.3) and (E12.4.4) are parabolas. However, this is because we assumed the friction factor to be independent of the flow rate, which is strictly not the case. We note that when the valve is partially closed the resulting increase of the loss coefficient makes the system curve a steeper parabola (see Fig. 12.5).

Example 12.5 Water at 20°C enters the pipe network depicted in Fig. E12.5.1. The diameters and lengths of the pipe sections are indicated in the figure. The total head at 1 is 15 m above atmospheric pressure. The pressures at the exits 4 and 6 are atmospheric. Calculate the water flow rates at 4 and 6.

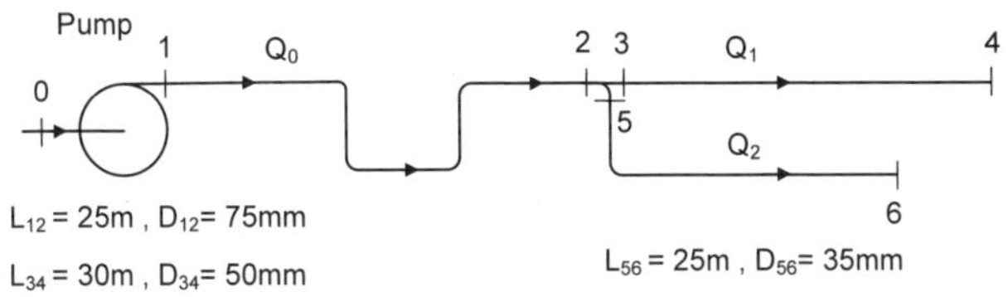

Fig. E12.5.1 Pipe network

Solution The total head loss from 1-4 consists of: (i) the friction losses in the duct sections 1-2 and 3-4, (ii) the loss in the *straight* section of the tee-junction 2-3, and (iii) the loss in the four 90°-elbows in section 1-2. Hence we have

$$H_{t1} - H_{t4} = \frac{8f_{12}L_{12}{Q_o}^2}{g\pi^2 {D_{12}}^5} + \frac{4\times 8k_{el1}{Q_o}^2}{g\pi^2 {D_{12}}^4} + \frac{8k_{s23}{Q_o}^2}{g\pi^2 {D_{12}}^4} + \frac{8f_{34}L_{34}{Q_1}^2}{g\pi^2 {D_{34}}^5} \quad \text{(E12.5.1)}$$

where the symbols have their usual meaning.

The total head loss from 1-6 consists of: (i) the friction losses in the duct sections 1-2 and 5-6, (ii) the loss in the *turning* section of the tee-junction 2-5, (iii) the loss in the four 90°-elbows in section 1-2 and (iv) the loss in the 90° elbow in section 5-6. Hence we have

$$H_{t1} - H_{t6} = \frac{8f_{12}L_{12}{Q_o}^2}{g\pi^2 {D_{12}}^5} + \frac{4\times 8k_{el1}{Q_o}^2}{g\pi^2 {D_{12}}^4} + \frac{8k_{t25}{Q_o}^2}{g\pi^2 {D_{12}}^4} + \frac{8f_{56}L_{56}{Q_2}^2}{g\pi^2 {D_{56}}^5} + \frac{8k_{el2}{Q_2}^2}{g\pi^2 {D_{56}}^4}$$

(E12.5.2)

Since the total head loss across the duct sections 1-4 and 1-6 are equal.

$$H_{t1} - H_{t4} = H_{t1} - H_{t6} = \Delta H = 15 \text{ m} \quad \text{(E12.5.3)}$$

Assuming that the density of water is constant, mass balance gives

$$Q_o = Q_1 + Q_2 \quad \text{(E12.5.4)}$$

Equation (E12.5.1) may be expressed in the compact form:

$$Q_1 = \left(\frac{\Delta H - a_1 Q_o{}^2}{b_1}\right)^{1/2} \tag{E12.5.5}$$

Equation (E12.5.2) may be expressed in the compact form:

$$Q_2 = \left(\frac{\Delta H - a_2 Q_o{}^2}{b_2}\right)^{1/2} \tag{E12.5.6}$$

The constants a_1, b_1, a_2, b_2 in Eqs. (E12.4.5) and (E12.5.6) involve the constant terms in Eqs. (E12.5.1) and (E12.5.2). Substituting from Eqs. (E12.5.5) and (E12.5.6) in Eq. (E12.5.4) we have

$$Q_o = \left(\frac{\Delta H - a_1 Q_o{}^2}{b_1}\right)^{1/2} + \left(\frac{\Delta H - a_2 Q_o{}^2}{b_2}\right)^{1/2} \tag{E12.5.7}$$

The total flow rate Q_o is obtained by solving Eq. (E12.4.7). A trial and error method has to be used because of the nonlinear form of the equation. An initial guessed value for Q_o is substituted in Eq. (E12.5.7). This value is adjusted iteratively until the LHS and the RHS of Eq. (E12.5.7) are equal. This gives the required flow rate.

We assume the following initial values for the friction coefficients: $f_{12} = 0.02$, $f_{34} = 0.02$, $f_{56} = 0.02$. The following loss coefficients for the fittings are obtained from Table 12.1: $k_{s23} = 0.9$, $k_{t25} = 1.23$, $k_{el1} = 0.82$, $k_{el2} = 1.26$.

The trial and error solution of Eq. (12.5.7) gives the following water flow rates:

$$Q_o = 11.92 \text{ Ls}^{-1}, \quad Q_1 = 8.35 \text{ Ls}^{-1}, \quad Q_2 = 3.57 \text{ Ls}^{-1}.$$

The computed flow rates are then used to obtain more accurate estimates of the friction coefficients using the MATLAB program in Appendix A12.1. Hence we have: $f_{12} = 0.0194$, $f_{34} = 0.0207$, $f_{56} = 0.0228$. These friction coefficients are then used to obtain better estimates of the water flow rates. Hence we have

$$Q_0 = 11.7 \text{ Ls}^{-1}, \quad Q_1 = 8.3 \text{ L}^3\text{s}^{-1}, \quad Q_2 = 3.4 \text{ m}^3\text{s}^{-1}$$

(Also, see worked example 11.3 for additional details)

Example 12.6 (a) Obtain the equation for the system curve for the two-branch pipe network shown in Fig. E12.5.1. (b) The pump has an impellor diameter of 180 mm and the performance characteristic shown in Fig. 12.4. Determine the flow rate delivered by the pump.

Solution We repeat the calculation outlined in worked example 12.5 for different values of the head, ΔH and hence obtain the data to develop the system curve, shown in Fig. E12.6.1, for the two-branch pipe network.

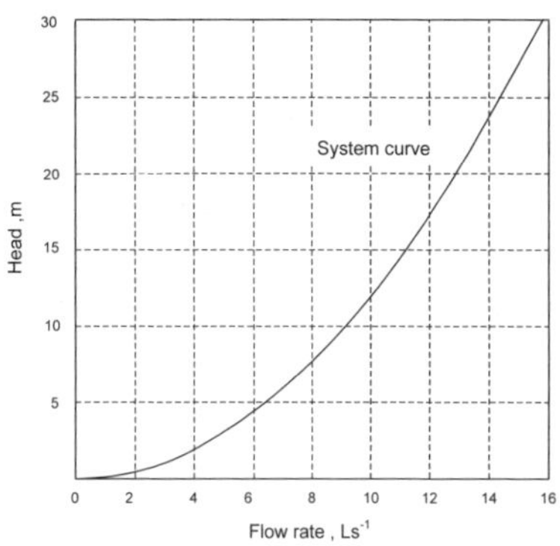

Fig. E12.6.1 System curve

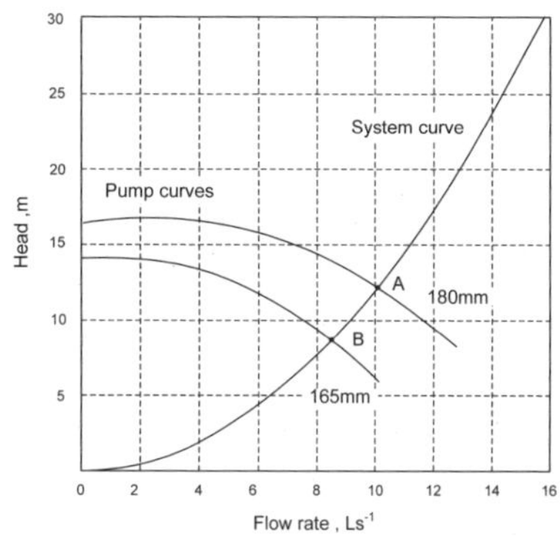

Fig. E12.6.2 Pump and system curves

(a) We recall that for the single-branch pipe network in Fig. E12.4.1, the parabolic equation (E12.4.4) of the system curve, was obtained analytically. It is clear from worked example 12.5 that such a direct approach is not possible for the two-branch pipe network depicted in Fig. E12.5.1. The pressure loss data for the two-branch system was obtained by solving the nonlinear equation, (E12.5.7), using an iterative method. Despite the numerical solution procedure used, the resulting data were found to be well-represented by the following parabolic equation:

$$\Delta H = 120.5 \times 10^3 {Q_o}^2$$

Here the total head, ΔH is in m and the total flow rate, Q_o is in m^3s^{-1}.

(b) We plot the pump characteristic and the system curve on the same graph as shown in Fig. E12.6.2. The point of intersection A of the two curves gives the operating point of the pump–piping system. Hence we obtain the flow rate as 10.05 Ls^{-1}.

Example 12.7 A variable speed centrifugal pump running at 1500 rpm delivers 6.5 Ls^{-1} of water at a head of 30.5 m. The power input is 3.75 kW. If the speed is increased to 2500 rpm, calculate the flow rate, the head, and the power input.

Solution The dimensionless numbers given by Eqs. (11.23)–(11.25) for dynamically similar fans are also applicable to centrifugal pumps. Let subscripts 1 and 2 denote quantities at the initial and final speeds. Hence we have

The flow rate, $Q_2 = Q_1\left(\frac{N_2}{N_1}\right) = \frac{6.5\times2500}{1500} = 10.83\ Ls^{-1}$

The head, $H_2 = H_1\left(\frac{N_2{}^2}{N_1{}^2}\right) = \frac{30.5\times2500^2}{1500^2} = 84.7\ m$

The power, $W_2 = W_1\left(\frac{N_2{}^3}{N_1{}^3}\right) = \frac{3.75\times2500^3}{1500^3} = 17.36\ kW$

Example 12.8 A water pumping system uses a pump with an impeller diameter of 165 mm, whose characteristics are shown in Fig. 12.4. (i) Calculate the power input to the pump when it delivers 9 Ls^{-1} of water. (ii) If the pump is replaced with one having an impeller diameter of 180 mm, calculate the power input when it delivers 9 Ls^{-1}.

Solution (i) From the pump characteristics (D = 165 mm) in Fig. 12.4, we obtain the total head for a flow rate of 9 Ls^{-1} as 7.8 m. The pump efficiency at this operating point is 55%. The pressure rise is given by

$$P_o = \Delta H\rho g = 7.8\times998\times\frac{9.81}{10^3} = 76.36\ \text{kPa} \qquad \text{(E12.8.1)}$$

The ideal power input is

$$\dot{W}_i = P_o\dot{Q} = 76.36\times\frac{9}{10^3} = 0.687\ \text{kW}$$

The actual power input is

$$\dot{W}_a = \frac{\dot{W}_i}{\eta_{pump}} = \frac{0.687}{0.55} = 1.25\ \text{kW}$$

(ii) If an impeller of diameter 180 mm is used, we obtain from Fig. 12.4, the pump head at a flow rate of 9 Ls^{-1} as 13.2 m. The corresponding pump efficiency is 62.5%. Following the above calculation procedure we obtain the new power input as 1.86 kW.

Example 12.9 The pump included in the pipe network shown in Fig. E12.4.1 has an impeller diameter of 180 mm and the performance curves of the pump for three different speeds are depicted in Fig. E12.9.1. If the pump is operated at 2000 rpm, calculate the total water flow rate, the head, and the ideal power input to the pump when (i) the valve is fully open, and (ii) when the valve is partially closed.

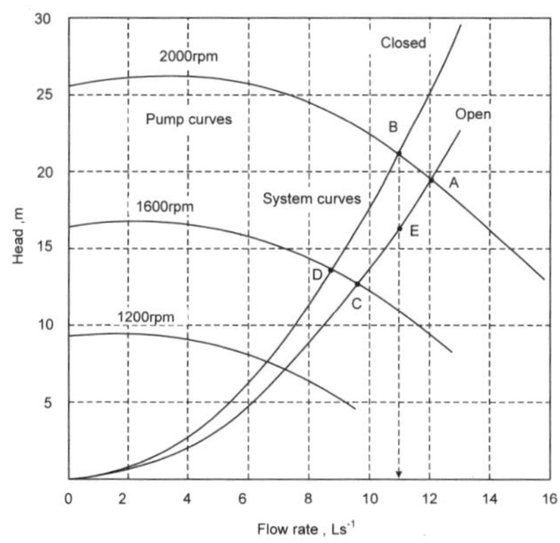

Fig. E12.9.1 Pump and system characteristics

Solution In worked example 12.4 we obtained the equation for the system curve when the valve is fully open as

$$\Delta H = 135 \times 10^3 Q_o{}^2 \tag{E12.9.1}$$

When the valve is partially closed the system equation is

$$\Delta H = 175 \times 10^3 Q_o{}^2 \tag{E12.9.2}$$

where Q_0 is the flow rate in m^3s^{-1}.

In Fig. E12.9.1 we have plotted the two system curves and the pump curves for the three speeds on same graph. The point of intersection A of the pump curve for 2000 rpm and the system curve when the valve is *open*, gives the balance point for the open position. We obtain the flow rate and the pump head at A as: 12.1 Ls^{-1} and 19.5 m. The ideal power input is given by

$$\dot{W}_A = (H\rho g)\dot{Q} = (19.5 \times 998 \times 9.81)\left(\frac{12.1}{10^3}\right) = 2.3 \text{ kW}$$

When the valve is partially closed while maintaining the pump speed at 2000 rpm, the new balance point B is the point of intersection of the pump curve at 2000 rpm and the new system curve. We obtain the flow rate and the pump head at B as: 11Ls^{-1} and 21.1 m. The ideal power input is given by

$$\dot{W}_B = (H\rho g)\dot{Q} = (21.1 \times 998 \times 9.81)\left(\frac{11}{10^3}\right) = 2.27 \text{ kW}$$

Note that the reduction in power input is relatively small.

Example 12.10 The performance curves of the pump included in the pipe network, shown in Fig. E12.4.1, are depicted in Fig. E12.9.1. The variation of head, H (m) with flow rate, Q (Ls^{-1}) at 2000 rpm is well represented by the quadratic equation:

$$H = 25.42 + 0.5288Q - 0.084Q^2 \qquad \text{(E12.10.1)}$$

(a) When the pump is operated at 1600 rpm, calculate the flow rate and power input (i) when the valve is open, and (ii) when the valve is partially closed.

(b) Calculate the required pump speed to reduce the flow rate from 12.1 Ls^{-1} to 11Ls^{-1}, keeping the valve open. What is the power input at this speed?

Solution (a) In Fig. E12.9.1 we have plotted the two system curves and the pump curves for the three speeds on same graph. The point of intersection C of the pump curve for 1600 rpm and the system curve when the valve is open, gives the balance point for the open position.

We obtain the flow rate and the pump head at C as: 9.3 Ls^{-1} and 12.7 m. The ideal power input is given by

$$\dot{W}_C = (H\rho g)\dot{Q} = (12.7 \times 998 \times 9.81)\left(\frac{9.3}{10^3}\right) = 1.16 \text{ kW}$$

When the valve is partially closed while maintaining the pump speed at 1600 rpm, the new balance point D is the point of intersection of the pump curve at 1600 rpm and the new system curve. We obtain the flow rate and the pump head at D as: 8.75Ls^{-1} and 13.7 m. The ideal power input is given by

$$\dot{W}_D = (H\rho g)\dot{Q} = (13.7 \times 998 \times 9.81)\left(\frac{8.75}{10^3}\right) = 1.17\text{ kW}$$

(b) The vertical line through B for a flow rate of 11Ls^{-1} intersects the original system curve at E, which is the balance point at the reduced pump speed. We obtain the flow rate and the pump head at E as: 11Ls^{-1} and 16.2 m. The ideal power input is given by

$$\dot{W}_E = (H\rho g)\dot{Q} = (16.2 \times 998 \times 9.81)\left(\frac{11}{10^3}\right) = 1.74\text{ kW}$$

In worked example 12.9 the flow reduction was achieved by partially closing the valve. The power input then was 2.27 kW. When the flow is reduced by decreasing the pump speed, the power input is 1.74 kW, which is a 23% reduction. We note that the use of variable speed pumps is a more energy efficient method of flow control.

Let the required pump speed be $2000f$, rpm where f is the fractional reduction in speed. Using the pump laws, we obtain the head and flow rate at a speed of 2000 rpm as

$$H = \frac{H_E}{f^2} = \frac{16.2}{f^2} \quad \text{and} \quad Q = \frac{Q_E}{f} = \frac{11}{f} \tag{E12.10.2}$$

Substituting from Eq. (E12.10.2) in Eq. (E12.10.1), the curve pump for 2000 rpm, we have

$$\frac{16.2}{f^2} = 25.42 + 0.5288 \times \frac{11}{f} - 0.084 \times \left(\frac{11}{f}\right)^2 \tag{E12.10.3}$$

Solution of the quadratic Eq. (E12.10.3) gives $f = 0.91$. Hence the required pump speed is given by, $2000f = 1820$ rpm.

Pump characteristics are usually presented graphically in manufacturer's catalogues. The analytical form given by Eq. (12.10.1) was obtained by curve fitting such graphical data. If an equation is not available, then the graphical approach described in worked example 11.8 may be used to determine the required pump speed.

Example 12.11 Two *identical* pumps , arranged in parallel, circulate water through a pipe network, as shown in Fig. E12.11.1. The characteristics of one of the pumps is depicted in Fig. E12.11.2. The system curve for the pipe network is given by

$$\Delta H = 58.6 \times 10^3 {Q_o}^2 \qquad \text{(E12.11.1)}$$

(a) Obtain the composite curve for the two pumps operating in parallel. Hence determine the flow rate through the pipe network when both pumps are running. What is the total ideal power input?

(b) If one of the pumps is switched off calculate the flow rate through the network. What is the ideal power input to the pump?

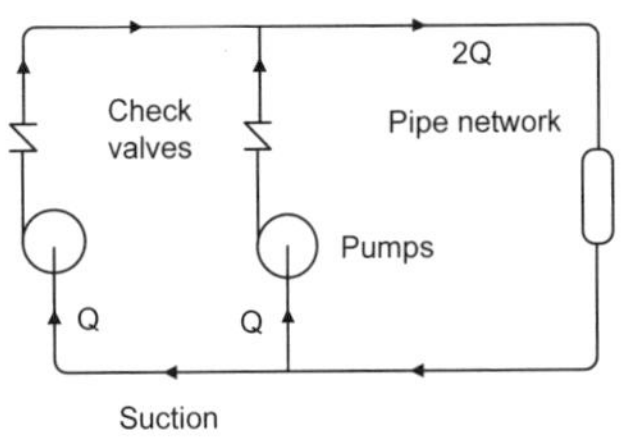

Fig. E12.11.1 Parallel pumps

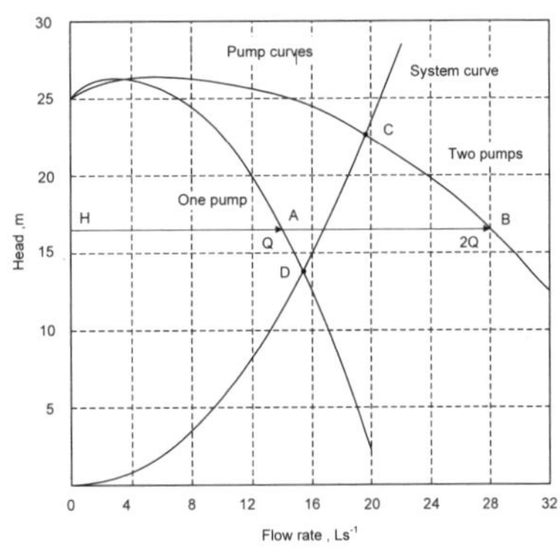

Fig. E12.11.2 Pump and system curves

Solution We use the given data for a single pump to construct the performance curve for two identical pumps operating in parallel, as shown in Fig. E12.11.1. When two pumps operate in parallel, the flow rate through the network for a given pump head is doubled.

Therefore a point A (Q,H) on the single-pump curve corresponds to the point B (2Q,H) on the composite curve as indicated in Fig. 12.11.2. By applying this procedure to a series of points on the single-pump curve we obtain the points needed to plot the two-pump curve.

On the same graph we have plotted the system curve, given by Eq. (E12.11.1), which is a parabola. The point of intersection C of the system curve and the two-pump curve gives the balance point when both pumps are running.

(a) We obtain the flow rate and the pump head at C as: 19.7Ls^{-1} and 22.7 m respectively. The ideal power input is given by

$$\dot{W}_C = (H\rho g)\dot{Q} = (22.7 \times 998 \times 9.81)\left(\frac{19.7}{10^3}\right) = 4.38 \text{ kW}$$

(b) When one of the pumps is switched off, the balance point shifts to D where the single-pump curve intersects the system curve. Note that the system curve is the same because no changes are made to the pipe network.

We obtain the flow rate and the pump head at D as: 15.5Ls^{-1} and 13.9 m respectively. The ideal power input is given by

$$\dot{W}_D = (H\rho g)\dot{Q} = (13.9 \times 998 \times 9.81)\left(\frac{15.5}{10^3}\right) = 2.11 \text{ kW}$$

It is important to note that although the two pumps are identical, switching one of them off results only in a 21% reduction in the total flow rate through the pipe network. This is due to the nonlinear forms of the pump and system curves.

Example 12.12 A centrifugal pump is used to pump water from a sump tank with its free surface located 4 m below the pump entrance. The total head loss between the free surface and the pump entrance, including friction losses and fitting losses, has been estimated as 2.1 m. The atmospheric pressure is 101 kPa. The water temperature is 40°C. (i) Calculate the available NPSH. (ii) How much further can the pump entrance be moved vertically before the available NPSH becomes just equal to the manufacturer-recommended NPSH of 2.2 m for the given conditions.

Solution (i) We shall express all pressures in meters. The atmospheric pressure is

$$h_o = \frac{101 \times 10^3}{998 \times 9.81} = 10.3 \text{ m}$$

The vapor pressure at 40°C, obtained from property tables in Ref. [2], is 7.38 kPa. Therefore

$$h_v = \frac{7.38 \times 10^3}{998 \times 9.81} = 0.75 \text{ m}$$

The available *NPSH* is given by Eq. (12.14) as

$$NPSH_{ava.} = (h_o + z_o) - (z_i + h_f + h_v)$$

Substituting numerical values we have

$$NPSH_{ava.} = (10.3 + 0) - (4 + 2.1 + 0.75) = 3.45 \text{ m}$$

(ii) We apply Eq. (12.14) with the available NPSH just equal to the manufacturer-recommended value of 2.2 m. Hence we have

$$NPSH_{ava.} = 2.2 = (10.3 + 0) - (z_i + 2.1 + 0.75)$$

From the above equation we obtain the allowable elevation of the pump entrance above the free surface as: $z_i = 5.25$ m. Therefore the pump entrance can be moved 1.25 m vertically from the present position.

Example 12.13 In the air conditioning system, shown schematically in Fig. E12.13.1, the evaporator supplies chilled water at 6°C to two cooling coils A and B with cooling capacities of 32 kW and 25 kW respectively. The rise in temperature of the chilled water in each coil is 3°C. The pipe lengths of the different sections of the network are indicated in the figure. Schedule-40 steel pipes with threaded fittings are used in the design. The pressure losses through coils A and B are 1.8 m and 1.5 m respectively. The pressure loss through the evaporator is 3.6 m. (i) Size the pipes for this system. (ii) Determine pump flow rate and the head required.

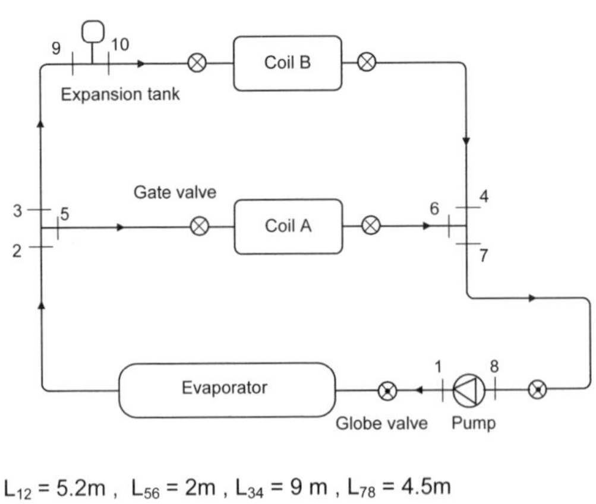

Fig. E12 13.1 Chilled water supply system

Solution We first calculate the chilled water flow rates through the two cooling coils A and B by applying the energy balance equation. Hence we have

$$V_A \rho c_w \Delta t_A = Q_A \tag{E12.13.1}$$

$$V_A \times 10^3 \times 4.2 \times 3 = 32$$

The volume flow rate through coil A is 2.54 Ls^{-1}. Similarly, the flow rate through coil B is obtained as 1.98 Ls^{-1}. Therefore the total flow rate through the evaporator is 4.52 Ls^{-1}.

Based on the above volume flow rates we now select the pipe diameters to satisfy the guidelines from Ref. [1], given in section 12.6.2. For this purpose we could either use the friction chart in Fig. 2.2 or the MATLAB code in Appendix A12.1.

For the flow rate of 4.52 Ls^{-1}, a pipe of diameter 65 mm gives a velocity of 1.36 ms^{-1} and a unit pressure loss of 306 $Pa.m^{-1}$. Since the diameter is larger than 50 mm, this pipe diameter satisfies the unit pressure loss limit of 400 $Pa.m^{-1}$ recommended in Ref. [1].

For the flow rate of 1.98 Ls^{-1}, we select a pipe of diameter 50 mm which gives a velocity of 1 ms^{-1} and a unit pressure loss of 242 $Pa.m^{-1}$. This diameter satisfies the limit on velocity as well as the limit on pressure loss.

For the flow rate of 2.54 Ls^{-1} we select a pipe of diameter 55 mm which gives a velocity 1.07 ms^{-1} and a pressure 240 $Pa.m^{-1}$. This diameter satisfies the limit on velocity as well as the limit on pressure loss.

Having selected the pipe diameters we now calculate the total pressure loss across each of the pipe sections through which the flow rate is constant. These are the sections 1-2, 7-8, 5-6, and 3-4 as seen in Fig. E12.12.1. The relevant loss coefficients for the fittings, obtained from Table 12.1, are listed below:

(i) For D = 65 mm, k-globe valve = 6.5, k-90° elbow = 0.85, k-tee-junction-straight = 0.9, k-tee-junction-branch = 1.3.
(ii) For D = 55 mm, k-gate valve = 0.17
(iii) For D = 50 mm, k-gate valve = 0.17, k-90° elbow = 1.0.

The head losses in the different pipe sections are given by:

$$\Delta H_{27(A)} = \Delta H_{25} + 2\Delta H_{gav} + \Delta H_{cA} + \Delta H_{f56} + \Delta H_{67}$$

$$\Delta H_{27(B)} = \Delta H_{23} + 2\Delta H_{gav} + 2\Delta H_{elb} + \Delta H_{cB} + \Delta H_{9,10} + \Delta H_{f34} + \Delta H_{47}$$

$$\Delta H_{12} = \Delta H_{evp} + \Delta H_{glv} + \Delta H_{elb} + \Delta H_{f12}$$

$$\Delta H_{78} = 3\Delta H_{elb} + \Delta H_{glv} + \Delta H_{f78}$$

The head developed by the pump is

$$\Delta H_{pump} = H_1 - H_8 = \Delta H_{18}$$

Under steady operating conditions

$$\Delta H_{pump} = \Delta H_{18} = \Delta H_{12} + \Delta H_{27} + \Delta H_{78} \qquad \text{(E12.13.2)}$$

The friction loss through a pipe may be obtained directly from friction chart in Fig. 12.2 and the loss through a fitting can be calculated using Eq. (12.6). The MATLAB code in Appendix A12.1 offers a convenient alternative to obtain the friction loss through the pipes as well as the losses through the fittings. The computed head losses are summarized in Table E12.13.1.

Table E12.13.1 Computed head losses

Component	Length/ number	Head loss, m
Pipes 1-2 and 7-8	9.7 m	0.3038
Evaporator	1	3.6
90°-Elbows	4	0.3215
Globe valves	2	1.2294
	($\Delta H_{7\text{-}8} + \Delta H_{1\text{-}2}$)	5.4547
Pipe 5-6	2 m	0.0489
Gate valves	2	0.0198
Cooling coil A	1	1.8
Tee-branches 2-5, 6-7	2	0.2459
	($\Delta H_{2\text{-}7}$ via 5-6)	2.1146
Pipe 3-4	9 m	0.2228
Gate valves	2	0.0176
Cooling coil B	1	1.5
90°-Elbows	2	0.1037
Tee-straights 2-3, 4-7, 9-10	3	0.2553
	($\Delta H_{2\text{-}7}$ via 3-4)	2.0994

We notice that the total head loss from 2 to 7 through the two paths 2-5-6-7 (coil A) and 2-3-4-7 (coil B) are nearly equal. Therefore the desired flow rates will be delivered to the two cooling coils. If these pressure losses were unequal, then the gate valves in the circuits could be used to provide the additional head loss to balance the flow. The total head developed by the pump is given by Eq. (E12.13.2). Therefore

$$\Delta H_{pump} = 5.455 + 2.115 = 7.57 \text{ m}$$

and the required flow rate is 4.52Ls^{-1}.

Example 12.14 A water-cooled condenser of a water chiller plant rejects 145 kW of heat to the atmosphere through a cooling tower as shown schematically in Fig. E12.14.1. The rise in temperature of the cooling water in the condenser is 6°C. The total length of piping from the pump exit 1 to the spray header 2 of the cooling tower is 14 m. The length of piping from the cooling tower sump 3 to the pump inlet 4 is 12 m.

The vertical distance between the spray header and the water surface in the sump is 1.2 m. The pipe inlet in the sump has a square inlet. The head loss through the condenser is 2.2 m. The head loss through the spray nozzles in the cooling tower is negligible. The pipe network uses schedule -40 steel pipes with threaded fittings. (i) Size the pipes for this system. (ii) Calculate the rise in total head required at the pump.

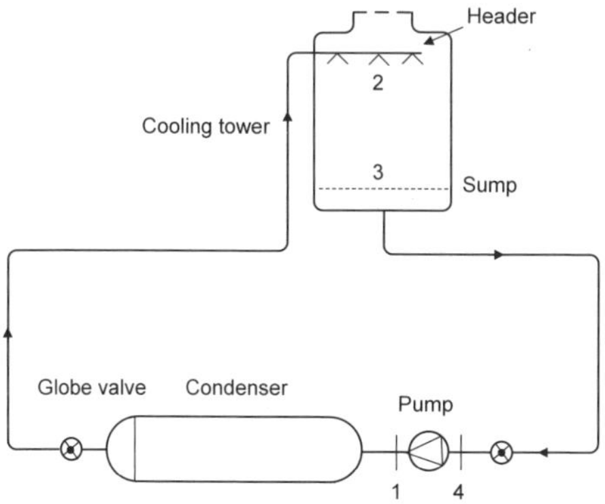

Fig. E12.14.1 Cooling water piping system

Solution Applying the energy balance equation to the condenser we have

$$\dot{m}_w \times 4.2 \times 6 = 145$$

The mass flow rate of water is 5.75 kgs^{-1}. Assuming the density to be 1000 kgm^{-3}, the volume flow rate of water is 5.75 Ls^{-1}.

Based on the above volume flow rate we now select the pipe diameters to satisfy the guidelines given in Ref. [1], listed in section 12.6.2. For this purpose we could either use the friction chart in Fig. 12.2 or the MATLAB code in Appendix A12.1.

For the flow rate of 5.75 Ls^{-1}, a pipe of diameter 65 mm gives a velocity of 1.75 ms^{-1} and a unit pressure loss of 481 $Pa.m^{-1}$. Since the diameter is larger than 50 mm, this pipe diameter is unacceptable because the unit pressure exceeds the limit of 400 $Pa.m^{-1}$. A pipe diameter of 70 mm gives a velocity of 1.49 ms^{-1} and a unit pressure loss of 333 $Pa.m^{-1}$, which is below the recommended pressure loss limit. Hence we choose a pipe of diameter of 70 mm for the entire network.

The relevant loss coefficients for the fittings, obtained from Table 12.1 are listed below: For D = 70 mm, k-globe valve = 6.3, k-90° elbow = 0.83, k-square inlet = 0.5 [1].

The loss through a fitting is given by Eq. (12.6), and the flow velocity is computed using Eq. (12.5). The computed head losses are summarized in Table E12.14.1.

Table E12.14.1 Computed head losses

Component	Length/ number	Head loss, m
Pipes 1-2	14 m	0.476
Condenser	1	2.2
90°-Elbows	4	0.378
Globe valves	1	0.717
	$(H_1 - H_2)$	3.77
Pipe 3-4	12 m	0.408
Globe valves	1	0.717
90°- Elbows	3	0.283
Square pipe entrance	1	0.057
	$(H_3 - H_4)$	1.465

The pressure in the cooling tower between 2 and 3 is atmospheric and (H_2-H_3) = 1.2 m. Now from the data listed in Table E12.14.1 we have

$$(H_1 - H_4) - (H_2 - H_3) = 3.77 + 1.465 = 5.235 \text{ m}$$

$$H_1 - H_4 - 1.2 = 5.235 \text{ m}$$

Therefore the required pump head, $(H_1\text{-}H_4) = 6.44$ m.

Example 12.15 Hot water is supplied to the three heating coils shown in Fig. E12.15.1 using a reverse-return piping network. The water flow rate through each coil is 0.13 Ls^{-1}. The three identical flow circuits from (2-3), (4-5) and (6-7) consist of 4 m of straight pipe, a gate valve, and a heating coil with a head loss of 0.85 m. The lengths of the different pipe sections are as follows: $L_{12} = 0.7$ m, $L_{24} = L_{35} = 3.2$ m, $L_{46} = L_{57} = 3.2$ m, $L_{89} = 7.4$ m. The pipe network uses schedule-40 steel pipes with threaded fittings. (i) Size the pipes for this system. (ii) Calculate the total head loss from 1 to 9.

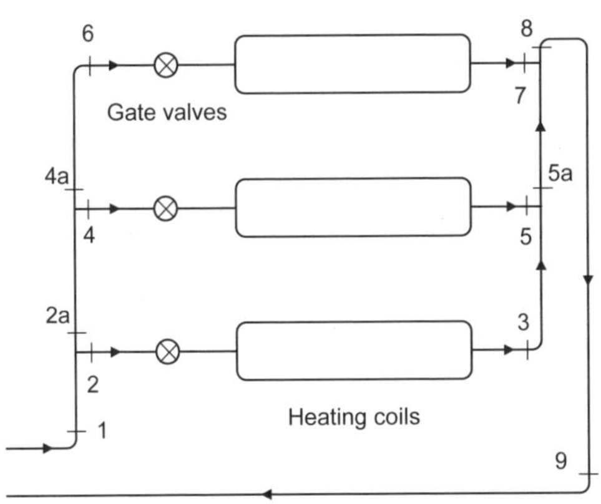

Fig. E12.15.1 Reverse-return pipe network of a heating system

Solution The water flow rates through the section 1-2, 2-4, and 4-6 are 0.39 Ls^{-1}, 0.26 Ls^{-1} and 0.13 Ls^{-1} respectively. Based on the above volume flow rates we now select the pipe diameters to satisfy the guidelines given in Ref. [1], and listed in section 12.6.2.

For this purpose we could either use the friction chart in Fig. 12.2 or the MATLAB code in Appendix A12.1. We shall select the pipe diameters of section 1-2, 2-4, and 4-6 as 25 mm, 20 mm and 15 mm respectively. The computed flow velocity, unit pressure loss, friction factor, and velocity head for the different sections are listed in Table E12.15.1.

Table E12.15 .1 Computed quantities

Section	D_{pipe} mm	Q Ls^{-1}	Velocity ms^{-1}	$\Delta P/\Delta L$ Pa. m^{-1}	Friction factor, f	Velocity Head, m
1-2	25	0.39	0.790	373	0.0297	0.0322
2-4	20	0.26	0.828	535	0.0313	0.0349
4-6	15	0.13	0.736	623	0.0347	0.0276
2-3	15	0.13	0.736	623	0.0347	0.0276
3-5	15	0.13	0.736	623	0.0347	0.0276
5-8	20	0.26	0.828	535	0.0313	0.0349
8-9	25	0.39	0.790	373	0.0297	0.0322
4-5	15	0.13	0.736	623	0.0347	0.0276
6-8	15	0.13	0.736	623	0.0347	0.0276

We see from the data in Table E12.15.1 that the pipe diameters selected are less than 50 mm. In all pipe sections the velocity is below $1.2ms^{-1}$, which is the limiting value recommended in Ref. [1]. However, in several sections the unit pressure losses are larger than 400 $Pa.m^{-1}$.

Now the head loss from 1 to 9 can be calculated using three different flow paths 1-2-3-8-9, 1-4-5-8-9 and 1-6-7-8-9. If the head loss computed along these paths are equal then the desired flow rates will be delivered to the three coils. Otherwise the gate valves may have to be used to balance the flows through the coils.

In order to compute the head loss through the three paths we first obtain the following loss coefficients from Table. 12.1.

(i) For D = 15 mm, k-gate valve = 0.33, k-90° elbow = 2.1, k-tee-junction-straight = 0.9, k-tee-junction-branch = 2.4.

(ii) For D = 20 mm, k-90° elbow = 1.7, k- tee-junction-straight = 0.9, k-tee-junction-branch = 2.1.

(iii) For D = 25 mm, k- 90° elbow = 1.5, k-tee-junction-straight = 0.9, k-tee-junction-branch = 1.8.

The head losses through the various fittings are calculated using Eq. (12.6), and the required velocity heads are given in Table E12.15.1. Each of the coil sections 2-3, 4-5 and 6-8 includes a gate valve, a heating coil and a pipe of length 12 m. The head loss for each path is obtained as 1.11 m. Therefore

$$\Delta H_{23} = \Delta H_{45} = \Delta H_{68} = 1.11 \text{ m}$$

The section 8-9 includes two elbows and a pipe length of 7.4 m. The head loss is obtained as 0.379 m. Therefore

$$\Delta H_{89} = 0.379 \text{ m}$$

Similarly, we obtain the following head losses (m) in the other sections:

$$\Delta H_{12} = 0.0846, \quad \Delta H_{1,2a} = 0.0557, \quad \Delta H_{2a,4} = 0.2049,$$

$$\Delta H_{2a,4a} = 0.1877, \quad \Delta H_{4a,6} = 0.2619, \quad \Delta H_{3,5a} = 0.2867$$

$$\Delta H_{5a,8} = 0.2063, \quad \Delta H_{5,5a} = 0.0579, \quad \Delta H_{78} = 0.0579$$

The total head loss (m) through flow path 1-2-3-8-9 (p1) is

$$\Delta H_{p1} = 0.0846 + 1.11 + 0.2867 + 0.2063 + 0.379 = 2.07$$

The total head loss (m) through flow path 1-4-5-8-9 (p2) is

$$\Delta H_{p2} = 0.0557 + 0.2049 + 1.11 + 0.0579 + 0.2063 + 0.379 = 2.01$$

The total head loss (m) through flow path 1-6-7-8-9 (p3) is

$$\Delta H_{p3} = 0.0557 + 0.1877 + 0.2619 + 1.11 + 0.0579 + 0.379 = 2.05$$

The maximum difference in the computed head losses through the three paths is about 3%. Hence we could expect the actual flow rates through the heating coils to be close to the desired the flow rates. Any minor adjustments required could be made by using the gate valves.

Problems

P12.1 A pump delivers 0.55 Ls^{-1} of water from a sump tank, with its free surface 1 m below the entrance of the pump, to a storage tank with its free surface 2 m above the pump. The total length of piping on the suction side of the pump is 2 m and the pipe has three 90°-elbows.

On the delivery side the pipe is 4 m long and there are four 90° elbows. The diameters of the suction and delivery pipes are 25 mm. The system uses schedule-40 steel pipes with threaded fittings. The

atmospheric pressure is 101 kPa. Calculate (i) the static head, the velocity head, and the total head at the entrance of the pump, (ii) the pressure rise at the pump, and (iii) the ideal work input to the pump.
[*Answers*: (i) 8.8 m, 0.064, 9.86, (ii) 40.4 kPa, (iii) 22.2 W]

P12.2 A pump circulates water through a closed pipe network connecting a water heater and a heating coil (heat exchanger) in a room. The total length of the pipe is 28 m and the diameter is 32 mm. There are six 90°-elbows and a globe valve in the network. The head loss through the water heater and the heating coil are 1.6 m and 1.2 m respectively. The water flow rate is 0.88 Ls^{-1}. The system uses schedule-40 steel pipes with threaded fittings. (i) Calculate the flow velocity and the unit pressure loss in the pipe. (ii) Calculate the increase in head across the pump. (iii) If the efficiency of the pump is 70%, calculate the power input to the pump.
[*Answers*: (i) 1.09 ms^{-1}, 493 $Pa.m^{-1}$, (ii) 5.2 m, (iii) 64 W]

P12.3 A pump delivers water to a pipe network of diameter 40 mm and length 25 m. The pipe section has six 90°-elbows and a gate valve at the exit. The system uses schedule-40 steel pipes with threaded fittings. (i) Obtain the system curve when the valve is fully open. What is the ideal work input to the pump when the flow rate through the pipe network is 3.4 Ls^{-1}? (ii) If the loss coefficient of the gate valve when it is partially closed is 1.2, obtain the new system curve with the valve partially closed.
[*Answers*: (i) $\Delta H = 720 \times 10^3 Q_o^{\ 2}$, 278 W, (ii) $\Delta H = 750 \times 10^3 Q_o^{\ 2}$]

P12.4 Water entering a pump at atmospheric pressure is delivered through a pipe network with two branches as shown in Fig. P12.4.1. The pressures at 3 and 4 are atmospheric. The diameters and lengths of sections 1-2, 2-3 and 2-4 are respectively (50 mm, 25 m), (32 mm, 30 m) and (25 mm, 25 m). The pressure head at the exit of the pump is 12 m. Calculate the water flow rates through the two branches 2-3 and 2-4.
[*Answers*: 2.37 Ls^{-1}, 1.3 Ls^{-1}]

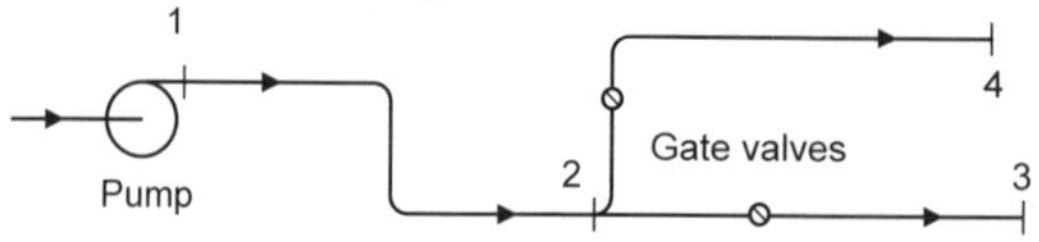

Fig. P12.4.1 Pipe network

P12.5 The performance curve of a pump at 1800 rpm is well represented by the quadratic equation:

$$H = 24.2 + 0.54Q - 0.092Q^2$$

where H (m) is the total head, and Q (Ls^{-1}) is the flow rate.

(a) Obtain the performance curve at 2000 rpm.

(b) This pump is installed in a pipe network with a system curve given by the equation:

$$H = 138 \times 10^3 Q^2$$

where H is the head loss in m and Q is the water flow rate in m^3s^{-1}.

Obtain the water flow rate and the ideal power input when the pump speed is 1800 rpm and 2000 rpm.

[*Answers*: (b) 11.5 Ls^{-1}, 2.06 kW and 12.7Ls^{-1}, 2.8kW]

P12.6 The performance curve of a pump at 2000 rpm is well represented by the quadratic equation:

$$H = 24.2 + 0.54Q - 0.092Q^2$$

where H (m) is the total head and, Q (Ls^{-1}) is the flow rate.

Two such pumps, installed in parallel, circulate water through a closed pipe network as depicted in Fig. E12.11.1. The system curve for the pipe network has been obtained as

$$H = 152 \times 10^3 Q^2$$

where H is the head loss in m, and Q is the water flow rate in m^3s^{-1}.

(i) Determine the flow rate when both pumps are running, and the ideal power input to the pumps. (ii) If one of the pumps is switched off, determine the flow rate through the pipe network, and the power input.

[*Answers*: (i) 12.56Ls^{-1}, 2.95kW, (ii) 11.13Ls^{-1}, 2.06 kW]

P12.7 The entrance of a centrifugal pump is located 4.4 m above the free surface of a sump tank from which water is pumped to a storage tank at the rate of 0.6 Ls^{-1}. The diameter of suction pipe is 25 mm and its length is 7.2 m. There are three 90° elbows in the pipe. The atmospheric pressure is 101 kPa and the water temperature is 40°C. (i) Calculate the available NPSH. (ii) How much further can the pump entrance be moved vertically before the available NPSH becomes just equal to the manufacturer-recommended NPSH of 2.2 m for the given conditions. [*Answers*: (i) 4.2 m, (ii) 2 m]

P12.8 Two identical heating coils are supplied with hot water at the rate of 0.25 L/s using a reverse-return pipe network as depicted in Fig. P12.8.1. Each of the coil sections 2-4 and 3-5 consists of 10 m of piping, a gate valve and a heating coil. The head loss through a coil is 0.8 m. The lengths of the other pipe sections are indicated in the figure. (i) Size the pipes for this system. (ii) Calculate the total head loss from 1 to 7 along the two flow paths 1-2-4-6-7 and 1-3-6-5-6-7.
[*Answers*: (ii) 2.26 m]

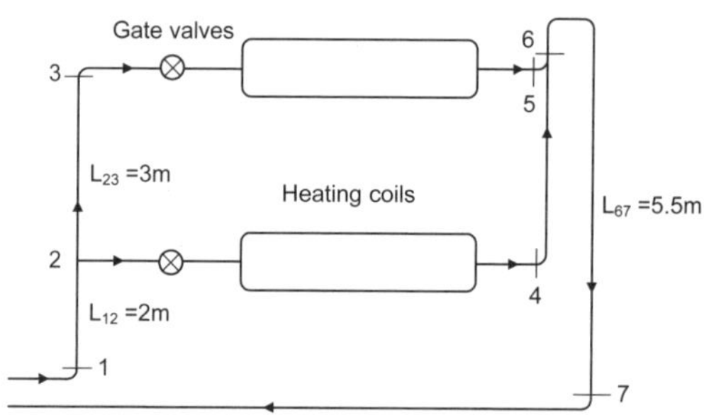

Fig. P 12.8.1 Reverse-return pipe network

P12.9 The heat removed from a condenser of a refrigeration plant is rejected to the atmosphere using two identical cooling towers as shown in Fig. P12.9.1. The temperature rise of the cooling water in the condenser is 6°C and the heat removal rate is 176 kW. The total length of pipe from 5 to 2 through the pump is 22 m, and the total length of pipe from 2 to 5 through the cooling tower is 8 m. The spray nozzles of the cooling towers are 4 m above the water surface in the sumps. The head loss through the condenser is 3.2 m of water.

Based on the required water flow rate through the condenser, the diameter of the pipe 5-2 through the pump has been selected as 75 mm and the diameter of pipe from 2-5 through the cooling tower has been selected as 60 mm. (a) Check whether the selected pipe diameters satisfy the guidelines listed in section 12.6.2. (b) Calculate the required rise of total head across the pump. (c) If the efficiency of the pump is 65%, calculate the power input.

[*Answers*: (b) 9.175 m, (c) 970 W]

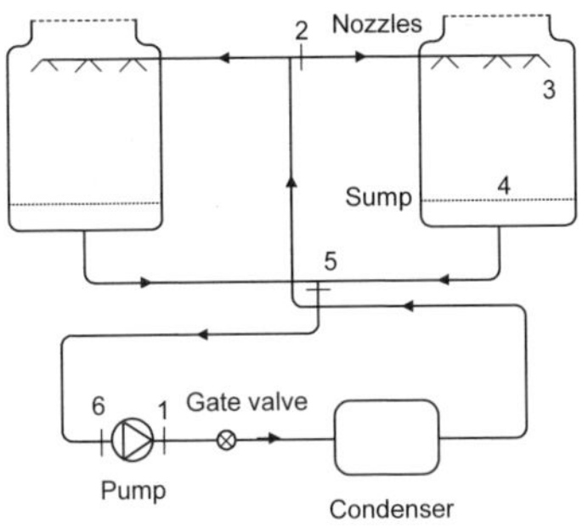

Fig. P12.9.1 Cooling tower pipe network

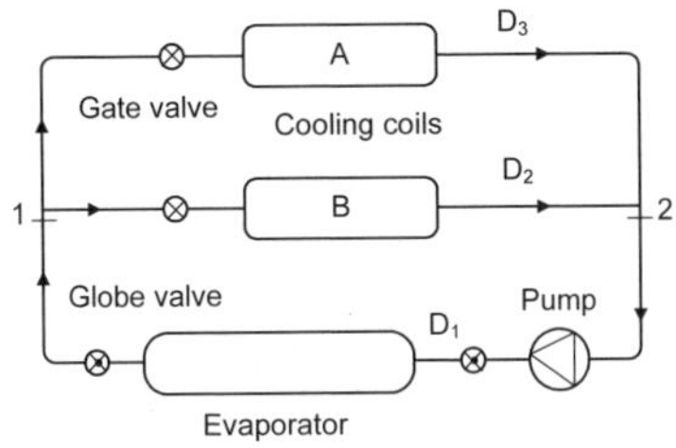

Fig. P12.10.1 Cooling coils

P12.10 In the air conditioning system, shown schematically in Fig. P12.10.1, chilled water is pumped to two cooling coils A and B with cooling capacities of 45 kW and 60 kW respectively. The temperature rise of water in each coil is 5°C. The flow circuit from 2 to1 through the pump includes 10 m of piping, two globe valves, and the evaporator with a head loss of 4 m. The flow circuit from 1 to 2 through coil A has 5.1 m of piping, a gate valve, and the cooling coil with a head loss of 1.7 m. The flow circuit from 1 to 2 through coil B has 1.5 m of piping, a gate valve, and the cooling coil with a head loss of 2 m.

Based on the chilled water flow rates through the three flow circuits the following pipe diameters have been selected (see Fig. P12.10.1): $D_1 =$ 65 mm, $D_2 =$ 52.5 mm, $D_3 =$ 48 mm. (i) Calculate the required water flow rate through the evaporator. (ii) Check whether the selected pipe diameters satisfy the guidelines listed in section 12.6.2. (iii) Calculate the required rise of total head across the pump. (iv) If the efficiency of the pump is 70%, calculate the power input.
[*Answers*: (i) 5Ls^{-1}, (iii) 8.4 m, (iv) 590 W]

References

1. *ASHRAE Handbook* - 2013 *Fundamentals*, American Society of Heating, Refrigeration and Air Conditioning Engineers, Atlanta, 2013.
2. Kuehn, Thomas H., Ramsey, James W. and Threlkeld, James L., *Thermal Environmental Engineering*, 3rd edition, Prentice-Hall, Inc., New Jersey, 1998.
3. Mitchell, John W. and Braun, James E., *Heating, Ventilation, and Air Conditioning in Buildings,* John Wiley and Sons, Inc., New York, 2013.
4. Stoecker, Wilbert F. and Jones, Jerold W., *Refrigeration and Air Conditioning*, International Edition, McGraw-Hill Book Company, London, 1982.

Appendix A12.1 - MATLAB Code for Head Loss in Pipe Sections

```
% head loss in a circular pipe length with fittings and coils
% tee-joint is at the end of the pipe length
% other fittings are within the pipe length
pi=3.141592;      % value of pi
g=9.81            % acceleration due to gravity, m per s^2
den= 998          % density of water kg per cu.m
vis=1.008*10^(-3) % viscosity of water Pa.s at 20 C
roug=0.000046     % pipe roughness
kentr=0        % loss coefficient of entrance to pipe
kvlv=0         % loss coefficient of valves
hcoil=0        % constant head loss of coils, m
ktees=0        % loss coefficient for straight section of tee-joint
kteeb=0        % loss coefficient for branch of tee-joint
kelb= 0        % loss coefficient of elbows
dcl= 5.5       % pipe length, m
dhmm=25            % pipe diameter, mm
dh=dhmm/1000       % pipe diameter, m
qalt=0.5           % water flow rate in, liters per s
qa = qalt/1000     % water flow rate, cu.m per s
for j=1:1:1        % loop if multiple flow rates are considered
```

```
ren=4*den*qa/(pi*vis*dh);   % Reynolds number
f=64/ren;                   % initial value for iterative solution
x=dh/roug;
% iterative solution of Colebrook's equation
for i=1:1:1000
y=1/f^0.5;
lhs=0.5*(y-1.14);
fu1=1+(9.3*x*y/ren);
rhs=log10(x/fu1);
com=[f,lhs,rhs];
if rhs >= lhs
   fr=f;
   break
end
f=f+0.0001;
end
% compute losses in pipe and fittings
plpipe=8*fr*dcl*den*(qa^2)/((pi^2)*(dh^5))  % pipe friction loss, Pa
pipunit=plpipe/dcl      % unit pressure loss in pipe, Pa per m
hpipe=plpipe/(den*g)  % head loss in pipe, m
vel=4*qa/(pi*dh^2)      % flow velocity, m/s
hdvel=vel^2/(2*g)       % velocity head, m
plelb=8*kelb*den*(qa^2)/((pi^2)*(dh^4))  % pressure loss in elbow, Pa
helb=plelb/(den*g)    % head loss in elbow, m
pentr=8*kentr*den*(qa^2)/((pi^2)*(dh^4)) % entrance pressure loss, Pa
hentr=pentr/(den*g)   % head loss in entrance, m
plvlv=8*kvlv*den*(qa^2)/((pi^2)*(dh^4))  % pressure loss in valves, Pa
hvlv=plvlv/(den*g)  % head loss in valves, m
pltees=8*ktees*den*(qa^2)/((pi^2)*(dh^4))  %  tee-line loss, Pa
htees=pltees/(den*g)   % tee-line head loss, m
plteeb=8*kteeb*den*(qa^2)/((pi^2)*(dh^4))  % tee-branch loss, Pa
hteeb=plteeb/(den*g) % tee-branch head loss, m
plcoil=hcoil*(den*g) % pressure loss through coil, Pa
ptloss=plpipe+plelb+plvlv+pltees+plteeb+plcoil+pentr % total loss, Pa
hloss=hpipe+helb+hvlv+htees+hteeb+hcoil+hentr % total head loss, m
end
```

Chapter 13

Building Energy Estimating and Modeling Methods

13.1 Introduction

Designing and operating buildings in an energy efficient manner has become a priority due to rising energy costs and concerns on climate change. This in turn has provided an incentive for manufacturers to develop building components like wall sections, windows, and doors with improved energy efficiency. Moreover, owners are undertaking extensive modifications to existing buildings to reduce energy costs.

To design energy efficient buildings it is necessary to develop quantitatively accurate methods to predict the long-term energy consumption rates in buildings. Although some methods for computing the energy flow rates through building envelopes were presented in chapters 8, 9 and 10, their main purpose was to determine the heating and cooling loads of the building.

The procedures available for predicting the energy performance of buildings can be divided into two main categories, those that can be carried out manually, and those requiring the use of main frame computers. There are also manual procedures involving extensive numerical calculations that are best performed on a computer.

The simplest manual procedure assumes that the energy input required to maintain indoor comfort conditions is a function mainly of the outdoor dry-bulb temperature. More accurate methods include the effects of transmitted solar radiation, internal heat gains, thermal storage in structural components and infiltration. The most sophisticated methods are based on hourly weather parameters, and incorporate detailed physical models of the different zones of the building and the various equipment of the HVAC system.

In this chapter we shall consider two manual methods called the degree–day method and the bin method for predicting the annual energy consumption. The physical basis of the different computerized procedures will be presented briefly.

13.2 Degree–Day Method for Estimating Energy Use

The degree–day method is used widely to estimate the winter heating loads of building in temperate climates. In this method it is assumed that: (i) the heating load of the building depends only on the indoor and outdoor dry-bulb temperatures, (ii) the efficiency of the heating system is constant, (iii) the internal energy gains due to solar radiation, people, lights, and equipment are constant, (iv) air infiltration affects only the sensible heating load.

These assumptions are consistent with the conservative design approach for estimating the design heating load where only the steady heat loss through the building envelope and infiltration heat loss are included. Moreover, internal heat gains from people, lights and equipment, and solar gains are ignored because these energy inputs help reduce the load on the heating system.

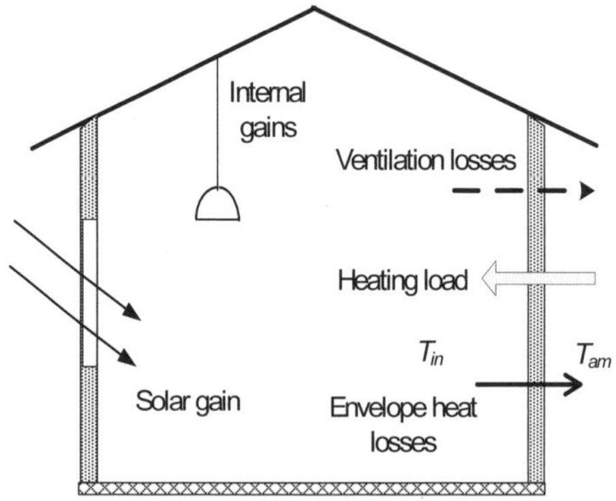

Fig. 13.1 Energy flows in a building

We shall develop the degree–day expression for the heating load by referring to the simplified physical model of a building, as shown in Fig. 13.1. Subject to the aforementioned assumptions, we can write the

following expressions for the various instantaneous energy flow rates in and out of the indoor conditioned space.
Heat flow rate across the envelope is given by

$$\dot{Q}_e = U_e A_e (T_{in} - T_{amb}) \tag{13.1}$$

where U_e and A_e are the heat transfer coefficient and area of the building envelope respectively. The indoor and ambient air temperatures are T_{in} and T_{amb} respectively. The indoor ambient temperature, T_{in} is usually controlled at a preset value by the space thermostat.
Sensible heat flow due to infiltration is given by

$$\dot{Q}_{inf} = \dot{m}_i c_p (T_{in} - T_{amb}) \tag{13.2}$$

where $\dot{m}_i$ is the infiltration rate and c_p is the specific heat of air.

Adding Eqs. (13.1) and (13.2) the total energy loss due to envelope heat transfer and infiltration may be expressed in the *equivalent form* [4]:

$$\dot{Q}_{loss} = U_e A_e (T_{in} - T_{amb}) + \dot{m}_i c_p (T_{in} - T_{amb})$$

$$\dot{Q}_{loss} = \left(U_e A_e + \dot{m}_i c_p\right)(T_{in} - T_{amb}) = U_o A_o (T_{in} - T_{amb}) \tag{13.3}$$

where $U_o A_o$ is an *effective* heat transfer conductance accounting for the sensible heat transfer across the building envelope and the sensible heat required to heat the infiltration air to the indoor temperature.

The rate of heat gain due to solar energy input, people, lights and equipment may be expressed as

$$\dot{Q}_{int} = \dot{Q}_{sol} + \dot{Q}_{people} + \dot{Q}_{lights} + \dot{Q}_{equip.} \tag{13.4}$$

Applying the first law of thermodynamics to the indoor air we have

$$\frac{dE_i}{dt} = \dot{L}_h + \dot{Q}_{int} - \dot{Q}_{loss} \tag{13.5}$$

Neglecting indoor energy storage rate, given by the LHS of Eq. (13.5), we can express the heating load as

$$\dot{L}_h = \dot{Q}_{loss} - \dot{Q}_{int} \tag{13.6}$$

Substituting from Eq. (13.3) in Eq. (13.5) we obtain

$$\dot{L}_h = U_o A_o (T_{in} - T_{amb}) - \dot{Q}_{int} \tag{13.7}$$

We now rearrange Eq. (13.7) as a simple rate equation by defining a *balance temperature*, or a *base temperature*. Hence we have

$$\dot{L}_h = U_o A_o (T_{bal} - T_{amb}) \tag{13.8}$$

The balance temperature is defined as

$$T_{bal} = T_{in} - \frac{\dot{Q}_{int}}{U_o A_o} \tag{13.8a}$$

The variations of the total heat loss (Eq. 13.3) and the heating load (Eq. 13.8) with ambient temperature are depicted in Fig. 13.2.

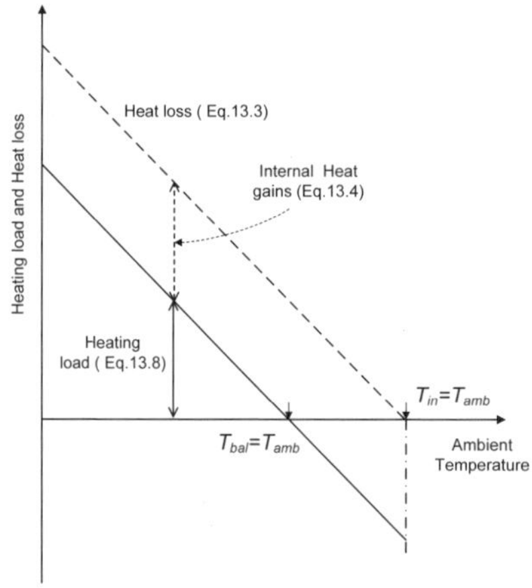

Fig. 13.2 Variation of heating load and heat loss with ambient temperature

These graphs are parallel straight lines with slope (U_oA$_o$). The constant vertical separation between them is the *internal heat gain*. The total heat loss is zero when the indoor temperature, T_{in} is equal to the ambient temperature, T_{amb}. The heating load, on the other hand, is zero when T_{amb} is equal to the balance temperature, T_{bal}, given by Eq. (13.8a). The required heat input by the heating system is negative (cooling) between T_{bal} and T_{in} because the internal heat gains more than compensate for the total heat loss.

The heating *degree–day* method is based on the observation that the heating system needs to supply heat only when T_{amb} is below T_{bal}. Therefore the total amount of heat to be supplied by the heating system during the heating season may be found by integrating Eq. (13.8), assuming that the effective heat transfer conductance and the balance temperature are constant. Hence we have

$$E_{he} = U_o A_o \int_0^{t_s} (T_{bal} - T_{amb})^+ dt \tag{13.9}$$

The superscript '+' indicates that only positive values of $(T_{bal} - T_{amb})$ should be included in the integration because heating is needed only when T_{amb} is lower than T_{bal}.

For a period, such as a month, Eq. (13.9) may be expressed in the form

$$E_{he,m} = U_o A_o (DD)_m \tag{13.10}$$

where $(DD)_m$ is the total heating *degree–days* for the month.

We use Eq. (13.10) to determine the amount of fuel energy required during the month by knowing the efficiency of the furnace, η_f. Hence we have

$$E_{fu,m} = \frac{U_o A_o (DD)_m}{\eta_f} \tag{13.11}$$

The integration of Eq. (13.9) to determine the number of degree–days for a month is illustrated graphically in Fig. 13.3.

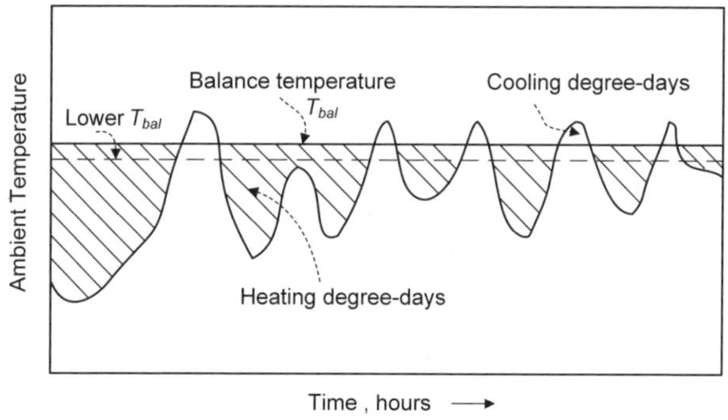

Fig. 13.3 Determination of degree–days

The hourly temperature distribution over the period of interest, typically a month, is plotted against time. The area of the curve below balance temperature, usually 18.3°C, gives the total heating degree–days for the month.

Three measures to reduce the heating load of a building are: (i) use better insulation to decrease U_oA_o, (ii) lower the thermostat setting, T_{in}, and (iii) allow more solar radiation to enter by installing transparent sections like skylights. From Eqs. (13.8) and (13.8a) we notice that these three changes lower the balance temperature, T_{bal}, which in turn decreases the monthly degree–days as seen from Fig. 13.3.

In Table 13.1 we have listed the monthly heating (HDD) and cooling (CDD) degree–days for two locations, Toronto and Singapore, extracted from Ref. [1]. Data included are for balance temperatures of 10°C and 18.3°C. Heating and cooling are both required in Toronto due to its temperate weather, while in Singapore with its tropical weather, only cooling is needed. The data also shows that there is a large reduction in the *heating degree–days* when the balance temperature is decreased from 18.3°C to 10°C. The converse is true for *cooling degree–days*.

Table 13.1 Monthly heating (HDD) and cooling (CDD) degree–days for balance temperatures of 10°C and 18.3°C for Toronto and Singapore. (*data extracted from the weather data CD accompanying the ASHRAE Handbook* - 2013 *Fundamentals* [1]).

Location	Toronto				Singapore	
Month	HDD 10	HDD18.3	CDD10	CDD18.3	CDD10	CDD18.3
Jan.	457	715	0	0	534	276
Feb.	407	640	0	0	503	269
Mar.	302	557	3	0	567	308
Apr.	110	336	26	1	560	310
May	16	167	120	13	585	327
Jun.	0	39	268	57	562	312
Jul.	0	6	358	107	567	308
Aug.	0	13	332	87	564	306
Sep.	3	85	194	26	544	294
Oct.	61	271	50	2	564	306
Nov.	189	434	6	0	531	281
Dec.	371	629	1	0	534	276

13.3 Bin Method for Estimating Energy Use

13.3.1 *Generation of bin data*

The degree–day method described above is a relatively straightforward procedure to estimate monthly and seasonal energy use. However it has the following limitations: (i) a single indoor temperature set-point is assumed to generate degree–day data, (ii) a constant heat gain rate is assumed, and therefore changes in occupancy rate cannot be included, (iii) variation in outdoor humidity is not included, and therefore latent heat loads have to be neglected.

In the *bin method* these limitations are partially overcome by compiling outdoor temperature data as a frequency distribution over a series of outdoor temperature intervals. As a further refinement, the above frequency distributions are developed for several fixed time periods of a single day, which are usually 3 consecutive 8-hour periods [2].

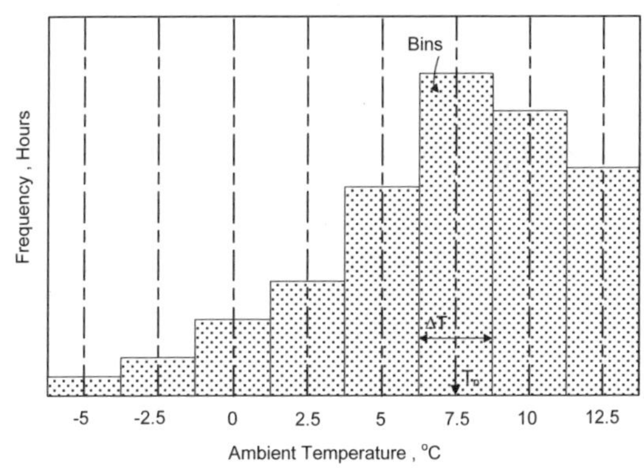

Fig. 13.4 Hourly bin data for ambient temperature

A typical distribution of hourly bin data is shown graphically in Fig. 13.4. Each bin in Fig. 13.4 is centered about a mean temperature, say T_n for the representative bin, *n*. A bin includes the number of hours the outdoor temperature is between (T_n -Δ $T_n/2$) and (T_n +Δ $T_n/2$) during a month or a year. The numerical values are for bins with temperature intervals, ΔT_n of 2.5°C.

An analytical method to compute monthly bin data using the number of days of the month, the monthly average ambient temperature, and statistical measures of the variability of weather is available in the recent book by Mitchell and Braun [4]. The original work was presented by Erbs *et al.* [2]. We shall now outline the main steps of this procedure.

The *daily average* ambient temperature at a location, $T_{av,d}$ for each month is used to obtain the *monthly average* temperature, $T_{av,m}$. Its standard deviation, σ_m given by

$$\sigma_m = \left[\frac{\Sigma(T_{av,d}-T_{av,m})^2}{N_{dm}}\right]^{1/2} \tag{13.12}$$

where N_{dm} is the number of days of the month.

The monthly average temperatures, $T_{av,m}$ is used to obtain the *yearly average* temperature at the location, $T_{av,y}$. Its standard deviation, σ_y is given by

$$\sigma_y = \left[\frac{\Sigma(T_{av,m}-T_{av,y})^2}{12}\right]^{1/2} \tag{13.13}$$

The standard deviation, σ_m has been correlated with the monthly average temperature $T_{av,m}$ and the standard deviation σ_y using the linear relation:

$$\sigma_m = 1.451 - 0.029T_{av,m} + 0.0369\sigma_y \tag{13.14}$$

The cumulative fraction, $Q(T_b)$ of the number of hours in a month below a *base temperature* T_b has been correlated with a dimensionless parameter H_m using the equation:

$$Q(T_b) = [1 + exp(-3.396H_m)]^{-1} \tag{13.15}$$

The parameter H_m is given by

$$H_m = \frac{(T_b-T_{av,m})}{\sigma_m(N_{dm})^{1/2}} \tag{13.16}$$

For any month *m*, the number of hours in bin *n* centered around temperature T_n in Fig. 13.4, may be obtained by using Eq. (13.15). Hence we have

$$N_{bin}(m,n) = 24N_{dm}\left[Q\left(T_n + \frac{\Delta T}{2}\right) - Q\left(T_n - \frac{\Delta T}{2}\right)\right] \tag{13.17}$$

The MATLAB code in Appendix A13.1, based Eqs. (13.12) to (13.17), may be used to obtain monthly and annual bin data for any location by knowing the monthly average ambient temperatures. An extensive tabulation of monthly average ambient temperatures for different locations is available in the weather data DC accompanying the *ASHRAE Handbook* - 2013 *Fundamentals* [1].

The yearly bin data for Toronto, given in Table 13.2, were generated using the MATLAB code in Appendix A13.1.

Table 13.2 Yearly bin data for Toronto

Temp. °C	−27.5	−25	−22.5	−20	−17.5	−15
No. Hours	1	2	5	10	23	51
Temp. °C	−12.5	−10	−7.5	−5	−2.5	0
No. Hours	106	208	356	510	604	624
Temp. °C	2.5	5	7.5	10	12.5	15
No. Hours	612	605	605	597	607	666
Temp. °C	17.5	20	22.5	25	27.5	30
No. Hours	766	801	593	276	93	28
Temp. °C	32.5	35				
No. Hours	8	2				

The above procedure may be extended to compute monthly degree–days using the following expression:

$$(DD)_m = \sum_{n=1}^{n=r} N_{bin}(m,n)(T_{bal} - T_n)^+/24 \qquad (13.18)$$

The summation in Eq. (13.18) is terminated ($n=r$) when the bin temperature, T_n exceeds the balance temperature, T_{bal}. The MATLAB code in Appendix A13.1 includes an optional section to compute the degree–days using the calculated bin data for a location.

We have used the above procedure to calculate the monthly heating and cooling degree–days for Toronto and Singapore for balance temperatures of 10°C and 18.3°C. The results agree well with the degree–days listed in Table 13.1, which were obtained directly from Ref. [1].

13.3.2 *Applications of the bin method*

The bin method may be used for energy estimation in situations where the heating system performance depends on environmental conditions such as the ambient dry-bulb and wet-bulb temperatures. Practical examples of such applications are: (i) air-source heat pumps where the rate of heat input to the building by the heating system is a strong function of the ambient dry-bulb temperature, (ii) cooling towers used in central air conditioning plants where the power input to the chiller depends on the outdoor wet-bulb temperature.

The performance degradation of furnaces and heat pumps due to cycling can be included in the bin method. Moreover, the bin method may be used to account for changes in occupancy that invariably occurs in most buildings.

13.3.3 *Cycling of furnaces*

The furnaces used for heating buildings are sized to meet the design heating load of the building. However, much of the time the actual load is below this design value. To deal with this practical situation the heating system usually operates in a cycling mode. It is switched on for the duration necessary to bring the indoor air temperature to the desired value and then switched off until the temperature again falls below the preset value. This type of cycling of a furnace effects the overall efficiency of the heating system and it may be accounted for by introducing a *part-load factor* (PLF) [3,4].

The PLF is defined as

$$PLF = 1 - c_d(1 - LF) \tag{13.19}$$

where LF the part-load ratio and c_d is the *degradation coefficient*.

The part-load ratio is given by

$$LF = \frac{\dot{Q}_{load}}{\dot{Q}_{cap}} \tag{13.20}$$

The coefficient c_d usually lies between 0.15 and 0.25 [3].
The actual heating system capacity during cycling is

$$\dot{Q}_{act} = \dot{Q}_{cap}PLF \tag{13.21}$$

The 'on-time' of the heating system is given by

$$\Delta\tau_{on} = \frac{\Delta\tau\dot{Q}_{load}}{\dot{Q}_{cap}PLF} \tag{13.22}$$

where $\Delta\tau$ is the time interval of interest, typically an hour. The required energy input into the heating system is given by

$$\Delta E = \frac{\dot{Q}_{cap}\Delta\tau_{on}}{\eta_{fur}} = \frac{\dot{Q}_{load}\Delta\tau}{PLF\eta_{fur}} \tag{13.23}$$

where η_{fur} is the rated efficiency of the furnace.

The part-load factor in Eq. (13.19) may also be used to account for the losses due to cycling of heat pumps and chillers [4].

The bin method allows us to include the *PLF* in the energy estimation procedure. We shall illustrate its application in worked examples 13.7 and 13.9.

13.3.4 *Air-source heat pumps*

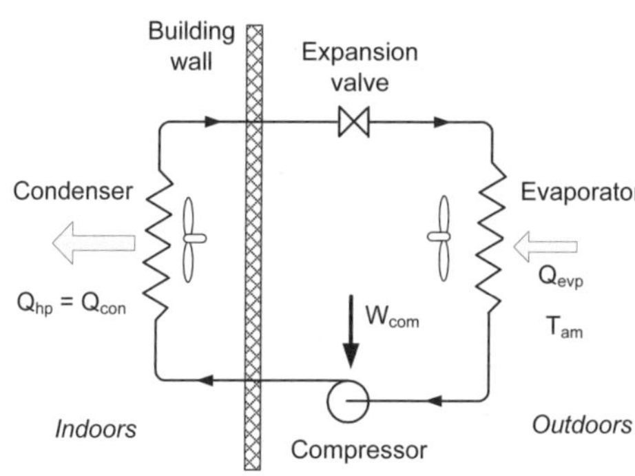

Fig. 13.5 Air-source heat pump

The principle of operation of reversible heat pumps for HVAC applications was explained in section 3.5. An *air-source* heat pump system used for space heating is shown schematically in Fig. 13.5. It operates essentially on the vapor compression refrigeration cycle studied in detail in chapter 3.

The evaporator of the cycle, located outdoors, absorbs heat from ambient air while the condenser located inside the building rejects heat to indoor air. The latter heat input by the heat pump is a strong function of the outdoor temperature. If the heat pump is unable to meet the heating load of the indoor space, then an auxiliary heat source such as an electric resistance heater is used to supplement the heat input of the heat pump.

Applying the energy balance equation to the heat pump cycle we obtain the total rate of heat input to the building as (see Fig. 13.5):

$$\dot{Q}_{hp} = \dot{Q}_{cond} = \dot{Q}_{evp} + \dot{W}_{comp} \tag{13.24}$$

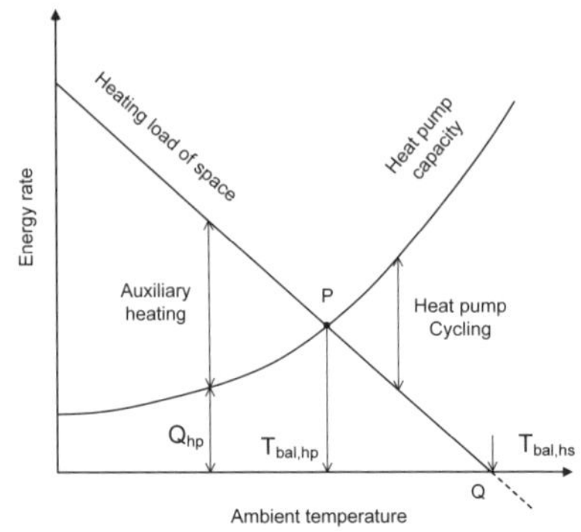

Fig. 13.6 Variation of heat input rates with temperature

In Fig. 13.6 we have shown the variations of Q_{hp} and the heating load of the building with ambient temperature. We notice that Q_{hp} decreases rapidly as the ambient temperature decreases.

At lower ambient temperatures the evaporating temperature and pressure of the refrigeration cycle are correspondingly lower. However, the condenser pressure, which is function of the indoor air temperature, is nearly constant. Therefore as the ambient temperature decreases, the pressure differential across the compressor rises, resulting in an increase in the compressor power input.

Moreover, at lower ambient temperatures the heat absorption rate per unit mass of refrigerant in the evaporator is lower and so is the refrigerant flow rate through the cycle [3]. The net impact of these various effects is a sharp decrease in the heat input to the building by the heat pump as seen in Fig. 13.6, and a corresponding decrease in the COP of the heat pump.

The heat pump capacity curve intersects the building heating load curve at point P, usually called the *heat pump balance point*, where the ambient temperature is $T_{bal,hp}$ (see Fig. 13.6). The point of intersection, Q of the heating load curve and the ambient temperature axis is the *house balance point*, $T_{bal,hs}$. When the ambient temperature is below $T_{bal,hp}$, the heat supplied by the heat pump is not sufficient to meet the heating load of the building and therefore an auxiliary heat input is required. This auxiliary energy source could be an electric resistance heater or a fuel-fired furnace.

When the ambient temperature is higher than $T_{bal,hp}$, the heat input of the heat pump is larger than the heating load of the building. Under these conditions the heat pump usually operates in a cycling mode where the room thermostat maintains the indoor temperature at the desired value by periodically switching the heat pump on and off. When the ambient temperature exceed $T_{bal,hs}$ the building has a negative heating load and therefore cooling is required.

The total rates of energy supply by the heat pump and the auxiliary source may be estimated using the bin method. The heating load of the building is given by Eq. (13.8) as

$$\dot{L}_h = U_o A_o \left(T_{bal,hs} - T_{amb}\right) \tag{13.25}$$

The heat supply rate and the COP of the heat pump, both functions of the ambient temperature, may be obtained from manufacture's catalogues as

$$\dot{Q}_{hp} = f_{hp}\,(T_{amb})$$

and

$$COP_{hp} = f_{cop}\,(T_{amb}) \tag{13.26}$$

When the ambient temperature is less than the heat pump balance temperature (see Fig. 13.6), the heat pump capacity is not sufficient to meet the heating load. The rate of heat supply by the auxiliary source under these conditions is given by

$$\dot{Q}_{aux} = \dot{L}_h - \dot{Q}_{hp}(T_{amb}) \tag{13.27}$$

When ambient temperature exceeds the heat pump balance temperature (see Fig. 13.6), the heat pump capacity is larger than the heating load. Under these conditions the heat pump operates in a cycling mode.

We define the part-load ratio as [4]

$$LR = \frac{\dot{L}_h}{\dot{Q}_{hp}}$$

The part-load factor is given by

$$PLF = 1 - c_d(1 - LR) \tag{13.28}$$

where c_d is the *degradation coefficient.*

The rate of electrical energy input to the heat pump is [4]

$$\dot{E} = \frac{\dot{L}_h}{COP_{hp} \times PLF} \tag{13.29}$$

For each bin temperature, $T_{bin} = T_{amb}$, we can determine the total heating load, the rates of heat input by the heat pump and the auxiliary energy source, and the electrical energy input to the heat pump, by using Eqs. (13.25) to (13.29). These quantities are multiplied by the number of hours in the bin to compute the total energy rates for the particular bin. By adding the individual bin-energy quantities we then estimate the total energy inputs to the heat pump and the auxiliary source for the entire heating period. We shall illustrate the calculation procedure in worked examples 13.14.

13.3.5 *Cooling towers*

A central air conditioning system using chilled water and air as energy distribution media is shown schematically in Fig. 3.8. Additional details of a cooling tower including cooling water and chilled water loops are depicted Fig. 6.6.

The condenser of the refrigeration plant, usually called a chiller, is cooled by water circulating between the condenser and the cooling tower. In section 6.5 we presented the modeling and performance evaluation of cooling towers in some detail. A MATLAB program to model cooling towers is given in Appendix A6.1.

The performance of a cooling tower depends on the local ambient wet-bulb temperature. It is the lowest temperature to which the water from the condenser of the chiller could be cooled. Therefore as the ambient wet-bulb temperature changes, the cooling rate of the condenser is affected. This in turn affects the power input to the compressor and the COP of the chiller plant. However, the heat absorbed in the evaporator is equal to the cooling load of the building.

The bin method can be used to estimate the long-term energy consumption of chillers. In this instance, the required bin data are the number of hours during which the ambient wet-bulb temperature is between (T_{wb} -$\Delta T_{wb}/2$) and (T_{wb} +$\Delta T_{wb}/2$), where T_{wb} and ΔT_{wb} are the mean wet-bulb temperature and the temperature interval respectively.

Using the cooling tower manufacturer's data it is possible to correlate the cooling tower water inlet temperature, which is equal to the condenser outlet water temperature, (see Fig. 6.6) with the ambient wet-bulb temperature [4].

Moreover, using the chiller manufacturer's data the condenser outlet water temperature can be correlated with the COP of the chiller. Thus for any wet-bulb temperature we can calculate the corresponding electrical power input to the compressor of the chiller. Hence the total electrical energy for each bin is obtained as

$$E_{comp} = N_{hr} W_{comp} \tag{13.30}$$

where N_{hr} is the number of hours in the bin and W_{comp} is the hourly work input to the compressor at the average bin temperature. The application of the bin method to estimate the chiller energy input is illustrated in worked example 13.12.

13.3.6 *Variable occupancy rates*

The occupancy rates in buildings affect the internal heat gains directly. Furthermore, the thermostat set-point is usually lower during unoccupied periods compared to occupied periods. The effect of these two factors on the balance temperature of the building is evident from Eq. (13.8a).

We mentioned in section 13.3.1 that bin data are sometimes available for specific time periods of a day. For example, Table B.3 in Ref. [3] gives monthly ambient temperature bin data for three daily time intervals or hour-groups, namely, 1 a.m. to 8 a.m., 9 a.m. to 4 p.m. and 5 p.m. to 12 p.m. This data can be used to estimate the total energy input to a building for which the occupancy rates during the different hour-groups are known.

A typical office building is usually unoccupied during the two days of the weekend. During the 5 working days it will be fully occupied from about 8 a.m. to 7 p.m. and unoccupied during the remaining hours. By assuming that the above weekly occupancy pattern is repeated over the heating season, we can determine the fractional number of hours that the building is occupied during each of the three hour-groups mentioned above.

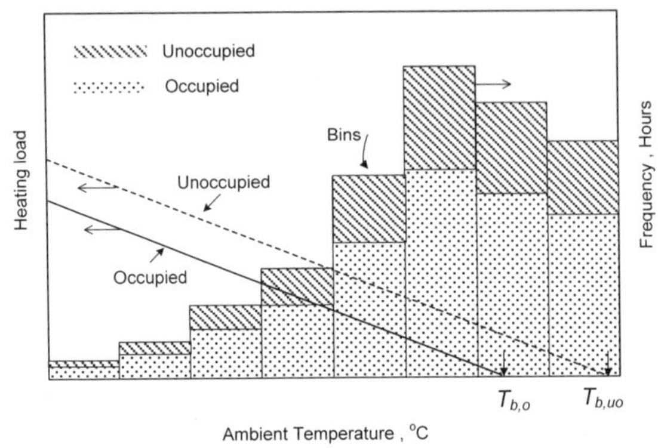

Fig. 13.7 Variable occupancy rates

For purposes of illustration we have shown in Fig. 13.7 the bin data for *one of the hour-groups*. For the same ambient temperature, the heating load when the building is occupied is shown to be lower than when it is unoccupied. This may happen due to the increase in internal heat gain from people, lights and appliances when the building is occupied.

Each temperature-bin consists of two areas, one indicating the occupied hours and the other the unoccupied hours. The ratio of these areas is the same for all bins in a single hour-group.

The total heat input for a *single hour-group* may be expressed as

$$E_{hg} = \sum_{n=1}^{N_{bt}} \left[f_o N_n A_o U_o \left(T_{b,o} - T_n \right) + f_{uo} N_n A_o U_o \left(T_{b,uo} - T_n \right) \right]$$

The occupied and unoccupied time fractions are f_o and f_{uo} respectively. The number of hours in bin-n at ambient temperature T_n, is N_n. The respective balance temperatures when the building is occupied and unoccupied are $T_{b,o}$ and $T_{b,uo}$, and the area–overall heat transfer coefficient product is A_oU_o. The total energy input is obtained by adding the energy inputs, E_{hg} for each hour-group. The calculation procedure is illustrated in worked example 13.13.

13.4 Simulation Methods for Estimating Energy Use

The energy estimation methods like the degree–day method and the bin method, presented in the preceding sections, are convenient to implement. However, they are unable to incorporate all the dynamic interactions occurring between the various subcomponents of an HVAC system.

An approach that has become increasingly popular in recent years is the use of component models to simulate the behavior of HVAC systems. These simulations are essentially computer-based and their complexity has increased over the years due to the advent of powerful and inexpensive computing facilities.

The widely available computer software programs listed in the *ASHRAE Handbook* - 2013 *Fundamentals* [1] include TRANSYS, DOE-2, EnergyPlus, and ESP-r. The simulation methodology used in these software programs is called *forward modeling*. This is a classical approach of simulation where each subcomponent constituting the complete HVAC system is described using a physical model [5].

We developed a number of such models for the subcomponents of typical HVAC systems in earlier chapters of this book. Each subcomponent is usually subject to one or more inputs which are often the output of some of the other components of the system. In addition, there are inputs like the weather conditions and the occupancy rates, that are external to the HVAC system.

13.4.1 *Central HVAC systems*

The forward modeling procedure is best illustrated by referring to a realistic HVAC system like the central air conditioning plant shown schematically in Fig. 13.8. The operation of such systems was outlined in chapters 1 and 3.

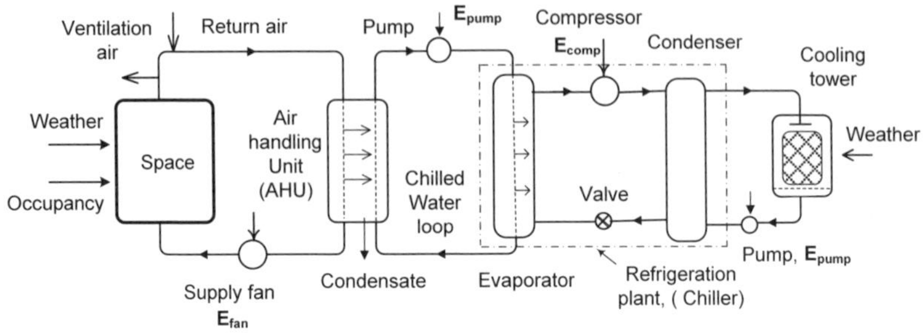

Fig. 13.8 Energy simulation of a central air conditioning system

Table 13.3 Summary of subcomponent-models, inputs and outputs

	Component: Conditioned space or building zone
Models	Methods in chapters 8 and 10, Heat Balance method, Transfer function method, Thermal network method [4].
Inputs	*Weather data*: Solar radiation, ambient temperature and humidity, wind speed; *Internal gains*: infiltration rate, heat input from people, lights and equipment; *HVAC system inputs*: supply air flow rate, ventilation air flow rate, air temperature and humidity.
Outputs	Space or zone temperature and humidity.
	Component: Air handling unit (AHU)
Models	Physical models as in chapter 7; AHU manufacturer's test data in the form of tables or fitted-curves (see Ref. [6]).
Inputs	Chilled water flow rate and inlet temperature; Return air mass flow rate, temperature and humidity.
Outputs	Chilled water flow rate and outlet temperature; Delivery air temperature and humidity; Condensate flow rate and temperature.
	Component: Refrigeration cycle, chiller
Models	Refrigeration cycle analysis in chapter 3; Condenser and evaporator models as in chapter 7; Chiller manufacturer's test data in the form of tables or fitted-curves (see Ref. [5,6]).
Inputs	Condenser cooling water flow rate and inlet temperature; Return chilled water flow rate and temperature; Power input to chiller.
Outputs	Chilled water flow rate and supply temperature.
	Component: cooling tower
Models	Cooling tower model in chapter 6; Cooling tower manufacturer's test data in the form of tables or fitted-curves (see Ref. [4]).
Inputs	Condenser cooling water flow rate and leaving temperature; Air flow rate through cooling tower; Ambient air dry-bulb and wet-bulb temperatures; Power input to cooling tower pump.
Outputs	Condenser cooling water flow rate and return temperature.

In Table 13.3 we have listed a few of the main subcomponents of the system depicted in Fig. 13.8. Typical physical models for each subcomponent, the likely inputs it may experience, and its outputs are also tabulated.

Furthermore, the fan–duct systems and the pump–piping systems that transfer air, water and refrigerants between the subcomponents listed above, were modeled in chapters 11 and 12 respectively. These models may be used to estimate the energy consumption of the fans and the pumps.

We recall that in developing physical models for equipment like wet cooling coils (AHUs) and cooling towers several simplifying assumptions had to be made. These assumptions could limit the capability of the models to accurately predict the behavior of commercially available equipment. In such cases it would be more accurate and convenient to model the component using manufacturer's test data directly, either in the form of tabulated data or fitted-curves.

Once the component models are developed, they can be assembled by identifying the outputs and inputs that are common to different components. The resulting block diagram is called an *information flow diagram* and the details of its construction are described in Ref. [5].

The complete simulation of a HVAC system has to include the control strategies used to operate the different components. For example, for the single-zone system depicted in Fig. 13.8, a zone thermostat may be used to shut off the air supply to the zone when the indoor temperature attains the preset value. Due to the interconnectedness of the different components, this control action would result in changes in the operating conditions of some of the other components. Therefore the models listed in Table 13.3 have to be appropriately modified to reflect these changes in operating conditions. The control strategies for HVAC systems and their implementation are discussed in detail in Ref. [4].

13.4.2 *Simulation of multi-chiller systems*

The variations of the COP with cooling load fraction of two typical water chillers used in central air conditioning systems, are depicted in Fig. 13.9. The change in COP with load fraction is relatively small for chiller

1, whose capacity is controlled using a variable speed drive. The COP is a maximum at a load fraction of about 0.6. However, for chiller 2 whose capacity is controlled by adjusting the inlet guide vanes, the change in COP at part-load is much larger, and the COP is a maximum at a load fraction of about 0.9.

In most buildings the cooling load varies during the day. Therefore it is more common to install several smaller chillers as opposed to a single large chiller to meet the varying cooling load. These chillers are switched on and off to match their combined part-load capacity to the prevailing cooling load. Such a multi-chiller system with 3 chillers and series of AHUs is shown schematically in Fig. 13.10.

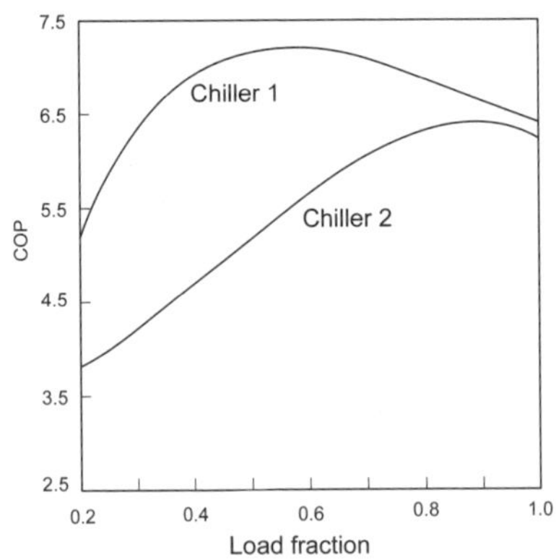

Fig.13.9 Variation of COP with load fraction

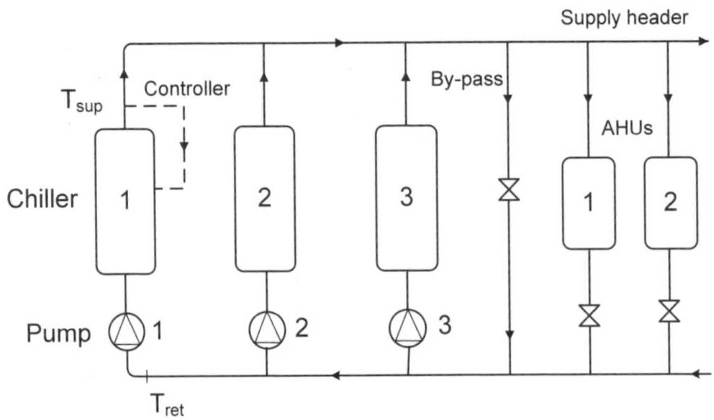

Fig. 13.10 Operation of multi-chiller systems

The supply water temperature, T_{sup}, typically about 6.7°C, from each chiller is maintained constant by a controller and the pumps maintain the chilled water flow rates constant. The chilled water flowing through the supply header pipe is distributed to the different AHUs where the air from the spaces are cooled and dehumidified.

When the building cooling load is less than the design cooling load, that is during part-load operation, the total water flow rate through the AHUs will be less than the total supply flow rate of the chillers. This difference in flow rates is balanced by the flow rate through the by-pass pipe. The return water temperature, T_{ret} changes with the total cooling load of the building.

Consider the operation of this multi-chiller system under *full-load* conditions. The total *rated* cooling capacity of the three chillers is equal to the maximum cooling load of the building. Therefore

$$Q_{c,1} + Q_{c,2} + Q_{c,3} = (\dot{m}_1 c_w + \dot{m}_2 c_w + \dot{m}_3 c_w)\Delta T_{fl} = Q_{fl} \qquad (13.31)$$

where $Q_{c,i}$ is the *rated cooling capacity* of chiller i (i = 1, 2 and 3). The water flow rate through chiller i is $\dot{m}_i$ and the specific heat capacity of water is c_w. The change in temperature of the water flowing through chiller i is ΔT_{fl}, and the total cooling load of the building is Q_{fl}.

Under *part-load* conditions (subscript *pl*) Eq. (13.31) becomes

$$(\dot{m}_1 c_w + \dot{m}_2 c_w + \dot{m}_3 c_w)\Delta T_{pl} = Q_{pl} \qquad (13.32)$$

From Eqs. (13.31) and (13.22) we have

$$\frac{Q_{pl}}{Q_{fl}} = \frac{\Delta T_{pl}}{\Delta T_{fl}} = f_l \qquad (13.33)$$

where f_l is the cooling load fraction. Now the capacity of an individual chiller, i under part-load conditions is

$$Q_{c,i(pl)} = \dot{m}_i c_w \Delta T_{pl} \qquad (13.34)$$

From Eqs. (13.32) and (13.34) we obtain the part-load capacity of chiller i as

$$Q_{c,i(pl)} = \left(\frac{\dot{m}_i c_w}{\dot{m}_1 c_w + \dot{m}_2 c_w + \dot{m}_3 c_w}\right) Q_{pl} = \left(\frac{Q_{c,i}}{Q_{c,1} + Q_{c,2} + Q_{c,3}}\right) Q_{pl} \qquad (13.35)$$

The electrical energy input to chiller i under part-load conditions is

$$W_{i,pl} = \frac{Q_{c,i(pl)}}{COP_i(f_l)} \qquad (13.36)$$

where $COP_i(f_l)$ is the COP of chiller i when the load fraction is f_l (see Fig. 13.9). The total electrical energy input is given by

$$E_{pl} = \sum \frac{Q_{c,i(pl)}}{COP_i(f_l)} \qquad (13.37)$$

We need to combine the three chillers in such a manner that for any particular cooling load the total electrical energy input to the chillers, given by Eq. (13.37), is a minimum. Since the variation of the COP with load fraction, f_l (Fig. 13.9) is dependent on the type of chillers used, the optimal combination of the chillers to meet a particular cooling load has to be determined by trial and error using Eq. (13.37). This is called chiller sequencing and it is illustrated in worked examples 13.10 and 13.11.

13.4.3 *Simulation of water-loop heat pump system* (*WLHPS*)

We shall now apply the forward modeling method to the water-loop heat pump system (WLHPS) depicted schematically in Fig. 13.11. These are central heating and cooling systems, commonly installed in large buildings like hotels. The WLHP system consists of a number of individual reversible heat pumps located in different zones of a building, and they exchange heat with a common water loop often incorporating a water storage tank.

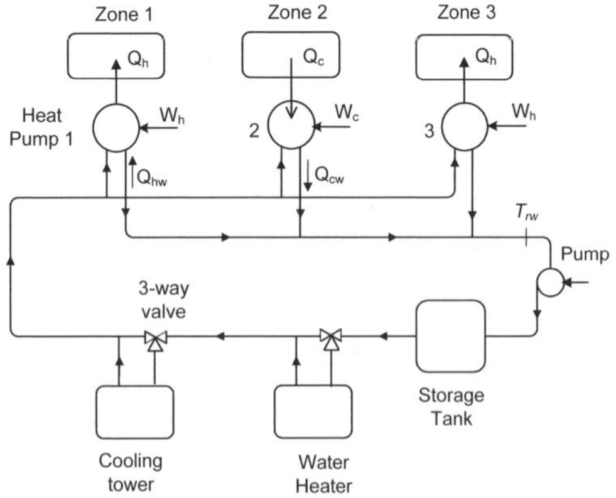

Fig. 13.11 Water-loop heat pump system (WLHPS)

For purposes of illustration we have shown three heat pumps serving three different zones of a building. Heat pumps 1 and 3 are in the heating mode while heat pump 2 is in the cooling mode. The water loop with the storage tank is a closed water source with which the heat pumps exchange energy.

In the water loop the temperature of the water returning from the heat pumps, T_{rw} is maintained between about 15°C (T_{min}) and 35°C (T_{max}). This is achieved by sending the water through the heater, if T_{rw} falls below T_{min}, and alternatively, through the cooling tower, if T_{rw} exceeds, T_{max}, using the two three-way valves.

We shall now write the governing energy equations for the system when only heat pumps 1 and 2 are in operation.

For heat pump 1 in the heating mode in zone 1 we have

$$Q_{hw} = Q_h \left(1 - \frac{1}{\varepsilon_h}\right) \tag{13.38}$$

$$W_h = \frac{Q_h}{\varepsilon_h} \tag{13.39}$$

where Q_h and Q_{hw} are the heating load of the zone and the heat absorbed from the water loop; W_h and ε_h are the electrical energy input and the COP of the heat pump in the *heating mode*. Typical value of ε_h is about 4.5.

For heat pump 2 in the cooling mode in zone 2 we have

$$Q_{cw} = Q_c \left(1 + \frac{1}{\varepsilon_c}\right) \tag{13.40}$$

$$W_c = \frac{Q_c}{\varepsilon_c} \tag{13.41}$$

where Q_c and Q_{cw} are the cooling load of the zone and the heat rejected to the water loop; W_c and ε_c are the electrical energy input and the COP of the heat pump in the *cooling mode*. Typical value of ε_c is about 3.5.

When both heat pumps 1 and 2 are in operation, the net heat transferred to the water loop is

$$\Delta Q = (Q_{cw} - Q_{hw})\Delta t \tag{13.42}$$

where Δt is the time interval, *typically one hour*.

The change in temperature of the water tank, ΔT is given by

$$\Delta T = \frac{3600\Delta Q}{\rho_w V_s c_w} \tag{13.43}$$

where ρ_w and c_w are the density and the specific heat capacity of water. The volume of the water storage tank is V_s.

We need to consider three phases of operation of the system. In the first phase, when $T_{rw} > T_{max}$, the water passes through the cooling tower. The total electrical energy consumption rate of the heat pumps is

$$E_{1hp} = \frac{(W_h + W_c)}{\eta_e} \tag{13.44}$$

The energy consumption rate of the cooling tower is

$$E_{1ct} = \frac{(Q_{cw} - Q_{hw})}{\eta_{ct}} \tag{13.45}$$

where η_e and η_{ct} are efficiencies of the electrical energy supply system to the heat pumps and the cooling tower respectively.

During the second phase of the system, when $T_{rw} < T_{min}$, the return water is heated by passing through the heater. The total electrical energy consumption rate for the heat pumps is given by Eq. (13.44) and the energy consumption rate of the water heater is

$$E_{2b} = \frac{(Q_{hw} - Q_{cw})}{\eta_{wh}} \tag{13.46}$$

η_{wh} is the efficiency of the heater.

During the third phase of operation of the system, when $T_{min} < T_{rw} < T_{max}$, both the heater and the cooling tower are bypassed. The total electrical energy consumption rate for the heat pumps is given by Eq. (13.44).

The energy performance of the WLHP-system can be simulated using Eqs. (13.38) to (13.46), if the hourly values of the heating and cooling loads of the different zones are known. We shall illustrate the numerical procedure in worked example 13.15.

13.5 Worked Examples

Example 13.1 An office building is occupied for 12 hours from 7 a.m. to 7 p.m. and unoccupied during the rest of the day. During the occupied period the building is maintained at 22°C and the occupants, lights and equipment generate heat at the rate of 14 kW. During the unoccupied period the internal heat gain is 1.2 kW and the indoor temperature is 18°C.

The effective heat transfer coefficient-area product, $(UA)_{eff}$ of the building is 1.6kW°C^{-1}. The hourly ambient temperatures T_A on a certain day are given in Table E13.1.1. Calculate (i) the balance temperatures for the occupied and unoccupied periods, (ii) the ambient temperature for which the heat loss is zero for the two periods, and (iii) the total heat input to the building if the furnace supplying the heat has a constant efficiency of 90%.

Table E13.1.1 Hourly ambient temperature

Hour	1	2	3	4	5	6	7	8	9	10	11	12
T_A °C	2.5	2.4	2.1	2.0	1.7	1.9	2.9	5	6.7	8.3	10	12.2
Hour	13	14	15	16	17	18	19	20	21	22	23	24
T_A °C	13.8	15	15	14.4	13.3	11.7	10	7.8	6.7	5	3.3	2.6

Solution We shall assume that the thermal storage effects in the building are negligible. Now the balance temperature at which the heating load is zero is given by Eq. (13.8a) as

$$T_{bal} = T_{in} - \frac{\dot{Q}_{int}}{U_o A_o} \tag{E13.1.1}$$

Substituting the given numerical values for the occupied and unoccupied periods in Eq. (E13.1.1) we have

$$T_{bal,occ} = 22 - \frac{14}{1.6} = 13.25°\text{C}$$

$$T_{bal,uoc} = 18 - \frac{1.2}{1.6} = 17.25°\text{C}$$

The heat loss is zero when the ambient temperature is equal to the indoor temperature. For the occupied and unoccupied periods these values are 22°C and 18°C respectively. It is clear from the tabulated ambient temperatures that the heat loss will always be positive.

The rate of heat input is given by

$$\dot{Q}_h = \frac{U_o A_o (T_{bal} - T_{amb})}{\eta_{fu}} \tag{E13.1.2}$$

Apply Eq. (E13.1.2) to the occupied period from 7 a.m. to 7 p.m. considering one hour at a time. Hence the total energy input is

$$E_{t,occ} = \sum_{r=7}^{r=18} 1.6(13.25 - T_{Ar})^{+} \times \frac{3600}{0.9} = 230 \times 10^3 \text{ kJ}$$

The '+' sign implies that the terms in the summation for which the ambient temperature T_{Ar} is greater than 13.25°C are ignored because the heat loss is negative (i.e. it is a heat gain).
Similarly, for the unoccupied 12 hours we have

$$E_{t,uoc} = \sum_{r=1}^{r=24} 1.6(17.25 - T_{Ar})^{+} \times \frac{3600}{0.9} = 1017.6 \times 10^3 \text{ kJ}$$

The total energy input is 1247 MJ.

Example 13.2 As a result of an energy audit of the building considered in worked example 13.1 the following measures are proposed to reduce the daily energy consumption: (a) decrease $(UA)_{eff}$ of the building to 1.4kW °C^{-1} by adding insulation, (b) lower the thermostat setting during the occupied and unoccupied periods to 20°C and 16°C respectively. (i) Calculate the new heat input to the building when the proposed changes are made. (ii) Show the impact of the proposed changes on a graph of heat input rate versus ambient temperature. Assume that the internal heat gains and the furnace efficiency are unchanged.

Solution We notice from Eqs. (E13.1.1) and (E13.1.2) that decreasing $(UA)_{eff}$ of the building lowers both the balance temperature and the required heat input. Substituting the given numerical values for the occupied and unoccupied periods in Eq. (E13.1.1) we have

$$T_{bal,occ} = 20 - \frac{14}{1.4} = 10°\text{C}$$

$$T_{bal,uoc} = 16 - \frac{1.2}{1.4} = 15.1°\text{C}$$

Apply Eq. (E13.1.2) to the occupied period from 7 a.m. to 7 p.m. considering one hour at a time. Hence the total energy input is

$$E_{t,occ} = \sum_{r=7}^{r=18} 1.4(10 - T_{Ar})^{+} \times \frac{3600}{0.9} = 95.8 \times 10^3 \text{ kJ}$$

The '+' sign implies that the terms in the summation for which the ambient temperature T_{Ar} is greater than 10°C are ignored because the heat loss is negative (i.e. it is a heat gain).

Similarly, for the unoccupied 12 hours we have

$$E_{t,uoc} = \sum_{r=1}^{r=24} 1.4(15.1 - T_{Ar})^{+} \times \frac{3600}{0.9} = 300 \times 10^3 \text{ kJ}$$

The individual effects of lowering the $(UA)_{eff}$ and the thermostat setting are depicted graphically in Figs. E13.2.1 (a) and (b) respectively. These same trends are applicable to both the occupied and the unoccupied periods. Now the slopes of the graphs according to Eq. (E13.1.2) are equal to $(UA)_{eff}$, and therefore when $(UA)_{eff}$ is reduced, the slopes are decreased. Moreover, the balance temperature, according to Eq. (E13.1.2) is also lowered. These effects combine to reduce the required heat input.

When the thermostat setting is lowered, keeping $(UA)_{eff}$ constant, the slopes of the graphs are unaffected. However, the balance temperature, according to Eq. (E13.1.2), is lowered.

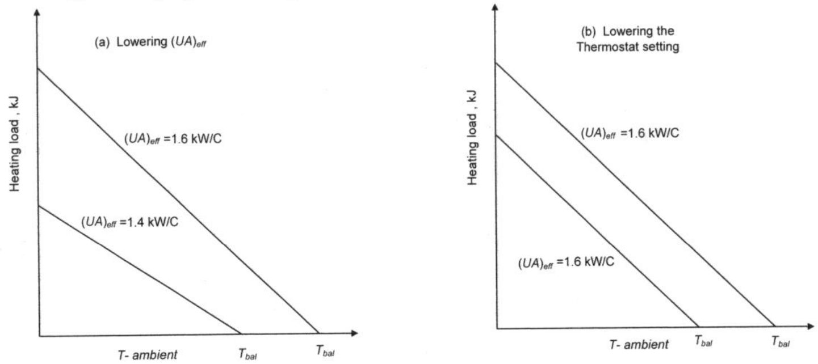

Fig. E13.2.1 (a) Effect of lowering $(UA)_{eff}$ (b) Effect of lowering thermostat setting

Example 13.3 A heat pump is used to heat a building maintained at 20°C. The occupants, lights and equipment generate heat at the rate of 4 kW. The effective heat transfer coefficient-area product, $(UA)_{eff}$ of the building is 0.9kW °C^{-1}.

The variations of the capacity, Q_{cap} (kW) and COP with ambient temperature, T_a (°C) of the heat pump (see Fig. 13.5) are well represented by the following equations:

$$Q_{cap} = 10.46 + 0.59T_a + 0.0092{T_a}^2$$

and $$COP = 2.27 + 0.082T_a + 0.0037{T_a}^2$$

The auxiliary heat source is a furnace with an efficiency of 88%.

The hourly ambient temperatures, T_A on a certain day are given in Table E13.3.1 below. Calculate (i) the balance temperature for the building, (ii) the heat pump balance temperature, (iii) the hourly heat input rate by the auxiliary energy source, and (iv) the hourly electrical energy input rate to the heat pump.

Table E13.3.1 Hourly ambient temperature

Hour	1	2	3	4	5	6	7	8	9	10
T_A °C	-9	-6	-3	0	3	6	9	12	15	18

Solution We shall assume that the thermal storage effects in the building are negligible. Now the balance temperature for the building at which the heating load is zero is given by Eq. (13.8a) as

$$T_{bal} = 20 - \frac{4}{0.9} = 15.55°\text{C}$$

The balance point for the heat pump, at which the capacity of the heat pump is equal to the heating load, is given by the point of intersection B of the heat pump capacity curve and the building heat load curve. This value is obtained graphically as shown in Fig. E13.3.1. Alternatively, we can solve simultaneously Eq. (13.8) and the heat pump capacity relationship given above. Hence we obtain the heat pump balance temperature as 2.34°C.

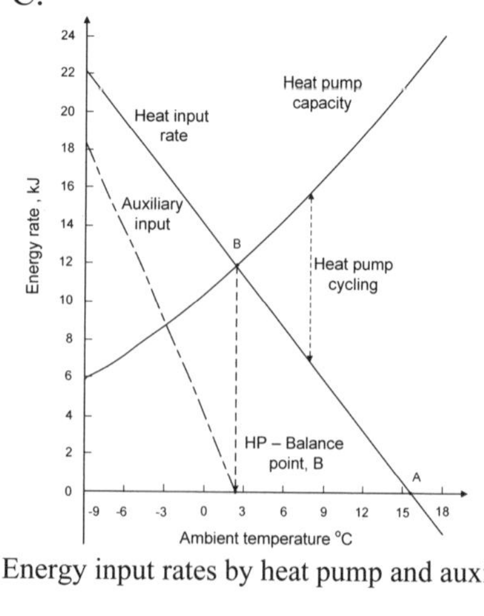

Fig. E13.3.1 Energy input rates by heat pump and auxiliary source

The required heat input rate is given by Eq. (13.8) as

$$\dot{L}_h = U_o A_o (T_{bal} - T_{amb})$$

$$\dot{L}_h = 0.9(15.55 - T_{amb}) \quad \text{(E13.3.1)}$$

The hourly heat rates computed using Eq. (E13.3.1) are listed in Table E13.3.2.

The rate of heat input by the heat pump is given by

$$Q_{cap} = 10.46 + 0.59T_a + 0.0092{T_a}^2$$

The values for different ambient temperatures, T_a are listed in Table E13.3.2.

The auxiliary heat input when $T_{amb} < T_{bal}$ is given by

$$Q_{aux} = \frac{\dot{L}_h - Q_{cap}(T_a)}{\eta_{eff}}$$

The electrical energy input is

$$E_{hp} = \frac{Q_{cap}}{COP(T_a)}$$

The electrical energy input rate to the heat pump when it is operating in a cycling mode ($T_{amb} > T_{bal}$) is given by

$$E_{hp} = \frac{\dot{L}_h}{COP(T_a)}$$

The hourly energy input rates (kW) are given in Table E13.3.2. These values are used to plot the graphs in Fig. E13.3.1.

Notice that for ambient temperatures above the heat pump balance point, the capacity of heat pump is larger than the heating load of the building. Under these conditions, the heat pump operates in a cycling mode where it is switched on for a period of time until the desired indoor temperature is reached, and then switched off.

Table E13.3.2 Hourly values of energy quantities (kW) and COP

Hour	1	2	3	4	5	6	7	8	9	10
$\dot{L}_h$	22.1	19.4	16.7	14	11.3	8.6	5.9	3.2	0.5	0
Q_{cap}	5.9	7.25	8.77	10.5	12.3	14.3	16.5	18.9	21.4	24.1
COP	1.83	1.91	2.06	2.27	2.55	2.9	3.31	3.79	4.33	4.9
Q_{aux}	18.4	13.8	9	4	0	0	0	0	0	0
E_{hp}	3.2	3.8	4.26	4.61	4.43	2.97	1.78	0.85	0.12	0

Example 13.4 A heat pump is used to heat a building that is occupied for 12 hours from 7 a.m. to 7 p.m. and unoccupied during the rest of the day. During the occupied period the building is maintained at 20°C, and the occupants, lights and equipment generate heat at the rate of 4 kW. During the unoccupied period the internal heat gain is 0.8 kW and the indoor temperature is 16°C. The effective heat transfer coefficient-area product, $(UA)_{eff}$ of the building is 0.9 kW°C^{-1}.

The variations of the capacity, Q_{cap} (kW) and COP with ambient temperature, T_a (°C) of the heat pump (see Fig. 13.5) are well represented by the following equations:

$$Q_{cap} = 7.85 + 0.44T_a + 0.007T_a^{\ 2}$$

$$COP = 2.27 + 0.082T_a + 0.0037T_a^{\ 2}$$

The auxiliary energy source is a natural gas furnace with an efficiency of 88%. The hourly ambient temperature, T_A on a certain day is given in Table E13.4.1 below. Calculate (i) the balance temperatures for the building during the occupied and unoccupied periods, (ii) the heat pump balance temperatures for the occupied and unoccupied periods, and (iii) the hourly heating load, the hourly heat inputs by the heat pump and the furnace for the two periods.

Table E13.4.1 Hourly ambient temperatures

Hour	1	2	3	4	5	6	7	8	9	10	11	12
T_A °C	2.5	2.4	2.1	2.0	1.7	1.9	2.9	5	6.7	8.3	10	12.2
Hour	13	14	15	16	17	18	19	20	21	22	23	24
T_A °C	13.8	13.8	13.5	13.6	13.3	11.7	10	7.8	6.7	5	3.3	2.6

Solution Using Eq. (E13.1.1) we obtain the balance temperatures for the occupied and unoccupied periods as

$$T_{bal,occ} = 20 - \frac{4}{0.9} = 15.56\text{°C}$$

$$T_{bal,uoc} = 16 - \frac{0.8}{0.9} = 15.1\text{°C}$$

We determine the heat pump balance points for the two periods by solving simultaneously Eq. (13.8) and the given heat pump capacity equation. Hence at the heat pump balance point we have

$$U_o A_o (T_{bal} - T_a) = 7.85 + 0.44T_a + 0.007{T_a}^2$$

For the occupied period we have

$$0.9(15.56 - T_a) = 7.85 + 0.44T_a + 0.007{T_a}^2$$

Therefore the heat pump balance temperature is 4.4°C

For the unoccupied period we have

$$0.9(15.1 - T_a) = 7.85 + 0.44T_a + 0.007{T_a}^2$$

Therefore the heat pump balance temperature is 4.2°C

The auxiliary heat input from the furnace is required only when the ambient temperature is below the respective heat pump balance temperatures for the two periods.

We use the same equations as in worked example 13.3 to calculate the hourly energy quantities (kW) for the occupied and unoccupied periods. These quantities are listed in Table E13.4.1.

Table E13.4.1 Hourly energy quantities (kW)

Hour	1	2	3	4	5	6	7	8	9	10	11	12
Q_{load}	11.4	11.4	11.7	11.8	12.1	11.9	11.4	9.5	8	6.5	5	3
Q_{hp}	9.0	8.95	8.8	8.75	8.6	8.73	9.2	10.2	11.1	12	12.9	14.3
Q_{furn}	2.68	2.83	3.3	3.5	3.9	3.6	2.51	0	0	0	0	0
E_{hp}	3.6	3.6	3.6	3.6	3.6	3.6	3.62	3.42	2.67	2.04	1.45	0.79
Hour	13	14	15	16	17	18	19	20	21	22	23	24
Q_{load}	1.6	1.6	1.85	1.76	3.5	5	7.6	6.6	7.6	9.1	10.6	11.3
Q_{hp}	15.3	15.3	15.1	15.1	14	13	13	11.7	11.1	10.2	9.4	9.0
Q_{furn}	0	0	0	0	0	0	0	0	0	0	1.4	2.5
E_{hp}	0.38	0.38	0.46	0.43	0.93	1.45	1.33	2.1	2.53	3.28	3.63	3.6

Example 13.5 A building in Toronto has an effective *UA* value of 2.25 kW°C^{-1}. Its heating system consists of a furnace with an average efficiency of 88%. The balance temperature is 18.3°C. (i) Use the degree–day data in Table 13.1 to estimate the monthly fuel energy input. (ii) If the balance temperature is decreased to 16°C by lowering the thermostat setting, obtain the monthly fuel energy input.

Solution The monthly fuel energy input to the furnace is given by

$$Q_{fm} = \frac{N_{dd}(UA)(24\times 3600)}{\eta_{fur}} \quad \text{(E13.5.1)}$$

where N_{dd} is the number of heating degree–days for the month, UA is the loss coefficient-area product and η_{fur} is the efficiency of the furnace.

The monthly heating degree days for Toronto for a balance temperature of 18.3°C are given in Table 13.1. Substituting numerical values in Eq. (E13.5.1) we compute the monthly fuel energy input. The results are given in Table E13.5.1.

Table E13.5.1 Monthly fuel energy inputs

Month	January	February	March	April	May	June
Deg.–days	715	640	557	336	167	39
$Q_{furn.}$ (GJ)	158	141	123	74	37	9
Month	July	August	Sept.	October	Nov.	Dec.
Deg.–days	6	13	85	271	434	629
$Q_{furn.}$ (GJ)	1	3	19	60	96	139

The annual energy input is 860GJ.

The degree-day data tabulated in Ref. [1] are valid only when the balance temperature is 18.3°C. When the balance temperature changes to 16°C due to the new setting of the thermostat, we need to compute the resulting degree-days for the months.

For this purpose we use the MATLAB program listed in Appendix A13.1. It incorporates the equations given in section 13.3.1 for computing bin data for a series of ambient temperatures. The bin data are integrated to obtain the degree days using Eq. (13.18). The only location-specific data required as input are the monthly average ambient temperatures, which are tabulated in Ref. [1]. The monthly fuel energy inputs for a balance temperature of 16°C are tabulated below.
The annual energy input is 698 GJ.

Table E13.5.2 Monthly fuel energy inputs

Month	January	February	March	April	May	June
Deg.–days	641	574	484	264	81	0
$Q_{furn.}$ (GJ)	142	127	107	58	18	0
Month	July	August	Sept.	October	Nov.	Dec.
Deg.–days	0	0	0	195	363	558
$Q_{furn.}$ (GJ)	0	0	0	43	81	123

Example 13.6 A building in Singapore has an effective *UA* value of 1.85 kW°C^{-1}. Its cooling system consists of a chiller with an average COP of 3.6. The balance temperature is 18.3°C. (i) Use the degree–day data in Table 13.1 to estimate the monthly electrical energy input to the chiller. (ii) If the balance temperature is increased to 20°C by raising the thermostat setting, obtain the monthly electrical energy input.

Solution The monthly electricity energy input to the chiller may be obtained using the cooling degree–days. However, it should be noted that this procedure is not considered accurate for cooling load estimation. Unlike the heating load, the cooling load depends on several time-varying inputs, in addition to the ambient temperature. These were discussed in chapter 10.

The monthly electrical energy input to the chiller is given by

$$E_{chi} = \frac{N_{dd}(UA)(24\times3600)}{COP} \tag{E13.6.1}$$

where N_{dd} is the number of cooling degree–days for the month, *UA* is loss coefficient-area product and *COP* is the average COP of the chiller.

The monthly cooling degree days for Singapore for a balance temperature of 18.3°C is given in Table 13.1. Substituting numerical values in Eq. (E13.6.1) we compute the monthly electrical energy input. The results are given in Table E13.6.1.

Table E13.6.1 Monthly electrical energy inputs

Month	January	February	March	April	May	June
Deg.–days	276	269	308	310	327	312
E_{chil} (GJ)	12.3	11.9	13.7	13.8	14.5	13.9
Month	July	August	Sept.	October	Nov.	Dec.
Deg.–days	308	306	294	306	281	276
E_{chil} (GJ)	13.7	13.6	13.1	13.6	12.5	12.3

The annual energy input is 159 GJ.

The data tabulated in Ref. [1] are valid only when the balance temperature is 18.3°C. When the balance temperature changes to 20°C due to the new setting of the thermostat, we need to compute the new degree–days for the months.

For this purpose we use the MATLAB program listed in Appendix A13.1. It incorporates the equations given in section 13.3.1 for computing bin data for a series of ambient temperatures. The bin data are integrated to obtain the degree–days using Eq. (13.18). The only location-specific data required as inputs are the monthly average ambient temperatures, which are tabulated in Ref. [1]. The monthly electrical energy inputs for a balance temperature of 20°C are tabulated below. The annual energy input is 132 GJ.

Table E13.6.2 Monthly electrical energy inputs

Month	January	February	March	April	May	June
Deg.–days	223	224	257	261	275	261
E_{chil} (GJ)	9.9	9.95	11.4	11.6	12.2	11.6
Month	July	August	Sept.	October	Nov.	Dec.
Deg.–days	257	254	243	254	231	223
E_{chil} (GJ)	11.4	11.3	10.8	11.3	10.3	9.9

Example 13.7 The furnace of a building in Toronto has a rated capacity of 30 kW and a rated efficiency of 85%. The *UA*-value of the building is 0.62 kW°C^{-1} and the indoor temperature is 22°C. The continuous internal heat gain of the building is 2.6 kW. The degradation coefficient of the furnace due to cycling is 0.25. Estimate the annual heating energy requirement for the building.

Solution We shall use the bin method to estimate the annual energy requirement. The bin data for Toronto are obtained using the MATLAB program in Appendix A13.1.
The balance temperature of the building is

$$T_{bal} = 22 - \frac{2.6}{0.62} = 17.8°\text{C}$$

The heat load at a bin temperature, T_{bi} is given by

$$\dot{Q}_{bi} = UA(T_{bal} - T_{bi}) \quad (T_{bal} > T_{bi})$$

$$\dot{Q}_{bi} = 0.62(17.8 - T_{bi})$$

The load factor, *LF* of the furnace is

$$LF = \frac{\dot{Q}_{bi}}{Q_{cap}}$$

The part-load factor, *PLF* is given by Eq. (13.19) as

$$PLF = 1 - c_d(1 - LF)$$

where the degradation coefficient, $c_d = 0.25$.
The heat input by the furnace for the bin temperature T_{bi} is given by Eq. (13.23) as

$$\dot{Q}_{i,fur} = \frac{\dot{Q}_{bi}}{\eta_{fur} PLF}$$

The yearly energy input for bin, *i* is

$$E_{bi} = 3600 N_{bi} \dot{Q}_{i,fur}$$

where N_{bi} is the number of hours per year in bin, *i*.

The numerical data obtained for the different bins are tabulated below. The annual heat input required is 200 GJ and the fuel energy input to the furnace is 278 GJ.

Table E13.7.1 Summary of yearly energy inputs in bins

N-bin	T_{mean}, C	No. hours	Q_{load}, GJ	*PLF*	Q_{fur}, GJ
1	−27.5	1	0.093	0.98	0.112
2	−25	2	0.198	0.97	0.240
3	−22.5	5	0.42	0.96	0.515
4	−20	10	0.881	0.95	1.096
5	−17.5	23	1.827	0.93	2.305
6	−15	51	3.705	0.92	4.74
7	−12.5	106	7.194	0.91	9.34
8	−10	208	12.88	0.89	16.96
9	−7.5	356	20.12	0.88	26.87
10	−5	510	25.95	0.87	35.17
11	−2.5	604	27.36	0.85	37.65
12	0	624	24.81	0.84	34.66
13	2.5	612	20.91	0.83	29.67
14	5	605	17.29	0.82	24.93
15	7.5	605	13.91	0.8	20.37
16	10	597	10.41	0.79	15.49
17	12.5	607	7.190	0.78	10.88
18	15	666	4.172	0.76	6.42
19	17.5	766	0.524	0.75	0.82
20	20	801	0	-	0

Example 13.8 The building described in worked example 13.7 is to be heated with a heat pump and an auxiliary heat source. The variations of the capacity, Q_{cap} (kW) and COP with ambient temperature, T_a (°C), of the heat pump are well represented by the following equations:

$$Q_{cap} = 20.92 + 1.18T_a + 0.0184{T_a}^2$$

and

$$COP = 2.27 + 0.082T_a + 0.0037{T_a}^2$$

Estimate (i) the annual electrical energy input to the heat pump, and (ii) the annual energy input to the auxiliary heat source.

Solution We shall use the bin method to estimate the annual energy requirement. The bin data for Toronto are obtained using the MATLAB program in Appendix A13.1.

The balance temperature of the building is

$$T_{bal} = 22 - \frac{2.6}{0.62} = 17.8\text{°C}$$

The heat load at a bin temperature, T_{bi} is given by

$$\dot{Q}_{bi} = UA(T_{bal} - T_{bi}) \quad (T_{bal} > T_{bi})$$

$$\dot{Q}_{bi} = 0.62(17.8 - T_{bi})$$

We determine the heat pump balance point by solving simultaneously Eq. (13.8) and the given heat pump capacity equation. Hence at the heat pump balance point we have

$$0.62(17.8 - T_a) = 20.92 + 1.18T_a + 0.0184{T_a}^2$$

Hence the heat pump balance temperature is −5.8°C.

The heat input from the auxiliary source is required only when the ambient temperature is below the heat pump balance temperature.

The yearly energy input required for bin, *i* is

$$E_{b,i} = 3600N_{bi}\dot{Q}_{bi}$$

where N_{bi} is the number of hours per year in bin, *i*.

The yearly energy supplied by the heat pump for bin, *i* is

$$E_{hp,i} = 3600N_{bi}\dot{Q}_{hp,i}$$

The yearly electrical energy input to the heat pump is

$$W_{hp,i} = \frac{3600 N_{bi} \dot{Q}_{hp,i}}{COP_i}$$

The yearly energy supplied by the auxiliary source for bin, *i* is

$$E_{aux,i} = E_{b,i} - E_{hp,i}$$

The numerical results obtained using the above equations are given in Table E13.8.1. The yearly heat input required is 200GJ. The heat input by heat pump is 183 GJ and the electrical energy input to the heat pump is 79 GJ. The auxiliary energy input is 17 GJ.

Table E13.8.1 Summary of yearly energy inputs in bins

N-bin	T_{bin}, C	No. hr.	E_{load}, GJ	E_{hp}, GJ	W_{hp}, GJ	E_{aux}, GJ
1	−27.5	1	0.093	0.0079	0.0028	0.0855
2	−25	2	0.198	0.0218	0.0086	0.177
3	−22.5	5	0.42	0.0619	0.0269	0.358
4	−20	10	0.881	0.1759	0.0833	0.705
5	−17.5	23	1.827	0.4928	0.2504	1.334
6	−15	51	3.705	1.331	0.7159	2.364
7	−12.5	106	7.194	3.463	1.899	3.731
8	−10	208	12.88	8.190	4.50	4.693
9	−7.5	356	20.12	16.80	9.02	3.314
10	−5	510	25.95	25.95	13.29	0
11	−2.5	604	27.36	27.36	13.10	0
12	0	624	24.81	24.81	10.93	0
13	2.5	612	20.91	20.91	8.37	0
14	5	605	17.29	17.29	6.24	0
15	7.5	605	13.91	13.90	4.5	0
16	10	597	10.41	10.41	3.0	0
17	12.5	607	7.190	7.19	1.86	0
18	15	666	4.172	4.17	0.96	0
19	17.5	766	0.524	0.52	0.11	0
20	20	801	0	0	-	0

Example 13.9 An air conditioner has a full load capacity of 12 tons and its COP is 3.8. It is installed in a building with a cooling load of 25 kW. Calculate (i) the electrical energy input rate, and (ii) the condenser heat rejection rate of the air conditioner.

Solution The full load capacity of the air conditioner is 12 tons which is equal to 42.2 kW. The actual refrigeration load is equal to the cooling load of the building which is 25 kW. We define the part-load ratio as

$$LR = \frac{\dot{Q}_{cl}}{Q_{cap}} = \frac{25}{42.2} = 0.592$$

As was discussed in section 13.3.4 the performance of the air conditioner decreases at part-load due to cycling. The effect of cycling on the COP may be accounted for by using a part-load factor defined as [3,4]

$$PLF = 1 - c_d(1 - LR) \qquad \text{(E13.9.1)}$$

where c_d is called a degradation coefficient which has a typical value of about 0.25. Substituting in Eq. (E13.9.1) we have

$$PLF = 1 - 0.25(1 - 0.592) = 0.9$$

Therefore COP under part-load conditions is

$$(COP)_{pl} = PLF(COP)_{fl} = 0.9 \times 3.8 = 3.42$$

The electrical energy input (work) to the air conditioner under part-load conditions is

$$\dot{W}_{pl} = \frac{\dot{Q}_{pl}}{(COP)_{pl}} = \frac{25}{3.42} = 7.31 \text{ kW}$$

The heat rejection rate in the condenser is

$$\dot{Q}_{con} = \dot{Q}_{pl} + \dot{W}_{pl} = 25 + 7.31 = 32.3 \text{ kW}$$

Note that if the part-load factor is not included, i.e. PLF =1, the electrical energy input is 6.58 kW and the heat rejection rate is 31.6 kW.

Example 13.10 The estimated hourly cooling loads of a building are given in Table E13.10.1. The air conditioning system consists of two similar chillers A and B, each of rated capacity 3500 kW. The variation of the COP of the chillers with load factor f_l is well represented by the equation (chiller 1 in Fig. 13.9):

$$cop(f_l) = 11.255f_l^3 - 28.14f_l^2 + 21.08f_l + 2.243$$

When the cooling load of the building is below 3500 kW, only one chiller is operated. The second chiller is switched on when the cooling load exceeds 3500 kW. Calculate the hourly electrical energy input to the two chillers.

Table E13.10.1 Hourly cooling loads (kW)

Hour	1	2	3	4	5	6
Q_l, kW	1758	1696	1570	1507	1633	1633
Hour	7	8	9	10	11	12
Q_l, kW	2010	2512	2763	3642	4396	5150
Hour	13	14	15	16	17	18
Q_l, kW	6154	6280	6029	5777	5652	5401
Hour	19	20	21	22	23	24
Q_l, kW	5150	4647	3894	3266	2638	2261

Solution The two chillers are switched on based on the prevailing cooling load conditions. When the load exceeds 3500, the second chiller is switched on and the two chillers share the load equally as discussed in section 3.4.2.

The operating conditions of the two chillers are as follows:

If $Q_{cooling}$ < 3500 kW, only chiller A is switched on.

The cooling load fraction is $f_l = \frac{\dot{Q}_{cl}}{3500}$

The work inputs to the two chillers are:

$$W_A = \frac{3500 f_l}{COP(f_l)}, \quad W_B = 0$$

If $Q_{cooling}$ >= 3500 kW, both chillers A and B are switched on.

The cooling load factor is $f_l = \frac{\dot{Q}_{cl}}{3500+3500}$

The work inputs to the two chillers are:

$$W_A = \frac{3500 f_l}{COP(f_l)}, \quad W_B = \frac{3500 f_l}{COP(f_l)}$$

We substitute the given numerical values in the above equations to obtain the hourly electrical energy input rates to the two chillers. The computed results are summarized in Table E13 10.1.

Table E13.10.1 Computed energy inputs to chillers A and B

Hour	1	2	3	4	5	6
Q_l, kW	1758	1696	1507	1507	1633	1633
W_{ca}, kW	246	238	223	215	230	230
W_{cb}, kW	0	0	0	0	0	0
Hour	7	8	9	10	11	12
Q_l	2010	2512	2763	3642	4396	5150
W_{ca}	279	357	401	254	306	368
W_{cb}	0	0	0	254	306	368
Hour	13	14	15	16	17	18
Q_l	6154	6280	6029	5777	5652	5401
W_{ca}	461	473	449	425	413	390
W_{cb}	461	473	449	425	413	390
Hour	19	20	21	22	23	24
Q_l	5150	4647	3893	3266	2638	2261
W_{ca}	368	326	270	498	379	316
W_{cb}	368	326	270	0	0	0

The total daily energy input to the two chillers is 12.6 MWh.

Example 13.11 The air conditioning system for the building described in worked example 13.10 is to include three chillers A, B and C of rated capacities 1750 kW, 1750 kW and 3500 kW respectively. When the load is below 1750 kW only chiller A is switched on. When the load is between 1750 kW and 3500 kW, chiller A and B are switched on. When the load is between 3500 kW and 5250 kW chillers A and C are switched on. When the load is between 5250 kW and 7000 kW all three chillers are operated. Obtain the hourly electrical energy input to the three chillers.

The variation of the COP of chillers A and B with the load fraction, f_l is well represented by the equation (see chiller 2 in Fig. 13.9):

$$COP_A(f_l) = -8.557f_l^3 + 11.72f_l^2 - 0.434f_l + 3.534$$

The variation of the COP of chiller C with the load fraction is well represented by the equation (see chiller 1 in Fig. 13.9):

$$COP_C(f_l) = 11.255\,f_l^3 - 28.14f_l^2 + 21.08f_l + 2.243$$

Solution The three chillers are switched on based on the prevailing cooling load. This type of matching of the total output of multiple chillers to the cooling load is commonly called chiller sequencing which can be used to reduce the total energy input to the chillers.

The operating conditions of the three chillers are as follows:

If $Q_{cooling} < 1750$ kW, only chiller A is switched on.

The cooling load factor is $f_l = \frac{\dot{Q}_{cl}}{1750}$

The work inputs to the three chillers are:

$$W_A = \frac{1750 f_l}{COP_A(f_l)}, \quad W_B = 0, \quad W_C = 0$$

If 1750 kW < $Q_{cooling}$ <= 3500 kW, only chillers A and B are switched on.

The load factor is $f_l = \frac{\dot{Q}_{cl}}{1750+1750}$

The work inputs to the three chillers are:

$$W_A = \frac{1750 f_l}{COP_A(f_l)}, \quad W_B = \frac{1750 f_l}{COP_B(f_l)}, \quad W_C = 0$$

If 3500 kW < $Q_{cooling}$ <= 5250 kW, only chillers A and C are switched on.

The load factor is $f_l = \frac{\dot{Q}_{cl}}{1750+3500}$

The work inputs to the three chillers are:

$$W_A = \frac{1750 f_l}{COP_A(f_l)}, \quad W_C = \frac{3500 f_l}{COP_C(f_l)}, \quad W_B = 0$$

If 5250 kW < $Q_{cooling}$ <= 7000 kW, all three chillers A, B and C are switched on.

The load factor is $f_l = \frac{\dot{Q}_{cl}}{1750+1750+3500}$

The work inputs to the three chillers are:

$$W_A = \frac{1750 f_l}{COP_A(f_l)}, \quad W_B = \frac{1750 f_l}{COP_B(f_l)}, \quad W_C = \frac{3500 f_l}{COP_C(f_l)}$$

We substitute the given numerical values in the above equations to obtain the hourly electrical energy input rates to the three chillers. The computed results are summarized in Table E13 11.1.

Table E13.11.1 Computed energy inputs to the chillers A, B and C

Hour	1	2	3	4	5	6
Q_l, kW	1758	1696	1507	1507	1633	1633
W_{ca}, kW	169	268	245	236	256	256
W_{cb}, kW	169	0	0	0	0	0
W_{cc}, kW	0	0	0	0	0	0
Hour	7	8	9	10	11	12
Q_l	2010	2512	2763	3642	4396	5150
W_{ca}	182	206	220	202	230	272
W_{cb}	182	206	220	0	0	0
W_{cc}	0	0	0	343	433	531
Hour	13	14	15	16	17	18
Q_l	6154	6280	6029	5777	5652	5401
W_{ca}	240	245	236	227	224	216
W_{cb}	240	245	236	227	224	216
W_{cc}	461	473	449	425	413	390
Hour	19	20	21	22	23	24
Q_l	5150	4647	3893	3266	2638	2261
W_{ca}	272	242	210	256	213	194
W_{cb}	0	0	0	256	213	194
W_{cc}	531	465	371	0	0	0

The total energy input to the three chillers is 13.63 MWh. Comparing this result with that obtained in worked example 13.10, we note that there is an increase in the total energy consumed by the chillers. This is because chillers A and B have lower COPs than chiller C. However, we could simulate alternative combinations of the three chillers to meet the hourly cooling load and hence determine the optimal combination. For instance, using chiller C when the cooling load is between 1750 kW and 3500 kW.

Example 13.12 A space with a constant cooling load of 155 kW is cooled by a chiller system where the condenser heat is rejected to the atmosphere through a cooling tower. The variation of the COP of the chiller with the ambient wet-bulb temperature is well represented by the equation

$$COP = 7.44 - 0.142t_{wb} + 0.0017t_{wb}^{\ 2}$$

The seasonal bin data for the local wet-bulb temperature is given in Table E13.12.1. Estimate the seasonal energy input to the chiller.

Table E13.12.1 Bin data for wet-bulb temperature

t_{wb}, °C	10	12.8	15.6	18.3	21.1	23.9	26.7
N-bin	672	650	840	1020	980	430	13

Solution In central air conditioning systems (see Fig. 13.8), the condenser cooling water rejects heat to the atmosphere in the cooling tower. The COP of the chiller depends on the temperature of water returning from the cooling tower, which in turn depends on the local wet-bulb temperature (see section 6.4).

The electrical energy input to the chiller is given by:

$$W_{chil} = \frac{Load_c}{COP(t_{wb})} = \frac{155}{COP(t_{wb})} \text{ kW} \qquad \text{(E13.12.1)}$$

The total energy input to the chiller for bin, *i* is

$$E_{cooling} = W_{chil} N_{bin}(t_{wb}) \text{ kWh} \qquad \text{(E13.12.2)}$$

The COP at any value of the wet-bulb bin temperature is calculated using the given expression. The energy input at any bin temperature and the total energy input are calculated using Eqs. (E13.12.1) and (E13.12.2) respectively. The computed results are given in Table E13.12.2.

Table E13.12.2 Computed energy inputs for bins

t_{wb}, °C	10	12.8	15.6	18.3	21.1	23.9	26.7
N-bin	672	650	840	1020	980	430	13
COP	6.19	5.91	5.65	5.41	5.2	5.02	4.87
W_{chil}, kW	25.02	26.24	27.45	28.64	29.79	30.87	31.86
E_{chil}, kWh	16816	17056	23059	29213	29191	13273	414

The total seasonal energy input to the chiller is 129.0 MWh

Example 13.13 An office building has the following occupancy schedule during a typical week. (i) From Monday to Friday it is fully occupied from 7 a.m. to 7 p.m. (12 hours) and unoccupied during the rest of the day. (ii) It is unoccupied on Saturday and Sunday. The *UA*-value for the building is 1.75 kW°C^{-1}. The internal heat gains and the indoor temperatures during the occupied and unoccupied periods are respectively, 14.8 kW, 22°C and 1.3 kW, 18°C. The building is heated by a fuel-fired furnace of capacity of 80 kW and efficiency 88%. Assume a degradation coefficient of 0.25 due to cycling of the furnace. The annual bin data for the three consecutive 8-hour periods, obtained from Ref. [3] are given in Table 13.13.1. Estimate the annual heating loads and heating energy inputs to the furnace for the occupied and unoccupied periods.

Table E13.13.1 Annual bin data for 3 hour - groups

Bin Temp, °C	N_{bin} (1h-8h)	N_{bin} (9h-16h)	N_{bin} (17h-24h)
−25	0	0	0
−22.22	0	0	0
−19.44	1	0	0
−16.67	7	0	1
−13.89	23	4	8
−11.11	47	10	26
−8.33	83	30	56
−5.55	125	65	89
−2.77	189	107	157
0	270	175	252
2.78	298	249	295
5.56	272	269	288
8.33	254	259	258
11.11	262	237	252
13.88	272	241	267
16.66	285	254	288
19.44	293	273	299
22.22	197	324	270
25	35	274	93
27.77	6	113	14
30.55	1	32	2
33.33	0	8	0
36.11	0	1	0

Solution We shall use the given information on the occupancy schedule of the building during a typical week to determine the occupied and unoccupied time fractions for each of the three hour-groups. Each hour-group has 56 hours per week.

For the first hour-group (1h-8h) the occupied and unoccupied fractions are: 5/56 = 0.09 and (1-0.09) = 0.91 respectively.

For the second hour-group (9h-16h) the occupied and unoccupied fractions are: 40/56 = 0.71 and (1-0.71) = 0.29 respectively.

For the third hour-group (17h-24h) the occupied and unoccupied fractions are: 15/56 = 0.268 and (1-0.268) = 0.732 respectively.

We assume that the above weekly occupancy pattern is repeated throughout the year.

Now the balance temperature at which the heating load is zero is given by Eq. (13.8a) as

$$T_{bal} = T_{in} - \frac{\dot{Q}_{int}}{U_o A_o} \tag{E13.13.1}$$

Substituting the given numerical values for the occupied and unoccupied periods in Eq. (E13.13.1) we have

$$T_{bal,occ} = 22 - \frac{14.8}{1.75} = 13.54°\text{C}$$

$$T_{bal,uoc} = 18 - \frac{1.3}{1.75} = 17.26°\text{C}$$

The heat load for a bin temperature, T_{bi} is given by

$$\dot{Q}_{bi} = UA(T_{bal} - T_{bi}) \quad (T_{bal} > T_{bi})$$

The load factor, *LF* of the furnace is

$$LF = \frac{\dot{Q}_{bi}}{Q_{cap}} = \frac{\dot{Q}_{bi}}{80}$$

The part-load factor, *PLF* is given by Eq. (13.19) as

$$PLF = 1 - c_d(1 - LF)$$

where the degradation coefficient $c_d = 0.25$.

The energy input rate to the furnace for bin temperature T_{bi} is

$$\dot{Q}_{i,fur} = \frac{\dot{Q}_{bi}}{\eta_{fur} PLF} = \frac{\dot{Q}_{bi}}{0.88 PLF}\ \text{kW}$$

The yearly energy input to the furnace for bin *i* during the occupied period is

$$E_{bi} = f_{occ} N_{bi} \dot{Q}_{i,fur}\ \text{kWh}$$

where N_{bi} is the number of hours in the bin, and f_{occ} is the occupied fraction.

The yearly energy input for bin *i* during the unoccupied period is

$$E_{bi} = f_{uocc} N_{bi} \dot{Q}_{i,fur} \text{ kWh}$$

where N_{bi} is the number of hours in the bin, and f_{uocc} is the unoccupied fraction.

The above equations are applied to each hour-group to obtain the total yearly heating load Q_{in} and the total energy input to the furnace E_{fur} for the occupied and unoccupied periods. The computed results are given in Table E13.13.2.

Table E13.13.2 Computed heating loads and furnace energy inputs

Hour-Group	Occupied period		Unoccupied period	
(time periods)	Q_{in} (MWh)	E_{fur} (MWh)	Q_{in} (MWh)	E_{fur} (MWh)
1 (1h-8h)	3.18	4.35	44.95	60.74
2 (9h-16h)	16.41	22.83	9.65	13.24
3 (17h-24h)	8.07	11.16	31.44	42.82
Total	27.66	38.34	86.04	116.8

The annual heating load is (27.66+86.04) = 113.7 MWh, and the annual energy input to the furnace is (38.34+116.8) = 155.4 MWh. Hence the average efficiency of the furnace is 113.7/155.5 = 73%.

Example 13.14 A house in Toronto has an effective *UA*-value of 1.3 kW°C^{-1} and it is heated using a ground-source heat pump, of capacity 40 kW and average COP 3.5. The average ground temperature for heat absorption by the heat pump is 6.5°C. The internal heat gains and the indoor temperature of the house are 5.4 kW and 22°C respectively. The degradation coefficient for cycling of the heat pump during part-load operation is 0.25. Calculate (i) heat pump balance temperature, (ii) the heating load, and (iii) the heat supplied by the heat pump and the auxiliary energy source.

Solution We shall assume that the heat source temperature for the heat pump is constant and equal to the average ground temperature of 6.5°C. At the heat pump balance temperature its capacity is equal to the heating load of the building.

When the external ambient temperature is below the balance temperature the heat pump is unable to meet the required heating load and therefore an auxiliary energy input is required. However, when the ambient temperature exceeds the balance temperature, the heat pump operates in a cycling mode because its capacity exceeds the heating load of the building. These scenarios are illustrated graphically in Fig. E13.14.1.

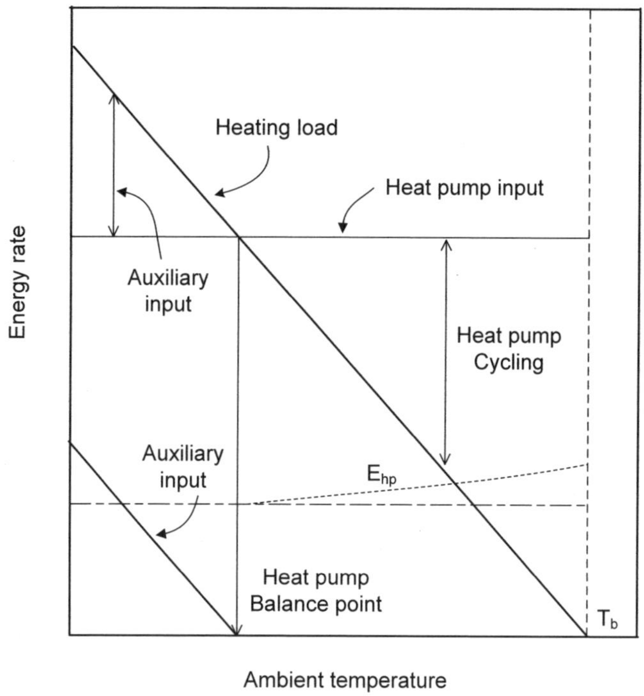

Fig. E13.14.1 Energy input rates by heat pump and the auxiliary source

The building balance temperature is given by Eq. (13.8a) as

$$T_{bal} = 22 - \frac{5.4}{1.3} = 17.85°\text{C}$$

The heat load for a bin temperature T_{bi} is given by

$$Q_{load} = UA(T_{bal} - T_{bi}) \quad (T_{bal} > T_{bi})$$

The heat pump balance temperature is given by

$$1.3(T_{bal} - T_{hp}) = 1.3(17.85 - T_{hp}) = 40$$

Hence the heat pump balance temperature is −13°C.

The auxiliary heat input when $T_{amb} < T_{bal}$ is

$$Q_{aux} = Q_{load} - Q_{cap}$$

The electrical energy input to the heat pump is

$$E_{hp} = \frac{Q_{cap}}{COP}$$

When the ambient temperature exceeds −13°C the heat pump operates in a cycling mode. The degradation in performance due to cycling is accounted for by using the degradation coefficient.

The load factor, *LF* of the heat pump is

$$LF = \frac{Q_{load}}{Q_{cap}} = \frac{Q_{load}}{40}$$

The part-load factor, *PLF* is

$$PLF = 1 - c_d(1 - LF)$$

where the degradation coefficient $c_d = 0.25$.

The electrical energy input rate to the heat pump when it is operating in a cycling mode ($T_{amb} > T_{bal}$) is given by

$$W_{hp} = \frac{Q_{load}}{PLF{\times}COP} = \frac{Q_{load}}{3.5PLF}$$

The bin data for Toronto are obtained using the MATLAB code listed in Appendix A13.1.

For each bin temperature we compute the heating load, the heat supplied by the heat pump, the electrical energy input to the heat pump, and the heat supplied by the auxiliary source using the above equations. The annual values are obtained by multiplying by the number of hours in the bin. The results are summarized in Table E13.14.1.

Table E13.14.1 Summary of yearly energy quantities (GJ) in bins

N-bin	T_{mean}, C	No. hours	Q_{load}	Q_{hp}	Q_{aux}	W_{hp}
1	−27.5	1	0.196	0.133	0.063	0.038
2	−25	2	0.416	0.299	0.117	0.085
3	−22.5	5	0.881	0.672	0.209	0.192
4	−20	10	1.849	1.503	0.346	0.429
5	−17.5	23	3.834	3.338	0.497	0.954
6	−15	51	7.777	7.285	0.492	2.082
7	−12.5	106	15.10	15.10	0	4.330
8	−10	208	27.05	27.05	0	7.017
9	−7.5	356	42.25	42.25	0	12.63
10	−5	510	54.50	54.50	0	16.64
11	−2.5	604	57.48	57.49	0	17.94
12	0	624	52.13	52.13	0	16.64
13	2.5	612	43.96	43.96	0	14.36
14	5	605	36.37	36.38	0	12.16
15	7.5	605	29.27	29.27	0	10.23
16	10	597	21.93	21.93	0	7.701
17	12.5	607	15.19	15.19	0	5.469
18	15	666	8.87	8.87	0	3.279
19	17.5	766	1.24	1.24	0	0.471
20	20	801	0	0	0	0

Hence we obtain the following total yearly energy quantities:

Total heating load = 420.31 GJ, Heat supplied by heat pump = 418.59 GJ, Heat supplied by auxiliary source = 1.72 GJ, Electrical energy input to heat pump = 133.35 GJ.

Example 13.15 Four zones of a building are heated and cooled by individual heat pumps that exchange heat with a common water loop (see Fig. E13.15.1). If the water temperature in the loop falls below 15°C, a gas-fired furnace in the loop is activated to maintain the temperature at 15°C. If the temperature rises above 35°C, the water is sent through a cooling tower which maintains the water temperature at 35°C (see Fig. 13.11).

The heating and cooling mode COPs of the heat pumps at 15°C are 4.5 and 8.5; at 35°C they are 9 and 3.5 respectively. The cooling (positive) and heating (negative) loads of the zones for four different time periods are given in Table E13.15.1. Calculate (i) the energy input rates to the heat pumps, (ii) the heat input rate to the furnace, and (iii) the heat rejection rate in the cooling tower for the different load conditions. Neglect any effects due to storage.

Table E13.15.1 Water-loop heat pump data

H/C -Loads (kW)	Zone 1	Zone 2	Zone 3	Zone 4
A	−14	−9	−12	−20
B	−14	−9	−10	18
C	−12	10	12	15
D	−10	10	14	20

Solution

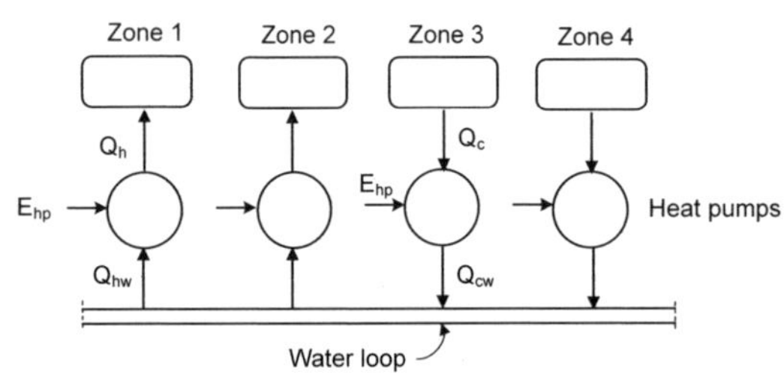

Fig. E13.15.1 Water-loop heat pump system

As seen in Fig. E13.15.1, the heat pumps operating in the heating mode extract heat from the water loop while those in the cooling mode reject heat to the water loop. If the net heat supplied to the water loop is positive, the water loop temperature will increase steadily and eventually exceed the limiting value of 35°C. When this happens water in the loop is sent through the cooling tower to maintain the loop temperature at 35°C as shown in Fig. 13.11.

The reverse occurs if there is a net heat out flow from the loop, in which case, the loop temperature will steadily decrease and eventually drop below the limiting value of 15°C. The water is then sent through a water heater to maintain the temperature at 15°C.

For each of the 4 load conditions listed in Table E13.15.1, we shall first determine whether the net heat flow to the loop is positive or negative.

For condition A, all four heat pumps are in the heating mode and are therefore extracting heat from the water loop. The heat extraction rate by a heat pump is given by

$$Q_{out} = Q_h\left(1 - \frac{1}{cop_h}\right)$$

The equilibrium temperature of the water loop is 15°C. The heating mode COP is 4.5. The total heating load

$$Q_h = 14 + 9 + 12 + 20 = 55 \text{ kW}$$

The rate of heat removal from the water loop is

$$Q_{out} = 55\left(1 - \frac{1}{4.5}\right) = 42.8 \text{ kW}$$

Therefore the furnace has to supply heat at the rate 42.8 kW to maintain the water temperature at 15°C.

The electrical energy supply rate to the heat pumps is

$$W_{hp} = \frac{55}{4.5} = 12.2 \text{ kW}$$

For condition B, zones 1, 2 and 3 are heated and zone 4 is cooled by the respective heat pumps. The heat extraction rate by heat pumps 1, 2 and 3 from the water loop is given by

$$Q_{out} = 33\left(1 - \frac{1}{cop_h}\right) \qquad \text{(E13.15.1)}$$

The rate of heat rejection to the water loop by heat pump 4 is given by:

$$Q_{in} = Q_c\left(1 + \frac{1}{cop_c}\right) = 18\left(1 + \frac{1}{cop_c}\right) \qquad \text{(E13.15.2)}$$

The *net heat extraction rate* from the water loop is

$$Q_{o,net} = 33\left(1 - \frac{1}{cop_h}\right) - 18\left(1 + \frac{1}{cop_c}\right) \qquad \text{(E13.15.3)}$$

We shall use Eq. (E13.15.3) to determine the equilibrium temperature of the water loop by substituting the COP values at the two limiting temperatures of 15°C and 35°C.

For 15°C, and 35°C the *net rates of heat out flow* are 5.55 kW, 6.2 kW respectively. Hence it is clear that due to continuous heat extraction, the water loop temperature will decrease steadily and eventually reach 15°C. The furnace has to supply heat at the rate 5.55 kW to maintain the water temperature at 15°C.

The total electrical energy supply rate to the heat pumps is given by

$$W_{hp} = \frac{33}{4.5} + \frac{18}{8.5} = 9.45 \text{ kW}$$

For condition C, zones 1 and 2 are heated and zones 3 and 4 are cooled by the respective heat pumps. The total heat extraction rate by heat pumps 1 and 2 from the water loop is given by

$$Q_{out} = 22\left(1 - \frac{1}{cop_h}\right) \quad \text{(E13.15.3a)}$$

The total rate of heat rejection to the water loop by heat pumps 3 and 4 is given by

$$Q_{in} = 27\left(1 + \frac{1}{cop_c}\right) \quad \text{(E13.15.4)}$$

Hence the *net heat extraction rate* from the water loop is

$$Q_{o,net} = 22\left(1 - \frac{1}{cop_h}\right) - 27\left(1 + \frac{1}{cop_c}\right) \quad \text{(E13.15.5)}$$

We shall use Eq. (E13.15.5) to determine the equilibrium temperature of the water loop by substituting the COP values at the two limiting temperatures of 15°C and 35°C.

For 15°C, and 35°C the *net rates of heat inflow* are 13 kW, 15.1 kW respectively. Therefore it is clear that due to continuous heat *inflow*, the water loop temperature will increase steadily and eventually reach 35°C. The cooling tower has to remove heat at the rate 15.1 kW to maintain the water temperature at 35°C.

The total electrical energy supply rate to the heat pumps is given by

$$W_{hp} = \frac{22}{9} + \frac{27}{3.5} = 10.2 \text{ kW}$$

For condition D, zone 1 is heated and zones 2, 3 and 4 are cooled by the respective heat pumps. The total heat extraction rate by heat pump 1 from the water loop is given by

$$Q_{out} = 10\left(1 - \frac{1}{cop_h}\right) \quad \text{(E13.15.6)}$$

The total rate of heat rejection to the water loop by heat pumps 2, 3 and 4 is given by

$$Q_{in} = 44\left(1 + \frac{1}{cop_c}\right) \quad \text{(E13.15.7)}$$

The *net heat extraction rate* from the water loop is

$$Q_{o,net} = 10\left(1 - \frac{1}{cop_h}\right) - 44\left(1 + \frac{1}{cop_c}\right) \qquad \text{(E13.15.8)}$$

We shall use Eq. (E13.15.8) to determine the equilibrium temperature of the water loop by substituting the COP values at the two limiting temperatures of 15°C and 35°C.

For 15°C, and 35°C the *net rates of heat inflow* are 41 kW, and 47.7 kW respectively. Hence it is clear that due to continuous heat *inflow*, the water loop temperature will increase steadily and eventually reach 35°C. The cooling tower has to remove heat at the rate 47.7 kW to maintain the water temperature at 35°C.

The total electrical energy supply rate to the heat pumps is given by

$$W_{hp} = \frac{10}{9} + \frac{27}{3.5} = 8.8 \text{ kW}$$

An extension of the above analysis is included in problem 13.12 below.

Problems

P13.1 A building with an effective heat conductance – *UA* of 0.68 kW $°C^{-1}$ is occupied for 14 hours from 7h to 20h and unoccupied during the rest of the day. The rates of internal heat gain during the occupied and unoccupied periods are 2.85 kW and 0.9 kW respectively. The corresponding indoor temperatures are 22°C and 16°C respectively.

The building is heated using a fuel-fired furnace of capacity 14 kW and rated efficiency 88%. The part-load degradation factor of the furnace is 0.25. The hourly ambient temperatures for a particular day are given in Table P13.1.1. Calculate (i) the balance temperatures, and (ii) the energy input to the furnace, during the occupied and unoccupied periods. Plot the energy flow rates versus temperature curves.

[*Answers*: (i) 17.8°C, 14.68°C, (ii) 96.6 kWh, 100.2 kWh]

Table P13.1.1 Hourly ambient temperatures

Hour	1	2	3	4	5	6	7	8	9	10	11	12
T_A °C	3	2	2	1	2	3	4	5	6	7.7	9.5	12
Hour	13	14	15	16	17	18	19	20	21	22	23	24
T_A °C	14	15	15	13	12	11.5	10	7	6	5	4	3

P13.2 The building described in problem 13.1 is to be heated using a heat pump whose capacity Q_{cap} (kW), and COP vary with ambient temperature t_a (°C) according to the equations:

$$Q_{cap} = 7 + 0.4t_a + 0.00615{t_a}^2$$

$$COP = 2.16 + 0.078t_a + 0.00357{t_a}^2$$

The conditions for the occupied and unoccupied periods are the same as in problem 13.1. The degradation factor due to cycling of the heat pump is 0.25. Calculate (i) the heat pump balance temperatures, (ii) the electrical energy input to the heat pump, and (iii) the auxiliary energy input required, for the occupied and unoccupied periods. Plot the energy flow rates versus temperature curves.
[*Answers*: (i) 4.53°C, 2.72°C, (ii) 56.7 kWh, 34.8 kWh, (iii) 0.7 kWh, 4.28 kWh]

P13.3 The rated capacity and COP of an air conditioner are 75 tons and 3.5 respectively. It is used to cool a building whose hourly cooling loads are given in Table P13.3.1. The degradation coefficient for part-load operation of the air conditioner is 0.25. Estimate (i) the hourly energy input rate to the air conditioner, (ii) the total energy input, and (iii) the average COP.
[*Answers*: (ii) 1014 kWh, 3.15]

Table P13.3.1 Hourly cooling loads

Hour	1	2	3	4	5	6
Q_l, kW	65	62	58	55	60	60
Hour	7	8	9	10	11	12
Q_l, kW	73	92	101	133	160	188
Hour	13	14	15	16	17	18
Q_l, kW	225	230	220	211	206	197
Hour	19	20	21	22	23	24
Q_l, kW	188	170	142	119	96	82

P13.4 The monthly average ambient temperatures for Vancouver are given in Table P13.4.1[1]. (i) Using the MATLAB code in Appendix A13.1 or otherwise, calculate the bin data for Vancouver. (ii) Calculate the degree–days for base temperatures of 18.3°C and 12°C.

[*Answer:* use the MATLAB code in Appendix A13.1 and compare degree–day data with data in Ref. [1]]

P13.4.1 Monthly average temperatures for Vancouver from Ref. [1]

Month	January	February	March	April	May	June
T_{am} °C	4.2	4.9	6.8	9.5	12.8	15.6
Month	July	August	Sept.	October	Nov.	Dec.
T_{am} °C	17.9	18	15	10.4	6.6	3.9

P13.5 A fuel-fired furnace of rated capacity of 30 kW and efficiency 90% is used to heat a house in Vancouver (see problem 13.4). The balance temperature for the house is 18.3°C. (i) Calculate the yearly energy input to the furnace using the bin method, (a) neglecting the performance degradation due to cycling, and (b) assuming a degradation coefficient of 0.25. (ii) Use the degree–day method to estimate the yearly energy input to the furnace. (iii) Obtain the answers to (a) and (b) if the balance temperature decreases to 12°C due to design changes.
[*Answers*: (i) (a) 176 GJ, (b) 216GJ, (ii) 176GJ, (iii) (a) 76GJ, (b) 95GJ]

P13.6 The full-load capacity and COP of an air conditioner are 25 tons and 3.5 respectively. The part-load degradation factor is 0.3. The air conditioner is installed in a building whose cooling load varies from 8 kW to 85 kW. (i) Plot a graph of the power input to the air conditioner versus the cooling load. (ii) If two such air conditioners are to be operated in an optimal manner in a building whose cooling load varies from 10 kW to 170 kW, plot the variation of the power input to the chillers with the cooling load.
[*Answer*: see worked examples 13.9 and 13.10]

P13.7 The building in worked example 13.10 is to be cooled using 4 chillers A, B, C and D with capacities of 750 kW, 1500 kW, 1500 kW, and 3500 kW respectively. The variation of the COP of the chillers A, B and C with load fraction is given by:

$$cop(f_l) = -8.55{f_l}^3 + 11.7{f_l}^2 - 0.43f_l + 3.8$$

The variation of the COP of chiller D with load fraction is given by:

$$cop(f_l) = 11.2\, f_l^{\,3} - 28.1 f_l^{\,2} + 21.0 f_l + 2.2$$

The chillers are to be operated in an optimal manner to satisfy the hourly cooling load given Table E13.10.1. Calculate (i) the hourly input of electrical energy to the chillers, (ii) the total electrical energy input, and (iii) the average COP of the cooling system.
[*Answers*: simulate different combinations and compare the total energy inputs]

P13.8 A building has a constant cooling load of 180 kW. It is cooled by a chiller system where the condenser heat is rejected to the atmosphere through a cooling tower. The variation of the COP of the chiller with the ambient wet-bulb temperature is well represented by the equation

$$COP = 6.8 - 0.13 t_{wb} + 0.0015 t_{wb}^{\,2}$$

The seasonal bin data for the local wet-bulb temperature is given in Table P13.8.1. Estimate (i) the seasonal energy input to the chiller, (ii) the average COP, and (ii) the total heat rejected in the cooling tower.
[*Answers*: (i) 600 GJ, (ii) 5.02, (iii) 3611 GJ]

Table P13.8.1 Bin data for wet-bulb temperature

t_{wb}, °C	11	12.9	15.5	18.2	21	23.5	26.5
N-bin	675	680	795	995	1010	480	12

P13.9 A building has an effective heat conductance – *UA* of 0.78 kW °C^{-1}. From Monday to Friday, it is occupied for 14 hours from 7h to 20h, and unoccupied during the rest of the day. The building is unoccupied during Saturday and Sunday. The rates of internal heat gain during the occupied and unoccupied periods are 3.2 kW and 1.1 kW respectively. The corresponding indoor temperatures are 21°C and 16°C respectively.

The building is heated using a fuel-fired furnace of capacity 30 kW and rated efficiency 88%. The part-load degradation factor is 0.25. The bin data for three hour-groups for the location are given in Table E13.13.1. Calculate (i) the balance temperatures for the occupied and unoccupied periods, and (ii) the annual energy input to the furnace during the occupied and unoccupied periods.
[*Answers*: (i) 16.9°C, 14.6°C, (ii) 27.9 MWh, 36.8 MWh]

P13.10 A building in Vancouver is heated with a heat pump whose capacity Q_{cap} (kW), and COP vary with ambient temperature T_a (°C), according to the following equations:

$$Q_{cap} = 31.38 + 1.77T_a + 0.0276{T_a}^2$$

$$COP = 2.27 + 0.082T_a + 0.0037{T_a}^2$$

The *UA*-value of the building is 0.62 $kW°C^{-1}$ and the indoor temperature is 22°C. The rate of internal heat gain of the building is 2.6 kW. The part-load degradation coefficient of the heat pump is 0.25. Estimate (i) the annual electrical energy input to the heat pump, (ii) the annual energy input to the auxiliary heat source, and (iii) the average COP.
[*Answers:* (i) 65.7GJ, (ii) 0.15 GJ, (iii) 2.27]

P13.11 A house in Vancouver has an effective *UA*-value of 1.3 $kW°C^{-1}$. It is heated using a ground-source heat pump, of capacity 25 kW and average COP 3.5. The average ground temperature for heat absorption by the heat pump is 7.2°C. The internal heat gains and the indoor temperature of the house are 5.4 kW and 22°C respectively. The degradation coefficient due to cycling of the heat pump is 0.25.

Calculate (i) heat pump balance temperature, (ii) the heating load, the heat supplied by the heat pump and the auxiliary energy source, (iii) the electrical energy input to the heat pump, and (iv) the average COP. Sketch the energy flow rates versus ambient temperature graphs.
[*Answers*: (i) −1.38°C, (ii) 315 GJ, 311.7 GJ, 3.2GJ, (iii) 98.8 GJ, (iv) 3.15]

P13.12 A water-loop heat pump system is used to heat and cool a building with 4 zones (see Fig. E13.15.1). If the water temperature in the loop falls below 15°C, a gas-fired furnace in the loop is activated to maintain the temperature of the water at 15°C. If the temperature rises above 35°C, the water is sent through a cooling tower which maintains the water temperature at 35°C. (See Fig. 13.11.) The heating and cooling mode COPs of the heat pumps are well represented by the equations:

$$\varepsilon_h = 0.2t_w - 1$$

$$\varepsilon_c = 7.25 - 0.15t_w$$

where t_w (°C) is the temperature of the water loop.

The cooling (positive) and heating (negative) loads of the zones for four different time periods are given in Table P13.12.1. Calculate (i) the total energy input rate to the heat pumps, (ii) the heat input rate to the furnace, and (iii) the heat rejection rate in the cooling tower, for the different load conditions. Neglect any effects due to storage.
[*Answers*: (A) 31.7 kW, 0 kW, 0 kW, (B) 24.4 kW, 0 kW, 0 kW, (C) 43.3 kW, 0 kW, 103.3 kW, (D) 44 kW, 16 kW, 0 kW]

Table P13.12.1 Water-loop heat pump data

H/C -Loads (kW)	Zone 1	Zone 2	Zone 3	Zone 4
A	−30	−30	10	20
B	−20	−20	30	20
C	−20	25	25	30
D	−30	−25	−25	20

P13.13 Analyze the WLHP system in worked example 13.13 including a water storage tank of volume 2m^3, if each time period is 6 hours.

References

1. *ASHRAE Handbook* - 2013 *Fundamentals*, American Society of Heating, Refrigeration and Air Conditioning Engineers, Atlanta, 2013.
2. Erbs, D. G., Klein, S. A. and Beckman, W. A., 'Estimation of degree–days and ambient temperature bin data from monthly–average temperatures'. *ASHRAE Journal*, 25(6), 1983.
3. Kuehn, Thomas H., Ramsey, James W. and Threlkeld, James L., *Thermal Environmental Engineering*, 3rd edition, Prentice-Hall, Inc., New Jersey, 1998.
4. Mitchell, John W. and Braun, James E., *Principles of Heating, Ventilation and Air Conditioning in Buildings*. John Wiley and Sons, Inc., New York, 2013.
5. Stoecker, Wilbert F., *Design of Thermal Systems*, International Edition, McGraw-Hill Book Company, London, 1989.

6. Stoecker, Wilbert F. and Jones, Jerold W., *Refrigeration and Air Conditioning*, International Edition, McGraw-Hill Book Company, London, 1982.

Appendix A13.1 - MATLAB Code for Bin Data and Degree–Days

```
% computation of yearly bin data for ambient temperature
% uses computed bin data to calculate the degree–days
tbas=[-27.5,-25,-22.5,-20,-17.5,-15,-12.5,-10,-7.5,-5,-2.5,0,2.5,...
   5,7.5,10,12.5,15,17.5,20,22.5,25,27.5,30,32.5,35];
% mean temperature of each bin, 26 bins in all
for i=1:26
tbou(i)=tbas(i)-1.25; % boundary temperatures of bins, C
end
tbou(27)=35+1.25
% monthly average ambient temperatures, C; data for Toronto
tma=[-4.7,-4.5,0.4,7.2,13.4,18.9,21.6,20.7,16.4,9.7,3.9,-2]
% number of days of the 12 months - Jan. to Dec.
nd=[31,28,31,30,31,30,31,31,30,31,30,31];
tyav=sum(tma)/12;   % yearly average temperature, C
sum2=0;
for i=1:12
 sum2=sum2+((tma(i)-tyav)^2)/12;
end
sigy=sum2^0.5; % standard deviation of daily ambient temperatures
for j=1:12
stdm=1.451-0.0290*tma(j)+0.0369*sigy;
% correlation for monthly average standard deviation
for i=1:27
hp(i)=(tbou(i)-tma(j))/(stdm*(nd(j)^0.5)); % H-parameter
qf(i)=(1+exp(-3.396*hp(i)))^(-1);
% cumulative fraction of number of hours below T-base
aaq(i,j)=qf(i);
% monthly cumulative fractions, j = month, i = bin-number
end
end
```

```
for j=1:12
   for i=1:26
      nhr(i,j)=(aaq(i+1,j)-aaq(i,j))*nd(j)*24
 % monthly bin data , j = month , i= bin
end
end
% compute yearly bin data in hours for 26 - different bins
for i=1:26
   sumy=0;
   for k=1:12
   sumy=sumy+nhr(i,k);
end
 ynhr(i)=sumy % number of hours per year in bin number, i
end
% compute monthly heating degree–days for the 12 months using %
computed bin data
tbala=22    % input balance temperature of building
for i=1:12
   sumdd=0;
   for k=1:26
   sumdd=sumdd+nhr(k,i)*(tbala-tbas(k))/24;
   if tbas(k)>=tbala
   break
   end
end
 yndd( i)=sumdd  % number of degree–days
end
```

Index

absorption cycles, 84, 87
absorption of solar radiation, 406
fenestrations, 411
opaque surfaces, 406
adiabatic saturation, 126
air distribution, 529, 539
air washers, 224
air-source heat pumps, 9, 82, 643
air–water mixtures, 119, 121
angle of incidence, 400

beam radiation, 396, 400
below grade heat transfer, 351
bin method, 638
black surface, 35
bypass systems, 166, 177

Carnot refrigeration cycle, 66
centrifugal compressors, 81, 539
centrifugal pumps, 596, 598
clear-sky model, 404
coefficient of performance - COP, 67
Colebrook's equation, 532, 594
condensers, 272
condition line, 165, 170, 176, 290
conduction, 20, 463
cylindrical systems, 26
internal heat generation, 28
conduction time factors, 472, 474
cooling coils, 165, 176, 285
cooling load estimation, 447, 459, 478
conduction time factors, 472, 476
heat balance method, 468
radiant time series, 471, 476
cooling tower performance, 230
cooling towers, 229, 646
analysis, 230, 232
approach, 233
range, 234
simplified model, 232
counter-flow heat exchangers, 267
cross-flow heat exchangers, 267, 275
cycling of furnaces, 642
degradation coefficient, 642

Darcy–Weisbach equation, 532, 593
degree of saturation, 121
dehumidification, 163, 285
design of pipe networks, 601
diffuse radiation, 396
diffusion coefficient, 219, 363
direct-return systems, 602
direct-contact processes, 217, 221
direction of solar beam, 397
dual-duct systems, 181
duct design methods, 546
equal friction method, 547
static regain method, 549
dynamic losses in fittings, 534, 595
equivalent length, 595

effective temperature, 455
effectiveness–NTU design method, 270
efficiency of finned surfaces, 278, 282
emissive power, 37
energy estimation methods, 633, 638
enthalpy of moist air, 123
enthalpy potential, 221, 224
equation of time, 400
evaporative cooling, 169
evaporators, 272

fan characteristics, 541
fan efficiency, 541
fan laws, 542
fan–duct interaction, 543
Fick's law, 218, 362
fin-tube heat exchangers, 282, 286
forced convection, 30
Fourier's law, 20, 26
free convection, 33
friction chart, 533, 594
gas-filled cavities, 346
  fenestrations, 348
  overall heat transfer coefficient, 349
generation of bin data, 638
grade-level heat transfer, 355
gray surface, 37, 39

head loss in pipes, 593
  friction chart, 594
heat balance method, 468, 478
heat exchanger types, 267
heat pump balance point, 644
heat pump cycles, 9, 82, 643
heat transfer coefficient, 31, 34
heat transfer correlations, 32, 34
heat transfer from human body, 452
  MET units, 452
  metabolic heat generation, 453, 460
heating and cooling degree-days, 634
  balance temperature, 636
heating load estimation, 340, 477
heating load, 447, 477, 478
hour angle, 399
humidification, 168
humidifiers, 224
  efficiency, 228
  NTU, 228
humidity ratio, 121
HVAC system types, 4-9
indoor air quality, 457
  ventilation rate, 459

indoor design conditions, 455
infiltration, 355
  heating load, 355
  flow rates, 356, 360
  stack effect, 357
  wind effect, 359
  mechanical effect, 359
internal heat gains, 459
  equipment, 461
  lighting, 460
  people, 459

Kirchoff's law, 38

Lewis number, 129, 165, 223
lithium bromide–water systems, 86, 87
LMTD design method, 267, 270

mass convection, 220
mass diffusion, 218
mass transfer coefficient, 221
mixing process, 160
modes of heat transfer, 17
moisture transfer, 361
    permeability, 363
multi-chiller systems, 651
multilayered structures, 21, 340
    isothermal plane method, 342
    parallel flow method, 341
    zone method, 343
multi-zone systems, 180

net positive suction head (NPSH), 598

operative temperature, 453
optimization of duct systems, 551
outdoor design conditions, 450

parallel resistances, 23
pressure loss, 532, 593
    fittings, 534, 595
    straight ducts, 532, 593
protractors of chart, 134, 135
psychrometric chart, 129, 135
psychrometric processes, 159
pump curve, 598
    flow control, 600
    system–pump interaction, 599

radiant time factors, 472, 473
radiation exchange, 38
radiation heat transfer coefficient, 347
radiation heat transfer, 34
reciprocating compressors, 80
reflected radiation, 405
reflection, 37
refrigerants, 76
refrigeration ton,74
reheat systems, 176, 180
relative humidity, 121
reverse-return systems, 602
reversible heat pumps, 9, 82, 643
rotary compressors, 81

sensible cooling, 162
sensible heat ratio, 134
sensible heating, 162
series resistances, 22
shading of fenestrations, 417
simulation methods, 649
    central HVAC system, 7, 79, 650
simultaneous heat and mass transfer, 221, 231
single-zone systems, 174
solar altitude angle, 398
solar azimuth angle, 399
solar declination, 399
solar heat gain coefficient (SHGC), 415
solar radiation fundamentals, 396
sol-air temperature, 407
solar time, 399
standard longitude, 399
standard refrigeration cycle, 68
standard time, 399

thermal comfort, 452
    clo unit, 454
thermal networks, 21, 340
thermal resistance, 21
three-heat-reservoir model, 85

total head, 592
total pressure distribution, 530
transient heat transfer, 19, 463
    lumped capacitance model, 465
    transfer function method, 464
transmittance of radiation, 409
    multilayered fenestrations, 410
two-stage cycles, 74

vapor compression cycle, 72, 74
variable air volume systems, 183
variable occupancy, 647

water distribution systems, 591
    energy equation, 592
    hydronic systems, 593
water-loop heat pumps, 10, 83, 654
wet-bulb temperature, 126, 128
wet-coil heat exchangers, 266, 283
    analysis, 284, 285
    numerical models, 288
    physical processes, 284
Wien's law, 36

zone air distribution, 552, 555
    air diffusion performance index, 554
    diffusers, 555